Proceedings of the 6th International Conference
"Computational Mechanics and Virtual Engineering"
COMEC 2015

Transilvania University of Braşov

15 - 16 October 2015, Braşov, Romania

Karol JÁRMAI, Chairman
Sorin VLASE, President
Michael M. DEDIU, Editor

DERC Publishing House
Tewksbury (Boston), Massachusetts, U. S. A.

Published and printed in 2015 in the United States of America

American Mathematical Society
2010 Mathematical Subject Classification: 65-xx, 70-xx, 74-xx, 76-xx, 82-xx

Library of Congress Cataloging in Publication Data

Computational Mechanics and Virtual Engineering, the 6th International Conference, COMEC 2015, 15-16 October 2015, Braşov, Romania / Karol Jármai, Chairman, Sorin Vlase, President, Michael M. Dediu, Editor
      p. cm. – (Proceedings of the 6th International Conference "Computational Mechanics and Virtual Engineering" COMEC 2015)
      Includes bibliographical references

ISBN-13: 978-1-939757319

PREFACE

After receiving many favorable comments and having interesting discussions regarding the Proceedings of the 4th International Conference "Advanced Composite Materials Engineering" COMAT 2012, the Proceedings of the 5th International Conference "Computational Mechanics and Virtual Engineering" COMEC 2013, and the Proceedings of 5th International Conference "Advanced Composite Materials Engineering" and the 3rd International Conference "Research & Innovation in Engineering" COMAT 2014, we are glad to present these Proceedings of the 6th International Conference "Computational Mechanics and Virtual Engineering" COMEC 2015. This conference was organized with the important support of Transilvania University of Brasov, Romanian Academy of Technical Sciences and Romanian Society of Theoretical and Applied Mechanics, to whom we thank very much.

The research and innovation in engineering in general, and in particular in Computational Mechanics and Virtual Engineering, are impressive and ever more important in everyday activities. They include, between many other techniques, fundamental studies of multiscale phenomena and processes in civil engineering, from kilometer-scale problems to a much finer scale, including nano-scale, geometric models, simulation, optimization and decision making tools within an integrated computer-generated environment, using parallel supercomputing, that accelerates the product design, engineering, manufacturing and testing, which are the main purposes of any engineering work. The applications are very diverse and include municipal infrastructure, manufacturing, aerospace, civil engineering structures, mining, communication, geotechnics, flow problems, automotive engineering, electric vehicles, predictive surgery, transportation, geo-environmental modeling, biomechanics, electromagnetism, medicine, prediction of natural physical events, and renewable energy, just to mention a few.

These Proceedings of COMEC 2015 include 54 papers, which analyze many important practical applications. The topics range from the energy release rate evaluation in sandwich composite structures, to tactile sensing by using smart materials, and to a statistically relevant experiment concerning the colloidal silver influence on human body.

We want to thank very much Professor Sorin Vlase for his remarkable effort and dedication for the organization of this international conference, Professor Karol Jármai for his distinguished chairmanship, and the organizing committee for their continuous assistance.

We also thank all the conference participants for their interesting presentations, and Mrs. Sophia Dediu for her assistance in preparing this volume.

There is, certainly, much more that can be said about these engineering research domains, than we presented here. We hope that the papers included here will provide ideas for our audience, and will stimulate more research, development and applications.

We look forward to receiving comments and suggestions from our readers.

Michael M. Dediu, Ph. D.

Boston, U.S.A., December 9, 2015

Previously published in this series:

1. Ioan Goia, *Mechanics of Materials*
2. Ionel Staretu, *Gripping Systems*
3. Proceedings of the 4th International Conference *"Advanced Composite Materials Engineering"* COMAT 2012,
4. Proceedings of the 5th International Conference *"Computational Mechanics and Virtual Engineering"* COMEC 2013
5. Proceedings of the 5th International Conference *"Advanced Composite Materials Engineering"* COMAT 2014,
6. Sorin Vlase: Mechanical Identifiability in Automotive Engineering
7. Gabriel Dima: The Evolution of the Aerostructures – Concept and Technologies

6th International Conference
"Computational Mechanics and Virtual Engineering "
COMEC 2015
15-16 October 2015, Braşov, Romania

ENERGY RELEASE RATE EVALUATION IN SANDWICH COMPOSITE STRUCTURES BY USING THE DIC

Octavian Pop[1], Ioan Goia[2], Dorin Rosu[2], Sorin Vlase[2], Mihaela Violeta Munteanu[2]
[1]University of Limoges, France, ion-octavian.pop@unilim.fr
[2]Transilvania University of Braşov, Romania, i.goia@unitbv.ro

Abstract: In this paper a new formalism based on the coupling between the optical full field techniques and the integral invariants is proposed in order to evaluate the fracture parameters in cracked sandwich composite structures. The formalism allows identifying the fracture parameters in terms of energy release rate. From the experimental tests the displacement field is obtained by digital image correlation measurements. In this case the experimental displacement fields are employed to calculate the strain and stress fields by a numerical approach. Then using the mechanical fields defined on the surface of specimen the integral invariants can be used in order to characterize the fracture process. The present study is limited to the identification of the mechanical fields using the experimental displacement fields measured by the optical methods.

1. INTRODUCTION

Since some years the optical methods find more and more, their applications in the mechanical characterisation of materials and structures. Associated to the full fields techniques, the optical methods can be easily correlated with the energetically approaches such as the integral invariants. Among the optical methods, the digital image correlations seem to be the best to characterise the mechanical behaviour in the case of composite. Another particularity of this optical full fields' technique is the possibility to coupling this one with the numerical approach such as the Finite Element Method (FEM). Using the optical methods, the zone of interest (ZOI) can be discretized either by the subsets similar to the finite elements of the mesh, in the case of DIC. In this paper a feasibility study in order to characterize the fracture behavior in sandwich composite structures [1 2] by digital image correlation technique, is proposed.

2. OPTICAL FULL-FIELD TECHNIQUES [3, 4, 5, 6]

Related with the optical full field methods, digital image correlation is an optical method allows to measure in-plane displacement.
The basic principle of method is based on the comparison between two images of sample plane surface acquired at different states one before deformation and the other one after. Then, the displacements are estimated by comparing the degree of resemblance between these two images subsets. As is shown in Figure 1, in DIC the displacements are calculated into Zone Of Interest (ZOI) discretized by small areas with multiple pixel called subsets (where, m is the subset numbers). Note, that each subset is characterized by a unique light intensity distribution (i.e. gray level). Assuming that during the

test the light intensity distribution on the plane surface of sample does not change, the displacements are estimated by searching the subsets changes (translation + rotation + rigid body motion) between the undeformed and the deformed images. Finally the displacement field is build from the displacement vectors (u1, u2) corresponding to displacements of all subsets centers.

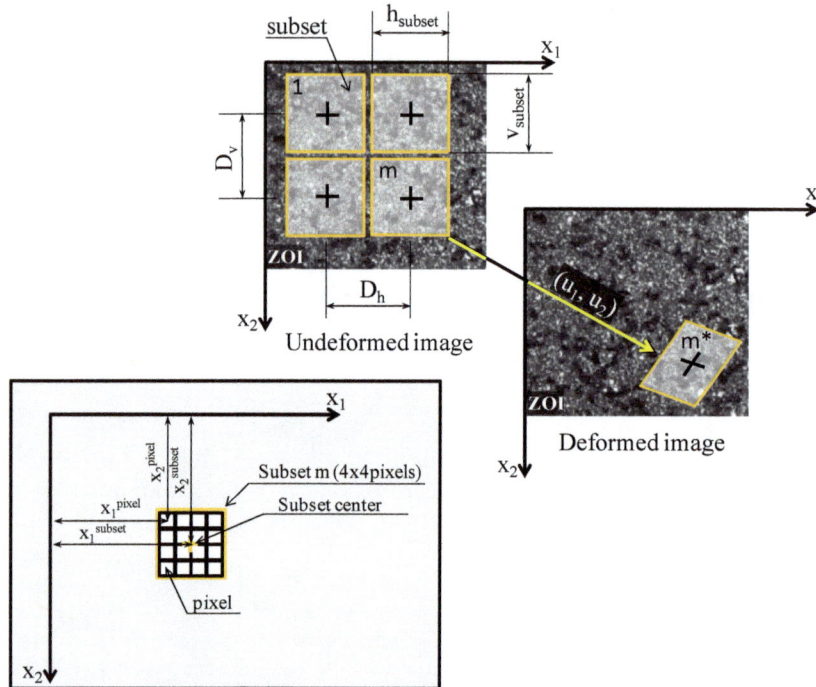

Figure 1: Principle of Digital Image Correlation

Another important aspect in DIC is the specimen plane surface preparation. So, in order to obtain a gray level distribution, it is necessary to create a characteristic speckle pattern a black speckle is deposited on the white background by spraying the black and white paints.

3. EXPERIMENTAL SETUP

The testing machine is an electromechanical press with a load capacity of 50kN. The test is run under displacement control and the velocity of the cross-head is fixed at 0.01 mm/s. As shown in Figure 2 the sandwich composite specimen is loaded in opening mode using a system which includes a steel pyramid and two rollers. The loading system allows perfect lips displacement symmetry.

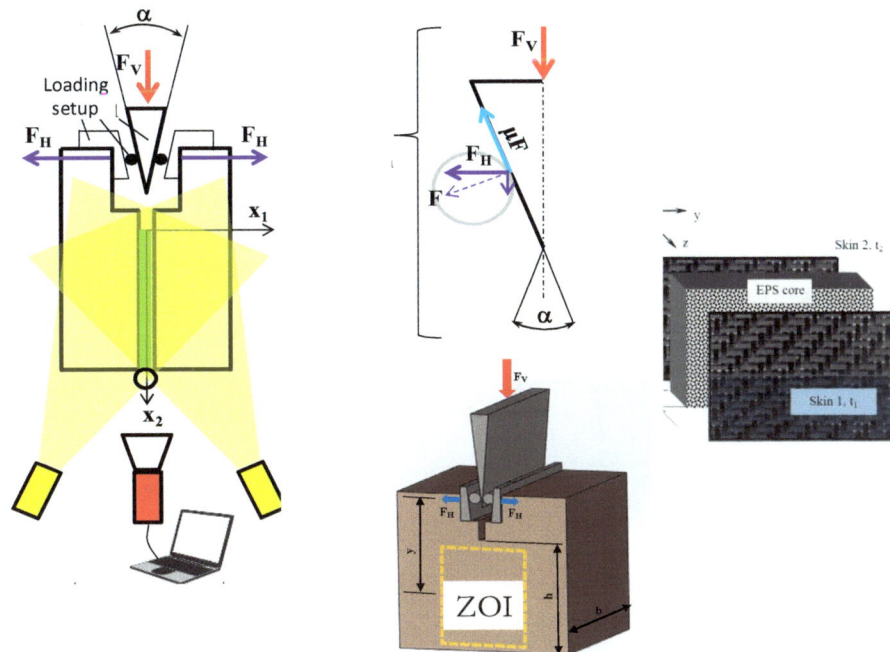

Figure 2: Experimental setup

Besides, Figure 3 puts in evidence the crack lips displacement symmetry during the test. In fact the plotted displacements correspond to the rollers displacement measured by mark tracking method.

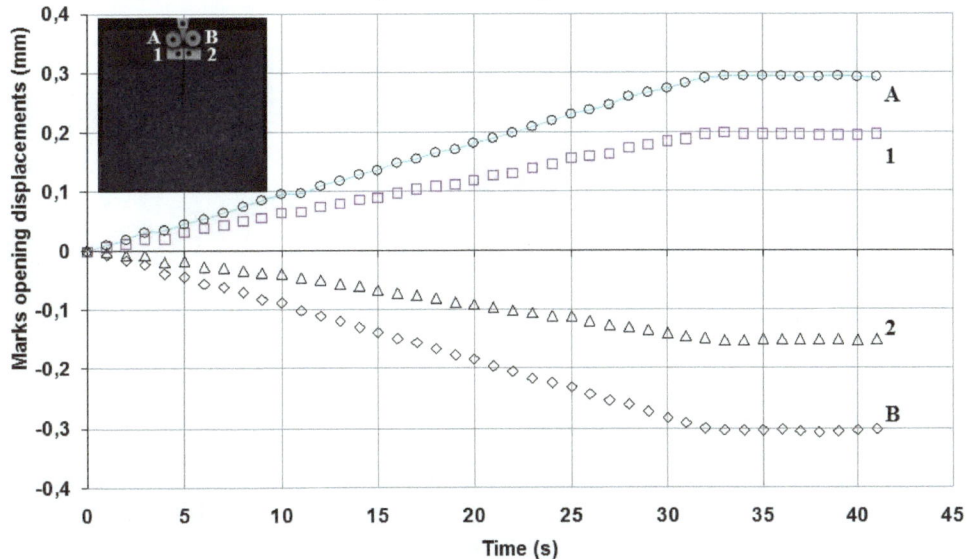

Figure 3: Experimental displacement of the loading rollers

As is show in Figure2, the specimen is a sandwich composite. The sandwich structure taken into account to accomplish the damping analysis presents two carbon/epoxy skins reinforced with a 0.3 kg/m2 twill weave fabric and an expanded polystyrene (EPS) 9 mm thick core with a density of 30 kg/m3 [1], [2]. The final structure's thickness is 10 mm. The carbon-fibre fabric used in this structure is a high rigidity one, that presents so called twill weave. The main feature of this weave is that the warp and the weft threads are crossed in a programmed order and frequency, to obtain a flat appearance. The skins were impregnated under vacuum with epoxy resin and sticked to the core.

During the test, the load-displacement evolution is recorded using a LVDT sensor and a load cell. Furthermore, an 8-bit Charge Coupled Device (CCD) camera (1392x1040 pixels^2) is synchronized with the testing machine to measure the displacement fields by DIC.

4. INTEGRAL INVARIANT J-INTEGRAL

In fracture mechanics the J-integral is associated with strain energy release rate or the work per unit of crack area. The theoretical concept of the J-integral was developed by Cherepanov [7,8] and Rice [9,10] who showed that the J-integral is independent of the path defined around the crack tip (Figure 4). The crack is oriented in x_1-direction, and the crack tip represents the origin of the coordinate system.

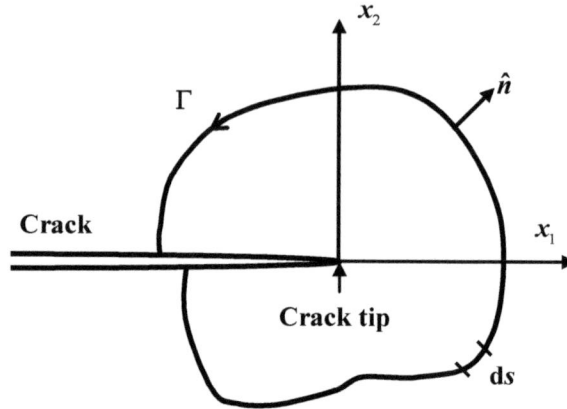

Figure 4: Domain of integration of J

The J-integral can be expressed in the following form:

$$J = \int_{\Gamma} \left(W \cdot n_1 - T_i \cdot \frac{\partial u_i}{\partial x_1} \right) \cdot ds \tag{4}$$

$$W = \frac{1}{2} \cdot \sigma_{ij} \cdot \varepsilon_{ij} \tag{5}$$

$$T_i = \sigma_{ij} \cdot n_j \tag{6}$$

Where: W is the elastic strain energy density, T_i is the traction along the contour Γ, n the unit normal of the contour path, u is the displacement vector, σ_{ij} and ε_{ij} are the stress and strain tensors, respectively.

By introducing Equations (5) and (6) in (4), we can see that the J-integral evaluation is based on the knowledge of the mechanical fields in terms of displacement, strain and stress. Therefore, the displacement vector (u) will be evaluated experimentally by DIC method, while the strain (ε_{ij}) and stress (σ_{ij}) tensors will be calculated from numerical approach.

5. EXPERIMENTAL RESULTS

As illustrated in Figure 5 the domain of integration can be assimilated by crowns defined by the centers of the correlation windows. By using the finite elements mesh generated from the experimental data, the J-integral is evaluated numerically via Equation (4) along different crowns (Γ_i) surrounding the crack tip.

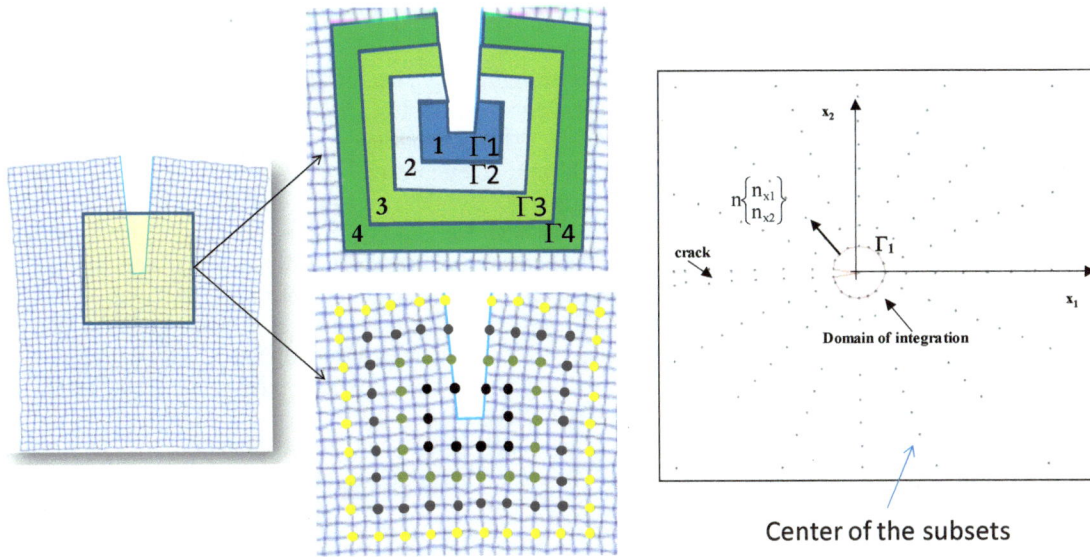

Figure 5: Domain of integration using the experimental results

The J-integral values versus loading amplitude and size of the domain of integration are presented in Table 1.

Load (N)	J –integral (J/m²)			
	Γ_1	Γ_2	Γ_3	Γ_4
280	25.33	25.01	24.99	25.42
560	55.23	55.03	55.02	54.98

The results presented in Table 1 show a low variation of the energy release according to integral path. These slight differences can be explained by experimental displacement field noises. In this case the displacement field optimization may be considered in order to increase the accuracy of the experimental fields.

6. CONCLUSION

In this study, a new experimental technique is proposed in order to characterize the fracture parameters in sandwich composites. The DIC is employed in order to measure the displacements fields on the specimen surface. Based on the experimental optical full field measurements the energetic approach is used to evaluate the energy release rate. Using the experimental displacement and the strain fields the stress tensor is calculated via a constitutive law. The stress fields are evaluated using the finite elements method. In fact the approach of this study consists of a combination of experimental and numerical techniques in order to evaluate the different mechanical fields employed in the energetic approach such as J or G-integral. For this purpose the experimental data are implemented in a finite element code in order to generate a finite element mesh. In this case the stress fields are calculated using the experimental data by imposing real boundary conditions. Now, using the mechanical fields measured or evaluated by coupling experimental with numerical approaches the J-integral is calculated for several loading steps and for different crowns defined around the crack tip. The J-integral evolution versus crown size shows the energy release rate invariance. The slight differences can be explained by the displacement field noises.

REFERENCES

[1] D.H. Teodorescu, S. Vlase - "Dynamic analysis of an ultra-lightweight sandwich structure for multiple applications". 3rd WSEAS International Conference on Dynamical Systems and Control, Arcachon, France, 13 – 15 October 2007, pp. 229-234.

[2] D.H. Teodorescu, S. Vlase, D.L. Motoc, I. Popa, R. Dorin, F. Teodorescu - "Mechanical behaviour of an advanced sandwich composite structure". WSEAS International Conference on ENGINEERING MECHANICS, STRUCTURES, ENGINEERING GEOLOGY (EMESEG '08), Heraklion, Greece, July 22-24, 2008, pp. 280-285.

[3] Sutton M.A., McNeill S.R., Helm J.D., Chao Y.J. - (2000) Advances in two-dimensional and three-dimensional computer vision, In: Rastogi P.K. (ed) Photomechanics, Springer, Berlin Heidelberg New York, pp 323–372.

[4] Sutton, M. A., Turner, J. L., Bruck, H. A. and Chae, T. A. - (1991) Full-field Representation of Discretely Sampled Surface Deformation for Displacement and Strain Analysis. Experimental Mechanics, 31(2), 168-177.

[5] Sutton, M. A., McNeill, S. R., Jang, J. and Babai, M - (1988) Effects of Subpixel Image Restoration on Digital Correlation Error. Journal of Optical Engineering, 27(10), 870-877.

[6] Sutton, M. A., Cheng, M. Q., Peters, W. H., Chao Y. J. and McNeill, S. R - (1986) Application of an Optimized Digital Correlation Method to Planar Deformation Analysis. Image and Vision Computing, 4(3), 143-151.

[7] Cherepanov GP - (1962) The stress in a heterogeneous plate with slits, in Russian, Izvestia AN SSSR, OTN, Mekhan. I Mashin. 1:131-137.

[8] Cherepanov GP - (1979) Mechanics of brittle Fracture, McGraw-Hill, New York.

[9] Budiansky B, Rice JR (1973) Conservation laws and energy-release rates, ASME, J. Appl. Mech. 40: 201-203.

[10] Rice JR - (1968) A path independent integral and the approximate analysis of strain concentration by notches and cracks, Journal of Applied Mechanics, 35:379-386.

**6th International Conference
"Computational Mechanics and Virtual Engineering"
COMEC 2015
15-16 October 2015, Braşov, Romania**

DOUBLE MAIN GIRDER DESIGN OF AN OVERHEAD TRAVELLING CRANE FOR MINIMUM COST

K. Jármai[1], J. Farkas[2]
[1] University of Miskolc, HUNGARY, jarmai@uni-miskolc.hu
[2] University of Miskolc, HUNGARY, altfar@uni-miskolc.hu

Abstract: A crane structure of two doubly symmetric welded box beams is designed for an overhead travelling crane for minimum cost. The following design constraints are considered: local buckling of web and flange plates, fatigue of the butt K weld under rail and fatigue of fil-let welds joining the transverse diaphragms to the box beams. The rails are placed over the inner webs of box beams. To increase the fatigue strength of the last mentioned welds, an efficient post welding treatment (PWT) is considered. For the formulation of constraints the relatively new standard for cranes EN 13001-3-1 [1] is used. The cost function consists of cost of material, assembly, welding and PWT. PWT is economic, since it is used only for diaphragms near the span centre of box beams, where the bending stresses are high. The optimization is performed by systematic search using a MathCAD program.
Keywords: crane girder, fatigue, post welding treatment, optimum design

1. INTRODUCTION

The main girder of overhead travelling cranes can be designed as a single or double box beam. The rail can be placed in the middle of the upper flange or over the inner web of the box beams. In our case we designed a double box beam with rails over the inner webs (Fig. 1). The research of post-welding treatments (PWT) does not give any data for these welds. PWT can cause a significant increase of fatigue strength for welds joining the transverse diaphragms to the upper flange, so we use these data. Our research shows that PWT can result in significant cost savings using them in welds joining the transverse diaphragms to the box or I-beams (Jármai et al. [2]).

2. DATA OF THE TREATED CRANE

The British Standard for cranes BS 2573-1 [3] is valid at present also. This BS gives characteristic parameters for crane groups. We select a workshop crane with a dynamic factor of $\psi_d = 1.3$, the governing number of cycles is $N = 4 \times 10^6$, the coefficient of spectrum is according to EN 13001-3-1 [1] $s_3 = 2$. The safety factor for fatigue is $\gamma_f = 1.25$.
Yield stress $f_y = 355$ MPa, according to EN 13001-3-1 the maximum design stress for plate thicknesses $t < 16$ mm is 323 MPa, for $16 < t < 40$ mm 314 MPa. We do not treat hybrid beams constructed with steels of two different yield stresses.
Span length is $L = 16.5$ m, hook load $P = 200$ kN, mass of the trolley $G_k = 42.25$ kN, distance of wheels $k = 1.9$ m, height of rail $h_s = 70$ mm, specific mass of the service-walkway and rail $p = 1900$ N/m, steel density $\rho = 7.85 \times 10^{-6}$ kg/mm³ or $\rho_0 = 7.85 \times 10^{-5}$ N/mm³, distance of transverse diaphragms $a = L/10 = 1650$ mm. The box beams are doubly symmetric.

2.1. BUCKLING CONSTRAINTS OF THE WEB UNDER THE RAIL

2.1.1. *Bending*

Stress from the vertical bending

$$\sigma_x = \frac{M_h}{W_x} \tag{1}$$

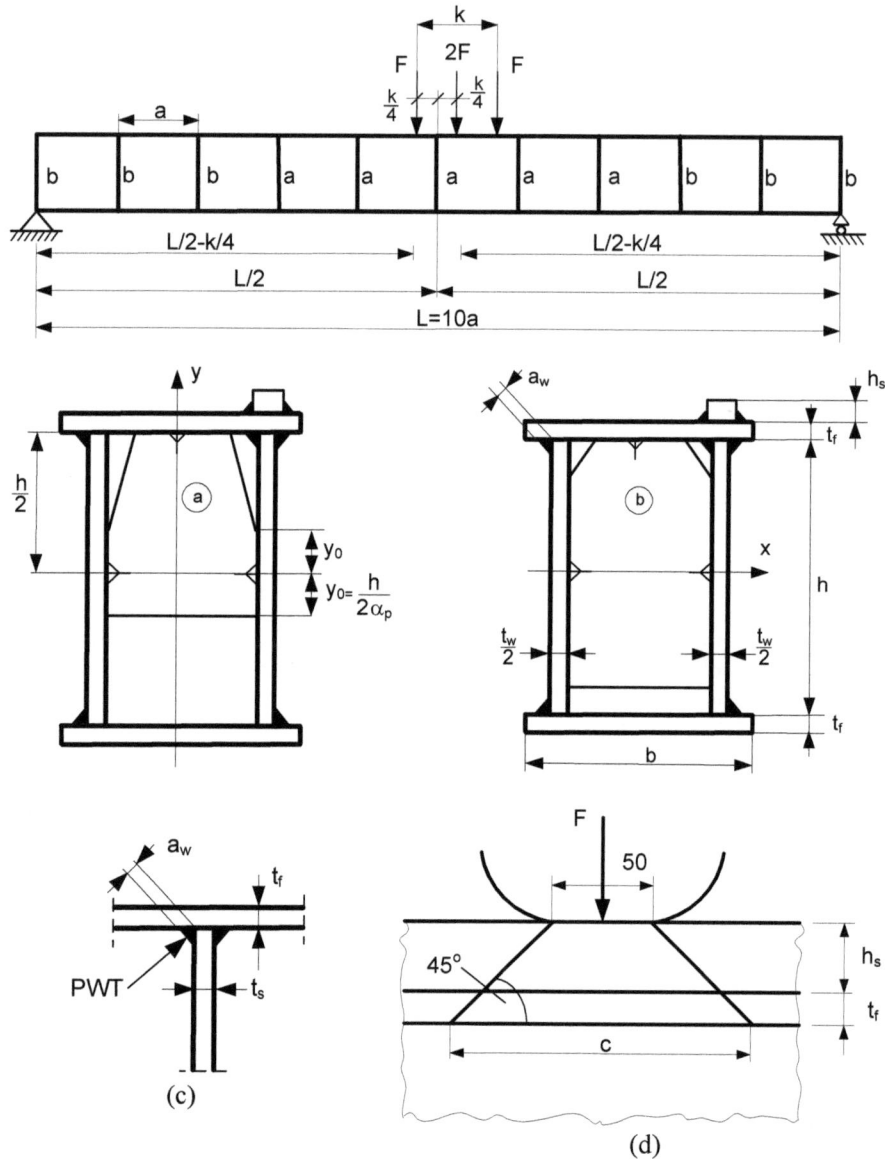

Figure 1: Data and cross-sections of the crane beams. Diaphragms (a) are used in the middle of beams for high bending stresses, PWT is used for the welds joining the diaphragms, diaphragms (b) are used near the beam ends, (c) shows the welds with PWT, (d) shows the load distribution in the beam web from the crane wheel.

Maximum bending moment in the case of the load position of two concentric forces

$$M_x = (1.05\rho_0 A + p)\frac{L^2}{8} + \frac{F}{2L}\left(L - \frac{k}{2}\right)^2, \quad F = \frac{\psi_d P + G_k}{4} \tag{2}$$

$$A = ht_{w0} + 2bt_{f0} \tag{3}$$

$$W_x = \frac{h^2 t_{w0}}{6} + b h t_{f0} \tag{4}$$

t_{w0} and t_{f0} are the rounded plate thicknesses.

Bending moment from the horizontal bending

$$M_y = 0.3 x 0.5 \left[\left(1.05 \rho_0 A + p \right) + \frac{G_k}{8L} \left(L - \frac{k}{2} \right)^2 \right] \tag{5}$$

The multiplier 0.5 expresses that two wheels are driven from four, 0.3 is the coefficient of mass force.

$$\sigma_y = \frac{M_y}{W_y}, \quad W_y = \frac{b^2 t_{f0}}{3} + \frac{h t_{w0} b}{2} \tag{6}$$

It is not necessary to calculate with effective width, when

$$\sigma_x \le k_x f_y, k_x = 1 \tag{7}$$

$$\lambda_x = \sqrt{\frac{f_y}{k_{\sigma x} \sigma_e}} \le 0.673, \quad k_{\sigma x} = 7.81 - 6.29 \psi_x + 9.78 \psi_x^2, \quad \psi_x = -\frac{\sigma_x - \sigma_y}{\sigma_x + \sigma_y} \tag{8}$$

$$\sigma_e = \frac{\pi^2 E}{12 \left(1 - v^2 \right)} \left(\frac{2 t_{w0}}{h} \right)^2, \quad E = 2.1 x 10^5 \text{ MPa}, v=0.3 \tag{9}$$

The required plate thickness

$$t_{w.req} = \frac{2h}{0.673 \, x 28.42 \, \varepsilon \sqrt{k_{\sigma x}}}, \quad \varepsilon = \sqrt{\frac{235}{f_y}} \tag{10}$$

2.1.2. Shear and torsion

From shear (approximately)

$$\tau_{ny} = \frac{V}{h t_{w0}}, \quad V = \left(1.05 \rho_0 A + p \right) + \frac{F}{2L} \left(L - \frac{k}{2} \right)$$

(11)

From torsion

$$\tau_t = \frac{2 M_t}{2 b h t_{w0}}, \quad M_t = \frac{F}{2L} \left(L - \frac{k}{2} \right) \frac{b}{2} + \frac{pLb}{4} \tag{12}$$

The constraint on shear buckling

if $\quad \tau = \tau_V + \tau_t \le k_{\tau 0} f_y / \sqrt{3}, k_{\tau 0} = 1 \tag{13}$

$$\lambda_\tau = \sqrt{\frac{f_y}{k_\tau \sigma_e \sqrt{3}}} \le 0.84, \quad k_\tau = 5.34 + \frac{4}{\alpha^2}, \quad \alpha = \frac{a}{h} = \frac{L}{10 h} \tag{14}$$

i.e. $\quad t_{w.req} = \frac{2h}{31 \varepsilon \sqrt{k_\tau}} \tag{15}$

2.1.3. Compression from a wheel

According to Figure 1d

$$\sigma_{y1} = \frac{2F}{c t_{w0}}, \quad c = 50 + 2 \left(h_s + t_{f0} \right) = 50 + 2 x 100 = 250 \text{ mm} \tag{16}$$

If $\quad \sigma_{y1} \le k_y f_y, k_y = 1 \tag{17}$

$$\lambda_y = \sqrt{\frac{f_y}{k_{\sigma y} \sigma_e \frac{a}{c}}} \le 0.831 \tag{18}$$

From the diagram of EN13001-3-1 [1] $c/a = 250/1650 = 0.15$ and $\alpha = a/h = 1650/620 = 2.7$ $k_{\sigma y} = 1$

$$t_{w.req} = \frac{2h}{60.97\,\varepsilon}$$ (19)

The complex check

$$\left(\frac{|\sigma_x|}{f_{bx}}\right)^{e_1} + \left(\frac{|\sigma_y|}{f_{by}}\right)^{e_2} - V_0\left(\frac{|\sigma_x\sigma_y|}{f_{bx}f_{by}}\right) + \left(\frac{\tau}{f_{b\tau}}\right)^{e_3} \leq 1,\ e_1 = 1+k_x^4, e_2 = 1+k_y^4, e_3 = 1+k_x k_y k_{\tau 0}^2$$ (20)

$$V_0 = \left(k_x k_y\right)^6 \text{ if } \sigma_x\sigma_y \geq 0,\ V_0 = -1 \text{ if } \sigma_x\sigma_y \leq 0$$ (21)

In our case $k_x = k_y = k_{\tau 0} = 1$ (22)

$$\sigma_{red} = \sqrt{(\sigma_x + \sigma_y)^2 + \sigma_{y1}^2 - (\sigma_x + \sigma_y)\sigma_{y1} + 3\tau^2} \leq f_y$$ (23)

3. BUCKLING CONSTRAINTS OF THE UPPER FLANGE

3.1. Vertical and horizontal bending

Similarly to the constraint on web buckling

$$t_{f.req} = \frac{b}{0.673\,x\,28.42\,\varepsilon\sqrt{k_{\sigma y}}},\ k_{\sigma y} = \frac{8.2}{1.05 + \psi_y},\ \psi_y = \frac{\sigma_x - \sigma_y}{\sigma_x + \sigma_y}$$ (24)

3.2. Torsion

Similarly to the web

$$t_{f.req} = \frac{b}{31\varepsilon\sqrt{k_{\tau b}}},\ k_{\tau b} = 5.34 + \frac{4}{\alpha_b^2},\ \alpha_b = \frac{a}{b}$$ (25)

4. FATIGUE CONSTRAINT FOR THE WELD UNDER THE RAIL

According to the EN 13001-3-1 [1] the fatigue strength of a K butt weld for the number of cycles $N = 4\text{x}10^6$ is $\Delta\sigma_C = 112$ MPa, the allowed stress for the spectrum factor $s_3 = 2$

$$\Delta\sigma_{Rd} = \frac{\Delta\sigma_C}{\gamma_f \sqrt[3]{s_3}} = 71.1 \text{ MPa}$$ (26)

and for shear

$$\Delta\tau_{Rd} = \frac{\Delta\tau_C}{\gamma_f \sqrt[3]{s_3}} = 50.8 \text{ MPa}$$

(27)

The complex constraint on fatigue is expressed as

$$\eta = \left(\frac{\sigma_x + \sigma_y}{\Delta\sigma_{Rd}}\right)^3 + \left(\frac{\sigma_{y1}}{\Delta\sigma_{Rd}}\right)^3 + \left(\frac{\tau_V + \tau_t}{\Delta\tau_{Rd}}\right)^5 \leq 1$$ (28)

5. FATIGUE CONSTRAINT FOR FILLET WELDS JOINING THE TRANSVERSE DIAPHRAGMS

The fatigue strength [4]

$$\Delta\sigma_C = \alpha_p 63 \text{ MPa}$$ (29)

α_P is the coefficient of the effect of PWT, for ultrasonic treatment 1.3, for HiFIT high frequency impact treatment 1.6.

The allowed stress

$$\Delta\sigma_{f.adm2} = \frac{\Delta\sigma_C}{\gamma_f \sqrt[3]{2}}$$

(30)

The constraint is given by

$$\sigma_x \leq \Delta\sigma_{f.adm2}$$

(31)

6. THE COST FUNCTION

The cost function is formulated according to the fabrication sequence (Farkas & Jármai books [5,6,7,8]).

(1) Welding of the upper flange, webs and transverse diaphragms, PWT of the welds joining the diaphragms. Two forms of diaphragms are used: the 5 diaphragms near the span centre are cut according to the Figure 1a, the other 6 diaphragms are constructed according to Figure 1b.
The structural volume for this fabrication phase is

$$V_1 = L(ht_{w0} + bt_{f0}) + 6bht_s + 2.5bht_s\left(1 + \frac{1}{\alpha_P}\right), \; t_s = 6 \text{ mm}, \; \alpha_P = 1.6$$

(32)

The number of the assembled structural elements is $\kappa_1 = 14$, the factor of the complexity of assembly is $\Theta_1 = 3$. The welding cost consists of four parts: GMAW-C welding of Butt K welds under the rail (K_{w11}), GMAW-C welding of the fillet welds joining the other web, welding of the diaphragms (K_{w12}) and PWT of the welds of 5 diaphragms (K_t).

$$K_{w1} = k_w(\Theta_1 \sqrt{\kappa_1 \rho V_1} + 1.3 x 0.3394\, x10^{-3} a_w^2 L + K_{w11}), \; k_w = 1.0 \text{ \$/min}$$

(33)

$$K_{w11} = k_w 1.3 x 0.1520\, x10^{-3} a_{w1}^{1.94} L, \qquad a_{w1} = t_{w0}/2$$

(34)

$$K_{w12} = k_w 1.3 x 0.7889\, x10^{-3} a_w^2 L_w, \quad a_w = t_{w0}/4, \quad L_w = 2\left[6(b+2h) + 5\left(b + \frac{h}{\alpha_P}\right)\right]$$

(35)

$$K_t = k_w L_t T_0, \; L_t = 10b, \; T_0 = 0.0033 \text{ min/mm}$$

(36)

(2) Welding of the lower flange with two GMAW-C fillet welds

$$K_{w2} = k_w\left(\Theta_2 \sqrt{\kappa_2 \rho V_2} + 1.3 x 0.3394\, x10^{-3} a_w^2 2L\right),$$
$$\Theta_2 = 2, \; V_2 = V_1 + bt_{f0}L, \; \kappa_2 = 2$$

(37)

Welding of the two webs from 11x1500 mm parts with GMAW-C butt K-welds

$$K_{w3} = k_w\left(\Theta_2 \sqrt{11 \rho V_3} + 1.3 x 0.152\, x10^{-3} x10h\left(\frac{t_{w0}}{2}\right)^{1.94}\right), \; V_3 = Lht_{w0}/2$$

(38)

Welding of the two flanges from 11x1500 mm parts with GMAW-C butt K-welds

$$K_{w4} = k_w\left(\Theta_2 \sqrt{11 \rho V_4} + 1.3 x 0.152\, x10^{-3} x10 bt_{f0}^{1.94}\right), \; V_4 = Lbt_{f0}$$

(39)

Material cost

$$K_m = k_m \rho V_2, k_m = 1.0 \text{ \$/kg}$$

(40)

Total cost

$$K = K_m + K_{w1} + K_{w11} + K_{w12} + K_t + K_{w2} + 2K_{w3} + 2K_{w4}$$

(41)

7. RESULTS OF OPTIMIZATION

The results are given in Table 1.

Table 1: Dimensions and deflection in mm, stresses in MPa, volume in mm^3, costs in $. Minima are marked by bold letters.

h	710	660	620	600
b	340	380	420	440
t_{w0}	30	28	26	26
t_{f0}	40	40	40	40
σ_x	61.95	62.6	62.7	62.8
Equation (19)	26.9	25.0	23.5	22.7
Equation (10)	20.0	18.4	17.2	16.6
w_{max}	9.3	10.0	10.5	10.7
Equation (28)	0.978	0.995	0.992	0.983
V_2x10^{-8}	**8.153**	8.222	8.367	8.547
K_t	11.2	12.5	13.9	14.5
K	14230	13890	**13690**	13930

8. CONCLUSIONS

The optimization has been performed by using a MathCAD program. Since the welding cost depends on the web thickness, the cost can be decreased by decrease of web thickness or web height. This decrease is stopped by the increase of cost caused by the increase of flange width. The web thickness is determined by the constraint on the maximal stress from the wheel load. In the systematic search we select a b and for this value h is searched, which fulfils the constraints. The web thickness is determined by the quality of the weld under the rail. Therefore, it is necessary to use high quality butt K weld. The governing constraints are the constraint on the compressive stress under rail and those on the fatigue. η should be smaller than 1 and σ_x should be smaller than $\Delta\sigma_{f.adm2} = 64.0$. The constraint of Equation 15 is passive.

ACKNOWLEDGEMENTS

The research was supported by the TÁMOP 4.2.4.A/2-11-1-2012-0001 priority project entitled 'National Excellence Program - Development and operation of domestic personnel support system for students and researchers, implemented within the framework of a convergence program, supported by the European Union, co-financed by the European Social Fund. The research was supported also by the Hungarian Scientific Research Fund OTKA T 109860 projects and was partially carried out in the framework of the Center of Excellence of Innovative Engineering Design and Technologies at the University of Miskolc.

REFERENCES

[1] EN 13001-3-1: 2010. Cranes – General design – Part 3-1: Limit states and proof competence of steel structure.
[2] Jármai,K. Pahlke,H. & Farkas,J. 2014. Cost savings using different post welding treatments on an I-beam subject to fatigue load. *Welding in the World* 58: 691-698.
[3] BS 2573-1: 1983.Rules for the design of cranes. Part 1. Specification for classification, stress calculations and design criteria for structures.
[4] Eurocode 3 -1-9: 2005. Design of steel structures. Fatigue strength of steel structures.
[5] Farkas,J. & Jármai,K. 2003. *Economic design of metal structures.* Rotterdam: Millpress.
[6] Farkas,J. & Jármai,K. 2008. *Design and optimization of metal structures.* Chichester, UK, Horwood Publishing.
[7] Farkas,J.& Jármai,K. 2013. *Optimum design of steel structures.* Heidelberg etc. Springer.
[8] Farkas,J. & Jármai,K. 2015. *Fémszerkezetek innovatív tervezése.* Miskolc, Gazdász-Elasztik Kiadó és Nyomda. (Innovative design of metal structures). (In Hungarian), 624 p.

6th International Conference
"Computational Mechanics and Virtual Engineering "
COMEC 2015
15-16 October 2015, Braşov, Romania

CTE'S POLYNOMIAL CURVES OF A THIN COMPOSITE SANDWICH

Horaţiu Teodorescu-Draghicescu[1], Sorin Vlase[1], Dana Luca Motoc[1], Mircea Mihalcica[1]

[1] Transilvania University of Brasov, Brasov, ROMANIA, e-mail: draghicescu.teodorescu@unitbv.ro

Abstract: *This paper presents coefficients of thermal expansion (CTE's) polynomial tendency curves of rank two in case of a thin composite sandwich structure with dissimilar skins. This structure presents following layers: 1 layer – RT500 glass roving fabric/ 2 layers – RT800 glass roving fabric/ 1 layer – CSM450 chopped strand mat/ 1 layer – nonwoven polyester mat/ 1 layer – CSM450 chopped strand mat/ 1 layer – gelcoat. Thermal expansions have been measured using a DIL 420 PC dilatometer from NETZSCH GmbH, on both glass fabric reinforced polyester skin and for the whole structure. The coefficients of thermal expansion have been experimentally determined only for the structure's upper skin.*
Keywords: *coefficients of thermal expansion, polynomial curves, thermal expansion, composite sandwich, dissimilar skins*

1. INTRODUCTION

Composite sandwich structures are widely used in many applications in which the ratio between strength and specific weight must be very high. One of the most important characteristics of a sandwich structure is represented by the bond between skins and core. This link is essential for the structure's subsequent loading. Stiffness represents also a very important feature of a sandwich structure. Three and four point bending tests as well as compression tests have been accomplished to obtain useful mechanical properties regarding the sandwich structures. General references [1], [5], [7], [9], [25] and [31] put into evidence the huge problematics regarding the composite materials. Behavior of composite structures subjected to the influence of temperature, modeling and analysis can be found in references [2], [15], [27] and [28]. Other important researches regarding theoretical as well as experimental approaches on various composites are highlighted in references [3], [4], [6], [8], [10-14], [16-24], [26], [29], [30], [32-37].

2. MATERIAL AND PROCEDURE

Thermal response of a sandwich structure with thin nonwoven polyester mat as core has been experimentally determined. The structure presents dissimilar skins from which one is a glass fabric reinforced polyester. A sandwich plate has been manufactured at Compozite Ltd., Brasov with the following structure:

- One layer RT500 glass roving fabric of 500 g/m^2 specific weight;
- Two layers RT800 glass roving fabric of 800 g/m^2 specific weight;
- One layer CSM450 chopped strand mat of 450 g/m^2 specific weight;
- One layer 4 mm thick nonwoven polyester mat as core;
- One layer CSM450 chopped strand mat of 450 g/m^2 specific weight;
- A usually used gelcoat layer which is a pigmented polyester resin.

From this sandwich plate, specimens have been cut for the experimental characterization. Thermal expansions have been measured using a DIL 420 PC dilatometer from NETZSCH GmbH, Germany, on both glass fabric reinforced polyester skin and for the whole structure. The coefficients of thermal expansion have been experimentally determined only for the structure's upper skin. For each sample, two successive heating stages, in order to size the influence of the thermal cycling and temperature interval from 20⁰C to 250⁰C at a heating rate of 1 K/min, have been used. To eliminate the system errors, the dilatometer has been calibrated by measuring a standard SiO_2 specimen under identical conditions.

3. RESULTS

Distributions of thermal expansion and coefficients of thermal expansion (also known as technical alpha) have been presented in Figs. 1-4.

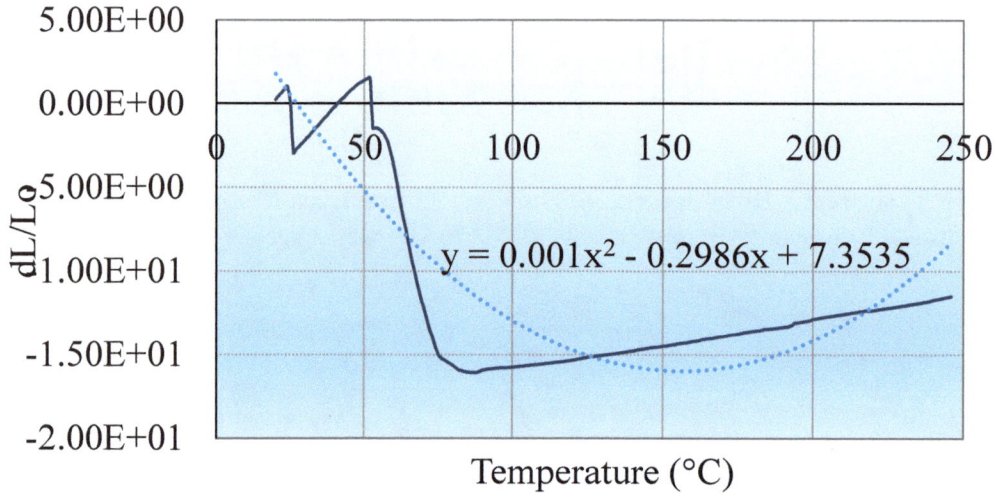

The plot is labeled with axis "dL/Lo" (y-axis) and "Temperature (°C)" (x-axis), with the equation $y = 0.001x^2 - 0.2986x + 7.3535$

Figure 1: Distribution of upper skin's thermal expansion

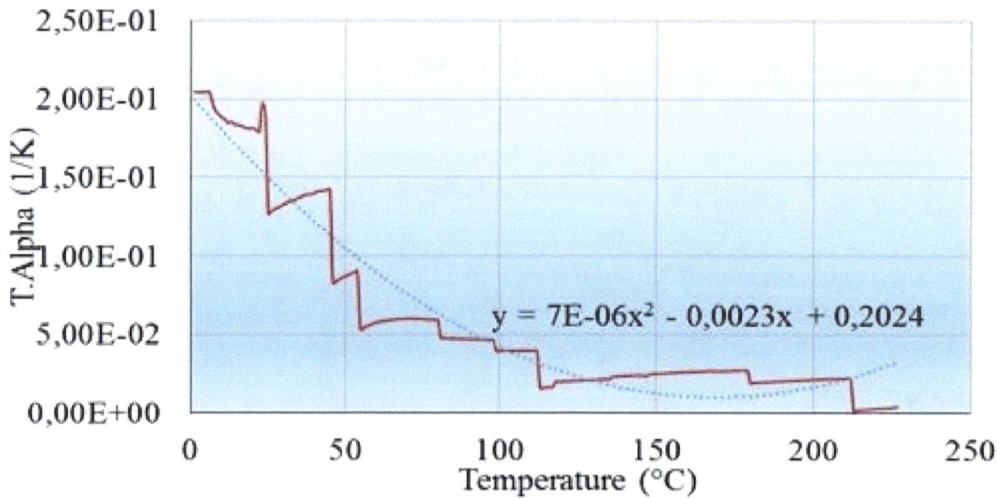

The plot is labeled with axis "T.Alpha (1/K)" (y-axis) and "Temperature (°C)" (x-axis), with the equation $y = 7E\text{-}06x^2 - 0,0023x + 0,2024$

Figure 2: Distribution of upper skin's coefficient of thermal expansion (technical alpha)

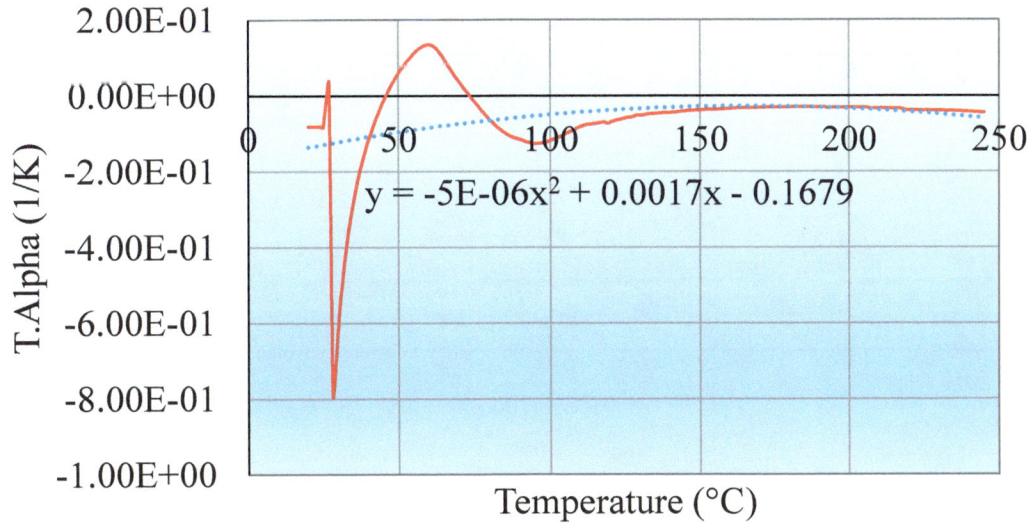

Figure 3: Distribution of sandwich structure's coefficient of thermal expansion in the first heating stage

The equation shown in Figure 3: $y = -5E-06x^2 + 0.0017x - 0.1679$

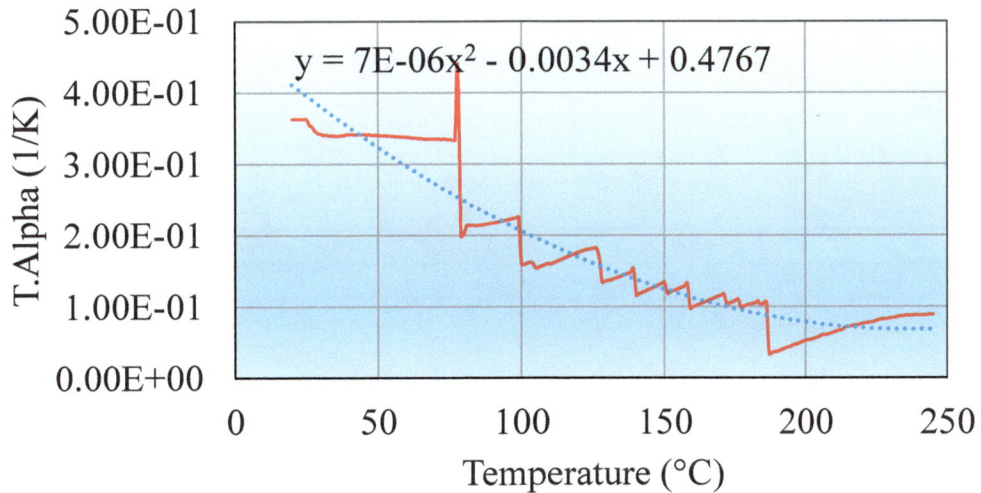

Figure 4: Distribution of sandwich structure's coefficient of thermal expansion in the second heating stage

The equation shown in Figure 4: $y = 7E-06x^2 - 0.0034x + 0.4767$

4. CONCLUSIONS

In Fig. 1, the negative thermal expansion in the first heating stage is due to the beginning of curing in the upper skin's structure. Regarding Fig. 4, in the second heating stage, the significant peak is due to the high shrinkage that took place in the sandwich structure. An application of this kind of structure is an underground large spherical cap shelter and formed by twelve curved shells bonded together. To withstand the soil weight, the wall structure present a variable thickness. This kind of structure can be used in outdoor applications also. For each distribution, a polynomial tendency curve of rank two has been generated. These polynomial tendency curves are important to model a specific experimental test and their parametric form is:

$$y = Ax^2 + Bx + C \tag{1}$$

For upper skin's thermal expansion, the parameters A, B and C are presented in table 1.

Table 1: A, B and C parameters in case of upper skin's thermal expansion

A	B	C
0.001	-0.2986	7.3535

For upper skin's coefficient of thermal expansion, these parameters are shown in table 2.

Table 2: A, B and C parameters in case of upper skin's coefficient of thermal expansion

A	B	C
$7 \cdot 10^{-6}$	-0.0023	0.2024

For the sandwich structure's coefficient of thermal expansion in the first heating stage, these parameters are presented in table 3.

Table 3: A, B and C parameters in case of sandwich structure's coefficient of thermal expansion in the first heating stage

A	B	C
$-5 \cdot 10^{-6}$	0.0017	-0.1679

In case of the sandwich structure's coefficient of thermal expansion in the second heating stage, the parameters A, B and C are shown in table 4.

Table 3: A, B and C parameters in case of sandwich structure's coefficient of thermal expansion in the first heating stage

A	B	C
$7 \cdot 10^{-6}$	-0.0034	0.4767

REFERENCES

[1] Ashbee KHG., Fundamental principles of fiber reinforced composites, Technomic Publishing Co. Inc., Lancaster-Basel, 1989.

[2] Teodorescu-Draghicescu H, Vlase S, Motoc DL, Chiru A., Thermomechanical Response of a Thin Sandwich Composite Structure. Engineering Letters, Sept. 2010, 18(3):EL_18_3_08.

[3] Doxsee LE, Rubbrecht P, Li L, Verpoest L, Scholle M., Delamination growth in composite plates subjected to transverse loads, Journal of Composite Materials, Vol.27, No.8, 1993, p. 764-781.

[4] Gao Z., A cumulative damage model for fatigue life of composite laminates, Journal of Reinforced Plastics and Composites, Vol. 13, 1994, p. 128-141.

[5] Gay D., Matériaux composites, Editions Hermes, Paris, 1991.

[6] Hashin Z., Failure criteria for unidirectional fiber composites, Journal of Applied Mech., No. 47, 1980, p. 329-334.

[7] Reddy JN., Mechanics of Laminated Composite Plates: Theory and Analysis, CRC, Press. Boca Raton, FL, 1997.

[8] Thesken JC., A theoretical and experimental investigation of dynamic delamination in composites, Fatigue and Fracture of Engineering Materials and Structures, Vol. 18, No. 10, 1995, p. 1133-1154.

[9] Cristescu ND, Craciun EM, Soos E., Mechanics of elastic composites, Chapman & Hall/CRC; 2003.

[10] Teodorescu-Draghicescu H, Vlase S., Homogenization and Averaging Methods to Predict Elastic Properties of Pre-Impregnated Composite Materials. Computational Materials Science, 50, 4, 2011, p. 1310–1314.

[11] Teodorescu-Draghicescu H, Vlase S, Scutaru ML, Serbina L, Calin MR., Hysteresis Effect in a Three-Phase Polymer Matrix Composite Subjected to Static Cyclic Loadings, Optoelectronics and Advanced Materials – Rapid Communications (OAM-RC), 5, 3, March 2011.

[12]   Vlase S, Teodorescu-Draghicescu H, Motoc DL, Scutaru ML, Serbina L, Calin MR., Behavior of Multiphase Fiber-Reinforced Polymers Under Short Time Cyclic Loading. Optoelectronics and Advanced Materials – Rapid Communications (OAM-RC), 5, 4, 2011.

[13]   Vlase S, Teodorescu-Draghicescu H, Calin MR, Serbina L., Simulation of the Elastic Properties of Some Fibre-Reinforced Composite Laminates Under Off-Axis Loading System. Optoelectronics and Advanced Materials – Rapid Communications, 5 (3–4), 2011, p. 424–429.

[14]   Teodorescu-Draghicescu H, Stanciu A, Vlase S, Scutaru ML, Calin MR, Serbina L., Finite Element Method Analysis Of Some Fibre-Reinforced Composite Laminates, Optoelectronics and Advanced Materials – Rapid Communications (OAM-RC), 5, 7, July 2011.

[15]   Marin M, Agarwal RP, Mahmoud SR., Modeling a Microstretch Thermoelastic Body with Two Temperatures. Abstract And Applied Analysis, Article No. 583464, 2013.

[16]   Teodorescu-Draghicescu H, Vlase S, Goia I, Scutaru ML, Stanciu A, Vasii M., Tensile Behaviour of Composite Specimens Made From Chopped Strand Mat Reinforced Polyester Resin. 24th Danubia – Adria Symposium on Developments in Experimental Mechanics, 19-22 September, 2007, Sibiu, Editura Universităţii Lucian Blaga Sibiu, p. 255–256.

[17]   Remy O, Wastiels J., Development of impregnation technique for glass fibre mats to process textile reinforced cementitious composites. Plastics, Rubber and Composites, 39 (3–5), 2010, pp. 195–199.

[18]   Owen MJ., Static and fatigue strength of glass chopped strand mat/polyester resin laminates. Short fiber Reinforced composite Materials, ASTMSTP. BA Sanders, Ed.; 1982, p. 64–84.

[19]   Marin M, Stan G., Weak solutions in Elasticity of dipolar bodies with stretch. Carpathian Journal Of Mathematics, Vol. 29 (1), 2013, p. 33–40.

[20]   Hill R, Cowking A, Carswell WS., An acoustic emission study of stress corrosion in a chopped strand mat GFRP composite. Composites, Elsevier 1989; 20(3):215:222.

[21]   Corum JM, Battiste RL, Ruggles MB, Ren W., Durability-based design criteria for a chopped-glass-fiber automotive structural composite. Composites Science and Technology, 61 (8), 2001, p. 1083–1095.

[22]   Marin M., On the minimum principle for dipolar materials with stretch. Nonlinear Analysis: R.W.A., Vol. 10 (3), 2009, p. 1572–1578.

[23]   Craig PD, Summerscales J., Poisson's ratios in glass fibre reinforced plastics. Composite Structures, 9 (3), 1988, p. 173–188.

[24]   Giroux C, Shao Y., Flexural and Shear Rigidity of Composite Sheet Piles. Journal of Composites for Construction, 7 (4), 2003, p. 348–355.

[25]   Berthelot JM., Composite Materials: Mechanical Behavior and Structural Analysis. Springer Verlag, New York, Berlin, Heidelberg, 1999.

[26]   Ionita A, Weitsman YJ., On the mechanical response of randomly reinforced chopped-fibers composites: Data and model. Composites Science and Technology, 66 (14), 2006, p. 2566–2579.

[27]   Segal L., The thermal expansion of reinforced nylon-6 composites through the matrix glass transition temperature. Polymer Engineering & Science., 19 (5), 1979, p. 365–372.

[28]   Wang J, Kelly D, Hillier W., Finite Element Analysis of Temperature Induced Stresses and Deformations of Polymer Composite Components. Journal of Composite Materials, 1 (43), 2009, p. 2639–2652.

[29]   Marin M, Agarwal RP, Mahmoud SR., Nonsimple material problems addressed by the Lagrange's identity. Boundary Value Problems, Article No.135, 2013, DOI: 10.1186/1687-2770-2013-135.

[30]   Naughton BP, Panhuizen F, Vermeulen VC., The Elastic Properties of Chopped Strand Mat and Woven Roving in G.R. Laminae. Journal of Reinforced Plastics and Composites, 4 (2), 1985, p. 195–204.

[31]   Kaw AK., Mechanics of composite materials. Taylor & Francis Group, New-York, 2006.

[32]   Modrea A, Vlase S, Teodorescu-Draghicescu H, Calin MR, Astalos C., Properties of advanced new materials used in automotive engineering. Optoelectronics and Advanced Materials – Rapid Communications (OAM-RC), 7 (5–6), 2013, p. 452–455.

[33]   Teodorescu-Draghicescu H, Vlase S, Chiru A, Purcarea R, Munteanu V., Theoretical and Experimental Approaches Regarding the Stiffness Increase of Fibre-Reinforced Composite Structures. Proc. of the 1st Int. Conf. on Manufacturing Engineering, Quality and Production Systems (MEQAPS'09), Transilvania University of Brasov, Romania, WSEAS Press, 2009, p. 449–452.

[34]   Vlase S, Purcarea R, Teodorescu-Draghicescu H, et al., Behavior of a new Heliopol/Stratimat300 composite laminate. Optoelectronics and Advanced Materials-Rapid Communications (OAM-RC). 7, (7-8), 2013, p. 569–572.

[35]   Vlase S, Teodorescu-Draghicescu H, Calin MR, et al., Advanced Polylite composite laminate material behavior to tensile stress on weft direction. Journal of Optoelectronics and Advanced Materials (JOAM). 14, (7-8), 2012, p. 658–663.

[36]   Niculita C, Vlase S, Bencze A, et al., Optimum stacking in a multi-ply laminate used for the skin of adaptive wings. Optoelectronics and Advanced Materials - Rapid Communications (OAM-RC), 5, (11), 2011, p. 1233–1236.

[37]   Modrea A, Vlase S, Calin MR, et al., The influence of dimensional and structural shifts of the elastic constant values in cylinder fiber composites. Journal of Optoelectronics and Advanced Materials (JOAM). 15, (3-4), 2013, p. 278–283.

6th International Conference
"Computational Mechanics and Virtual Engineering "
COMEC 2015
15-16 October 2015, Brașov, Romania

ON THE TACTILE SENSING BY USING SMART MATERIALS

Ligia Munteanu[1], Dan Dumitriu[1], Veturia Chiroiu[1], Cornel Brisan[2], Mircea Bara[2], Doina Marin[1]

[1] Institute of Solid Mechanics, Romanian Academy, Bucharest, e-mails: ligia_munteanu@hotmail.com, dumitri04@yahoo.com, veturiachiroiu@yahoo.com, marin_doina@yahoo.com
[2] Technical University of Cluj-Napoca, e-mails: Cornel.Brisan@mmfm.utcluj.ro, bmvbara@yahoo.com,

Abstract : *The aim of this work is to present a virtual experiment concerning the recognizing of the shape and texture of a 3D object performed by simulation the action of an array of nanopiezotronic transistors integrated into the skin. A flexible finger with the muscles made of Nitinol wires and the skin made of auxetic material is considered. The array of nanopiezotronic transistors makes possible the detection of the pressure-induced changes in the auxetic skin. An inverse problem is solved in order to find these parameters from the condition that the n-ellipsoid best fits the set of data points probed by touch with the finger*
Keywords : *Nitinol, auxetic material, tactile sensing, shape and texture*

1. INTRODUCTION

Robotic tactile sensing involves techniques for knowledge transfer from human to robot. The robotic tactile sensing in touching, grasping and manipulating of the objects is the base for exploring and differentiating the objects from one another with respect to shape, surface texture, stiffness, temperature etc. [1, 2].
Geometrically and structurally, the skin is a complex mechanical system supported by the deformable system of muscles and tendons. The stiffness of various skin layers significantly varies with epidermis being considerably stiffer than the dermis (the Young's modulus of base layer, i.e. the epidermis is 10–10000 times that of the dermis). Skin acts as a multilayered, nonlinear, nonhomogeneous and viscoelastic medium in order to convert the surface indentation into stress and strain fields. The thickness of the skin in adult humans varies between 0.6–0.8 mm and the Young's modulus is around $4 \times 105 \, \text{N/m}^2$.
In order to mimic the tactile sensing capabilities of the human skin, a flexible finger with the muscles and skin made of Nitinol (NiTi) wires and auxetic material, respectively, is considered in this paper, in the spirit of the article Munteanu et al. [3]. The robotic shape detection of the objects within the contact area (7–12 mm) of the fingertip may be realized by using the interface piezotronic effect. The piezotronic effect arises as a result of the polarization of non-mobile ions in the crystal, unlike the piezoresistive effect which results from a change in band gap, charge carrier density, or density of states in the conduction band of the strained semiconductor material. Therefore, the piezoresistive effect is a symmetric volume effect without polarity, whereas the piezotronic effect is an interface effect that asymmetrically modulates local contacts at different terminals of the device because of the polarity of the piezoelectric potential [4, 5].
As mentioned above, the muscles are made of NiTi wires. The phase change in the NiTi wires is achieved by heat exchange with a heat source and a heat sink. The actuation frequency of the NiTi wires is only dependent on the rate of heat transfer with its surroundings. The heat transfer mechanism for most Nitinol wires are based on resistive heating and cooling with forced convection or natural convection. Because this is an inefficient heat exchange mechanism which requires the use of electrical power, we chose the semiconductors for which Peltier effect has shown high actuation frequency. In other words, we use the forced convection heating and cooling to actuate the NTi wires. This can overcome

the low energy density resistive heating systems and the low efficiency of the thermoelectric heat transfer mechanism, even though it should need additional devices such as a pump and valves.

Conventional foam exhibits pores with an average diameter of around 1mm, while the auxetic foam has a possible average diameter of a few micrometers or even nanometers. Processing manufacturing techniques of auxetic materials can control various features of the pore shapes and is performed by a compression process [6-10].

2. INVERSE PROBLEM

Let us consider a flexible finger modeled as a cylindrically rod of length $L = 6$cm and radius $R = 0.5$cm, with three embedded Nitinol wires (yellow) in an aluminum matrix, at uniform absolute temperature T_0 (Figure 1). The NiTi wire has the length L and radius r, and are placed in a parallel arrangement to form the vertices of an equilateral triangle. The x-axis has the distal direction, the y-axis the radial (reference) direction, and z-axis the tangential direction. The rod is covered with a thin layer of auxetic material (red colour) representing the skin of the finger. Both the matrix and the NiTi wires are assumed to be initially straight at $t = 0$ and $T_0 = 33^0$ C. The NiTi wires are heated above the austenitic start temperature by passing an electrical current, and the deflected beam tends to return to the initial configuration. The NiTi alloy acts as an actuator transforming electrical energy into mechanical energy, annihilating the deformed shape of the rod. Topological view of the skin with hexagonal pores and the nanowires (red circle) is shown in Figure 2.

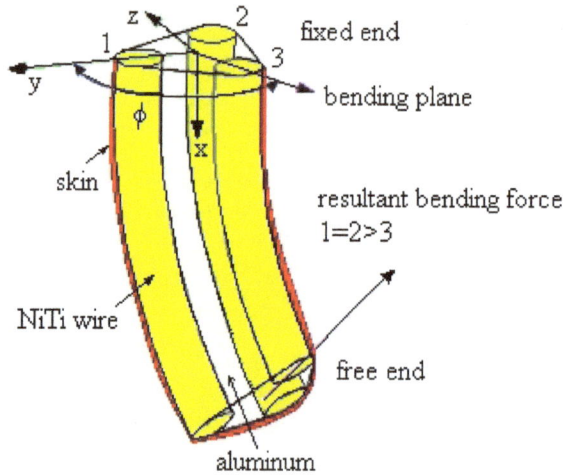

Figure 1: A flexible finger.

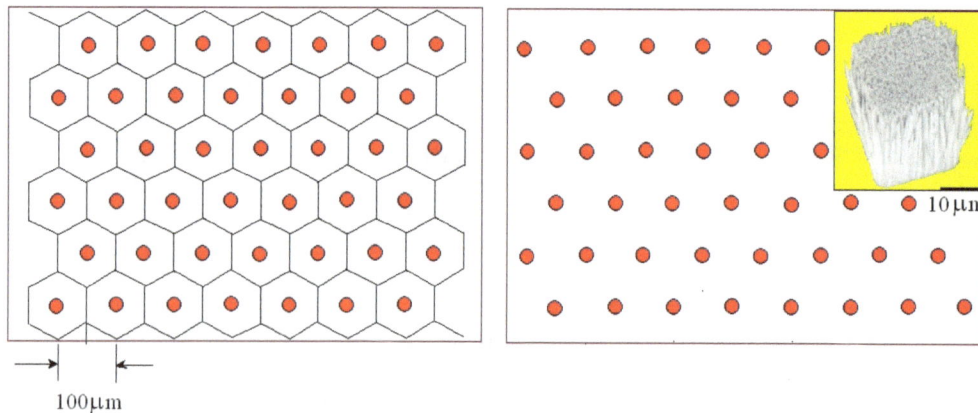

Figure 2: Topological view of the auxetic skin with hexagonal pores where the nanowires are positioned (red circle).

The operation of the gripper finger relies on the elastic deformation of three embedded NiTi wires (55% Ni, 45%Ti) in an aluminum matrix. Using a different force in each NiTi wire a range of extension forces causing the finger to bend according to the constraints provided by the end plate. The larger the force, the larger the resulting finger tip deflection. In addition to bending, the triangular arrangement enables the direction of fingertip movement to be controlled.

The finger motion is described by a complex set of equations containing the aluminum equations, the NiTi wires (muscles) equations, the equations of the auxetic skin coupled with the ZnO nanowires, the conditions on the interfaces between aluminum-NiTi wires, aluminum-auxetic material, NiTi wires-auxetic material, auxetic material- ZnO nanowires, boundary conditions and initial conditions. e assume that a finger is used to probe an object to detect its shape and the texture. The array of nanopiezotronic transistors makes possible the detection of the pressure-induced changes in the auxetic skin.

Although it is difficult to detect the shape of the object only with one finger, we still consider this variant for simulation the detection of the shape of objects with well-defined geometry such as balls or eggs. Instead, the texture can be detected using a single finger. To detect both the shape and texture, the finger can walk, rub and rotate on the surface of the object until a control parameter reaches a value proposed by algorithm.

3. RESULTS

Consider a damaged a graphite plate of length 9cm. height 5cm and thickness 1cm (figure 3). The material is strongly anisotropic and damages make the texture to contain shallows, cracks and bumps. Therefore the finger will have to touch the material along 25 arbitrary paths shown in figure 4. If results require, the path traveled by the finger will have to get thicker where the texture is difficult. The finger walks without press the surface of the object with the velocity of 1cm/sec. The surface irregularities, elevations and simples not exceed the limit of the spatial resolution of the sensor. Some arbitrary cross-sectional slices of the image of the object furnished by the inverse technique are shown in figure 5.

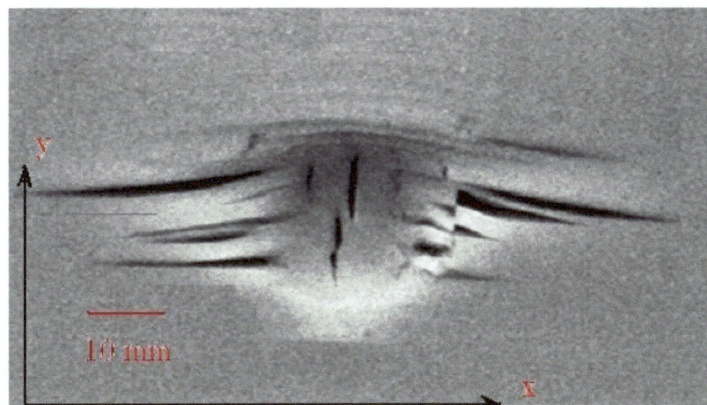

Figure 3: A damaged graphite plate.

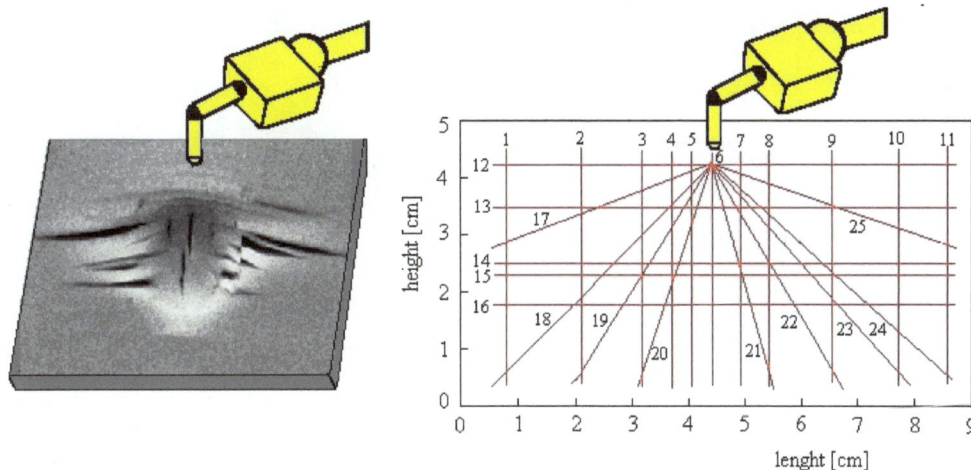

Figure 4: The finger tracking the plate. To the left the trajectories of the finger are shown.

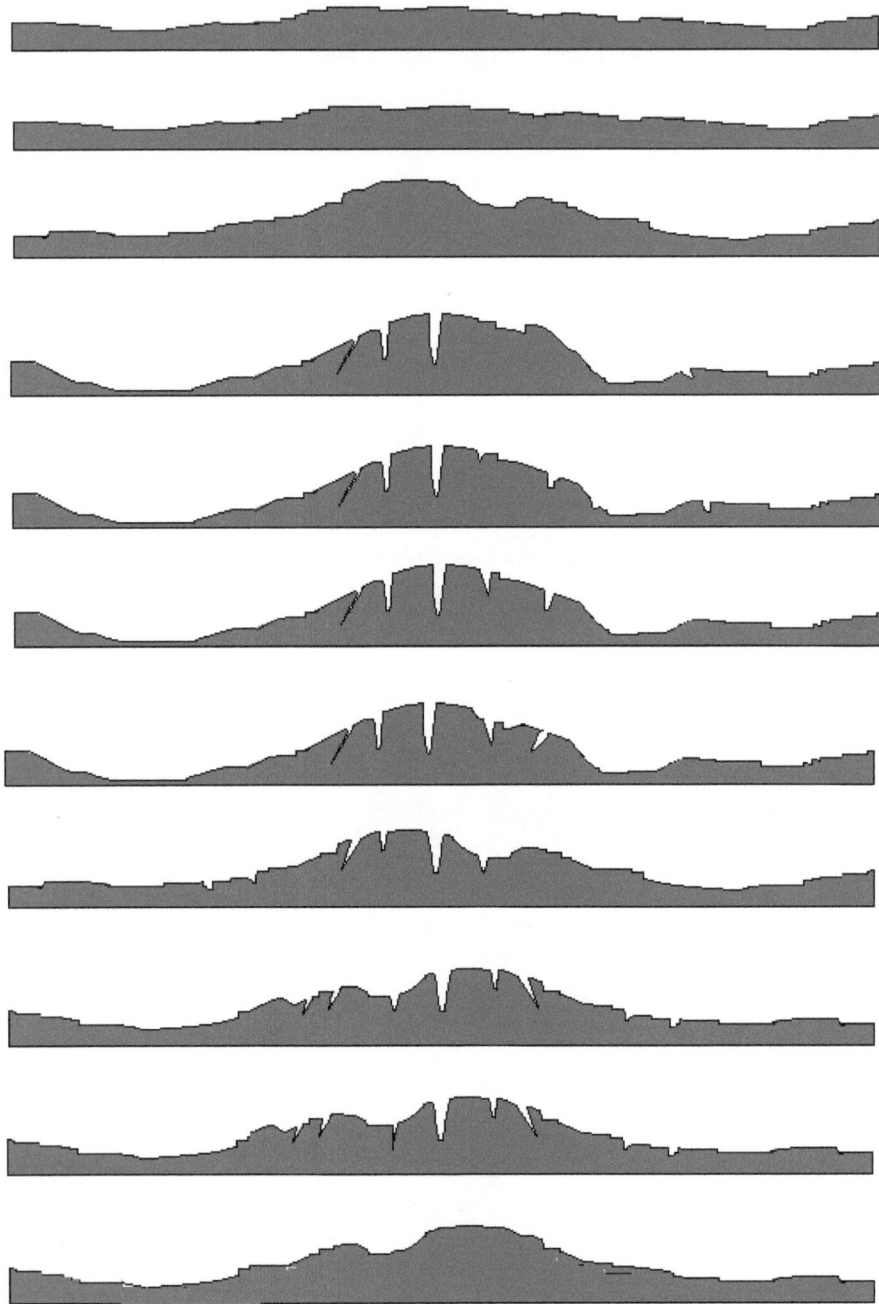

Figure 5: Some arbitrary cross-sectional slices of the image of the object.

4. CONCLUSION

The aim of this work is to present a virtual experiment concerning the recognizing of the shape and texture of a 3D object performed by simulation the action of an array of nanopiezotronic transistors integrated into the skin. A flexible finger with the muscles made of Nitinol wires and the skin made of auxetic material is considered. The array of nanopiezotronic transistors makes possible the detection of the pressure-induced changes in the auxetic skin. The shape and texture of the

3D object is best estimated by determining the surface and texture of the object as an n-ellipsoid defined by 12 parameters. An inverse problem is solved in order to find these parameters from the condition that the n-ellipsoid best fits the set of data points probed by touch with the finger.

ACKNOWLEDGEMENT.

The authors gratefully acknowledge the financial support of the National Authority for Scientific Research ANCS/UEFISCDI through the project PN-II-PT-PCCA-2011-3.1-0190, Contract nr.149/2012. The authors acknowledge the similar and equal contributions to this article.

REFERENCES

[1] Shikida, M., Shimizu, T., Sato, K., Itoigawa, K., Active tactile sensor for detecting contact force and hardness of an object. Sens. Actuators A, Phys. 103, 213–218, 2003.

[2] Fearing, R.S., Tactile sensing mechanisms. Int. J. Robot. Res., 9(3), 3–23, 1990.

[3] Munteanu, L., Dumitriu, D., Chiroiu, V., Brişan, C., Bara, M., Marin, D., *On the tactile sensing based on the smart materials*, CMC: Computers, Materials & Continua, 46(2), 79-103, 2015.

[4] Wang, Z.L., Piezotronics and Piezo-Phototronics. Springer, New York, 2013.

[5] Zhang, Y., Liu, Y., Wang, Z.L., Fundamental theory of piezotronics. Adv. Mater., 23(27), 3004-3013, 2011.

6] Chiroiu, V., Ionescu, M.F., Sireteanu, T., Ioan, R., Munteanu, L., On intrinsic time measure in the modeling of cyclic behavior of a Nitinol cubic block. Smart Materials and Structures, 24(3), 035022, 2015.

[7] Chiroiu, V., Munteanu, L., A flexible beam actuated by a shape memory alloy ribbon. Proc. of the Romanian Academy, Series A: Mathematics, Physics, Technical Sciences, Information Science 4(1), 2003.

[8] Munteanu, L., Chiroiu, V., Brişan, C., Dumitriu, D., Sireteanu, T., Petre, S., On the 3D normal tire/off-road vibro-contact problem with friction. Mechanical Systems and Signal Processing, 54-55, 377-393, 2015.

[9] Munteanu, L., Chiroiu, V., Serban, V., From geometric transformations to auxetic materials. CMC: Computers, Materials & Continua, 42(3), 175-203, 2014.

[10] Munteanu, L., Brişan, C., Donescu, St., Chiroiu, V., On the compression viewed as a geometric transformation. CMC: Computers, Materials & Continua, 31(2), 127–146, 2012.

6th International Conference
"Computational Mechanics and Virtual Engineering "
COMEC 2015
15-16 October 2015, Braşov, Romania

THE DETERMINATION OF THE ELASTIC CHARACTERISTICS OF UNIDIRECTIONAL PREPREG CARBON FIBER-REINFORCED EPOXY RESIN COMPOSITES

Mariana D. Stanciu, Ioan Curtu, Teodora Sturzu, Horatiu Teodorescu Draghicescu

Transilvania University of Brasov, Department of Mechanical Engineering, Brasov, Romania,
mariana.stanciu@unitbv.ro, curtui@unitbv.ro, sturzu_teodora@yahoo.com, hteodorescu@yahoo.com

Abstract: *The unidirectional prepreg carbon fiber-reinforced epoxy resin composites are used on a wide scale in various applications: automobile industry, recreation boat industry, chemical industry (pipes and tanks) etc. Depending on the stresses at which the composite parts are exposed to, they have different reinforcement structures (woven fabric, mats, unidirectional fibers). The axial loading of the structures having complex geometry implies unidirectional fibers reinforcement of the composite, therefore, it is of utmost importance to know the elasticity modulus, the fracture strength and the deformations induced. In the absence of these properties, the design and production of the structures would not be possible. This paper's objective is to determine the characteristic curve of the unidirectional prepreg carbon fiber-reinforced epoxy resin composites, as a result of the investigation upon two types of materials, belonging to different producers. The results suggested that between the two types of materials there are differences of approximately 9% among the values of the longitudinal elasticity modulus and of approximately 22% for the fracture strength.*
Keywords : *unidirectional prepreg carbon fiber-reinforced composite, longitudinal elasticity modulus, tensile test*

1. INTRODUCTION

The uniqueness and diversity of the characteristics by which composite materials may be described contributed to their growing importance at a global level. Starting from the lightest fly fishing rods, carbon fiber composite materials are used more often given their high value of the ratio strength – weight, as well as the easy methods of production. The goal of the composite materials is to obtain rigid and resistant components, yet having a low density. There materials contain reinforced fibers and polymer resins. Fibers are considered the basic component of composites, being the one which transmits forces. The polymer matrix transfers forces among fibers and offers resistance to corrosion, shock tolerance, thermal and environment stability [1, 2]. Some types of fibers may have identical chemical composition and similar mechanical properties, but they are different in structure, depending on the producer. Carbon fiber reinforced polymers (CFRP) are highly resistant and light. Carbon fibers are obtained by carbonization of polyacrylonitrile (PAN) fibers, pitch resins or the Rayon (through oxidation and thermal pyrolysis) at high temperatures. Then, by graphitization, the strength and elasticity of the fibers may be improved. Carbon fibers are obtained with diameters found between 9 and 17 *μm* and contain approximately 90% carbon. Research on unidirectional carbon fiber reinforced composites have emphasized values of the longitudinal modulus of elasticity E, varying between 136 GPa and 188 GPa depending on the type of the fibers, the volume ratio and the used matrix [3, 4]. This paper presents the experimental results determined on unidirectional, prepreg carbon fiber-reinforced epoxy resin composite specimens, and obtained by the void-pressure method.

2. TYPES OF SPECIMENS TESTED TO TRACTION

The studied composites were obtained from unidirectional carbon fibers having diameters between 0,008…0,1 mm; 1,9 g/cm^3 density; the ultimate tensile strength $\sigma_{rt} = 10,0 \dots 3,00\ GPa$, E=230…400 GPa and epoxy matrix. Two types of specimens were analyzed, encoded U and G, reinforced with unidirectional epoxy resin prepreg fibers (Fig.1). The shape and dimensions of the specimens tested to tensile are given in figure 2, a and b, having been obtained in conformity with the SR-EN ISO 527-2 specifications (Fig. 2, a and b).

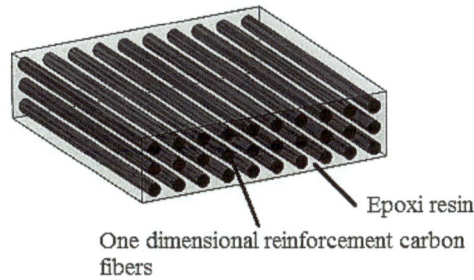

Figure 1: Reinforcement of composite with unidirectional fibers and epoxy resin

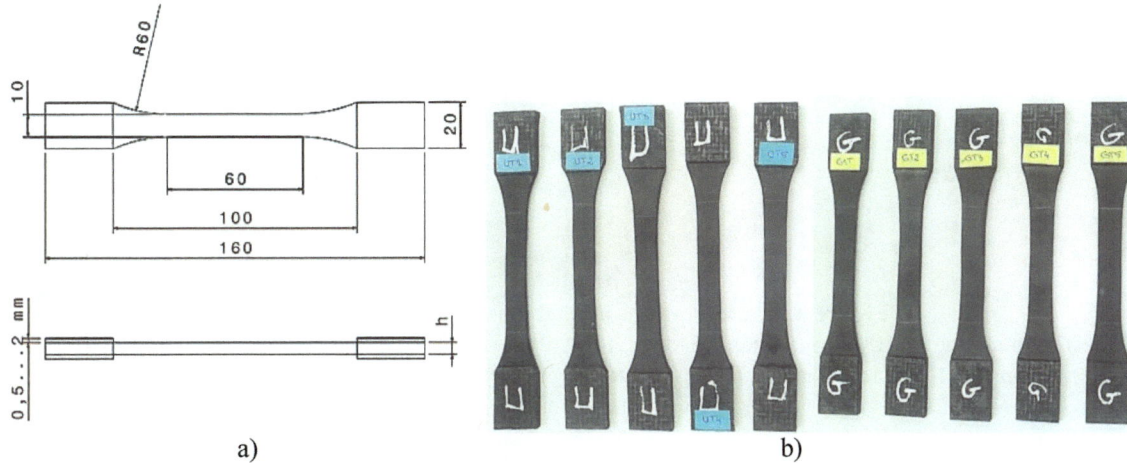

a)   b)

Figure 2: a) Geometry and dimensions of the specimens tested to tensile; b) Types of tested specimens

Before starting the experimental part, the dimensions of each specimen have been precisely measured: the thickness and the length of the cross section. These dimensions of the specimens, alongside specifications regarding the advancing speed of the machine, presented in Table 1, have been introduced as input data into the computer connected to the testing machine, which had a NEXYGEN soft to receive the experimental data from the machine and to process them statistically.

Table 1: Data regarding the dimensions of the tensile tested specimens

Samples	Thickness $h\ mm$	Width $b\ mm$	Area of cross section $A\ mm^2$	The speed of loading, mm/min	Reference lengths, mm
GT1	1,86	11,74	21,8364	1	50
GT2	1,91	11,81	22,5571	1	50
GT3	2,02	12,08	24,4016	1	50
GT4	1,94	11,47	22,2518	1	50
GT5	1,95	11,86	23,1270	1	50
UT1	1,89	10,53	19,9017	1	50
UT2	1,85	10,32	19,0920	1	50
UT3	1,88	10,79	20,2852	1	50
UT4	1,83	10,59	19,3797	1	50
UT5	1,76	10,93	19,2368	1	50

3. EQUIPMENT USED AND THE EXPRIMENTAL METHOD

The experiment has been accomplished by aid of the testing machine with constant tensile speed, in conformity with ISO 527, including: fixed part, with lower grip for fixing the specimen and a mobile part also having grip to fix the specimen and a loading cell [5,6]. The testing machine, an LS100 Lloyd's Instrument type, presented in figure 3, belonging to the Mechanical Engineering Department of Transylvania University of Brasov, is characterized by the following: maximum loading domain: 100 kN; maximum track: 840 mm; loading resolution: <0,01% of the loading cell used; extension resolution: <0,1 $microns$; loading cell: XLC-100K-A1; analysis software: NEXYGEN MT. The tensile testing machine allows the electronic recording of the experimental result, through the NEXYGEN Plus soft (Fig. 3, a). The extensometer, by SR EN ISO 527, has the capacity to determine the relative variation of the reference length of the specimen in every moment of the test (Fig 3, b).

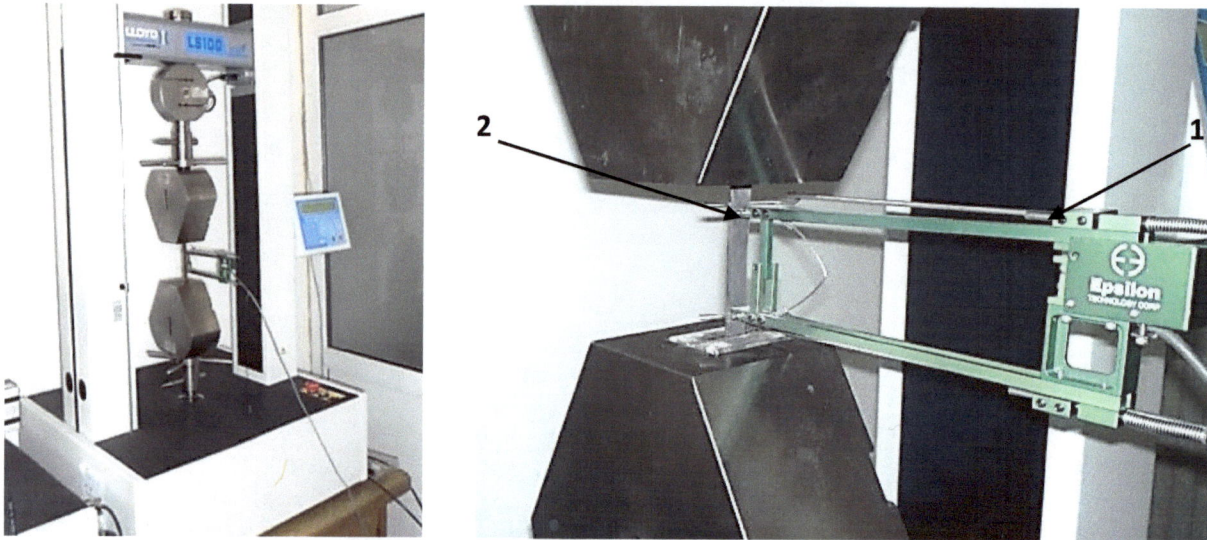

Figure 3: The equipment used at tensile stress:
a – tensile testing machine LS100 Lloyd's Instruments; b – Axial extensometer 1, specimen 2 (3542 Epsilon Technology Corp), belonging to the Mechanical Testing laboratory of Transylvania University of Brasov, Mechanical Engineerin Department

The specimens were fixed between the grips of the tensile testing machine, such that the axis of the specimen to coincide with the direction of the central tensile line of the assembly in the tightening system. The extensometer has been attached so as to prevent the slipping between the extensometer and the specimen.

4. RESULTS AND DISCUSSION

After testing, it has been observed that the obtained values differ from a type of specimens to another (U and G), as well as among the specimens of the same category (Table 2). The forces recorded at failure for the composite G are with 21,12 % lower than those of the material U (Fig.4). The mean value of the force at failure for specimens U is 18,48 kN, while for composite G is 22,41 kN.

Table 2: Values of elastic characteristics obtained through experimental test

Samples	Stiffness kN/m	Longitudinal modulus of elasticity MPa	The maximum stress at the maximum loading N	Ultimate tensile strength MPa
GT1	42376,93	106465,60	21436,03	1077,095
GT2	47165,31	123521,10	22469,83	1176,924
GT3	36330,79	89550,00	24814,55	1223,283
GT4	32732,51	84450,50	21162,11	1091,973
GT5	53416,97	138840,60	22192,74	1153,661
UT1	50544,28	115734,00	29695,68	1359,917
UT2	40241,78	89199,82	33022,59	1463,956
UT3	43718,09	89580,38	31746,28	1300,992
UT4	46691,63	104916,50	30918,57	1389,486
UT5	46056,84	99573,74	33987,68	1469,610

The specific strain ε encountered in the case of type G specimens are 4,59% higher than those of U composite (fig.5). The ultimate tensile strength σ_r for the type G material is higher with approximately 22,20% than that of U material (Fig 6). The unidirectional carbon fiber composite of type G is more elastic than the U type, the ratio between their longitudinal moduli of elasticity being E_G/E_U=0,9192 (Fig. 7).

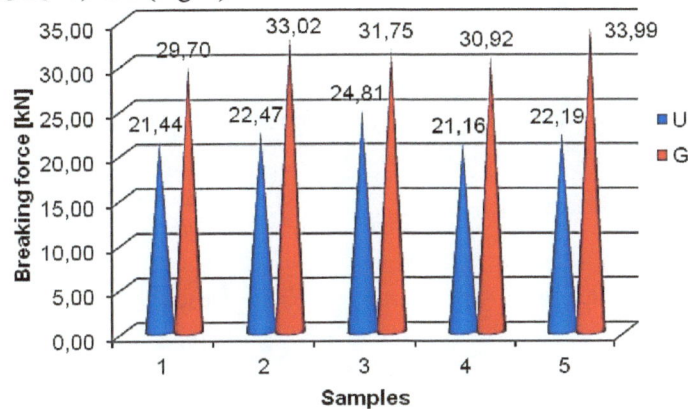

Figure 4: Variation of breaking force

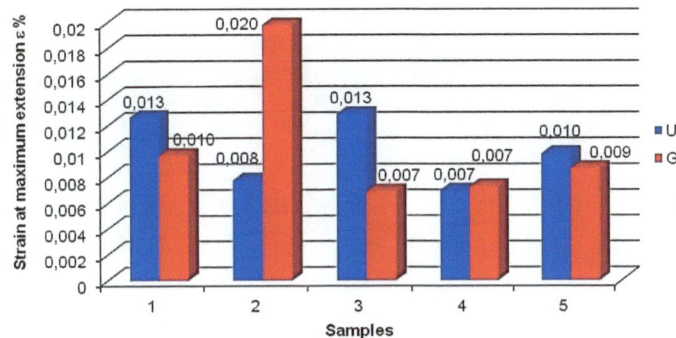

Figure 5: Specific strain ($\overline{\varepsilon_G} = 0,010469$; $\overline{\varepsilon_U} = 0,010009$)

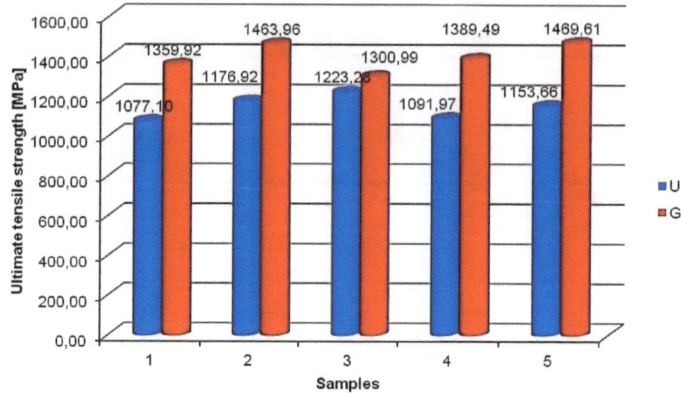

Figure 6: Variation of ultimate tensile strength

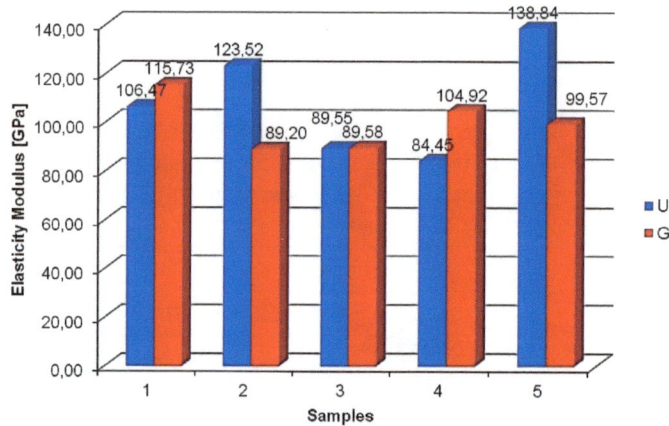

Figure 7: Variation of the longitudinal modulus of elasticity ($\overline{E_U} = 108570 \ MPa$; $\overline{E_G} = 99801 MPa$)

In figure 8, a, the characteristic curves of the U type of specimens are presented, where it can be observed that up until the value of 600 MPa in tensile strength the material has an elastic region. Around 650 MPa, the interlaminar stress is recorded, as the composite includes 4 layers of unidirectional fabric (Fig. 8, b). Between 650 and 1100 MPa, plastic deformations appear simultaneously with the failure of the layers which induces a rise in stress with approximately 24% for each layer in particular. A similar behaviour is observed in the case of the G type of composite, where the ultimate tensile strength of the layers is higher than that of the U composite, although the first interlaminar ultimate tensile strength appears around a 640 MPa value.

In figure 9 the elastic-visco-plastic model of the multilayered unidirectional carbon fiber, epoxy resin composite is presented. Elastic elements k determine the linear correlation stress (σ) - specific deformation (ε); plastic elements μ represents horizontal correlations σ-ε where the fracture of each lamina is recorded; the elements viscous η characterized slowly increase of tensile stress value with a low strain rate which is produced due to the interlaminar bonding and between fiber and matrix [8, 9].

a)

b)

c)

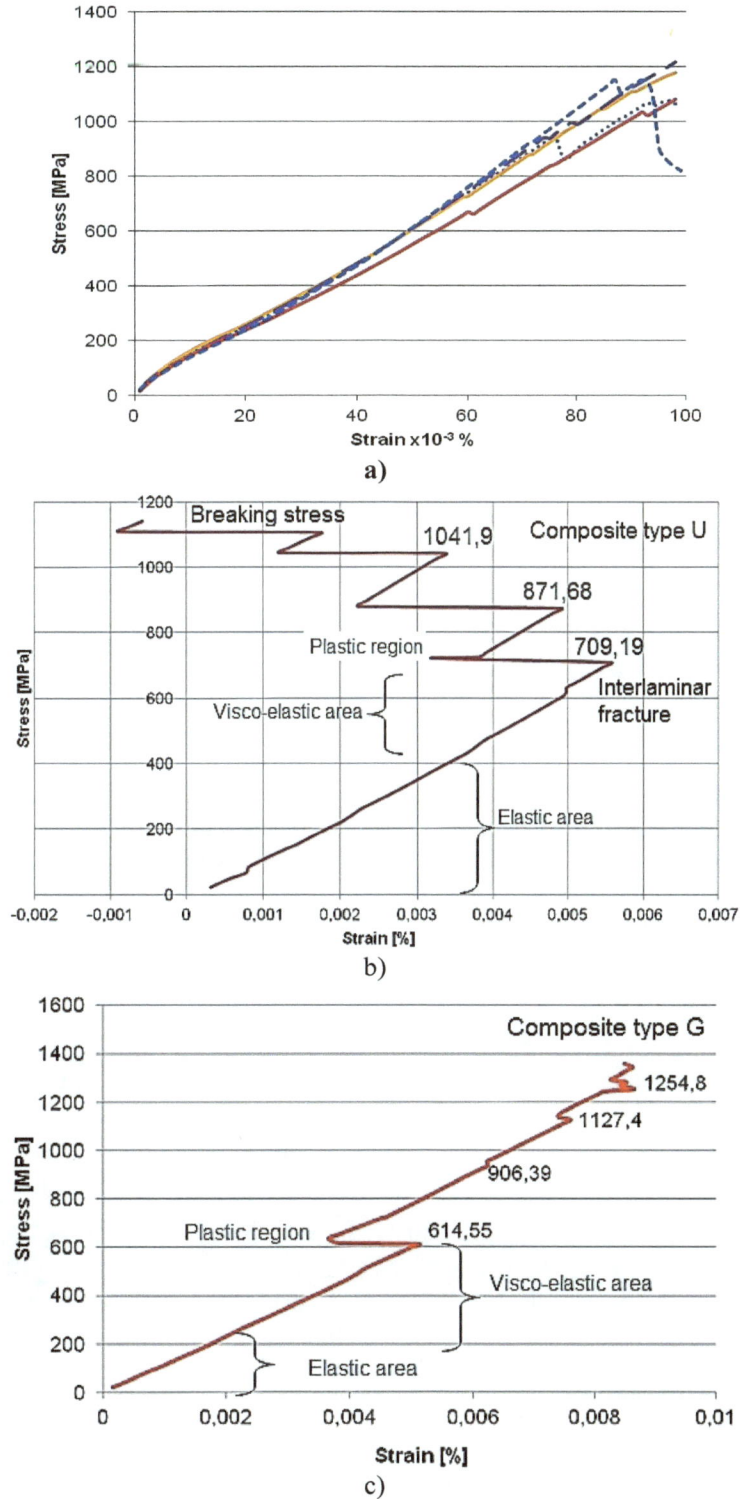

Figure 8: a) Characteristic curves of unidirectional composite; b) variation of interlaminar fracture stresses in case of U composite; c) variation of interlaminar fracture stresses in case of G composite;

33

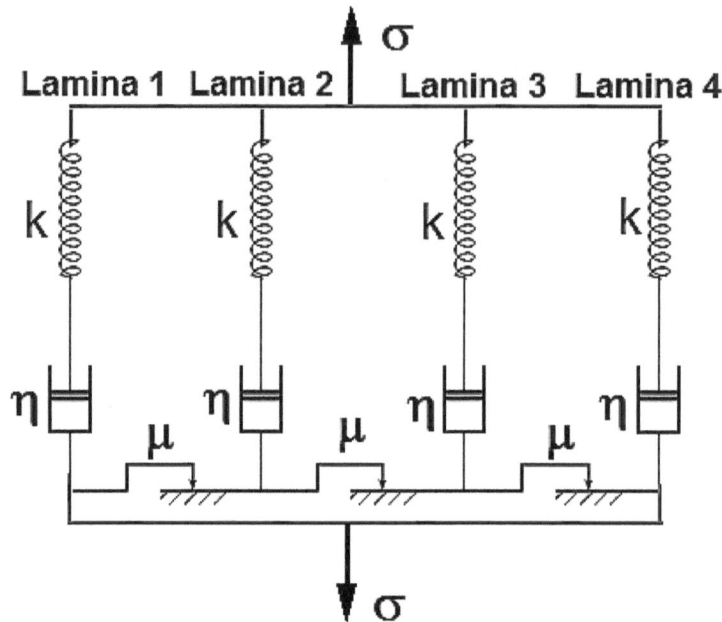

Figure 8: Elasto-visco-plastic model of composite (k - the elastic element type Hooke; η - characteristics of viscous element type Newton; μ - characteristics of plastic element type Saint Venant) [9]

5. CONCLUSION

In conclusion, the unidirectional carbon fiber reinforced composites present elastic characteristics which are different depending on three factors: the carbon fiber fabric producer, the imperfections developed when cutting the specimens, such that the loading axis does not coincide with the orientation of the fibers, the prepreg resin content and even production flaws. Yet, the advantages of the multilayered unidirectional carbon fabric composite are influenced by its capacity to store strain energy, its elastic-visco-plastic behaviour and its high tensile strength even in the case when cracks and interlaminar failures are developed.

ACKNOWLEDGEMENTS

This paper was supported by Program partnership in priority domains -PNII under the aegis of MECS -UEFISCDI, project No.PN-II-PT-PCCA-2013-4-0656.

REFERENCES

[1] Teodorescu-Draghicescu H., Sorin V., Stanciu M.D., Curtu I., Mihalcica M., Advanced Pultruded Glass Fibers-Reinforced Isophtalic Polyester Resin, Materiale Plastice 52 (1), 2015.
[2] Kormanikova E., Sejnoha M., SzavaI., Kotrasova K., Dogaru F., Sicakova A., Valenta R., Selected Chapters of Mechanics of Composite Materials I, Equilibria, Kosice, Slovakia, 2011.
[3] Stanciu M.D., Terciu O.M., Curtu I., Compozite Lignocelulozice – Aplicatii in Industria Automobilelor, Ed. Universitatii Transilvania din Brasov, 2014.
[4] Stoian V., Nagy Gyorgy T., Dan D., Gergely J., Daescu C., Materiale Compozite pentru Constructii, Ed. Politehnica, 2004.
[5] Stanciu M.D., Curtu I., Terciu O.M., Impact behavior of composite materials used for automotive interior parts, in Proceedings of the 10th HSTAM International Congress on mechanics, 25-27 May 2013, Chania Creta, Grecia, 2013.
[6] Terciu O. M., Curtu I., Cerbu C., Stanciu M.D., Testarea la tracţiune a materialelor compozite lignocelulozice cu aplicaţii în industria autovehiculelor, în Buletinul AGIR Creativitate, Inventică, Robotică, Anul XVII, nr. 1, Ed. AGIR, 2012.
[7] Soos, E., Resonance and stress concentration in a prestressed elastic solid containing a crack. an apparent paradox, International Journal of Engineering Science, 34 (3) 1996.
[8] Stanciu M.D, Curtu I., Reologie – suport de curs prima parte, Editura Universitatii Transilvania din Brasov, 2012.

6th International Conference
"Computational Mechanics and Virtual Engineering "
COMEC 2015
15-16 October 2015, Braşov, Romania

CONSIDERATIONS REGARDING THE HORIZONTAL FUEL CHANNELS IN THE CANDU 6 NUCLEAR REACTOR. PART 1 - PRESENTATION OF THE FUEL CHANNEL

Constantin D. Stănescu [1], Gabi Roşca-Fârtat [2], Constantin Popescu [3]

[1]Polytechnic University, Bucharest, ROMANIA, prof_cstanescu@yahoo.com.
[2,3]Polytechnic University, Bucharest, ROMANIA, rosca_gabi@yahoo.com, puiu_2001uss@yahoo.com.

Abstract: The aim of this study is to identify the fuel channel components based on which is made the installation into calandria of CANDU 6 nuclear reactor. The CANDU 6 is a 740 MW pressure tube reactor designed by Atomic Energy of Canada Limited (AECL) to provide safe and reliable nuclear power. The CANDU reactor design is based on the experience derived from preceding CANDU reactors and virtually every design feature of the latest CANDU reactor by ensuring its compliance with the latest Canadian nuclear regulations and the fundamental safety principles of the International Atomic Energy Agency (IAEA) Safety Standards. The design of the CANDU fuel channel is accordingly the result of continuing intensive engineering development of its major components. The fuel channel is designed to ensure a radiation exposure protection of workers and public, during the reactor operation. The reactor assembly of the CANDU 6 nuclear reactor consists of the horizontal, cylindrical, low-pressure calandria and the end-shield assembly. This enclosed assembly contains the heavy water moderator, the 380 fuel channels assemblies and the reactivity mechanisms. The fuel channels are one of the major distinguishing features of a CANDU reactor and their reliability is crucial to the performance of the reactor. Each fuel channel consists of four major components: the pressure tube, the calandria tube, the annulus spacers and the end fittings. Fuel bundles are enclosed in the fuel channels that pass through the calandria and the end-shield assembly. The fuel channels are assembled and installed into the calandria vessel at the reactor site following installation of the calandria. The fuel channel assembly is made according with the specific requirements of the tools and equipments, installation procedures and the quality assurance program.
Keywords: Candu reactor, Zirconium alloy, calandria tube, fuel channel, pressure tube, fuel bundle, end fitting, annulus spacer.

1. INTRODUCTION

The CANDU 6 is a 740 MW pressure tube reactor designed by Atomic Energy of Canada Limited (AECL) to provide safe and reliable nuclear power.
The nuclear reactors are designed and manufactured with respect of the specific requirements of codes and standards for the manufacture of components, equipment and systems required for the construction and operation of CANDU nuclear power plant.
The requirements for CANDU reactor design must comply with the codes of Canada Standards Association (CSA), Atomic Energy Control Board (AECB) of Canada and International Energy Agency (IAEA) which specify the specific and regulatory requirements.
The fuel channels, in number of 380, pressure tubes made of zirconium-niobium alloy, located inside the calandria tubes, chuck in the end fitting, are connected by the network pipes to the cooling system.
Each fuel channel consists of four major components: the pressure tube, the calandria tube, the annulus spacers and the end fittings. Fuel bundles are enclosed in the fuel channels that pass through the calandria and the end-shield assembly.

2. FUEL CHANNEL COMPONENTS

The CANDU reactor design is based on the experience derived from preceding CANDU reactors and virtually every design feature of the latest CANDU reactor is identical to, or is an evolutionary improvement of, an earlier proven design. The CANDU 6 reactors are the following general features of the fuel channels:
- 380 fuel channels;
- pressure tube made of Zi 2,5% NB, diameter 103,4 mm, thickness 4,19 mm;
- calandria tube made of Zircaloy 2, diameter 129,0 mm, thickness 1,37 mm;
- annulus spacers made of Incon.X750, coil diameter 4,83 mm, 4 pieces;
The time life of the fuel channel is for 30 years at 80% of its capacity and 24 years for full capacity functioning.

2.1. General presentation

The fuel channels are one of the major distinguishing features of a CANDU reactor, and their reliability is crucial to the performance of the reactor. The components of the fuel channel design are illustrated in Figure 1.

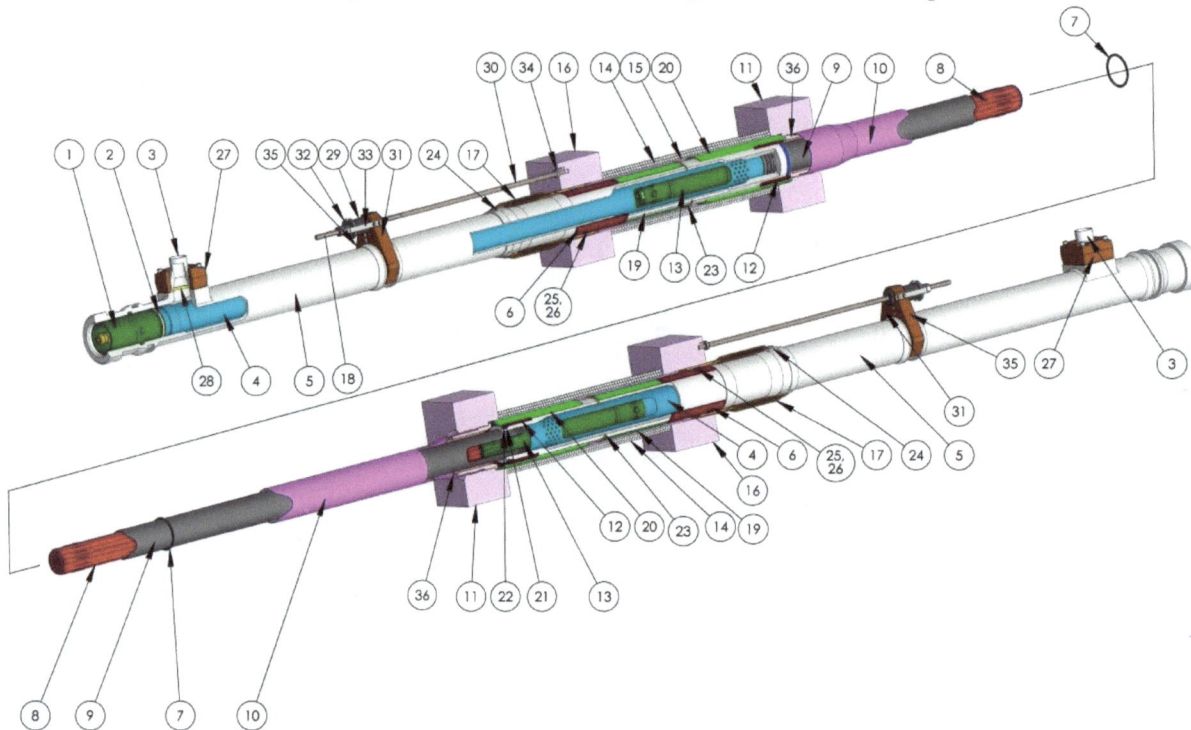

Figure 1: Representation of a CANDU fuel channel

1. Channel closure; 2. Closure seal insert; 3. Feeder coupling; 4. Liner tube; 5. End fitting body; 6. Outboard bearings; 7. Annulus spacer; 8. Fuel bundle; 9. Pressure tube; 10. Calandria tube; 11. Calandria tubesheet; 12. Inboard bearings; 13. Shield plug; 14. Endshield shielding balls; 15. Endshield lattice tube; 16. Fuelling tubesheet; 17. Channel annulus bellows; 18. Positioning assembly; 19. End fitting shielding sleeve; 20. Lattice tube shielding sleeve; 21. End fitting inner ring seal; 22. Elastic safety lock for end fitting inner ring seal; 23. Elastic safety lock for end fitting shielding sleeve; 24. Support ring for annulus bellows; 25. Annulus bellows outer ring seal; 26. Elastic safety lock for Annulus bellows outer ring seal; 27. Feeder coupling attachment; 28. Feeder gasket; 29. Rod positioning threaded part; 30. Rod positioning; 31. Right fastening piece for rod positioning; 32. Counter nut locking; 33. Safety lock for counter nut; 34. Lock pin for rod positioning; 35. Left fastening piece for rod positioning; 36. Crimping ring for calandria tube;

2.2. Calandria tube

A calandria tube surrounds each pressure tube. Calandria tubes have an internal diameter of about 129 mm and span the calandria vessel between the two end shields. The calandria tube is illustrated in Figure 2.

Figure 2: Representation of the calandria tube

These tubes provide access through the calandria for the pressure tube/end fitting assemblies. The calandria tubes help to support the fuel channel pressure tubes by means of four spacers per channel, as is illustrated in Figure 3.

Figure 3: Representation of calandria tube with pressure tube, end fitting and fuel bundle

2.3. Pressure tube

The pressure tubes are the most important part of the fuel channel as they pass through the calandria and contain the fuel bundles. They are zirconium alloy tubes (Zr-2.5% Nb) that are about 6 meters long, about 11 cm in diameter and have a wall thickness of about 4 mm.
The design of the pressure tube consists primarily of the determination of the length, the inside diameter and the wall thickness of a simple thin-walled cylinder which is illustrated in Figure 4.

Figure 4: Representation of pressure tube

One of the main requirements of the pressure tube design is to optimize wall thickness and to minimize neutron absorption for radiation exposure protection of workers.
The product specification requires that the following tests and examinations also be carried out on each pressure tube before it can be accepted: hydrostatic pressure test, chemical analysis, tensile testing and corrosion testing.

2.4. End fitting

The end fitting, manufactured from a modified AISI 403 stainless steel, is an out of core extension of the pressure tube that provides the connection for on power fuelling, the connection to the feeders coupling and the connection with the pressure tube, which is illustrated in Figure 5.

Figure 5: Representation of end fitting

The outboard end contains a removable closure plug and provides facilities on which a fuelling machine can clamp and make a high pressure seal to allow on-power refuelling. Near the outer end of each end fitting is a side port for connection of a feeder pipe connection.
The inboard end of each end fitting is connected to one end of a pressure tube by a rolled joint, which is illustrated in Figure 6.

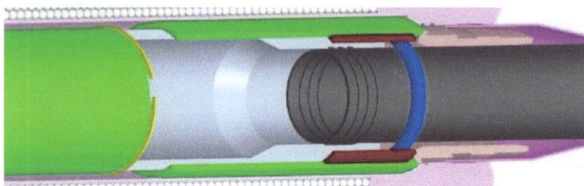

Figure 6: Representation of pressure tube rolled joint

2.5. Annulus spacer

Each pressure tube is separated from a calandria tube by means of four spacers. These spacers are positioned so that pressure tube sag will not allow the contact with the calandria tube. That is illustrated in Figure 7.

Figure 7: Representation of annulus spacers positioning on pressure tube

The spacers are made by forming Inconel wire into a close coiled helical spring, which is illustrated in Figure 8.

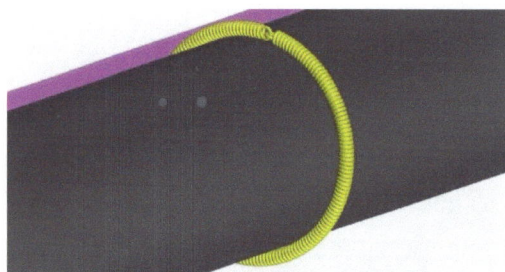

Figure 8: Representation of annulus spacer

The axial movement of the pressure tubes is allowed by a rolling motion of the annulus spacers, which results in almost no wear on the pressure and calandria tubes where they contract the spacers.

2.6. Feeder coupling

The feeder pipe connection located on the side of each end fitting is necessary for cooling system connection. The bolted feeder pipe connection has a metallic seal, as is illustrated in Figure 9.

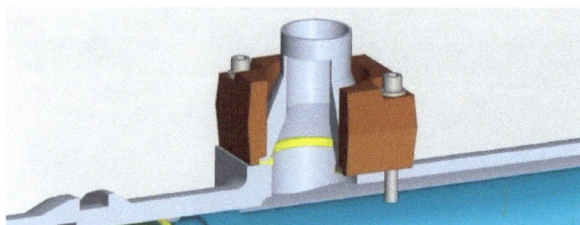

Figure 9: Representation of feeder coupling

Four bolts pass thorough a flange into holes tapped into the end fitting body to tighten this connection. The flange holds a hub welded to each feeder pipe tightly against the metal seal ring.

2.7. Positioning assembly

Each fuel channel is located axially within the reactor by a positioning assembly which is connected to one end shield, as is illustrated in Figure 10.

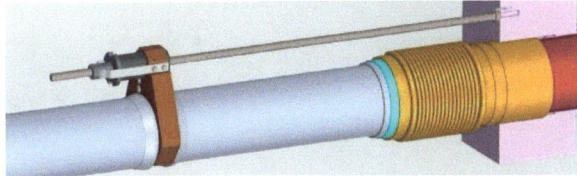

Figure 10: Representation of position assembly

A second positioning assembly is installed at the other end of the fuel channel but it is not attached to the end shield so axial motion resulting from thermal expansion, so that the pressure tube elongation to be permitted.

2.8. Annulus bellows

The annulus bellows, which is illustrated in Figure 11, connects between an end fitting and the reactor end shield, allows axial motion of the channels and also limits the torque imparted to the end fitting by the feeder piping. Each end of the bellows is welded to an end ring. One end ring is attached to the lattice tube/calandria tubesheet by welding and another is a shrink-fit onto the end fitting.

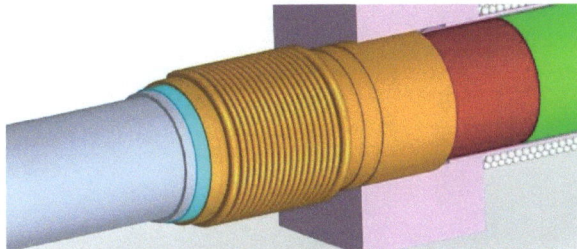

Figure 11: Representation of annulus bellows

2.9. Channel closure plug

The channel closures, illustrated in Figure 12, are located in each end fitting of a fuel channel to seal the primary coolant and to permit on-power access to the fuel channel by the fuelling machines. he channel closures can be remotely removed by a fuelling machine.

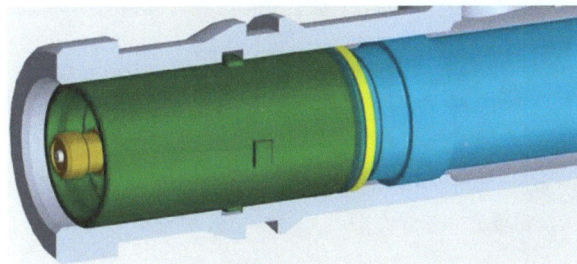

Figure 12: Representation of channel closures

2.10. Shield plug

The shield plugs, which provide shielding where the fuel channels pass through the reactor end shield, are latched into the end fitting, which is illustrated in Figure 13. They are also removed by the fuelling machine before the refueling of a channel can occur.

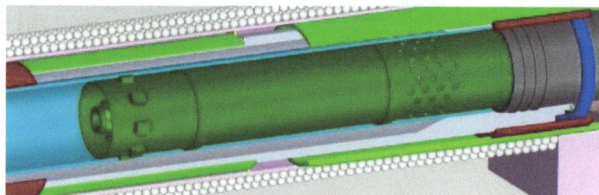

Figure 13: Representation of shield plugs

3. CONCLUSIONS

The design and the configuration characteristics of the fuel channel from the CANDU nuclear reactor are essentially in the design of device components. The fuel channel design has increased margins with extended operating life and is considered a fundamental part in the CANDU system.
The install operations to a new fuel channel must comply with the described requirements from the specified documents by AECL.
The current CANDU 6 reactors have a design life of 40 years at an average of 85% capacity. The pressure tube design life is 25 years at the reactor's 85% capacity factor.
The fuel channels from CANDU reactors, which use thin-walled zirconium alloy pressure tubes, represent a specialized application of pressure vessel design.
The fuel channels have made a significant contribution to the very high capacity factors attained in CANDU reactors since they allow on-power refueling.

REFERENCES

[1]   Cheadle B.A., Price E.G., *"Operating performance of CANDU pressure tubes"*, presented at IAEA   Techn. Comm. Mtg on the Exchange of Operational Safety Experience of Heavy Water Reactors, Vienna, 1989.
[2]   Roger G. Steed, *"Nuclear Power in Canada and Beyond"*, Ontario, Canada, 2003.
[3]   Venkatapathi S., Mehmi A., Wong H., *"Pressure tube to end fitting roll expanded joints in CANDU PHWRS"*, presented at Int. Conf. on Expanded and Rolled Joint Technology, Toronto, Canada, 1993.
[4]   AECB, *"Fundamentals of Power Reactors"*, Training Center, Canada.
[5]   AECL, *"CANDU Nuclear Generating Station"*, Engineering Company,   Canada.
[6]   ANSTO, *"SAR CH19 Decommissioning"*, RRRP-7225-EBEAN-002-REV0, 2004.
[7]   CANDU, *"EC6 Enhanced CANDU 6 - Technical Summary"*, 1003/05.2012.
[8]   CNCAN, *"Law no. 111/1996 on the safe deployment, regulation, authorization and control of nuclear activities"*, 1996.
[9]   CNCAN, *"Rules for the decommissioning of objectives and nuclear installations"*, 2002.
[10] IAEA, *"Assessment and management of ageing of major nuclear power plant components important to safety: CANDU pressure tube"*, IAEA-TEDOC-1037, Vienna 1998.
[11] IAEA, *"Assessment and management of ageing of major nuclear power plant components important to safety: CANDU reactor assemblies"*, IAEA-TEDOC-1197, Vienna 2001.
[12] IAEA, *"Decommissioning of Nuclear Power Plants and Research Reactors"* Safety Standard Series No. WS-G-2.1, Vienna 1999.
[13] IAEA, *"Nuclear Power Plant Design Characteristics, Structure of Power Plant Design Characteristics in the IAEA Power Reactor Information System (PRIS)"*, IAEA-TECDOC-1544, Vienna 2007.
[14] IAEA, *"Organization and Management for Decommissioning of Nuclear Facilities"*, IAEA-TRS-399, Vienna 2000.
[15] IAEA, *"Selection of Decommissioning Strategy: Issues and Factors"*, IAEA-TECDOC-1478, Vienna 2005.
[16] IAEA, *"State of the Art Technology for Decontamination and Dismantling of Nuclear Facilities"*, IAEA-TRS-395, Vienna 1999.

[17] IAEA, *"Water channel reactor fuels and fuel channels: Design, performance, research and development"*, IAEA-TEDOC-997, Vienna 1996.

[18] IAEA, *"Heavy Water Reactor: Status and Projected Development"*, IAEA-TEREP-407, Vienna 1996.

[19] Nuclearelectrica SA, *"Cernavoda NPP Unit 1&2, Safety features of Candu 6 design and stress test summary report"*, 2012.

[20] UNENE, Basma A. Shalaby, *"AECL and HWR Experience"*, 2010;

6th International Conference
"Computational Mechanics and Virtual Engineering "
COMEC 2015
15-16 October 2015, Braşov, Romania

MODELING BEHAVIOR OF BIOLOGICAL MEMBRANE COMPONENTS

Mihaela A.Ghelmez[1], Alexandru Mihai Rosu[1], Alireza Khastan[2]

[1]"Politehnica" University Bucharest, Physics Department, "Photonics and Advanced Materials Centre", Bucharest, ROMANIA
[2]Institute for Advanced Studies in Basic Sciences, Zanjan, IRAN
E-mail: mghelmez@yahoo.com, mghelmez@physics.pub.ro

*Abstract: This paper presents some of our results in modeling biological membrane behavior. Samples of very simple membrane models were built by using fatty acids in liquid crystal state, subjected to different external perturbations, like physical fields and impurities. Typical measurements were performed. Theoretical assumptions and some computer studies by using Table Curve3D capabilities and Matlab libraries allowed us to characterize changes occurred in the structure and ordering of the liquid crystal textures: local mechanical deformations in the systems, director **n** reorientation, and displacement of the electric charges. These changes determined the nonlinear feature of the the answer of the samples. Theoretical conclusions are experimentally validated and in agreement with other works.*
Keywords: biological membrane, fatty acids, liquid crystals, mechanical deformation, computer models

1. INTRODUCTION

Computers proved a useful tool for getting information on biological membrane structure and properties.

This study presents the computer help for modeling the changes induced by some external perturbations in some biological membrane simple models, realized by fatty acids (FA). These changes were experimentally evidenced by the nonlinear optical answer of the samples under non-destructive laser light, while in the liquid crystal (LC) state, a state involved in many mechanisms of the living matter. Many works proved that, in some conditions, the biological membrane turns in a cooperative manner from a "close-packed", gel-type structure, to a LC phase, responsible for many membrane mechanisms [1]. It is known that all the mesogenic compounds possess a high optical nonlinearity that can be pointed out even at low optical laser power [2]. In our experiments we used a Helium-Neon (He-Ne) laser light, incident on FA typical sandwich cells. Depending on the melting process speed (less than $5\,°C/min$), FA go to a disordered isotropic liquid phase via a LC state. Between some temperature values, our samples are *smectic* liquid crystals.

In electric field, they are nonlinear dielectrics with a weak conduction. In laser field, they generally are nonlinear optical materials, depending on the carbon atom number in the FA molecule, the saturation feature etc. [3]. The connection between the electric and optic behavior is the dielectric constant and refractive index of the material change under the respective external signals. The large change of the refractive index with the light intensity determines a nonlinear transmission of the light through the sample, the possibility of self-focusing of the laser beam (lens-like effect [4]), optical activity change with the intensity of the light, laser pulse width change etc. [5]. By adding small amount of other substances (impurities), they act as an external perturbation as well. Cholesterol (Ch) or some usual drugs (aspirin) proved their role on the membrane behavior, while in the LC state [5].

The experimental data were processed with TableCurve3D computer program that gives the possibility for obtaining the equations that model the answer of the samples and forecast the behavior of different fatty acids.

For simulating the dynamics of the phenomena inside the material, a model based on Runge-Kutta functions in Matlab is used.

2. EXPERIMENTAL: MATERIALS AND METHODS

To account for some of biological membrane mechanisms in view of the mosaic-fluid model developed by Singer [6], studies were carried out on saturated fatty acids (SFA) and some mixtures of them and with Ch. As it is known, these acids - butyric (4:0),caproic (6:0), caprilic (8:0), capric (10:0), lauric (12:0), myristic (14:0), palmitic (16:0), stearic (18:0),arachydic (20:0) are very important in animals and plants, too. The unsaturated fatty acids (UFA): linoleic (16:2, 9, 12 cis), elaidic (18:1, 9 trans), arachidonic (20: 5, 8, 11, 14 all cis) were studied as well. FA was sandwiched between two glass plates of transparent SnO_2 electrodes, about 2 cm long, with 20μm Mylar spacers at both ends. Each sample had electrical contacts and placed under a polarizing microscope equipped with a camera that allows visualization of microstructural images in polarized light. The temperature (4-150°C) and the electrical voltage applied on the sample (0-50V) were continuously monitored and controlled by a thermo stated setup built in our laboratory.

Hysteresis curves of the current I versus the applied voltage exhibited a nonlinear dielectric feature and a negative resistance of the samples [5]. This behavior is similar to the dielectrics with a spontaneous polarization [7]. I=I(t) plots at a constant temperature T, t being the time, when a step voltage U has been applied and then removed, allow for the determination of the mechanism of the electric conduction of the samples, of the internal resistance, the relaxation time of charge carriers, and of the space charge within the sample. The electric conduction and molecular arrangement of these systems depend on their chemical composition and present memory effects and other modifications under some stimuli from the environment [8, 9], by using a model of the system like in Figure 1:

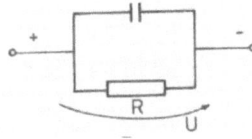

Figure 1: The electric system modeling the sample.

The current through the sample has two components [6]: the residual I_s stationary current:

$$I_s = U/R \tag{1}$$

where R is the internal resistance of the sample.
the absorption current I_0:

$$I_0 = US \, exp[-t/\tau] \tag{2}$$

where S is the conduction, and τ is the relaxation time (after the current decreased of "e" times).
The experimental dependencies I=I(t) for pure acids and mixture samples (see, for example, Figure 2), are quite similar with the theoretical ones, with some differences depending on the type of acid, showing an exponential decreasing of the current, typical for the strong dielectric materials.

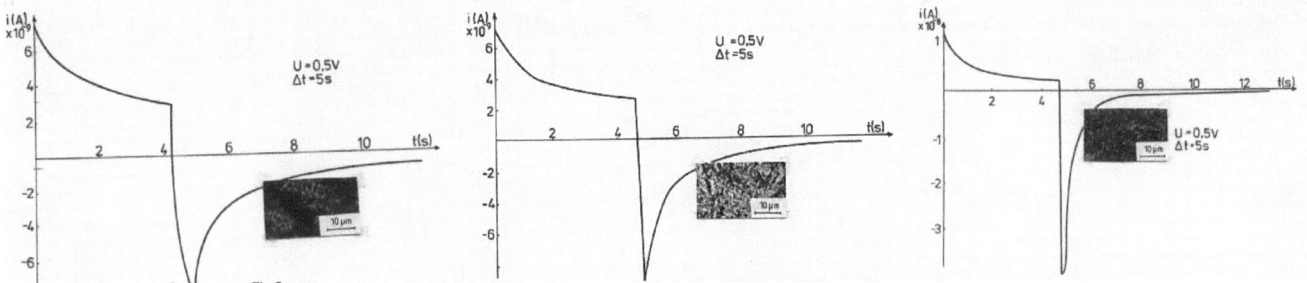

Figure 2: Dependency I=I (t) for U=0.5V and Δt=5s for mixture arrachidic-lauric-butyric-cholesterol in molar percentages (0.50÷0.25÷0.25÷0), (0.5÷0.25÷0.15÷0.15), (0.25÷0.25÷0.25÷0.25)

After electric field removal, the curves are similar with the ones for the currents limited by a space charge in dielectrics. The space charge is given by the surface delimited by the curve and the time axis [8]. The sample texture, inserted in Figure2 was obtained at the polarizing microscope, between crossed polarizers.

All the measurements took into consideration the Ch percentage and the discussion of the results has been done with respect with the hypothesis that Ch acts similarly with an external electric field, applied on the sample [8].

Emergent laser power from the samples (P_{out}) was registered at increasing and decreasing of the incident laser power (P_{in}). The experimental setup contained a He-Ne laser with a linear polarized c.w. beam, a variable attenuator for varying the input optical power, the sample, and an optical power meter. A feedback mirror (R = 96%) was added to reach the required value for self - focusing, by the return of the beam on the liquid crystal cell. A necessary condition is that the reflected beam touches the cell in the same point with the incident one. At this moment, the on-axis optical power decreases, the optical level being maintained even so the incident power is increased. Typical image is represented in Figure 3.

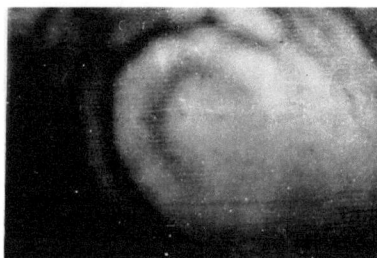

Figure 3: Mixture ($0.25 \div 0.25 \div 0.25 \div 0.25$), laser mode TEM_{00}

At the incident optical power decreasing, the molecules relax gradually to the initial positions and the output power takes more raised values, by the competition between the self -focusing effect - who is maintained until a certain power level - and the increased on-axis power. In almost all cases, the dependencies of P_{out} versus P_{in} were nonlinear and presented an optical hysteresis, similar with some of our previous results [9, 10].

3. COMPUTER STUDY OF THE EXPERIMENTAL RESULTS

The proposed computer models are in agreement with basic mathematical assumptions [11], and with the optical textures of the samples [12], as they were experimentally observed. The optical output power versus the input one and carbon atom number in FA was processed and analyzed by means of the TableCurve3D software, which gives also numerical information, including method/criteria of choosing equation, standard deviation and confidence limits for the fitted parameters, function extreme, an analysis of variance, and data table statistics. Precision Summary display was used to determine how much precision is preserved in the current equation. A residuals graph is also displaying the residuals for the current surface-fit. The data table can be weighted, if necessary. Thus, the program gives the possibility to choose other experimental conditions in terms of the future experiment requirements and purposes.

Figure 4 shows the dependencies of laser power (P_{max}=45mW), in terms of carbon C atoms number in SFA and Figure 5 – in UFA. One can observe the changes occurred in laser field, beginning with a small C number for SFA, but especially at a big C number for UFA; the nonlinear feature of the curves is different, increasing for UFA and decreasing for SFA. Working with a certain number of samples, graphs could forecast the behavior of other FA, with a different C number, by indicating the used equation.

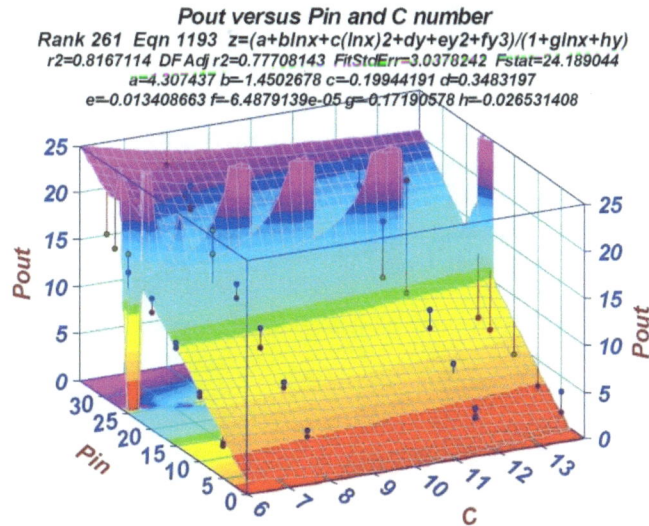

Pout versus Pin and C number
Rank 261 Eqn 1193 z=(a+blnx+c(lnx)2+dy+ey2+fy3)/(1+glnx+hy)
r2=0.8167114 DF Adj r2=0.77708143 FitStdErr=3.0378242 Fstat=24.189044
a=4.307437 b=-1.4502678 c=-0.19944191 d=0.3483197
e=-0.013408663 f=6.4879139e-05 g=0.17190578 h=-0.026531408

Figure 4: Output laser power versus input laser power and number of carbon atoms in molecule (SFA)

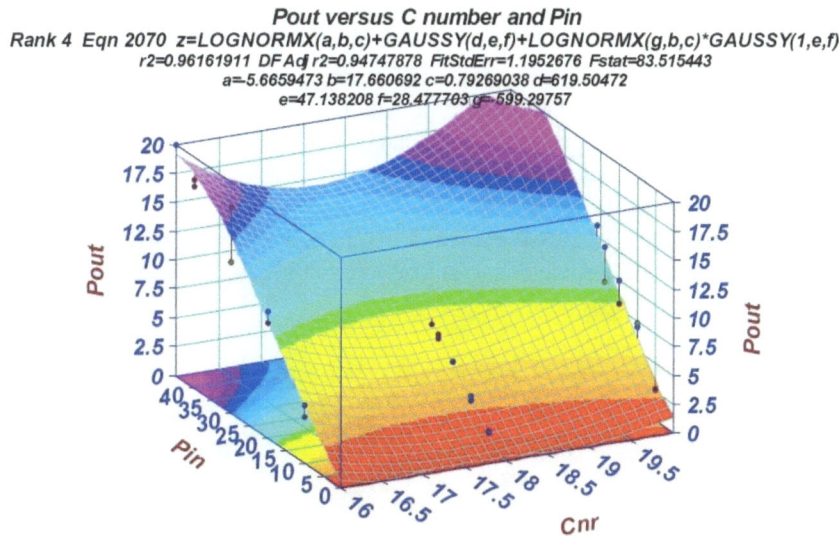

Pout versus C number and Pin
Rank 4 Eqn 2070 z=LOGNORMX(a,b,c)+GAUSSY(d,e,f)+LOGNORMX(g,b,c)*GAUSSY(1,e,f)
r2=0.96161911 DF Adj r2=0.94747878 FitStdErr=1.1952676 Fstat=83.515443
a=-5.6659473 b=17.660692 c=0.79269038 d=619.50472
e=47.138208 f=28.477703 g=-599.29757

Figure 5: Output laser power versus input laser power and number of carbon atoms in molecule (UFA)

These equations showed the nonlinear answer of the samples. Taking into account these results and previous results for other saturated and unsaturated FA and mixtures [13, 14], we believe that external interventions -physical fields or impurities - led to changes of the molecular arrangement and of the electric state of the samples. The length of the molecule is important in SFA and the carbon atom number in UFA; the distance between the double bound and the carboxyl group is also important [13]. These modifications can lead to some local mechanical deformations in the system, namely the director n orientation direction can be forced to modify by external perturbations. This reorientation of the director determines the observed changes of the refractive index, connected with the dielectric constant of the material. Relatively low laser power [15] can be useful for emphasizing the changes of the structure and properties in these membrane models, by their nonlinear interaction with these substances, in LC state- which represents a noninvasive method for analyzing real biological samples.

The mathematical model of the dynamics of phenomena inside the material, by using similar considerations as in [12], is based on Runge –Kutta functions in Matlab, by means of considering differential equations, adequate for the imprecise behavior which is inherent to many real processes [11]. To deal with periodic phenomena, by considering some derivatives with a switching point, providing sufficient conditions that the solutions match adequately and illustrate the real process.

The free term u (t) of the differential equation corresponds to the magnitude of the external perturbation; the working period is chosen about 0.2ms. The function f:

$$f^{(2)} = [(6\tau^4 - 2)/(\tau^2 - 1)^4] f + u(t) \qquad (3)$$

generated by the material under the influence of the external perturbation, can be integrated on this working time. For the external observer, the behavior of the material presents a slowly varying evolution in time, a single oscillation on the whole working interval, similar to the behavior noticed during the experiment.

For simulating the dielectric answer (z) of the material under the influence of an external electrical perturbation, we consider a differential equation able to generate an adequate truncated "test-function" φ similar to a Dirac pulse, in filtering and sampling procedures:

$$\varphi = exp\ [1/(\tau^2 - 1)] \qquad (4)$$

where $\tau = t - t_m$, t_m being the middle of the working period. This function has nonzero values only for $\tau \in$ [-1, 1]. The derivative $\varphi^{(2)}$ of this function related to τ is:

$$\varphi^{(2)} = [(6\tau^4 - 2)/(\tau^2 - 1)^4]\ \varphi + u(\tau) \qquad (5)$$

This corresponds to the magnitude of the external signal, where u (τ) = 1 on the working interval (-0.99; 0.99).

As it is known, such an equation of evolution, beginning to act at an initial moment of time, involves the derivative $f^{(n)}$ of certain order n jumps at the initial moment from the a null value to another one, in contradiction with the property of the test-functions to have continuous derivatives of any order on the whole real axis (time axis). Our truncated function φ has nonzero values on the interval $\tau \in$ [-1, 1] and a certain number of continuous derivatives on the whole time axis, and plays the role of the function f at a certain time moment, very close to the initial moment $\tau = -1$. Taking into account the expression of φ, the simplest differential equation satisfying these requirements, has the form:

$$f^{(1)} = [-2\tau/(\tau^2 - 1)] f \qquad (6)$$

and has the function φ as a possible solution. We use the initial moment of time is $\tau_0 = -1 + 0.01$ and the initial condition for f is $\varphi(\tau_0)$ for numerical simulation using equations Runge-Kutta of 4-5 order in Matlab.

The influence of the function φ is a sum of all effects, which can be mathematically represented as an integral of function φ multiplied by the progressive wave inside the material. Then one can write:

$$z^{(1)}(\tau) = \varphi(\tau)\ sin(\pi\tau - \Phi) \qquad (7)$$

where Φ represents an initial phase of the progressive wave. The function z (t) is represented in Figure 6, for u (τ) = 1 and $\Phi = \pi/12$, that models the experimental dielectric behavior presented in Figure 2.

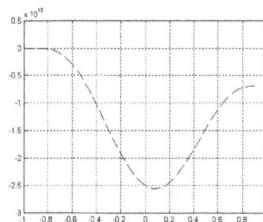

Figure 6: z versus t (u = 1, $\Phi = \pi/12$)

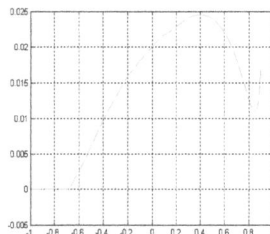

Figure 7:f versus t for u=sin($10^{11}\pi$t)

Figure 7 models the action of Ch, similar with the one of the electric field, but dependent on the Ch percentage added. The increase of this over 75% leads to some clusters formation inside the sample, and the fluidity of this decreased drastically.

Changes inside the material appeared due to the external optical pulse of the laser can be similarly taken into account. Let's consider null initial conditions for the system and add a free term in the differential equation – corresponding to the magnitude of the electrical field of the external optical signal, at a frequency of about 10^{15}Hz. The working period was chosen approximately equal to the period when the optical signal is received by the detector - about 0.2ms. The differential equation can be written as:

$$f^{(2)} = [(0.6\tau^4 - 0.36\tau^2 - 0.2)/(\tau^2 - 1)^4]\, f + u(t) \qquad (8)$$

where u is represented by an alternating function with a frequency 10^{11} times greater than the working period of 0.2 ms. By similar numerical simulations, we have obtained for f the results presented in Figure 8 for $u = cos(10^{11}\pi\tau)$.

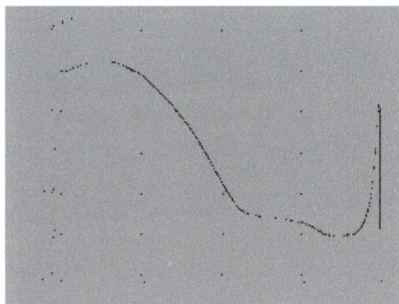

Figure 8: f versus t for $u = cos(10^{11}\pi\tau)$).

Similar considerations could be applied for other kind of perturbations, as for example some drugs (aspirin). We summarized in [13] some of these results.

4. CONCLUSIONS

In this paper, we present experimental and computer studies of behavior of substances involved in the structure and functions of the biological membrane, namely systems based on fatty acids, with a mesomorphic behavior.

Typical sandwich cells containing thin films of saturated and unsaturated fatty acids, components or forerunners of the biological membrane, were built in our laboratory and generally showed smectic liquid crystal textures during their evolution between some temperature values, characteristic for every of them. The electric conduction and molecular arrangement of these systems depend on their chemical composition and present modifications under some stimuli from the environment, as a result of the local mechanical deformations in the systems, director **n** reorientation, and displacement of the electric charges.

Relatively low electric fields and undistructive laser powers are useful for emphasizing the changes of the structure and properties of the samples, by their nonlinear interaction with these substances, easy to obtain in the liquid crystal state.

The experimental data can be processed by using TableCurve 3D computer program, and the adequate equation describing the sample behavior, precision intervals, confidence limit etc. are directly evidenced.

The experimental graphs and the theoretical ones are in agreement each other and with other results from the literature [16, 17]. Samples present a nonlinear dielectric feature, which evidenced the changes occurred in connection with the carbon number, the length of the molecule and the distance between the double bound and the carboxyl group is also important. Data processing and simulations were performed with computer help, modeled by Runge-Kutta functions in Matlab, with truncated test-functions generated by differential equations. The perturbation - let's be it electric, laser, or "impurities" with importance for biology and medicine (cholesterol, drugs, other acids etc.) – can be treated as a free term of the used equation, and adequate initial conditions and working intervals should be taken into account.

The obtained results and conclusions are important for the living processes and show once more our responsibility to proceed in a way that sustains life healthy.

REFERENCES

[1] A.C. Guyton, J.E. Hall, Textbook of Medical Physiology, W.B.Saunders Company, Ninth Edition, 1996
[2] I.C.Khoo, Y.R.Shen, Optical nonlinear effects in liquid crystals, Phys.Rev. Vol.A23, 2077-2089, 1981
[3] M.Dumitru (formerly Ghelmez), C.Motoc, M.Honciuc, R.Honciuc, Nonlinear Optical Properties of Some Fatty Acid-Cholesterol Mixtures, Mol. Cryst.Liq.Cryst., 215, 295,1992
[4].M. Dumitru (formerly Ghelmez), Doctor's Thesis, I.F.A./ICEFIZ, 1991
[5] M.Dumitru (Ghelmez), M.Honciuc, M.C.Piscureanu, C.Gheorghe, New Nonlinear Optical Materials for Optoelectronics, Optical Engin. 35(5), 1372-1376, 1996

[6] S.J. Singer and G. L. Nicholson, The Mosaic Model of the Cell Membranes, Science, 175, 720, 1972

[7] V.Tareev, Physics of Dielectric Materials, Mir Publisher, Moscow, 1975

[8] Mihaela Ghelmez (Dumitru), "Nonlinear Optical Effects in Biological Membrane Simple Models", Ed. Printech Bucuresti, 2000, ISBN 973-625-185-0

[9] M. Dumitru (formerly Ghelmez), Operating Conditions of Some Liquid Crystals in Interactions with Physical Fields, (Invited Paper ECLC'97, Cited in pref. vol., Zakopane, Poland), pag.326-334, 1997 SPIE Vol. 3319, M.Tykarska, R.Dabrowski, J.Zielinski, Editors, 326, 1998

[10] M.Ghelmez (Dumitru), B. Dumitru, M. Honciuc, C. Popa, A. Sterian, Studies in light field of the mesomorphic state of some biological membrane models, Synthetic Metals, 8590, 124, 1, 163, 2001

[11]. A. Khastan, J.J. Nieto, R. Rodríguez-López, Periodic boundary value problems for first-order linear differential equations with uncertainty under generalized differentiability, Information Sciences 02/2013; 222:544–558

[12] M.Ghelmez (Dumitru), C.Toma, M.Piscureanu, A.Sterian, Laser signals' nonlinear change in fatty acids, Chaos Soliton Fract., 17/2-3, 405, 2003

[13] M. Ghelmez (Dumitru), E. Slavnicu, Biological membrane simple models in physical fields, Ed.Printech, Bucuresti, 2005, ISBN 973-718-392-4

[14] M.Ghelmez, M.Berteanu, B.Dumitru, Fatty acids based biological membrane simple models, JOAM, Vol.10, Issue3, pp.707-712, 2008

[15] M.Dumitru (formerly Ghelmez), I.M.Popescu, M.Honciuc, Low Power Laser Light in New Nonlinear Optical Materials, Journal de Physique IV, Colloque C7, Vol.1, pag.C7-757 - C 7-760, 1991

[16] M. Dumitru (formerly Ghelmez), M. Honciuc, M. C. Piscureanu, Liquid crystalline state of some fatty acids and mixtures, Nonlinear Optics of Liquid and Photorefractive Crystals, pag.164-174, 1998, Ed.G.V.Klimusheva, ISBN 0-8194-2947-3

[17] L. N. Lisetski, V. Ya. Malikov, O. Ts. Sidletski, P. E. Stadnik, Thermo- and Photostimulated Currents in Bioequivalent Liquid Crystalline Structures, Molecular Crystals and Liquid Crystals, 12/1998; 324(1), pag.243-249

6th International Conference
"Computational Mechanics and Virtual Engineering "
COMEC 2015
15- 16 October 2015, Brasov, Romania

RANDOM VIBRATION OF DUFFING OSCILLATOR FOR N CORRELATION FUNCTIONS

Petre STAN[1], Marinică STAN[2]
[1]University of Pitesti , Romania, email: petre_stan_marian@yahoo.com
[2] University of Pitesti , Romania, email: stanmrn@yahoo.com

Abstract: *The response of a Duffing oscillator to narrow band random excitation is considered. Results obtained by applying the method linearization statistiques to random vibration problems are discussed. The equivalent linearization are found to give reasonable results only for very small non-linearities. This method is applicable to a variety of problems involving the response of lightly damped systems to broad-band random excitations. The theoretical analyses are verified by numerical results. Theoretical analyses and numerical simulations show that when the intensity of the random excitation increases.*
Keywords: *Duffing oscillator, random excitation, the power spectral density*

1.   INTRODUCTION

The present approximate representation of the spectrum is applied to a nonlinear oscillator in wich the non-linearity has pronounced on the response spectrum. The effect of non-linearities on the response power spectral density has been studied by a number of investigators This method is applicable to a variety of problems involving the response of lightly damped systems to broad-band random excitations. If however, any of the basic components behave nonlinearly, the vibration is called nonlinear vibration. The differential equations that govern the behaviour of vibratory non-linear systems are non-linear.

2 SYSTEM MODEL

Consider a Duffing oscillator of which the equation is

$$m\ddot{\eta}(t) + c\dot{\eta}(t) + k\eta(t) + \alpha k\eta^3(t) = W(t) \tag{1}$$

where m is the mass, c is the viscous damping coefficient, $W(t)$ is the external excitation signal with zero mean, α is the nonlinear factor to control the type and degree of nonlinearity in the system and $\eta(t)$ is the displacement response of the system.
Dividing the equation by m, the equation of motion can be rewritten as:

$$\ddot{\eta}(t) + 2\xi p\dot{\eta}(t) + p^2\eta(t) + \alpha p^2\eta^3(t) = f(t), \tag{2}$$

where ξ is the critical damping factor and p is the undamped natural frequency, for the system.
We want to act on this oscillator random excitations narrowband random excitations products through a number of n correlation functions containing it will introduce parameters $A_1,...A_n$, $\alpha_1,...\alpha_n$, $\gamma_1,...\gamma_n$ real, strictly positive.

$$R_F(\tau) = A_1 e^{-\alpha_1 \tau} \cos\gamma_1\tau + A_2 e^{-\alpha_2\tau}\cos\gamma_2\tau + \ldots A_n e^{-\alpha_n\tau}\cos\gamma_n\tau.$$
(3)

The parameter A_k influences directly proportional to the spectral density of initial excitation intensity and moderate printing relatively rapid variations. Increasing parameter α_k produces excitations with increases and decreases slow spectral density. Increasing parameter α_k widens excitation power and the drop was performed narrowing the spectrum. Excitation control parameters γ_k contribute to the excitation spectral density leves peak delayed. We say that the parameter maximum spectral density peaks moves to the right

The power spectral density of excitation W (t) is determined using the relationship

$$S_F(\omega) = \frac{1}{2\pi}\int_{-\infty}^{\infty} R_F(\tau)e^{-i\omega\tau}d\tau.$$
(4)

Solving this integral the relation sends us

$$S_F(\omega) = \frac{A_1\alpha_1}{\pi}\frac{\omega^2+\alpha_1^2+\gamma_1^2}{\left|(i\omega)^2+2\lambda(i\omega)+\alpha_1^2+\gamma_1^2\right|^2} + \frac{A_2\alpha_1}{\pi}\frac{\omega^2+\alpha_2^2+\gamma_2^2}{\left|(i\omega)^2+2\lambda(i\omega)+\alpha_2^2+\gamma_2^2\right|^2} + \ldots + \frac{A_n\alpha_n}{\pi}\frac{\omega^2+\alpha_n^2+\gamma_n^2}{\left|(i\omega)^2+2\lambda(i\omega)+\alpha_n^2+\gamma_n^2\right|^2}$$
(5)

The power spectral density of response is

$$S_\eta(\omega) = |H(\omega)|^2 S_F(\omega) = \frac{S_F(\omega)}{(k_e-m\omega^2)^2+c^2\omega^2} = \frac{1}{m^2}\frac{S_F(\omega)}{(p_e^2-\omega^2)^2+4\xi^2 p^2\omega^2},$$
(6)

Substituting equation (5) into (6), obtain

$$S_\eta(\omega) = \left\{ \frac{1}{\pi m^2}\frac{A_1\alpha_1(\omega^2+\alpha_1^2+\gamma_1^2)}{\left[\left(p^2-\omega^2+3\alpha p^2\sigma_\eta^2\right)^2+4\xi^2 p^2\omega^2\right]\left[\left(\alpha_1^2+\gamma_1^2-\omega^2\right)^2+4\alpha_1^2\omega^2\right]} + \right.$$

$$+ \frac{A_2\alpha_2(\omega^2+\alpha_2^2+\gamma_2^2)}{\left[\left(p^2-\omega^2+3\alpha p^2\sigma_\eta^2\right)^2+4\xi^2 p^2\omega^2\right]\left[\left(\alpha_2^2+\gamma_2^2-\omega^2\right)^2+4\alpha_2^2\omega^2\right]} + \ldots +$$

$$\left. + \frac{A_n\alpha_n(\omega^2+\alpha_n^2+\gamma_n^2)}{\left[\left(p^2-\omega^2+3\alpha p^2\sigma_\eta^2\right)^2+4\xi^2 p^2\omega^2\right]\left[\left(\alpha_n^2+\gamma_n^2-\omega^2\right)^2+4\alpha_n^2\omega^2\right]} \right\}$$
(7)

We start from the known formula

$$\sigma_\eta^2 = \int_{-\infty}^{\infty}|H(\omega)|^2 S_F d\omega.$$
(8)

We obtain

$$\sigma_\eta^2 = \sum_{k=1}^{n}\frac{A_k\alpha_k}{m\pi}\int_{-\infty}^{\infty}\frac{(\omega^2+\alpha_k^2+\gamma_k^2)}{\left\{\left[p^2-\omega^2+3\alpha p^2\sigma_\eta^2\right]^2+4\xi^2 p^2\omega^2\right\}\left[\left(\alpha_k^2+\gamma_k^2-\omega^2\right)^2+4\alpha_k^2\omega^2\right]}d\omega.$$
(9)

This formula has a integral type

$$\int_{-\infty}^{\infty}\frac{\omega^2+d}{\left|(i\omega)^2+2\lambda(i\omega)+d\right|^2\left|(i\omega)^2+b_1(i\omega)+b_0\right|^2}d\omega = \frac{\pi(b_o h_1 + h_1 h_2 - h_3)}{b_0(h_1 h_2 h_3 - b_o h_1^2 d - h_3^3)}$$
(10)

where

$$h_1 = b_1+2\lambda, \quad h_2 = b_0+2\lambda b_1+d, \quad h_3 = 2\lambda b_0+db_1.$$
(11)

We finally get a 4 degree equation with unknown σ_η^2

$$l\sigma_\eta^8 + n\sigma_\eta^6 + r\sigma_\eta^4 + s\sigma_\eta^2 + q = 0.$$
(12)

We can find a solution of the equation as a boundary of the next string

$$x_n = x_{n-1} - \frac{f(x_{n-1})}{f'(x_{n-1})}, \quad \forall n \geq 1.$$

(13)

3. NUMERICAL RESULTS

We consider a two component excitation and

$$m = 1kg, k = 30\frac{N}{m}, c = 3\varepsilon\frac{Ns}{m}, \varepsilon = 3m^{-2},$$

$$A_1 = 50N^2, A_2 = 50N^2, \alpha_1 = 1s^{-1}, \alpha_2 = 1,5s^{-1}, \gamma_1 = \gamma_2 = 3s^{-1}.$$

(14)

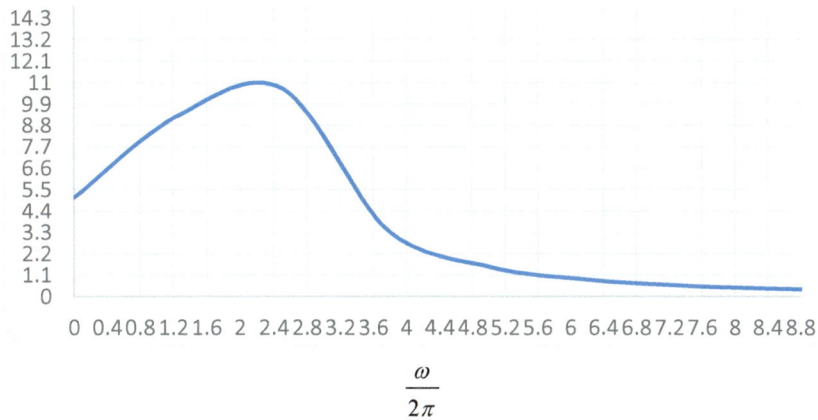

$$\frac{\omega}{2\pi}$$

Figure 1: The power spectral density of excitation S_F $[N^2 \cdot s]$ for $A_1 = A_2 = 40\ N^2, \alpha_1 = \alpha_2 = 1s^{-1}, \gamma_1 = \gamma_2 = 2s^{-1}, n = 2.$

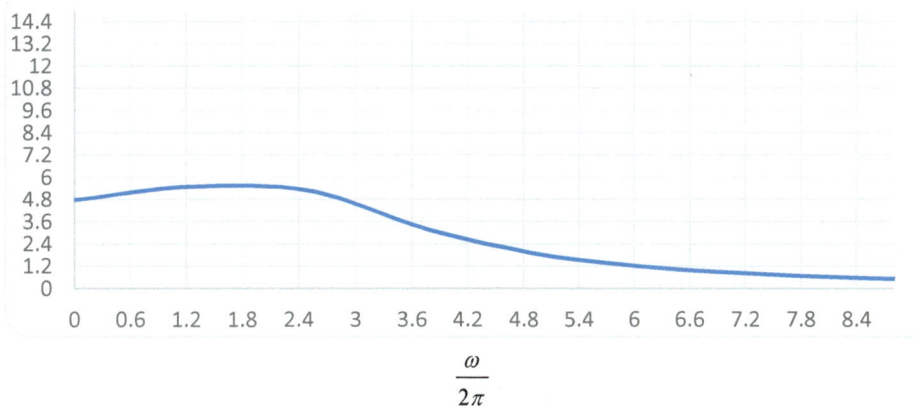

$$\frac{\omega}{2\pi}$$

Figure 2: The power spectral density of excitation S_F $[N^2 \cdot s]$ for $A_1 = A_2 = 30\ N^2, \alpha_1 = \alpha_2 = 2s^{-1}, \gamma_1 = \gamma_2 = 2s^{-1}, n = 2.$

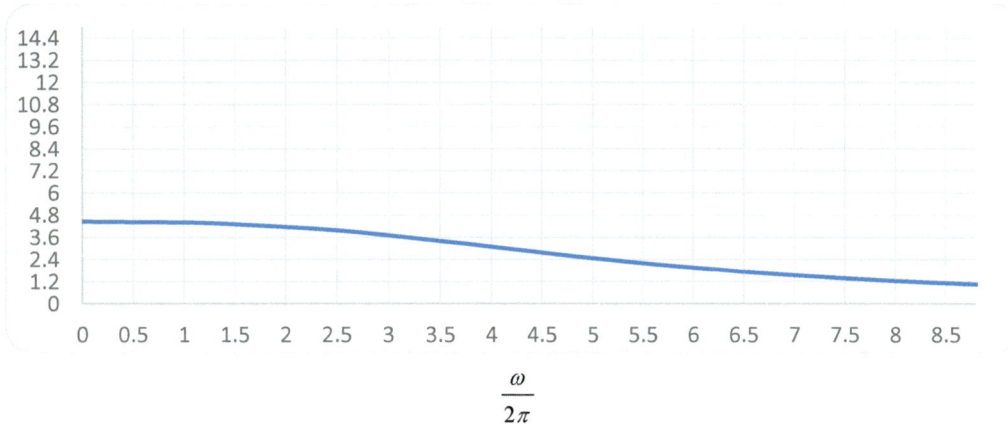

$$\frac{\omega}{2\pi}$$

Figure 3: The power spectral density of excitation S_F $[N^2 \cdot s]$ for $A_1 = A_2 = 20\ N^2, \alpha_1 = \alpha_2 = 3s^{-1}, \gamma_1 = \gamma_2 = 1s^{-1}, n = 2.$

We obtain

$$1458 \cdot 10^3 \sigma_\eta^8 + 6231,91 \cdot 10^4 \sigma_\eta^6 + 7254,52 \cdot 10^3 \sigma_\eta^4 + 1250 \sigma_\eta^2 - 143 = 0$$

(15)

which has the solution

$$\sigma_\eta^2 = 0,041 m^2.$$

(16)

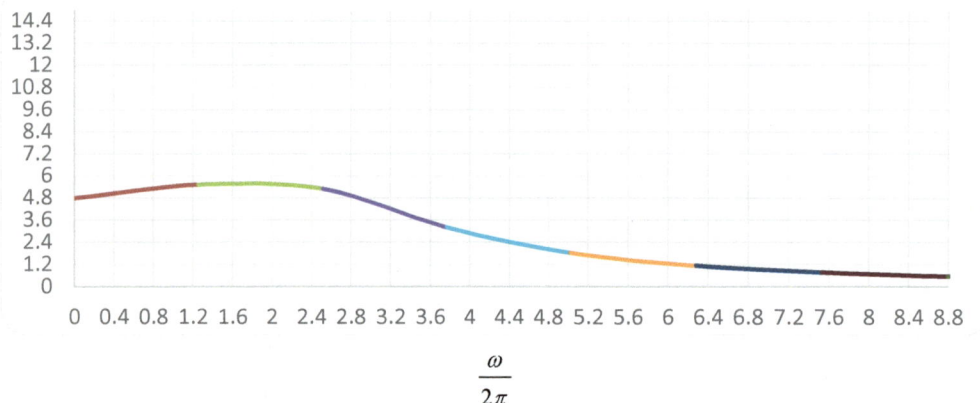

Figure 4: The power spectral density of excitation $S_F \ [N^2 \cdot s]$ for $A_1 = A_2 = 20 \ N^2, \alpha_1 = \alpha_2 = 2s^{-1}, \gamma_1 = \gamma_2 = 2s^{-1}, n = 3$.

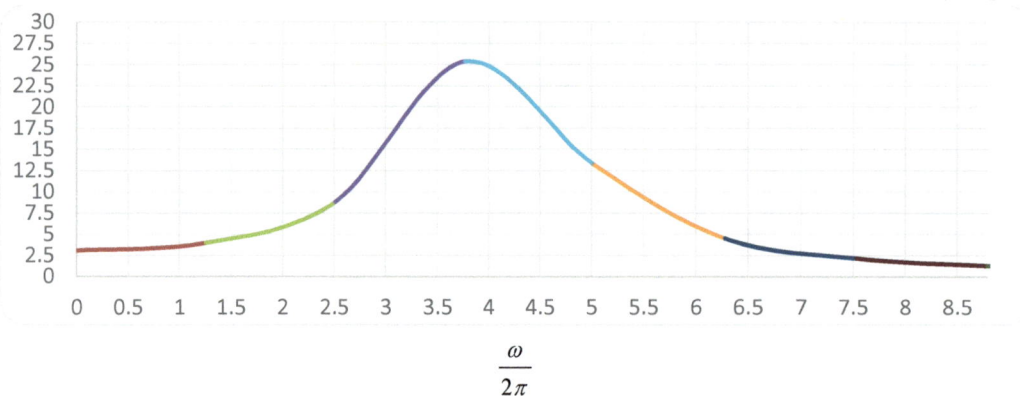

Figure 5: The power spectral density of excitation $S_F \ [N^2 \cdot s]$ for $A_1 = A_2 = 55 \ N^2, \alpha_1 = \alpha_2 = 1s^{-1}, \gamma_1 = \gamma_2 = 4s^{-1}, n = 3$.

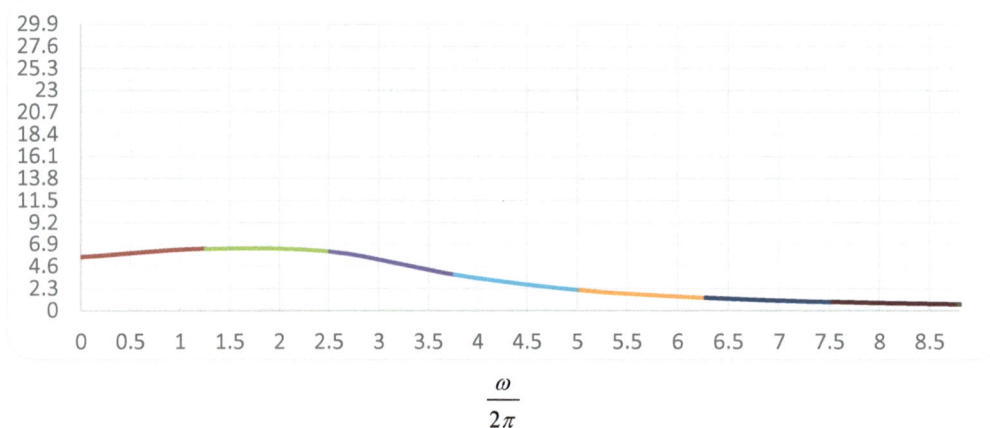

Figure 6: The power spectral density of excitation $S_F \ [N^2 \cdot s]$ for $A_1 = A_2 = 70 \ N^2, \alpha_1 = \alpha_2 = 2s^{-1}, \gamma_1 = \gamma_2 = 2s^{-1}, n = 1$.

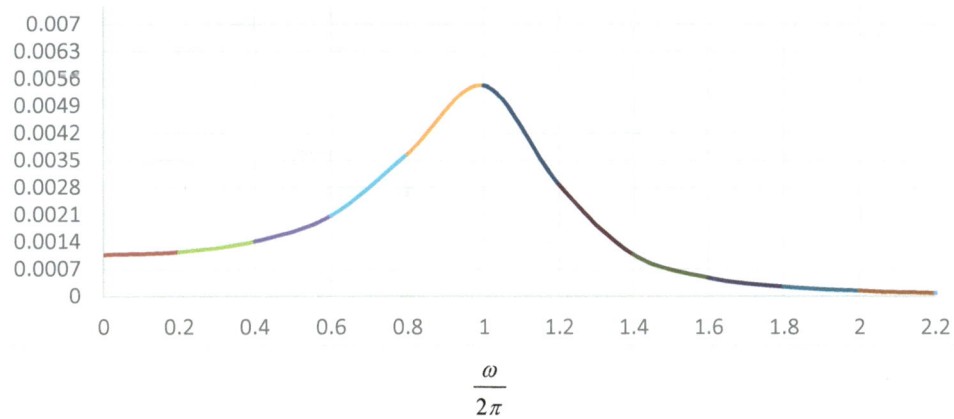

Figure 7: The power spectral density of response $S_\eta [m^2 \cdot s]$ for $m = 1kg, k = 30\dfrac{N}{m}, c = 3\dfrac{Ns}{m}, \alpha = 3m^{-2}$.

4. CONCLUSION

The theoretical analyses are verified by numerical results. Theoretical analyses and numerical simulations show that when the intensity of the random excitation increases. A second-order closure method is presented for determining the response of non-linear systems to random excitations. The random excitation is taken to be the sum of a deterministic harmonic component and a random component. The presence of the nonlinearity causes multi-valued regions where more than one mean-square value of the response is possible. Various applications of the theory to engineering problems are outlined. Using computer diagrams below its trend highlighted how the power spectral density given by equation (6). The parameter $A_k \left[N^2 \right]$ influences directly proportional to the spectral density of initial excitation intensity and moderate printing relatively rapid variations (fig. 1, 2, 3). Increasing parameter $\alpha_k \left[s^{-1} \right]$ produces excitations with increases and decreases slow spectral density. Increasing parameter α_k widens excitation power (fig. 4, 5, 6), and the drop was performed narrowing the spectrum. Excitation control parameters $\gamma_k \left[s^{-1} \right]$ contribute to the excitation spectral density levels peak delayed. (fig. 3, 4).

REFERENCES

[1] N. Pandrea, S. Parlac, Mechanical vibrations, Pitesti University (2000)

[2] A. Blaquiere, Nonlinear System Analysis, Academic Press, New York (1966)

[3] P. Stan, M. Stan, On response of random vibration for nonlinear systems, 5th International conference, Advanced Composite Materials Engineering, Brasov, (2014)

[4] P. Stan, Analysis of single-degree of freedom non-linear structure under Gaussian white noise ground excitatio, The book of University by Piteşti, Serie Appl. Mech. 6, 2007

[5] P. Stan., Response of Duffing Oscillator under narrow band random excitation, The book of University by Piteşti, Serie Appl. Mech. 6, (2007).

[6] P. Stan., "Random vibrations of non-linear oscillators", Second International Conference of Romanian Society of Acoustics on Sound and Vibration, (2004)

[7] P. Stan, M. Stan., The Random vibrations for the nonlinear oscillators with an elastic sine-like feature, 5th international conference, Thermal Systems and Environmental Engineering, Galati, 2014.

[8] P. Stan, M. Stan, On response of random vibration for nonlinear systems, 5th international conference Advanced Composite Materials Engineering, Brasov, (2014).

[9] S. Graham, Theory and problems of mechanical vibrations, (1993)

6th International Conference
"Computational Mechanics and Virtual Engineering"
COMEC 2015
15-16 October 2015, Braşov, Romania

BENDING BEHAVIOR ASSESMENT OF NDFEB MAGNET BASED SAMPLES

Ph.D. Stud. Eng. Melania Tămaş[1], Prof .Univ. Emerit. Ph.D. Eng.Constantin D. Stănescu[2], Ph.D. Stud.Eng. Doina Ene[3]

[1] Ion Neculce" National College, Bucharest, ROMANIA, melaniat2@yahoo.com
[2] Polytechnic University, Bucharest, ROMANIA, prof_cstanescu@yahoo.com
[3] I. Socolescu" Technical Architecture College, Bucharest, ROMANIA, doinaene@ymail.com

Abstract: The technological advancements impose new solutions for growing adding value. In this context, this paper presents some experimental researches concerning the bending behavior of steels magnets, NdFeB based. Neodymium magnet samples of different density and hardness values. Mechanical processing for cold plastic deformation by bending of technological samples is characterized by technical difficulties at bending low angle technological tests. The size and shape of the geometric deformations of materials depends on the size, type and mode of application of the stresses to which they were subjected and their properties. For technological samples with small thickness (up to 8 mm), present fractures, in the case of free bending test.
Keywords: NdFeB magnets, bending behavior, thermal characteristics, electrical characteristics

1. INTRODUCTION

The upward level of requirements and demands, constantly growing, regarding the performance of human activities, led to unprecedented development and continuous improvement of science and engineering technology [1,4].
Theoretical and experimental research undertaken in Romania and abroad, in the field of composite materials technology is an interdisciplinary field in which materials engineering, mechanical engineering and chemical technology contributes to achieving the final project: "the composite" [5-9].
Today, composite materials are used in all areas where technological progress requires a combination of properties that cannot be provided by conventional materials.
The article deals with possibilities and behavior of materials used to produce permanent magnets, NdFeB based (see figure 1), on machining of cold plastic deformation by bending [10, 11].

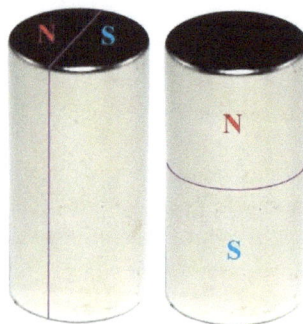

Figure 1: Rare earth magnets, NdFeB based

The size and shape of the geometric deformations of materials depends on the size, type and mode of application of the stresses to which they were subjected and their properties [12].

2. CHEMICAL COMPOSITION AND TECHNICAL FEATURES OF TECHNOLOGICAL TESTING SAMPLES

In the framework of experimental researches, encompassing plastic deformation mechanical processing, they were used eight types of samples, with diffrent densities and hardness with carbon steel 0,5% based matrix of composite material, and neodymium alloy, which incorporates complementary material composed of boron ferrite (FeB) short fibers, evenly spaced, with guided distribution.

The samples, acquired from the company Arca Hobber Chemicool SRL, have been designed and molded with the following dimensions (see Figure 2) [4-7]
- Length: 80 - 100 mm;
- Width: 30 - 40 mm;
- Thickness: 5 - 10 mm;

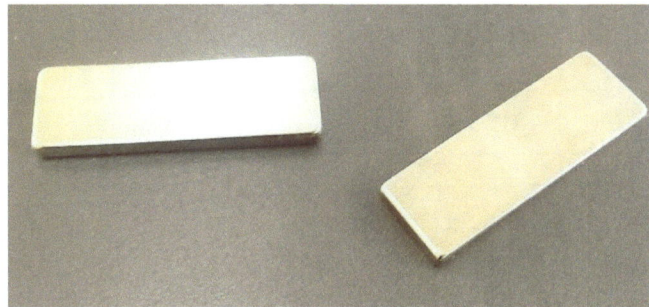

Figure 2: Geometrical configuration of the test samples used for experimental research by plastic bending deformation

Table 1 shows the chemical composition of experimental saples, based on NdFeB.

Table 1: Chemical composition of experimental semples based on NdFeB

Name	Chemical element components, %						
	C	Fe	B	Nd	Al	Nb	Dy
NdFeB	0,5	64,2-68,5	1-1,2	2,9-3,2	0,2-0,4	0,5-1	0,8-1,2

Where: Nb-niobiu, Dy-disprosiu;

Table 2 presents the mechanical characteristics of experimental samples, based on NdFeB.

Table 2: Main Mechanical characteristics of experimental samples

Sample	Density g/cm^3	Hardness			Resistance Nm/mm^2		Force		Young module	Rigidity N/m^2	Compresibility 10^{-12} mm^2/N	Poison ratio
		HB	HRC	HV			F_i	F_f				
W4	4.4-5.5	40-45	551	570	1980	780	8	9.8	1.6	0.64	9.8	0.24
W6	5.3-5.8	40-45	551	570	1980	780	8	9.8	1.6	0.64	9.8	0.24
W8	5.6-6.0	35-38	551	570	1980	780	8	9.8	1.6	0.64	9.8	0.24
W8H	5.6-6.0	35-38	551	570	1980	780	8	9.8	1.6	0.64	9.8	0.24
W10	5.8-6.1	35-38	551	570	1980	780	8	9.8	1.6	0.64	9.8	0.24
W10H	6.0-6.2	35-38	551	597	1980	780	8	9.8	1.6	0.64	9.8	0.24
W12	6.2-6.6	35-38	551	597	1980	780	8	9.8	1.6	0.64	9.8	0.24
W12D	6.2-6.6	35-38	551	597	1980	780	8	9.8	1.6	0.64	9.8	0.24

Where: Fi is tensile force;
      Ff –flexural force (10^{-12} m^2/N).
Table 2 presents the thermal characteristics of experimental samples, based on NdFeB [13, 16].

Table 3: Thermal characteristics of experimental samples

Sample/ characteristic	Symbol	U.M	W4	W6	W8	W8H	W10	W10H	W12	W12D
Thermal conductibility	K	kcal/mh ^{0}C	7.7	7.7	7.7	7.7	7.5	7.5	7.5	7.5
Specific heat capacity	C	kcal/mh ^{0}C	0.12	0.12	0.12	0.12	0.1	0.1	0.11	0.11
Melting point	T/C	^{0}C/K	1016/ 1289	1016/ 1289	1016/ 1289	1016/ 1289	1016/ 1289	1017/ 1290	1017/ 1290	1017/ 1290
Boiling point	Tf	^{0}C/K	3070/ 3343	3070/ 3343	3070/ 3343	3070/ 3343	3070/ 3343	3060/ 3340	3060/ 3340	3060/ 3340
Coefficient of thermal expansion	c	10^{-4}0C	3.4	3.4	3.4	3.5	3.5	3.5	3.5	3.5
Temperature resistivity coefficient	a	10^{-4}0C	2	2	2	2	2	2.1	2.1	2.1
Curie temperature	Tc	^{0}C	310	310	310	316	316	316	316	316
The heat of vaporization	cv	mΩ/cm	283.68	283.68	263.68	263.68	263.68	263.68	263.68	263.68

Table 4 presents the electrical characteristics of exterimental samples based on NdFeB.

Table 4: Electrical characteristics of experimental samples

Sample/ characteristic	Symbol	U.M	W4	W6	W8	W8H	W10	W10H	W12	W12D
Electrical conductibility	σ	10^6S/m	0,667	0,667	0,667	0,667	0,670	0,671	0,720	0,720
Electrical resistance	ρ	kcal/mh0C	150	150	150	152	153	158	158	158

The analysis of HV and HRC values of experimental specimens show that the materials used in these samples gives a very high hardness, mechanical strength and good tear.

3. THE ASSESSMENT OF MECHANICAL PROCESSING BEHAVIOR TO BENDING PLASTIC DEFORMATION OF EXPERIMENTAL SAMPLES

In order to produce of experiments, the technological samples has been demagnetized at a temperature of 110^{0}C, within an oven, model Nabertherm having the possibility of variation of the temperature in the range of 30-3000^{0}C. The laboratory equipment belongs to SIMAR INDUSTRIAL SA company from Bucharest (figure 3).

Figure 3: The oven for heating to demagnetize of technologic samples, model Nabertherm and temperature range 30-3000^{0}C

In order to acheive the mechanical test, has been used two diffrent experimental equipments, a mechanical driven one and a hydraulic driven another one.

The equipment belong to Laboratory of Plastic Deformation from TMS Department, and these are:

- AbKant apparatus, model 850, for free bending tests;
- Hydraulic universal machine, model WE60 (Hydraulic universal material testing machine) for mold bending.

The figure 4 presents free bending tests using the Abkant 850 apparatus.

Figure 4: Free bending tests of technological samples using the AbKant 850 apparatus

The bending tests has been acheived at two bending angle values: α=170^0 şi α=90^0. The experimental results has presented it Table 5.

Table 5: Experimental results of bent test samples, using AbKant 850 apparatus

Simbol of test sample	LxHxW (mm)	Bending angle (grade)	Recover angle (grade)	Assessment
W4	100x40x5	160-130	-	cracked
W6	100x40x5	160-130	-	cracked
W8	100x40x5	160-130	-	cracked
W8H	100x40x5	140-120	0.3	fissures and creases
W10	100x40x5	160-120	0.2	fissures and creases
W10H	100x40x5	140-110	0.2	accepted
W12	100x40x5	120-110	0.1	accepted
W12D	100x40x5	120-90	0.1	accepted

3. CONCLUSION

In the case of free bending technology, the mechanical processing of cold plastic deformation of test samples with small thickness (up to 8 mm) presents difficulties at bending low angles, resulting creases, fissures and cracks.

REFERENCES

[1] Arai K., Namikawa H., et al, Aluminum or phosphorus co-doping effects on the fluorescence and structural properties of neodymium-doped silica glass, J. Appl. Phys. 59, 3430 (1986).

[2] Fraden J., Handbook of Modern Sensors: Physics, Designs, and Applications, 4th Ed. USA: Springer. ISBN 1441964657, p. 73. (2010).

[3] *** What are neodymium magnets? wiseGEEK website. Conjecture Corp. 2011. Retrieved October 12, 2012.

[4] http://www.lpi.usra.edu/lpi/meteorites/The_Meteorite.shtml

[5] Bai G., Gao R.W., et al. Study of high-coercivity sintered NdFeB magnets, Journal of Magnetism and Magnetic Materials, Volume 308, Issue 1, January 2007, Pages 20–23.

[6] Yu L Q, Wen Y.H, et al. Effects of Dy and Nb on the magnetic properties and corrosion resistance of sintered NdFeB, Journal of Magnetism and Magnetic Materials, Volume 283, Issues 2–3, December 2004, Pages 353–356.

[7] https://www.kjmagnetics.com/neomaginfo.asp

[8] http://www.chemicool.com/elements/cerium.html

[9] Altintas Y., Spence A., End Milling Force Algorithms for CAD Systems, Annals of the CIRP, Vol. 40/1, pp. 31-34, 1991.

[10] Altintas Y., Spence A.D., A Solid Modeler Base Milling Process Simulation and Planning System Quality Assurance through Integration of Manufacturing Processes and Systems, PED-56 ASMtE, pp. 65-79, 1992.

[11] Arsecularatne, J.A., Jawahir I.S., Prediction and Validation of Cutting Forces in Machining with Chip Breaker Tools, Trans. NAMRI, Vol. XXIX, 2001, pp. 367-374.

[12] Athavale S., A Damage-Based Model for Predicting Chip Breakability for Obstruction and Grooved Tools, Ph.D. Thesis, North Carolina State University, Raleigh, North Carolina, USA, 1994.

[13] Stănescu. C.D., Căiniceanu L., Ionescu S., Tamaş M., Experimental research on improving power audio speaker devices performance using permanent NdFeB Magnets, Special Technology, in: Session of the Comission of Acoustics, Academia Română, 2013.

[14] Stănescu. C.D., Ene D., Tamaş M., Costner J., Research on modelling by finite elements method of Trying to impact using Buys and LS-DYNA Programs (First Part), in: Materials Engineering University of Chicago, USA, 2014.

[15] Zhang G.M., Kapoor S.G., Dynamic Generation of Machined Surface, part- i: Mathematical Description of the Random Excitation System, Journal of Engineering for Industry, Transaction of ASME, 1991.

[16] Zhang G.M., Kapoor S.G., Dynamic Generation of Machined Surface, part- ii: Mathematical Description of the Tool Vibratory Motion and Construction of Surface Topography, Journal of Engineering for Industry, Transaction of ASME, 1991.

6th International Conference
"Computational Mechanics and Virtual Engineering "
COMEC 2015
15-16 October 2015, Braşov, Romania

IMITATION-BASED ROBOT PROGRAMMING METHOD FOR PREDICTING THE MOTION

Aurel Fratu
"Transilvania" University of Brasov, ROMANIA, e-mail fratu@unitbv.ro

Abstract: One of the main barriers to automating a particular task with a robot is the amount of time needed to program the robot. In this paper the decrease in programming time is accomplished by combining off-line and on-line programming techniques .The method consists in using a programming platform that allows us to expediently compose robot programs. On the programming platform there is carried out the virtual prototype of the physical robotic arm to be programmed and the real working space wherein it is intended to work. In this paper, we propose a new approach to robot programming based on virtual robot prototype behavior in experimental scenery. The actions for the each robotic task are computed for virtual robot prototype and are transferred online, with a central coordination, to corresponding physical robot, which must imitate her virtual "homonymous".
Keywords: virtual robots, path learning by imitation, motion programming, behavioral control

1. INTRODUCTION

A characteristic feature of robot programming is that usually it is dealing with two different worlds; the real physical world to be manipulated and the abstract models - in particular virtual prototypes models - representing this world in a functional or descriptive manner by programs and data. In the simplest case, these models are pure imagination of the programmers; in high level programming languages, e.g. it may consist of CAD data.

Creating accurate robot path points for a robot application is an important programming task. It requires a robot programmer to have the knowledge of the robot's reference frames, positions, operations, and the work space.

In the conventional "lead-through" method, the robot programmer uses the robot teach pendant accessory to position and to orientation the robot joints and end-effector and record the satisfied robot pose [3].

The basic idea behind these approaches is to relieve the programmer from knowing all specific robot details and liberate him from coding every small motion.

Today's robot simulation software provides the robot programmer with the functions of creating virtual robot and virtual path points in an interactive virtual 3D design environment.

By the time a robot simulation design is completed; the simulation robot program is able to move the virtual robot end-effector to all desired virtual path points for performing the specified operation. By different scenarios one creates particular motion paths in the virtual space that are selected for particular motion paths in physical work space.

However, because of the inevitable dimensional differences of the components between the physical robot work space and the simulated robot work space, the virtual robot path points must be adjusted relative to the actual position of the components in the physical robot work space. This task involves the techniques of calibrating the position coordinates of the simulation device models with respect to the physical robot path points [1].

Learning by imitation based on our method represents a new research topic in robotics being a promising approach, towards effective robot programming.

This work follows a recent trend in Programming by Imitation. We present a new method to programming physical robot through the imitation based on the virtual robot prototype.

Movement imitation requires a demonstrator. In our approach the demonstrator is the virtual robot prototype and the imitator is the physical robotic manipulator. Conformity our method the dynamics of the motion of the virtual robots is reproduced by the physical robot.

In this paper one use the virtual robot prototypes and the motion capture systems to obtain the reference motion data, which typically consist of a set of trajectories in the Cartesian space.

We restrict our study to imitation of manipulative tasks. In particular, the ability to imitate virtual robots gestures and pursue task-relevant paths is essential skills for physical robotic manipulators, in the spirit of our method.

Inspired by the personal patent [8] and based on the motion imitation concept this paper develops a general policy for robot motion programming based on virtual robot prototypes.

The strong point of the proposed method is that it provides fast on-line re-planning of the motion in the face of spatial-temporal perturbations and it allows generalizing a motion to unseen context based on particular scenarios. The objective of the present paper is giving an overview of a new robot programming method based on the virtual prototypes.

2. PROGRAMMING FROM A SIMULATED VIRTUAL MODEL

Programming by imitation has appeared as one way to respond to growing need for intuitive control methods and is one of the most promising programming techniques for robotic manipulators. This technique allows even inexperienced users to easily program the robots based on the teaching by imitation paradigm.

We present a description of the theoretical aspects of the programming from a simulated virtual model and of the strategy which is at foundation of this approach. The advantages of such programming approach as an alternative to the classical methods (e.g. vision guided trajectory imitation) are on-line adaptation to the motion of the virtual prototype.

A solution to the above problem is to construct a virtual prototype model and transfer the virtual trajectory by interacting with the physical model.

Designing a model would be an option; however, the behavior of the robots is very difficult to model. Moreover, the use of system knowledge is contrary to our research aim. Therefore we focus on creating a virtual prototype model from experimental data obtained from the physical robot. Thus, we will prove that our method guarantees the motion optimization for each robotics tasks using a planning module and a visualization module.

The planning package communicates primarily with simulation environment. A planning module can send messages to the simulation system such as computed plans for the robots The planning module can further send trajectory and planning structure information to visualization; so users can see the results of this algorithm. The planning module also receives control signals from the simulation module, such as when to start planning joint trajectories [2].

The visualization module is in charge for visualizing any aspect needed by the programmer for the optimization process. Users interact with the simulation environment through the visualization. This includes, but not limited to, computer screen. The visualization provides an interface to develop interactive implementations based on imitation strategy.

In our work, we assume that learning of the deterministic part for description motion dynamics should be sufficient to design the corresponding robot control.

We particularly refer to the ability of the system to react to changes in the environment that are reflected by motion parameters, such as a desired target position and motion duration. Therefore, the system is able to manage with uncertainties in the position of a manipulated object, duration of motion, and structure limitation (e.g., joint velocity and torque limits).

3. ONLINE IMITATION METHOD OF THE VIRTUAL DYNAMICAL SYSTEMS

In robotics, one of the most frequent methods to represent movement strategy is by means of the learning from imitation. Imitation learning is simply an application of supervised learning. One goal of imitation of the dynamical systems is to use the ability of coupling phenomena to description for complex behavior [6].

In this paper, we propose a generic modeling approach to generate virtual robot prototype behavior in experimental scenery. The actions for the each task are computed for virtual robot prototype and are transferred online, with a central coordination, to corresponding physical robot, which must imitate her virtual "homonymous". Notice the similarity between moves of the virtual robot prototype in the virtual work space and the "homonymous" moves in the real work space of the physical robot. We assume to use the virtual robot prototypes and the motion capture systems to obtain the reference motion data, which typically consist of a set of trajectories in the operational space.

Our method consists in using a programming platform on which there is carried out the virtual prototype of the physical robotic arm to be programmed and the real working space wherein it is intended to work. The method combines off-line and on-line programming techniques

In the robot program there is written a source code intended to generate the motion paths of the virtual robotic arm prototype. The numerical values of the prototype articulation variables are sent to the data register of a port of the information system which, via a numerical interface, is on-line transferred into the data registers of the controllers of the actuator of the physical robotic arm.

Finally, there are obtained tracking structures due to which the moving paths of the virtual robotic arm joints are reproduced by the physical robotic arm joints, thereby generating motion within the real working space.

Imitation learning from our strategy demonstrates how to obtain dynamical virtual models with CAD systems. Those online adjusted virtual models are among the most important properties offered by a dynamical systems approach, and these properties cannot easily be replicated without the feed-back from physical robot of our proposed structure.

The objective of a movement is to generate a reaching movement from any start state to a goal state [8]. The discrete dynamical system is initialized with a minimum movement, which is frequently used as an approximate model of smooth movement.

The proposed structure uses a virtual demonstrator for planning the movements of the physical robot. We investigate the potential of imitation in programming robotic manipulators with multiples degrees of freedom when the associated joints must be coordinated concurrently.

The imitation strategy consists in a proportional real-time mapping between each virtual joint and the corresponding physical joint.

The system requires the programmer to perform an initial calibration routine to identify the range of motion for each movement. Each range is divided into as many intervals as the number of feasible discrete configurations of the corresponding joint.

3.1 Predicting Robot Control Latency

The imitation method supposes a delay between the action of the virtual prototype and physical robot. The time elapsed between making an action decision and perceiving the consequences of that action in the environment is called the control delay.

A predictor approximates a negative delay. It has access to the delayed robot state as well as to the un-delayed controller actions and is trained to output the un-delayed robot state. The predictor contains a forward model of the robot and provides instantaneous feedback about the consequences of action commands to the controller. If the behavior of the robot is predictable, this strategy can simplify controller design and improve control performance. All physical feedback control loops have a certain delay, depending on the system itself, on the input and output speed and, of course, on the speed at which the system processes and transfer the information from virtual environment to physical environment.

In our system we use the concept of motor prediction and the paper describes a method to reduce the effects of the system immanent control delay. It explains how we solved the task by predicting the movement of the robots, using a virtual prototype model. Just virtual robot positions and orientations as well as the most recent motion commands sent to the physical robot are used as input for the prediction.

In our system the global sensorial module is used to determine the positions of all robot joints. In order to control the behavior of the robot in a way that is appropriate for the situation of her virtual homonym on the virtual scene we need the exact positions of them at every moment. Because of the delay inherent in the control loop, however, the behavior control system actually reacts to the environment with finite delay (about few ms size).When moving faster the delay becomes more significant as the error between the real position and the virtual position used for control grows.

In order to correct this immanent error we have developed an informatics network which processes the positions and the orientations of each joint. We use recorded preprocessed data of moving virtual robot to train the network. It predicts the actual positions of the robots. These predictions are used as a basis for control. The action commands will be sent to the physical robots during the real time.

Using the concept of motor prediction one predicts the joints' position of the robot arm, rather than sensing it by sensorial system. In this model the predictions are based on a copy of the motor commands acting on the virtual joints. In effect, the virtual joints' position of the virtual robot arm is made available before physical sensory signals become available.

These predictions are used in an inner control loop to generate sequences of actions that guide the robot towards a target state. Since this cannot account for disturbances, the robot predictions are delayed by the estimated dead time and compared to the sensed robot state. The deviations reflect disturbances that are feedback into the controller via an outer loop. The fast internal loop is functionally equivalent to an inverse-dynamic model that controls a robot without feedback.

A simple approach to implement the predictor would be to use a mathematical method. This method is very effective to handle linear effects, for instance the motion of a free joint [5]. It is however inappropriate for robot arm that contain significant non-linear effects, e.g. caused by the slippage or by the behavior of its motion controller. For this reason, we must use a virtual prototype to predict non-linear systems. But this approach requires predicting first a good virtual model of the robot.

We then measure the time between sending the command to change the direction of motion and perceiving a direction change of the robot's movement. There are many possibilities to counter the adverse effects of the control delay. The easiest way would be to reduce precision requirements, but this would lead to unnecessary collisions and uncontrolled behavior. Control researchers have made many attempts to overcome the effects of delays.

3.2 System Architecture

Action commands are sent via a wireless link from virtual robot to the physical robot that contains only minimal local intelligence. Our imitation experimental system is illustrated in Figure 1.

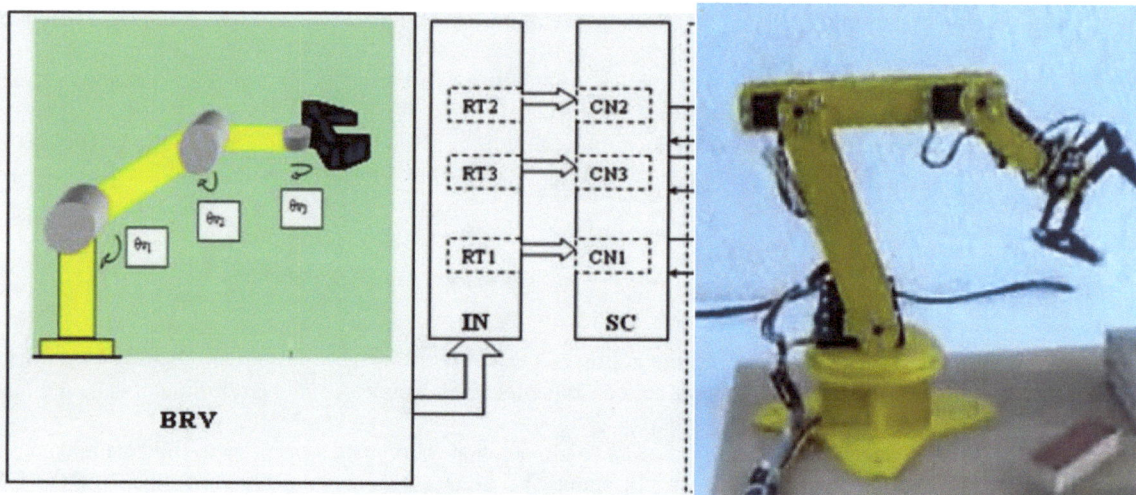

Figure 1 Experienced robotic manipulator

For behavior control of the physical robot is use one Video camera. It looks at the field from above and produces an output video stream, which is forwarded to the central PC. Images are captured by a frame memory and given to the vision module.

The global computer vision module analyzes the images, finds the paths of the physical robots and produces as output the positions and orientations of the robot's joints, as well as the position of the arm. It is described in detail in [3].

Based on the gathered information from virtual robot and the physical robot, the behavior control module then produces the commands for the physical robot: desired rotational velocity, driving speed and direction, as well as the activation of the logical device. The central PC then sends these commands via a wireless communication link to the physical robots. The hierarchical reactive behavior control system of the team virtual robot – physical robot is described in [5]. For each robot joint is needed a microcontroller for omni directional motion control. It receives the commands and controls the movement of the robot using PID controllers (see [4]). Feedback about the speed of the joints is provided by the pulse generators which are integrated in each servo-motor

4. EXPERIMENTAL CONFIGURATION FOR ROBOT PROGRAMMING BY IMITATION

The approach was implemented using a simulation software platform called Robot Motion Imitation Platform (RMIP) developed at University of Brasov [8], who its feasibility has been demonstrated in some simulation settings for manipulative works, using Virtual Reality.

We extended the simulation software platform to programming a physical robot. We see them as intelligent structure that can be used to transfer - in real time - the gesture of the virtual robot to the physical robots for generating complex movements in real work space. One transfers via intelligent interface, the virtual joint angles data from a motion capture system to a kinematic model for a physical robotic manipulator.

The RMIP is an architecture that provides libraries and tools to help software developers in programming.

Platform RMIP is focused on 3D simulation of the dynamics systems and on a control and planning interface that provides primitives for motion planning by imitation.

Also is a object-oriented infrastructure for the integration of controllers, as well as the integration of planners with controllers to achieve feedback based planning, In particular, RMIP is used to provide concrete implementations of motion planners.

The platform RMIP utilizes a framework which allows both high level and low level controllers to be composed in a way that provides complex interactions between virtual and physical robots.

In our strategy we just included the possibility of functional coupling virtual and physical robot model. The functional coupling consists to map of the motion of the demonstrator robot arm with the imitator robot arm. The former is displayed on the screen, giving the visual information to programmer and is connected to the latter through intelligent interface. In our experiments we have adapted as imitator a small robotic manipulator, named EXMAN and we start with a 3 degree-of-freedom (DOF) discrete movement system that models point-to-point attained in a 3D Cartesian space.

The arm moved in a free space without obstacles, had three active joints (shoulder, elbow and wrist) and was driven by electrical actuators.

Figure 2 shows our experiment involving the imitation learning for a physical robotic arm with 3 degrees-of-freedom (DOFs) for performing the manipulate tasks. We demonstrated the imitation of elbow, shoulder and wrist movements. Importantly, these tasks required the coordination of 3 DOFs, which was easily accomplished in our approach.

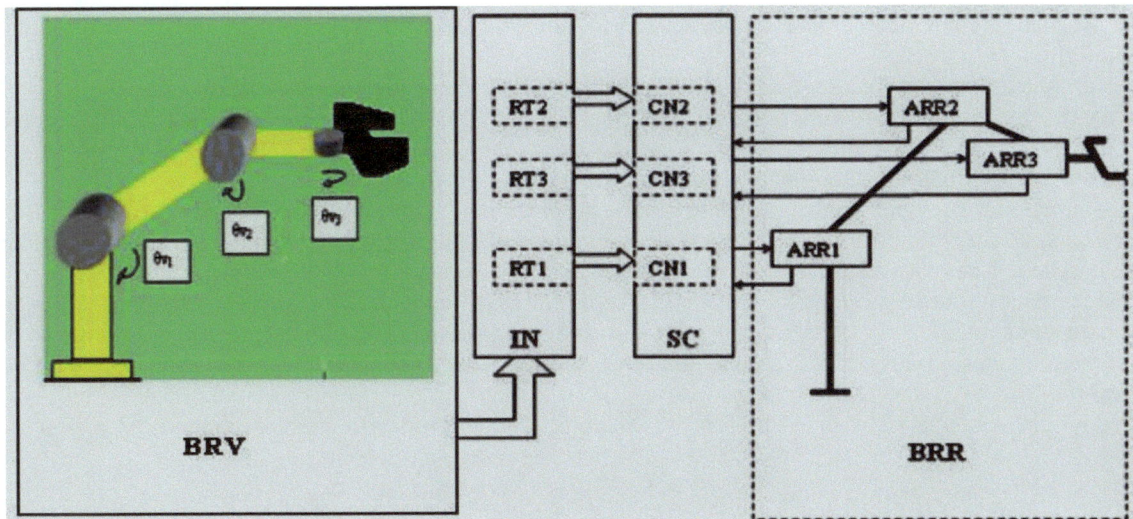

Figure 2 Imitation system architecture

The imitated movement was represented in joint angles of the robot. Indeed, only kinematic variables are observable in imitation learning. The robot was equipped with a controller (a PD controller) that could accurately follow the kinematic strategy (i.e., generate the torques necessary to follow a particular joint angle trajectory, given in terms of desired positions, velocities, and accelerations) [7].

Figure 2 also displays (left image) the user interface of a virtual robotic manipulator arm, which has been created which a dynamical simulator.

To generate the motion sequence for the real robot, one captures the motions from a virtual robot model and maps these to the joint settings of the physical robot.

Initially, a set of virtual desired postures is created to the virtual robotic arm BRV and the pictures' positions are recorded for each posture, during motion.

These recorded pictures' positions provide a set of Cartesian points in the 3D capture volume for each posture. To obtain the physical robot postures, the virtual pictures' positions are assigned as positional constraints on the physical robot. To obtain the physical joint angles one use standard inverse kinematics (IK) routines. The IK routine then directly generates the physical joint angles on the physical robot for each posture.

Referring the Figure 2 we comment the following: on programming platform, a robot program is carried out off-line, and one sends into the data registers of a port of the hardware structure, the numerical values of the joint variables of the virtual prototype of the robotic arm (BRV) and displays on a graphical user interface, the evolution of the virtual prototype during the carrying out of the robotic task. Via numerical interface (IN) the virtual joint dataset, from the data registers of the port of the hardware structure of the programming platform are transferred into the data registers of the numerical comparators of the controllers. These datasets are reference inputs of the pursue loops, resulting a control system (SC).

The symbolic spatial relations specifying the virtual work space can be used for the automatic pursuits of possible virtual path as well as for planning of appropriate behavior of the physical robot arm BRR, which may guide the motion process during task execution.

The easiest way to generate the spatial relations explicitly is the interactively programming of the behavior of the virtual prototype in his virtual environment, in order to specify suitable positions θ_{v1}, θ_{v2}, θ_{v3}. This kind of specification provides an easy to use interactive graphical tool to define any kind of robot path; the user has to deal only with a limited and manageable amount of spatial information in a very comfortable manner.

The robot programming system has to recognize the correct robot task type and should map it to a sequence of robot operations [7]. The desired pathways are automatically transferred and parameterized in the numerical interface IN, using the path planner.

The applicable robot tasks are designed and the desired pathways are programmed off-line and stored in the buffer modules RT1, RT2, RT3.

The comparative modules CN1, CN2, CN3 furnish, to the pursuit controllers, the datasets involving the expected state of the virtual robot prototype and the measured state of the physical robot.

While motion execution is in progress, the real robot joints ARR1, ARR2, ARR3 are activates into the real work space. Each actuator was connected by a sensor in the closed-loop. Each time, a skill primitive is executed by the robot control system SC; it changing the robot joints position. As no time limit for the motion is specified, the physical robot imitates the behavior of the virtual robot.

In our laboratory currently we are developing Cartesian control architecture able to interpret the physical robot commands in the above given form. The basis of our implementation is a flexible and modular system for robot programming by imitation.

In our experimental configuration in order to prove the correctness of the robot programming by imitation we have chosen an robotic manipulator equipped with electrical actuators, mounted on the physical robot's joints.

The robot's control unit is connected via TCP/IP to a PC equipped with the interface card; the PC is running the simulation and control process. The robot control system receives and executes each 16 ms, an elementary move operation.

Due to the kinematics limitations of physical robot, the resolution of the joints is quite limited. After the calibration phase, physical robot starts imitating the gestures of the virtual robot demonstrator. The user can decide to activate all the three degrees of freedom concurrently or with a restricted subset by deactivating some of the sensors attached to each joint of the physical robot.

Our system requires an essential step in that one converts the position errors into motor commands by means of the PD controller [4].

5. CONCLUSION

We propose to use a virtual robot as a predictor for the robot motion, because this approach doesn't require an explicit model and can easily use robot commands as additional input for the prediction. This allows predicting future movement changes before any of them could be detected from the visual feedback. As with all control systems, there is some delay between making an action decision and perceiving the consequences of that action in the environment. We have successfully developed, implemented, and tested a imitation system for predicting the motion of the physical robots. The prediction compensates for the system delay and thus allows more precise motion control. The virtual robot is trained with data recorded from real robots. We have successfully field-tested the system at several scenarios. The predictions using the virtual prototypes improve speed and accuracy of the physical robot motion.

The virtual tracking paths of the virtual joints are reproduced in the real work space by the actuators of the physical robotic arm. The reference datasets are obtained using a motion capture channel taking into account the joints motion range.. In future work we will investigate these idea and we will examine the possibility of how to implement them.

REFERENCES

[1] R. Diankov and J. Kuffner, "Openrave: A planning architecture for autonomous robotics". Technical report, CMU-RI-TR-08-34, The Robotics Institute, Carnegie Mellon University, 2008.

[2] M. Turpin, N. Michael, V. Kumar," Trajectory planning and assignment in multi robot systems". Proc. Workshop on Algorithmic Foundations of Robotics, 2012

[3] J. Snape, J. van den Berg, S. J. Guy, and D. Manocha, "Independent navigation of multiple mobile robots with hybrid reciprocal velocity obstacles", IEEE/RSJ Int. Conf. Intelligent Robots and Systems, St. Louis, Mo., pp. 5917–5922, 2009.

[4] M. Nicolescu, M. Mataric, *Natural methods for robot task learning- Instructive demonstrations, generalization and practice,* In Proc. Intl Joint Conf. on Autonomous Agents and Multi-agent Systems (AAMAS), Melbourne, Australia, pp. 241-248, 2003.

[5] S. Calinon, F. Guenter and A. Billard, *On learning, representing and generalizing a task in a humanoid robot,* In IEEE Trans. on Systems, Man and Cybernetics, Part B, vol. 37, no. 2, pp., 286-298, 2007

[6] M. Pardowitz, R. Zoellner, S. Knoop, and R. Dillmann, *Incremental learning of tasks from user demonstrations, past experiences and vocal comments,* In IEEE Trans. on Systems, Man and Cybernetics, Part B, vol. 37, no. 2, pp. 322-332, 2007.

[7] J. Kober, J. Peters, *Learning motor primitives for robotics,* In Proceedings of the IEEE International Conference on Robotics and Automation, pp. 2112–2118, 2009, Piscataway, NJ: IEEE

[8] A. Fratu, B., Riera and V., Vrabie : *Predictive strategy for robot behavioral control,* Proceedings in Manufacturing Systems, ISSN 2067-9238, Volume 9, Issue 3, 2014, pp. 125-130.

6th International Conference
"Computational Mechanics and Virtual Engineering "
COMEC 2015
15-16 October 2015, Braşov, Romania

ANALYSIS OF A COMPLEX MANUFACTURING SYSTEM MODELED WITH PETRI NETS

Elisabeta Mihaela Ciortea[1], Mihaela Aldea[2]
1 "1 Decembrie 1918" University of Alba Iulia, Str. Unirii, no. 15-17, cod 510009, Romania, ciortea31mihaela@yahoo.com
2 "1 Decembrie 1918" University of Alba Iulia, Str. Unirii, no. 15-17, cod 510009, Romania, maldea7@yahoo.com

Abstract : *The paper presents an example of monitoring and control for a complex manufacturing system modeled with Petri nets. It also proposes a method to monitoring and control in case of some repairs, or the maintenance according to the technical book of machines that are part of manufacturing process analyzed. The method uses a classical complex manufacturing model simulated with Petri network based on specifications, and implements the hierarchical and distributed control of the manufacturing chosen system. The paper is structured in two stages. First, the activities are analyzed in a conceptual level, and are defined by a Petri network model, having as support the contract documents, corresponding to manufacturing process. The second stage is dedicated to the procedure for carrying technological flow with the analysis of equipments that are involved in flux, without major losses occur. The study is appropriate for hierarchical and distributed control systems. The method can be extended to complex automated systems.*
Keywords: Petri Net, manufacturing system

1. INTRODUCTION

Due to the need of flexibility industrial system, were introduced industrial robots in manufacturing systems for automating the technological operations without recourse to a significant design. This flexibility resulted from the general structure of the manufacturing system, from the control of tracking and reprogrammable systems, and it can be exploited if the robot is programmed accordingly. A single robot cannot perform tasks in optimum conditions if this is used in systems with additional equipment. In this sense, usually in order to integrate the robot in manufacturing system, are used numerically controlled machine tools, conveyor belts or other machines with special purpose.

Because the robotic systems are made to work in parallel and are asynchronous, it cannot accurately predict when interest events for the robot program may appear. Thus the signal lines are supported by systems fitted with a robot that serves to coordinate more robots, machine tools or machines. So far there is an adequate computing language and strictly dedicated to a complete manufacturing system optimization, in parallel and in real time with specific constraints. In general are used relatively short programs with links between them, in order to ensure the continuity of manufacturing system operation.

The structure of a working system in a complex manufacturing system consists of one or more lines, each consisting of one or more equipments such as robots, intelligent machine tools. Inside the cell, the machine executes cooperation tasks such as processing, assembling, storing.

The task execution stages of a system consisting of a robot or a machine tool are provided as follows - the transfer of the object from the starting position into a position target is a sequence of next secondary tasks: it moves his arm in the starting position, grabbing the object, then it moves into the specified position, and put the object at the specified location. Thus the tasks are divided into a hierarchical system.

One effective method used to describe and control such systems is the library of Petri networks, which is a modeling tool for asynchronous and competing systems with discrete events. A Petri network model used for analyzing distributed and hierarchical control for manufacturing systems with discrete events includes control algorithms and it is accessed to control the manufacturing process.

The Petri networks used in this paper are analyzed based on oriented graph with which is associated manufacturing system. With the help of Petri networks it can model, simulate and control manufacturing systems with discrete events.

2. SYSTEMS DISCRETE EVENTS USING PETRI NETWORKS

A manufacturing process is characterized by the objects flow that passes through subsystems and receives adequate commands. Each subsystem executes manufacturing operations which are decomposed in machining operations, assembly and transportation operations, such as loading, transportation and unloading.

A characteristic of control for systems with discrete events is that a manufacturing process can be decomposed into a set of discrete events and conditions linked.

A working example for robot operations: the arm waits until it reaches a piece for processing; it takes the piece and sends it to the next operation. The steps are:

- it is conditioned the robot on hold;
- it is analyzed the event when the piece reach to the work area;
- it is analyzed the state of piece that is on hold;
- is launches the event and the robot arm begins manipulating; the arm is occupied with the work piece;
- the robot arm continues manipulation; it is conditioned – the manipulation was completed; it is analyzed the event with the work piece sent to another operation.

This system is one with state event. These events have the following characteristics:

- Asynchronies, the events occur when conditions are fulfilled;
- Conditioning, that is defined by conditions before and after the event;
- Parallel, that is into a system of two or more conditions can occur simultaneously events, and those that do not interact, can occur independently;
- Mistakenly, that is the overlapping the prior condition with the one that defined the event.

Because of these characteristics may occur some shortcomings such as: when the system reaches a state which it is not feasible to satisfy any event; in case of existence of one condition the previous event occurs, it may appear the striking effect (injury).

3. THE ANALYSIS OF PETRI NETWORKS

Petri network is intended to define the process-type behavior of the system network. The network captures the event appearance of the analyzed system. The graphical representation of the manufacturing process is based on the graphical representation of the appearance by labeling network elements.

Let be $\gamma = (\Sigma, M_o)$ a Petri net from [1]

$$\Sigma = (P, T, F, K, W) \tag{1}$$

that is called process of the network γ any couple $\pi = (N, \rho)$, where $N = (B, E, F')$ is a finite network of occurrences, and

$$\rho : B \cup E \to P \cup T \tag{2}$$

is a function that satisfies:

a) $\rho(B) \subseteq P$, $\rho(E) \subseteq T$

b) $M_o(p) = \left| \rho^{-1}(p) \cap {}^o N \right|$ for any $p \in P$

c) $W(p, \rho(e)) = \left| \rho^{-1}(p) \cap {}^o e \right|$

$W(\rho(e), p) = \left| \rho^{-1}(p) \cap e^o \right|$ for any $e \in E$ and $p \in P$

The network evolution P/T marked can be defined as the set of all processes in the network.

Starting from defining the processes [1] are introduced by N^o in Σ, M_{N^o}, according to the relation:

$$M_{N^o}(p) = \left| \left\{ b \in N^o \cap B \middle| \rho(b) = p \right\} \right|, \forall p \in P \tag{3}$$

To build the sets or to determine network processes it proceeds as [1]:

$\Pi_o(\gamma)$ contains only pairs of form (N, ρ) where $N = (B, \phi, \phi)$, $|B| = \Sigma_{p \in P} M_o(p)$ and B contains $M_o(p)$ different conditions b, $B = \rho^{-1}(p)$ for each $p \in P$,

It is assumed that $\Pi_i(\gamma)$ for $i \geq 0$ it was constructed. For each $\pi = (N, p) \in \Pi_i(\gamma)$, where $N = (B, E, F')$ and each $t \in T$ such that $M_{N^o}|t\rangle$, let's define $\pi' = (N', p')$ where $N' = (B', E', F'')$, satisfies the conditions:

- B', E', F'' include B, E, F',

- For any $x \in B \cup E$, $\rho'(x) = \rho(x)$,

- It adds a new element e to the network N with $\rho'(e) = t$. For each $p \in P$ with $W(p, t)\rangle 0$ it chooses $W(p, t)$ different conditions b labeled by $p(\rho(b) = p)$ and it adds the arcs (b, e) to F''. Similarly for each $p \in P$ with $W(t, p)\rangle 0$ it adds $W(t, p)$ different conditions b labeled by $p(\rho'(b) = p)$ and it adds the arcs (e, b).

Then $\Pi_{i+1}(\gamma)$ consist of $\Pi_i(\gamma)$ and all $\pi' = (N', \rho')$ constructed.

In Figure 1 is presented a Petri network Σ and it is illustrated a process π of the network. The locations and the transitions from the process are labeled as suitable locations and transitions for the manufacturing process. In figure it observes the conflict situations at various locations bordered by transitions, but these can be solved by building the model and so the conflict is resolved.

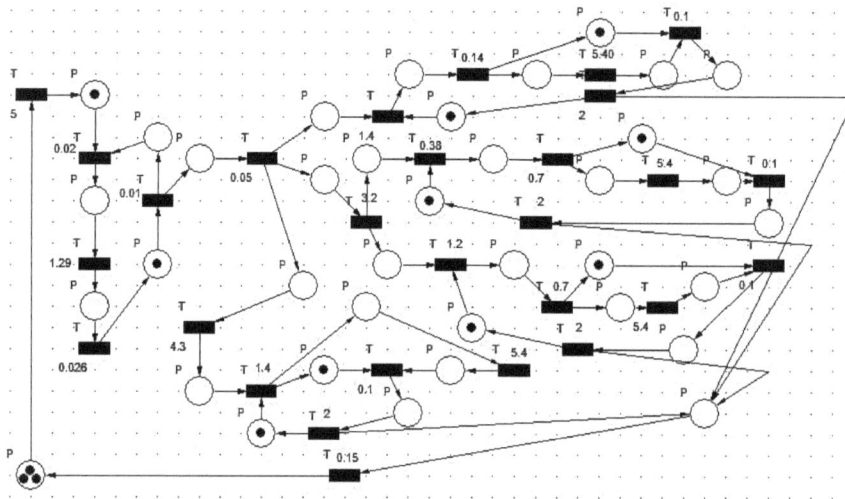

Figure 1. Manufacturing model constructed by Petri network

4. THE SYSTEM MODELING

The model used for modeling describes a manufacturing system consisting of a set of sorting or selection of matter. After that, using the conveyor belt properly for five workstations or flexible cells occur the robot arm that loads the material to be processed, and it takes after processing. The processed material is transported in mobile warehouses from where the process continues, but is not shown in the model. The processing times are real times selected from the specifications of processes and the transport times were selected following a modification of placement system of the machines to optimize transportation times, but without interfering waiting strings.

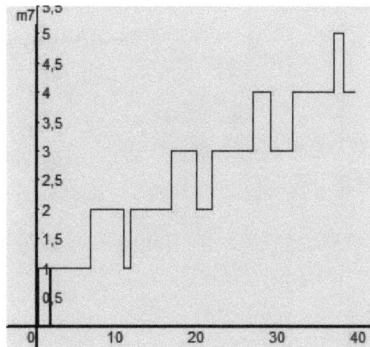

Figure 2: The graphical representation of the average duration for the first conveyor transport

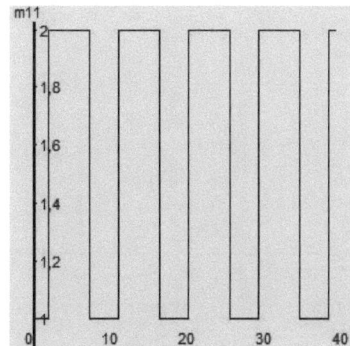

Figure 3: The graphical representation of the average duration occupancy of the machine

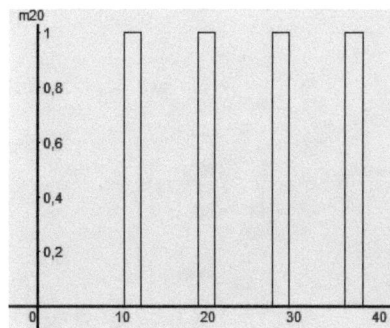

Figure 4: The graphical representation of the average duration occupancy of the machine through the robotic arm

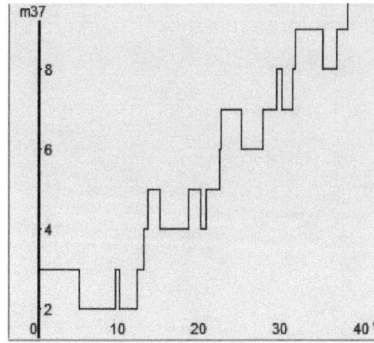

Figure 5: Graphic representation of the average number of conveyor thrugh the collection point

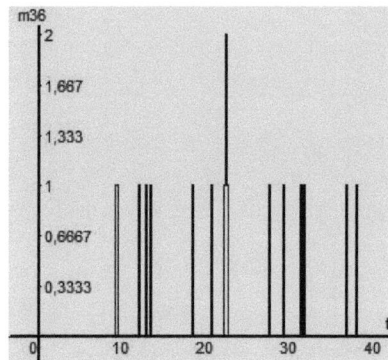

Figure 6: The graphical representation of the average number of conveyor through warehouse

According to the technical documentation, it is ensured the maintenance and in case of some errors, current repair occurs. To illustrate this, in the manufacturing model we assumed an intervention at first machine in the system.

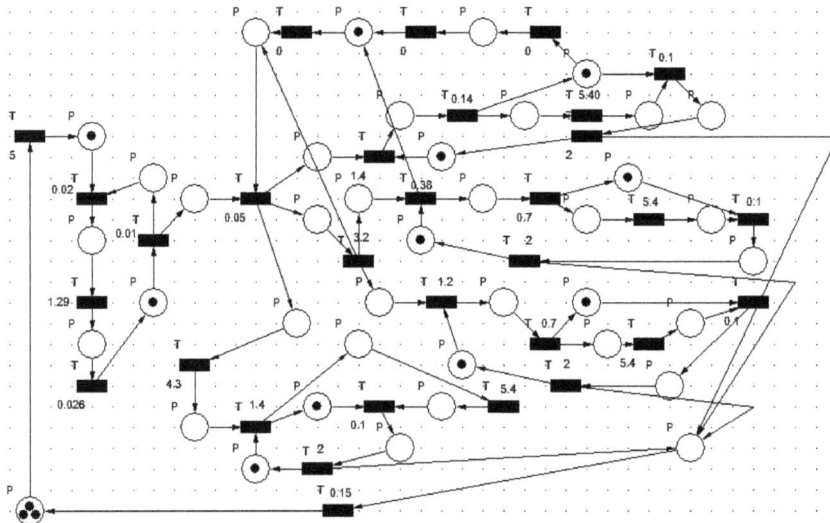

Figure 7: Network with intervention system

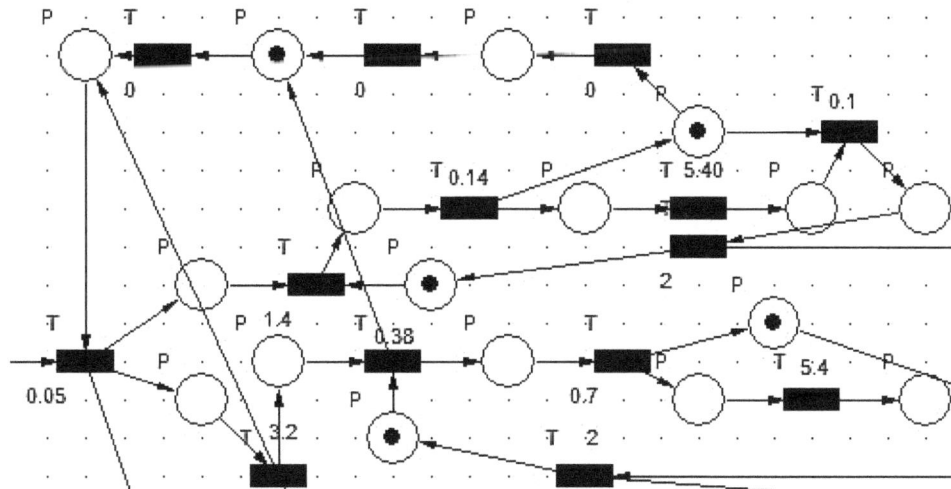

Figure 8: Illustration of the intervention system

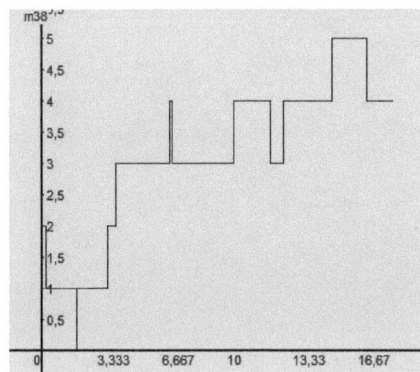

Figure 9: Graphical representation for tracking the intervention at the first machine.

5. CONCLUSION

Applying the product manufacturing times and the transport times obtained by applying the recommendations from the documents that accompany the manufacturing process, it is obtained graphics representations that highlight the average time of the transport activity, but also the evolution of the average manufacturing based on the transport activities times using as parameters lots of finished products

In analyzing the transport it is aimed the study of suggested simulation solutions. Initially the method was developed for a flexible system and in time it was developed also for complex systems even in the case of intervention for repairs systems such as not to be seriously economically affected the manufacturing process.

The graphs analyze applies in solving problems seeking a representation method that allows to study easily the entire problem and to highlight all the results.

It was elaborated the transportation system model that established the sizing, the internal structure and corresponding couplings. Besides workstations, in the system were included control stations and relevant logistics subsystems.

The Petri nets use as parameters the average value of the exponential distribution parameters assigned to the position which models the transport availability. Is obtained graphical representations regarding the evolution of the average times manufacturing, the average length of occupancy, the graphical representation of the average transport for each conveyor and graphical representation of the average number of conveyors from storage.

REFERENCES

[1] Camerzan I. „Proprietăţi structurale ale reţelelor Petri temporizate", Teza de doctor în informatică, Universitate de Stat din Tiraspol, Chişinău, 2007-12-15

[2] Păstrăvanu O. „Sisteme cu evenimente discrete. Tehnici calitative bazate pe formalismul reţelelor Petri", Editura MatrixRom, Bucureşti, 1997, ISBN 973-9254-61-6]

[3] Păstrăvanu O., Matcovschi M., Mahulea C. „Aplicaţii ale reţelelor Petri în studierea Sistemelor cu evenimente discrete", Editura Gh. Asachi, 2002, ISBN 973-8292-86-7

[4] Ciortea E. M., "Aspects of discrete events in manufacturing systems", Proceedings of the 14 th International Conference Modern Technologies, Quality and Innovation, Volume I, ISSN 2069-6736, 2010

[5] Ciortea E. M., " Analytical interpretation to optimize transport systems", Advanced Materials Research, vol.827, Modern Technologies in Industrial Engineering II, pag. 881-885 , ISSN

6th International Conference
"Computational Mechanics and Virtual Engineering "
COMEC 2015
15-16 October 2015, Braşov, Romania

THE 3D LIGHTWEIGHT OPTIMIZATION FOR HIGH LOADED PARTS – CASE STUDY

Gabriel. D. Dima[1], Marius Diaconu[2], Ion Balcu [3], Ionuţ Teşulă [4]

[1]Transylvania University, Brasov, ROMANIA, dumitru.dima@unitbv.ro
[2]NUARB Aerospace, Brasov, ROMANIA, m.diaconu@nuarb.ro
[3]Transylvania University, Brasov, ROMANIA, balcu@unitbv.ro
[4]Schaeffler, Brasov, ROMANIA, tesulinu@schaeffler.com

Abstract: The 3D structural optimization is the most difficult optimization process, being used in aerospace industry only in few applications. In accordance with the efforts to set up a methodology to help to the implementation of extensive structural optimization in the development teams, within the paper a case study from a helicopter pilot seat is presented. The optimization of the fitting targeted lightweight, with respect to the constraints of a high loading environment and the manufacturing issues. Iterations of optimization, together with conclusions and lessons learned are presented.
Keywords: lightweight, topology optimization, aerospace, design for manufacturing

1. INTRODUCTION

Aerospace structure and components have to comply with the major requirement of lightweight. This constraint emerged from the first aircrafts to fly and remains till nowadays on the spot. To comply with it, the manufacturers created internal lightweight methodologies and approach the design in an iterative process. Structural optimization may be also a solution, but its applicability is limited due to the resources involved vs. results. The objective of the paper is to present a case study of the structural optimization of a high loaded part, together with lessons learned and design recommendations.

2. TECHNICAL REQUIREMENTS

The lightweight allows to additional payload and/or extended range for an aircraft, therefore weight saving policies are close connected to the commercial issues as presented in [4], [5], [13].
Even the interest for the structural optimization can be reached since the 30's, due to calculation and manufacturing constraints, consistent results may be seen only after the 90's as published in [5], [6], [8], [3], [11]. These researches were focused on specific applications (parts or subassemblies) and not on working methodologies, this process being only at the beginning. The big manufacturers have no internal optimization methodologies, for specific applications just outsourcing work packages to the commercial solutions manufacturers or to R&D centers. Most of works are focused on case studies [3], or on methodology [2], [9], [10] reporting savings of 20 – 35%, within a time consuming and not very smooth optimization process. The stability, aero-elastic and manufacturing constraints are also critical, making the optimization process a multi-objective process [1], [6], [7].

This paper presents a machined part 3D optimization process, with a case study of a the floor attachment fitting of a helicopter crashworthy seat. The optimization results are post processed in three different designs to comply both with the manufacturing and the lightweight constraints.

As usual, the 3D structural optimization supposes an initial design, followed by a stress analysis as a start point to assess the optimization results. Structural optimization projects may be done on existing parts or to new developed parts. In the state of the art approach, after the design - stress iterations, the prototype is produced, tested, certified and released to operation. After the pre-serial production, up to three weight saving programs may follow-up, or fatigue upgrades addressing the critical stressed parts. This programs (stretched as usual in 10 – 20 years) lead to the upgraded structure of an aircraft. This may not be considered as an optimization program, but an improvement which requires high skilled engineers, and previous aircraft lifecycle experience.

Structural optimization brings some advantages as: less design iterations and shorter time to market, better results taking into account numerical optimization commercial solutions existing on the market.

3. THE WORKING METHODOLOGY

Within this case study, the parts subjected to structural optimization were designed by an experienced designer and based on the state of the art solutions. The static FE analysis was performed to provide a reference for the optimization results (mass vs. stress and displacements). The optimization output was used for the part redesign according to the manufacturing constraints. Different designs were proposed to check the progress of the output parameters.

The loading conditions are according helicopter's certification regulations (FAR27) consisting on inertial load on different axis as follows (g = 9.81 m/s^2):

- LC1 – FWD Crash, N_X = -18.4g
- LC2 – Down Crash, N_Z = -30.0g
- LC3 – Rear, N_X = 1.5g
- LC4 – Up, N_Z = 8.0g

For CAD design, the Catia V5R19 software was used, while for static stress analysis Hypermesh/ Radioss 10.0 was used. For topology optimization, the Hypermesh/ Optistruct 10.0 were used. The FE model consisted in 3D tetrahedral 2nd order elements, the mesh size being of 2.0 mm. The FE model of fiting was with all displacement restrainted in the screwing holes, and the loads applied in the hole containing the floor rails attachement pin.

4. THE INITIAL DESIGN

The case study refers to the floor panels attachment fittings of a helicopter crashworthy pilot seat (Fig. 1). The fittings are attaching the seat to the cockpit floor longerons below the plane of the honeycomb panels. The fittings are machined from aluminum alloy 2024T3, a very usual alloy used in aerospace structures.

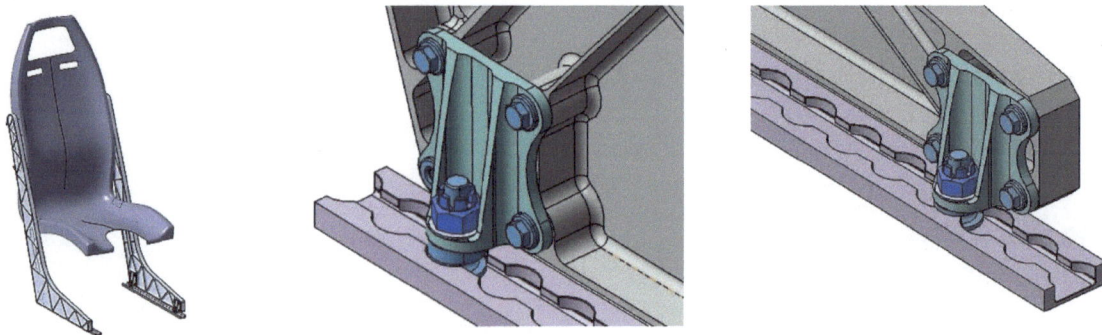

Figure 1: Isometric view of the crashworthy pilot seat and the attachment fittings

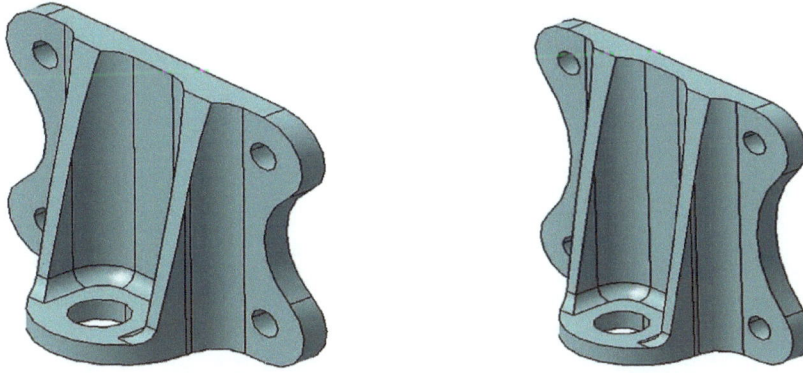

Figure 2: The front and rear attachment fittings

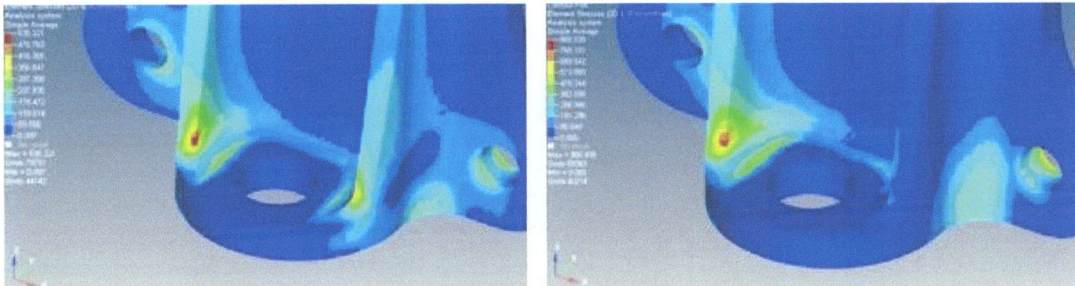

Figure 3: Front and rear attachment fitting – von Misses stress

The initial stress results of the fittings are indicated that the front attachment fitting presents higher von Misses loads (Fig. 2 and 3). Therefore, only the front fitting was considered for the optimization process, the rear fitting being redesigned according the output of the front fitting optimization process.

The results of the analysis of the initial design (Design start = DS) are shown in Table 1, the maximum von Misses stress being 863.7 MPa, displacement 0. 076 mm (load case 1, crash forward), for a part mass of 0.116 kg.

Table 1: The initial design results (DS)

Load Case	Stress [MPa]	Displacement [mm]
1. FWD	863,7	0,076
2. DOWN	361,4	0,032
3. REAR	145,7	0,013
4. UP	8,4	0,001

5. THE OPTIMIZATION RESULTS

The structural optimization consisted in a 3D topology optimization using the Optistruct 10.0 commercial solution. The optimization method consisted in choosing a volume with the permissible limits and defining the mounting points and force application, using a stress parameter and weight parameter; the constraints for these parameters was less then the initial values (700 MPa and 0.105Kg, respectively).

The optimization output consisted in a shape linking the load application area with the mounting holes areas.

The shape is organic, without the possibility of manufacturing with the existing state of the art technology (Fig. 4). Even the additive manufacturing technology may be suitable, a certified technology for aerospace industry had to be selected –

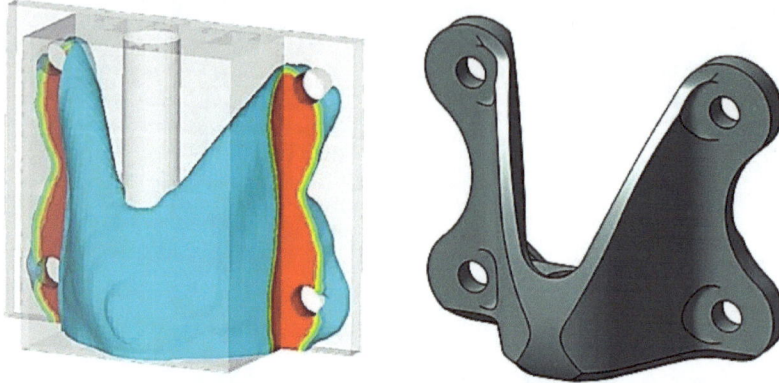

Figure 4: The optimization result and the design D00

the multi axis machining. The resulted design followed the output shape, taking into account the machining constraints (generation of surfaces, basement, corner and base radius, etc). Because the mass was 0.101 kg, further designs were generated to decrease the mass.

The design D01 was developed on the basis of the design D00. The inner cutout geometry and the curved central V - shaped edge were simplified, together with a consistent weight saving of 20%. The design D02 was based on idea that the load application area should consist of two levels in order to decrease the stresses due to the moment loads. The design is more robust, but the mass increased with 10% relative to the design D01.

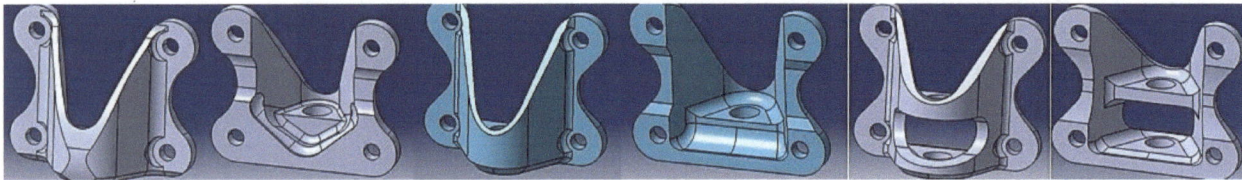

Figure 5: The design iterations D00, D01 and D02

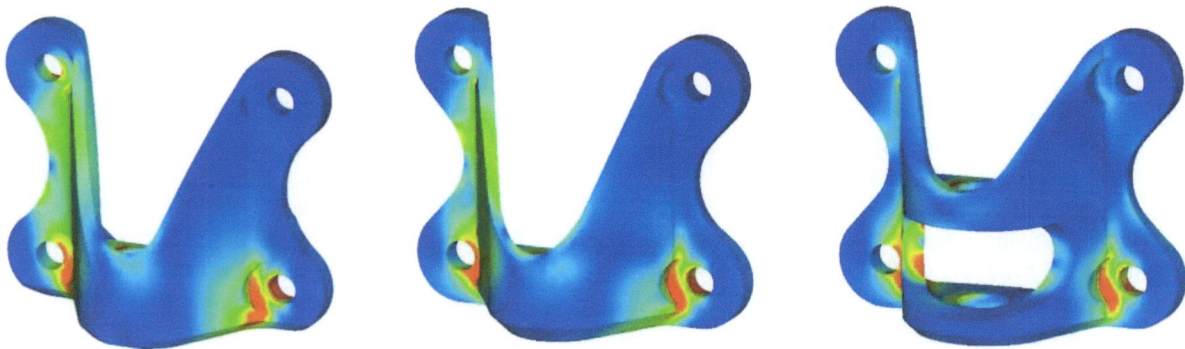

Figure 6: The von Misses stress distribution of the design iterations D00, D01 and D02

6. RESULTS AND DISCUSSIONS

The optimization output passed through a redesign process to allow machining of the resulted shape. The design D00 was the first design iteration to comply with the manufacturing constraints. As per graphs from figures 7 and 8, the design D00 presents a major stress reduction (about 50%) for a double rigidity than the initial design (DS). Taking into account the weight saving, the progress from variant DS to D00 is significant.

 For the design D01, the stresses are higher for a small variation of rigidity, together with a consistent improvement of the machineability. Taking into consideration the 20% of weight reduction, this variant may be considered for further product development. The design D02, comes up with a von Misses stress growth for a 10% weight saving and the same machineability as D01. Due to the high values of the von Misses stress, this variant was not developed anymore.

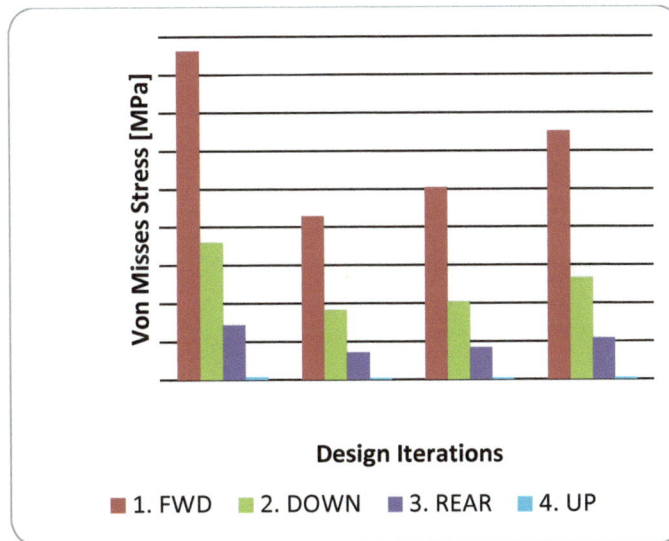

Figure 7: Maximum von Misses stress of the design iterations

The parameters of the fitting may be further adjusted depending of the parameter with the highest importance. Thus, if there is not a target for weight, by increasing the wall thickness/ decreasing fillet radii the stress level in hot spot stress areas may be decreased to a level required by the ultimate tensile stress of the material. Thus, a consistent weight saving may be obtained by changing material form steel to titanium or from titanium to aluminum. If the stress level is a target, then, the part will be refined by design/ stress iterations until the von Misses stress is near below the target, resulting the mass and the displacement.

Even the optimization process represents a step forward in the design process, an additional process of adjustment of the geometrical parameters may offer the final configuration of the part.

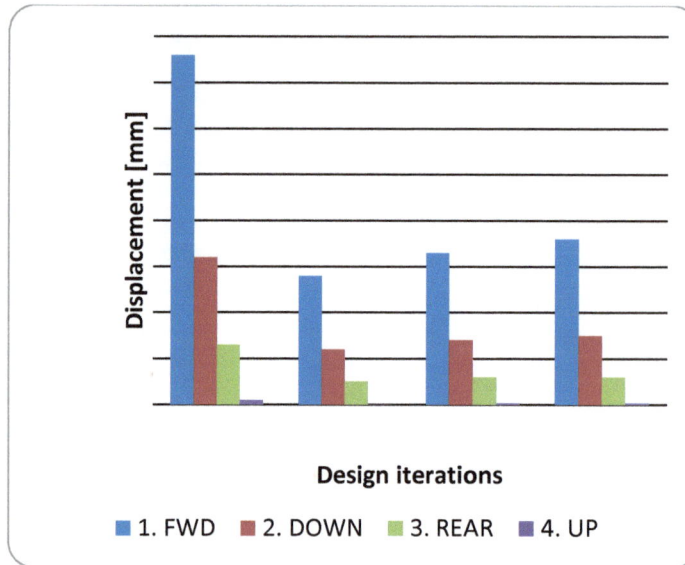

Figure 8: Maximum displacement of the design iterations

7. CONCLUSIONS

Within the paper a case study of structural optimization of a high loaded part was presented. After the stress and displacement assessment of the initial design, a 3D topology optimization was made using a commercial numerical solution. Based on the optimization output, three design variants were proposed. Using the static stress analysis, the new designs were assessed in order to select the best combination of the output parameters.

Based on the whole process results, the following conclusions may be formulated:

- only the critical load case/ cases have to be considered for the static stress/ displacement assessment, the approach being conservative;
- the output of the optimization process needs redesign by an experienced designer;
- the best results may be obtained by proposing few design to comply with the manufacturing constraints;
- finding the best configuration of the strength/ rigidity/ weight and machineability is not a linear process (the lowest stress were obtained for the first design iteration);
- the optimization process is multidisciplinary and there is no any available methodology to get a result to comply with both strength and machineability requirements.

Future investigations are worth to be done in the area of compliance with other requirements as maintainability and inspectability, considering also new manufacturing techniques as additive manufacturing.

REFERENCES

[1] Brackett D, Ashcroft I, Hague R. Topology optimization for additive manufacturing, Loughborough, Leicestershire, 2011

[2] Dima G., Balcu I., Zamfir M., Method for lightweight optimization for aerospace milled parts – case study for a helicopter pilot lightweight crashworthy seat side struts, 8th International Conference Interdisciplinarity in Engineering, INTER-ENG Proceedings, Tirgu-Mures 2014

[3] Gubisch M. New Lufthansa seat saves nearly 30% in weight, Flightglobal, Dec 2010, www.flightglobal.com

[4] Hertel H. Leichtbau., Bauelemente, Bemessungen und Konstructionen von Flugzeugen und anderen Leichtbauwerken, Springer, New York, 1980

[5] Kaufmann M. Cost/ Weight Optimization of Aircraft Structures, KTH School of Engineering Sciences, Stockholm, 2008

[6] Krog L, Tucker A, Rollema G. Application of Topology, Sizing and Shape Optimization Methods to Optimal design of Aircraft Components, Airbus UK Ltd, 2002,Bristol

[7] Maute K, Allen M. Conceptual design of aero-elastic structures by topology optimization, Structural and Multidisciplinary Optimization, vol.27, no.1-2, pp. 27 – 42, 2004

[8] Schuhmacher G. Optimizing Aircraft Structures, Concept to Reality, Altair Engineering, 2006.

[9] Teşulă I., Balcu I., Dima G., The Lightweight Optimization of a Composite Shell – Case Study of a Helicopter Crashworthy Seat , Advanced Composites Materials Engineering COMAT Proceedings, pp. 36-46, Derc Publishing House, USA, 2014

[10] Yancey R. Optimization and Modelling Technologies, Altair Engineering, 2006

[11] Wang Q, Liu Z., Gea H., New topology optimization method for wing leading-edge rib, Journal of Aircraft,vol.48, no. 5, 2011

[12] * * *, JAR/ FAR Part 27 – Airworthiness Standard: Normal category rotorcraft, Federal Aviation Administration, 2013

[13] * * *, "What's a few pounds here and there?", International Conference Innovative Aircraft Seating Proceedings, Hamburg, 2011

6ᵗʰ International Conference
"Computational Mechanics and Virtual Engineering "
COMEC 2015
15-16 October 2015, Braşov, Romania

STRUCTURES WITHOUT ADHESIVE WITH POTENTIAL APPLICATIONS IN FURNITURE DESIGN OBTAINED THROUGH LIGNIN ACTIVATION

Ramona-Elena Dumitraşcu[1], Loredana Anne-Marie Bădescu[2]

[1] Transylvania University of Brasov, Brasov, ROMANIA, r.dumitrascu@unitbv.ro
[2] Transylvania University of Brasov, Brasov, ROMANIA, loredana@unitbv.ro

Abstract: In the current conditions of full wood recovery, the problem of finding the most efficiency bonding solutions remains permanently open. At the same time market requirements as the product be environmentally friendly and easy to recycle lead the authors of this paper to looking for technologies that ensure these requirements. Wood welding technology as an alternative method of joining elements or wooden structures is the technology that we stopped with more so since has not been addressed so far in Romania. The paper is based on a theoretical studies performed by author and shows the experimental results of some wooden structures that can be achieved by activating lignin, as binder, without the use of adhesives.
Keywords: lignin activation, structures, furniture design

1. INTRODUCTION

Wood joints to obtain products is an indispensable process from wood industry. Wood industry is a sector that contributes to the economical competitiveness through the variety of products for import and export in countries like UK, Belgium, USA, and Slovakia. Finding efficient solutions of joining wood to obtain environmentally friendly and easy to recycle structure was a challenge for the author.

Activation lignin from wood, as binder, through friction has not been addressed so far in Romania. To contribute to the knowledge development in this field, research regarding wood joints without adhesive began in 2013, through a postdoctoral research project proposed by the author.

By addressing an unconventional method like this, could expand the product range and also could improve the production system for better productivity. Applying an unconventional method of obtaining the product, in general, involves rethinking their manufacturing process. Also, the product must be designed in according with requirements of the method. It is a complex process which requires a large study from product design to product manufacturing.

Furniture design is an area where Romania has potential, given existing resources. As traditional sector, this is one of the few areas that bring profit. In this sense, each stages of knowledge development through research represent an important step towards a safe and sustainable development.

During this period, according to development strategies, efficient use of wood in view of its future possibilities of reuse is a priority of international interest.

Wood recovered from various pieces of old furniture for reuse contributes to environmental requirements and beyond. In this context, beech wood has a particular importance, being often used as a feedstock in the production of furniture and interior decoration. It has a content of lignin in the amount of 24.75%. Studies in the literature show that the sustainability of the timber remains unchanged over time.

PAL is a wood-based product obtained from wood chips and also is frequently used for usual furniture pieces. Joints between two plates of PAL are frequently found to the furniture, particularly at products with minimalist design. Reintegration of the particleboards plates (PAL) in other products and in combination with wood are the current challenges.

Therefore, the study was conducted on wood of beech (*Fagus Sylvatica L.*) and of wood-based material (PAL) to analyse some possibilities of joints. In this context, the proposals of structures for tests are presented in figure 1.

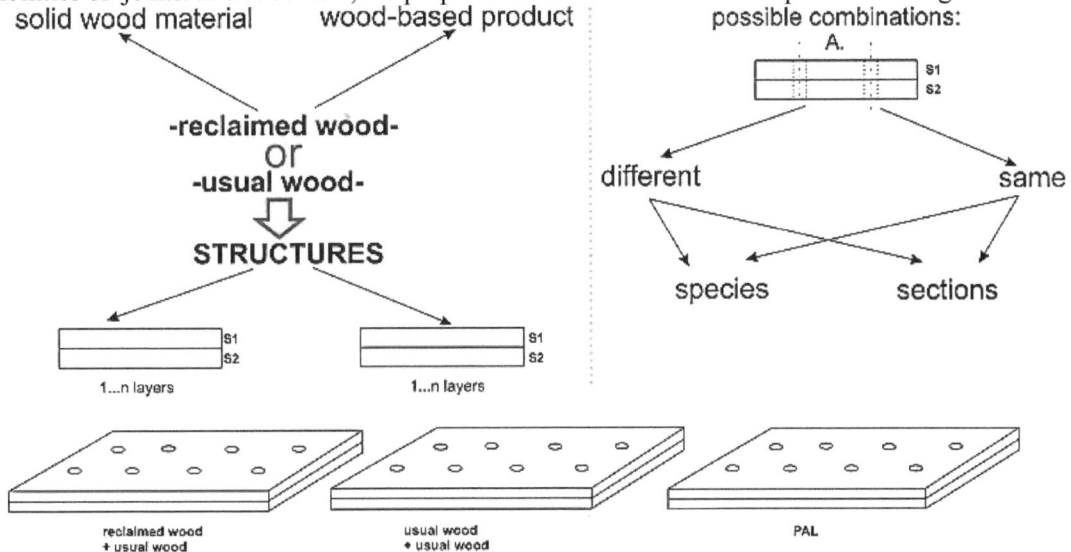

Figure 1: Type of structures proposed from:
solid wood material (beech) and wood-based product (PAL)

The paper presents the results obtained from combining the wood of the same species (beech and beech) and other species (beech and maple). At the same time, were made tests regarding joints of wood-based products through activating lignin, such as PAL's. This is a contribution of knowledge development for their further applications, in the context in which these are mainly used for furniture and also in various industries from Romania. It is known that in innovative furniture design, aesthetics and atypical structures and combination are often sought.

The lignin activation through a mechanical process of friction represent a opportunity for analize the posibilities of joints. As an alternative for classical joints, the structures achieved could be an solution for current requirements from the product design.

2. OBJECTIVE

Knowing the possibilities to activation lignin through rotational friction and structures behaviour can be an advantange for using this method in innovative furniture design as an alternative for clasical joints. The aim of this paper presents theoretical studies and shows the experimental results of some wooden structures that can be achieved by activating lignin, as binder, without the use of adhesives.

3. METHODS, MATERIAL AND EQUIPMENT

The method used for tests was rotational friction at high temperature. The method consists in activating chemical wood components (in special lignin) through a mechanical process of friction.

Joining wood through rotational friction can be achieved by maintaining fixed the support and rotating on its axis the dowel, under the action of external force.

The method choosed for this stage of study, facilitates the investigation of the interaction between wood dowels and other wood suport – in our case, solid wood material (*Fagus Sylvatica L.*) and other wood-based materials (particleboards -

PAL). The method require an common equipment, not very expensiv and efficient for usual applications. The study was carried out on a drilling machine adapted with an special clamping device, as is showed in figure 2.

Figure2: Drilling machine used to activate the lignin and insert the dowels by rotational friction

Material used for tests was PAL for furniture applications, beech wood. Maple *(Acer Platanoides)* was used for test the structures obtained from different species. Dimensions of samples tested were: 60x20x18 mm for wood-based products (PAL) and for solid wood material *(Fagus Sylvatica L. and Acer Platanoides)* 160x40x12 mm for each lamella. To joints the wood lamella, was used commercial ribbed wood dowels *(Fagus Sylvatica L.)* with diameter of 8 mm and length of 30 mm. The dowels were inserted by a fast movement, 1600 rpm, in a pre-drilled hole of 6 mm on the support (solid wood material or wood-based material). The structures design is showed in figure 1. The specimens were sectioned to analyze the connection from the contact area.

4. RESULTS AND DISCUTIONS

The structures obtained through lignin activation by rotational friction are presented in figure 2.

At joints between maple *(Acer Platanoides)* and beech wood lamella *(Fagus Sylvatica L.)* it was be observed that for some samples, the connection is not compact compared with the structures obtained only from beech wood (figure 6, figure 5).

When the material is pre-drilled, between the layers, in the contact area can appear an agglomeration of material which leads to gaps. These gaps affect the structure joints. As an observation, for good joints is necessary to have perfect linearity between the layers and of course a high clamping force. In this way, when the dowels are inserted in the solid wood material it can avoid gaps between layers, and the connection will be compact.

Figure 3: Examples of samples tested:
Solid wood material *(Fagus Sylvatica L. and Acer platanoides)* and wood-based material (particleboards -PAL)

Structures obtained from PAL presented good connection, even if between the layers is a melamina material. It can be estimated that the joints structures without melamina are more compact. Also, it is important to have a good force for clamping pieces during the pre-drilling process, to reduce the gaps between the layers.

Results obtained show that for good connection between layers, there is a correlation between dowels repartition, linearity of lamella, clamping force and fiber orientation (for solid wood material). On the other hands, specific wood characteristics has an important influence on the joints.

In general, the unconventional methods appeared as an alternatives to the clasical methods. Thus for uses in practical applications and in order to compete with standard adhesives, joining through lignin activation was tested and by international researchers (Ganne, 2008; Gerber, 2006). They shown that further investigations must be done in order to optimise the joint geometry (geometry & repartition of dowels) and to get reliable values for the design. In this sense, for the proposed structures, repartion of dowels were also investigated.

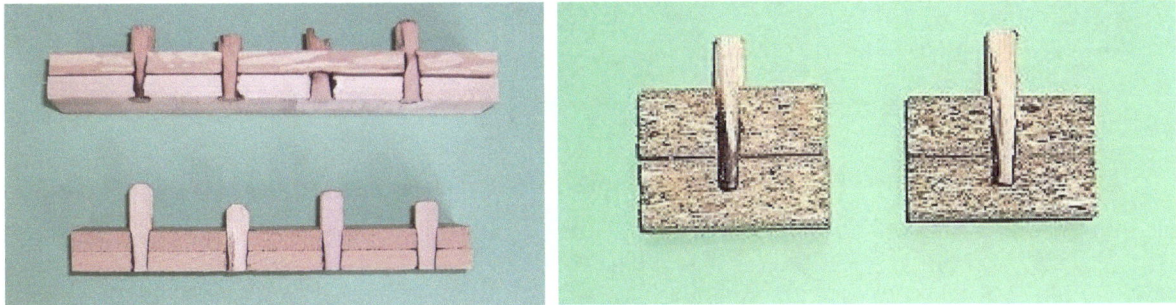

Figure 4: Sections on the samples obtained through lignin activation by rotational friction

Table 1: Main parameters of the process

Process	Wood Material	
	PAL (particleboards)	Solid wood material
Rotation speed	Density 610 - 700 kg/m3	Density: 720 kg/m3
Insertion speed	Lignin content: -	Lignin content : 24.75% and 27,30%
Dowels diameter	Fiber orientation: all directions	Fiber orientation: //
Hole diameter		
Time for joints	Moisture content: 8.5%	Moisture content: 9.6%

The dowels can be distributed on the surface in line. The distance between dowels depends of the dimensions of structure, for good results.

Structural changes that occur in wood at high rotation speed influences the quality of connection. It is known that when friction movement is stopped, the material begins to cool, it hardens to form a new structure with a solid contact layer (Pizzi 2004, Ganne 2008). If the rotation speed is lower, the connection quality colud be better – the wood fibers are interwine without degradation. Time for connection is around 1-2 seconds. That means efficiency and productivity of the method.

Figure 5: Section through structures obtained from: beech and beech wood *(Fagus Sylvatica L.)*

In the contact area, both joints in wood and in particleboards material (PAL) registered a uniform and compact connection. For some samples, in section, it can be seen that the joint is uniform but not perfect compact (figure 6). In sections, can be observed that the diameter of the dowel is not constant on all surface: the diameter decreases from the entrance in mterial to the end of insertion process. For good connection, an alternative insertions of the dowels could be a good choise.

Figure 6: Defects observed to the structures obtained from: beech (*Fagus Sylvatica L.*) and maple wood *(Acer platanoides)*

Where identified two types of defects which can appear: between layers and between support and wood dowels. Possible defects which can appear at joining layers and between dowels and wood support are presented in figure 7 and figure 6.

Figure 7: Possible defects at joining layers from particleboards (PAL)

Using joints obtained by rotational friction, without glue, come to complete the current requirements of product life. In this way, wood industry through products and processes involved plays a triple contribution in: economy – environment – society. The positive impact on them will lead to continue sustainable development.

5. CONCLUSION

The existing "binder" in the solid wood material can be opened and activated by pressure and rotational movement, during the friction procedure between the dowels and the support. Obtaining structures by this methods it could be an alternative for future products. Being a economical (no adhesive cost, simple equipment used) and ecological (the bonding of 2 samples without adhesive)method, rotational friction for joint wood could disrupt the traditional use of fasteners and adhesives in furniture manufacture. On the other hand, from this point of view, the pieces can be easily reused at the end of their life (no other material is used for connection), in according with the actual requirements. This method is one of the most (from the existing ones) efficient in terms of applicability because it requires no modern equipment for laboratory tests.

ACKNOWLEDGEMENT

This paper is supported by the Sectoral Operational Programme Human Resources Development (SOP HRD), financed from the European Social Fund and by the Romanian Government under the project number POSDRU/159/1.5/S/134378.

REFERENCES

[1] Christelle Ganne-Chedéville, Soudage linéaire du bois : étude et compréhension des modifications physico-chimiques et développement d'une technologie d'assemblage innovante. Thèse de doctorat Université Henri-PoinCaré, Nancy, France, p. 234, 2008.

[2] Dumitrașcu, R.E., Bădescu, L.A.M, Contribuții teoretice și experimentale la biofizica unor structuri inovative realizate prin sudarea lemnului - Grant Posdru-134378-2013, Brașov – Romania.

[3] Gerber, C., Gfeller B., Balz Joint connection with welded thermoplastic dowels & Wood Welding Technologies, CTI research, Swiss School of Engineering for the Wood industry, Biel-Bienne, Switzerland, 2005.

[4] Pizzi A., Leban J.-M., Kanazawa F., Properzi M., Pichelin F., Wood dowels bonding by high speed rotation welding, Journal of Adhesion Science and Technology. Vol. 18, (11), 1263-1278, 2004.

6th International Conference
"Computational Mechanics and Virtual Engineering"
COMEC 2015
15-16 October 2015, Braşov, Romania

VIRTUAL ENGINEERING ECO - INNOVATION LIBRARY FOR THE MANAGEMENT ISSUES OF THE WASTE OF ELECTRICAL AND ELECTRONIC EQUIPMENT

Irina Rădulescu[1], Florica Costin[2], Alexandru Valentin Rădulescu[1]
[1] POLITEHNICA University of Bucharest, Bucharest, ROMANIA, e-mail: irena_sandu@yahoo.com
[2] S.C. ICTCM S.A. Bucharest, Bucharest, ROMANIA, e-mail: coca.costin@gmail.com

Abstract: Research conducted in this paper are part of a national project, with the topic: Virtual eco-innovation hub to increase competitiveness in the field of waste electrical and electronic equipment recycling" (EcoInnEWaste). The project aims to respond to a challenge with profound economic, social and environmental implications: the increase of firms organizational competitiveness, that are operating in the field of electrical and electronic equipment recycling (e-waste, WEEE) in Romania and the increase of R & D private and public entities involvement in eco-innovation promoting for the green economy. One of the innovative tools that will be created for the traders in the e-waste recycling area, that will be a component of the hub, is the virtual engineering eco – innovation library. The execution of this documentation component of EcoInnEWaste platform allows the users access to relevant information resources for recycling domain.
Keywords: eco - innovation, library, WEEE, management

1. INTRODUCTION

Research conducted in this paper is part of a more ample national project, with the topic: *Virtual eco-innovation hub to increase competitiveness in the field of waste electrical and electronic equipment recycling*" (EcoInnEWaste). The project aims to respond to a challenge with profound economic, social and environmental implications: the increase of firms organizational competitiveness, that are operating in the field of electrical and electronic equipment recycling (e-waste, WEEE) in Romania and the increase of R & D private and public entities involvement in eco-innovation promoting for the green economy, [1].
One of the innovative tools that will be created for the traders in the e-waste recycling area, that will be a component of the hub, is the virtual engineering eco – innovation library.
The execution of this documentation component of EcoInnEWaste platform allows the users access to relevant information resources for recycling domain, [2], [3].
It enables users to consult a library of eco - innovation and the implementation model of eco – innovative technologies, which are useful for business representatives and companies that are involved or they wish to become involved in WEEE problem. The access of the hub information will be restricted, being paid or free, function the information category and the databases restrictions, but a lot of questions will find answers for free – by using the library or the Hub Forum. In terms of hub information management, they can be edited / added / deleted only by the platform operators.

2. THE VIRTUAL ENGINEERING ECO – INNOVATION LIBRARY FUNCTIONALITIES

To establish virtual engineering eco – innovation library functionalities it was necessary to analyze and clarify:
- The networking opportunities between library components;
- The demand – supply connection;
- The legislative information collection;
- The eco – innovation news;
- The participation in the eco-innovation events;
- Gathering of practical information regarding the collection, the waste transport, the landfills of waste, their treatment and / or recycling.

For the library achieving it is aimed to obtain quickly practical or theoretical information for the user.

Specifying library components, interfaces design and implementation of design specifications are designed to build an accessible virtual engineering eco - innovation library, easy to use, that will provide up-to-date and useful information.

The library conception shows its utility by its components (Figure 1), which offers information about:
- Environmental legislation;
- Actors involved in e-Waste management;
- Studies, research, basic knowledge of the WEEE guidelines;
- Eco - innovative technologies;
- News from eco-innovation domain, etc

.

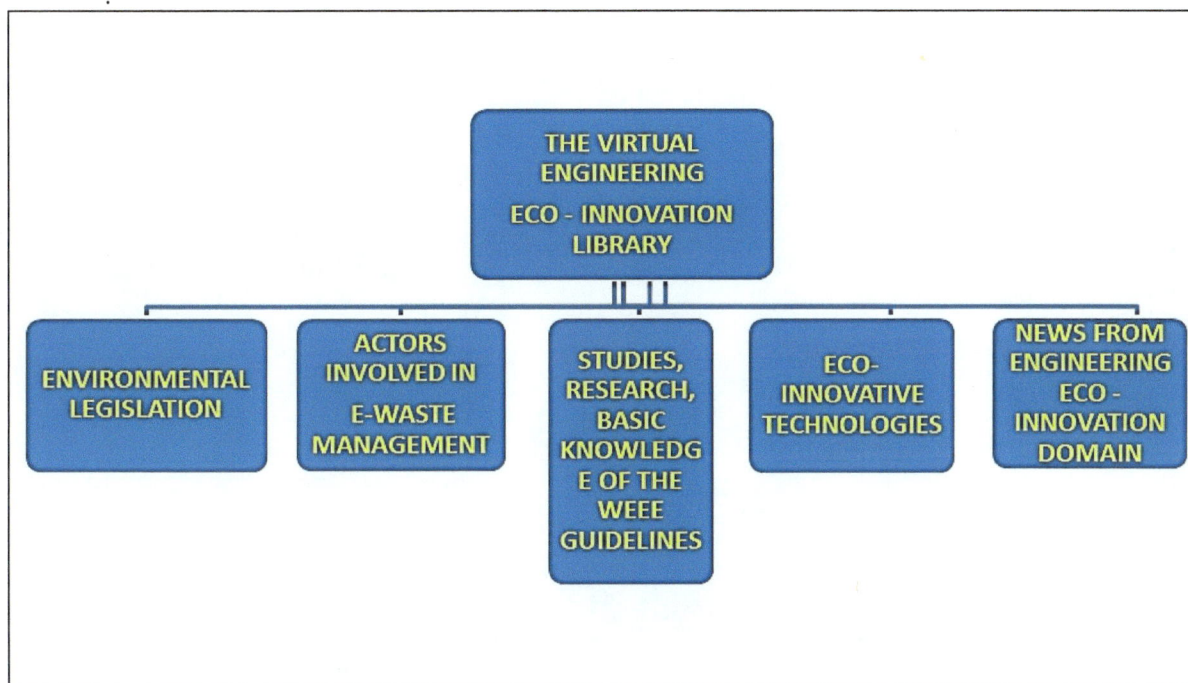

Figure 1: The structure of the virtual engineering eco – innovation library

The conception and the functionalities of the virtual engineering eco – innovation library have the quality of being suited to serve the purpose to inform the business environment in WEEE domain, [4].

There is also presented the range of operations that can be run on a computer to follow the necessary steps in order to obtain the required information.

The functional structure of the virtual engineering eco – innovation library and the information delivery required by the user must involve:
- The ability to manage the information (add, delete, change) by administrators and their selection criteria;
- The possibility of search terms customization by the user;
- The possibility of on WEEE legislation for the users;
- The possibility of documenting specialized users on businesses involved with the WEEE;
- The possibility of a specialized documentation on the WEEE studies, research, guides for the users;

- The possibility of a specialized documentation concerning the eco -innovative technologies for the users;
- The possibility of specialized documentation concerning the eco – innovation news for users;
- The possibility to save obtained information, after the documentation stage.

3. THE VIRTUAL ENGINEERING ECO – INNOVATION LIBRARY DEVELOPMENT

The development of the virtual engineering eco – innovation library involves the development of 5 modules, which functions were presented before. By accessing each one of these modules, the users can obtain information (paid or free, in function the hub restrictions), which can be saved.

So, the access of the library leads to the option of obtaining information from modules regarding: *Environmental legislation, Actors involved in e-Waste management, Studies, research, basic knowledge of the WEEE guidelines, Eco - innovative technologies, News from eco-innovation domain.* Once it was selected the domain name to be accessed, the user can proceed to the next step.

If the user selects the module providing information on *Environmental Legislation*, he must follow the next step, by selecting one of the three sub-modules regarding the legislation: *National Legislation, European Union Legislation* or *Other countries Legislation*, (Figure 2). By searching a key word, in Romanian or English, it can be obtain the required information, as a text file, a link or a pdf file, with the possibility to save it.

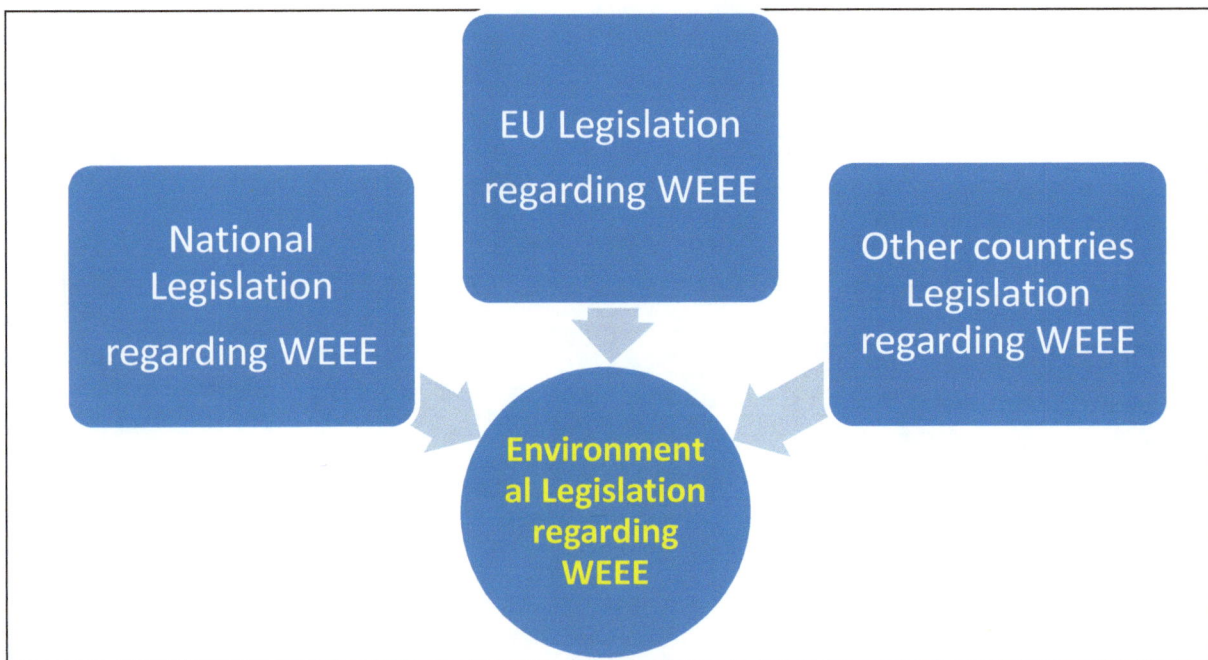

Figure 2: The module on the Environmental Legislation regarding WEEE

If the user selects the module providing information on *Actors involved in e-Waste management* (Figure 3), he can choose to obtain information about the business environment involved in WEEE, about NGOs and other organizations involved in WEEE or about Authorities. Each sub - module offers the possibility to search the information by a key word and to obtain a text file, a link or a pdf file, also with the possibility to save it.

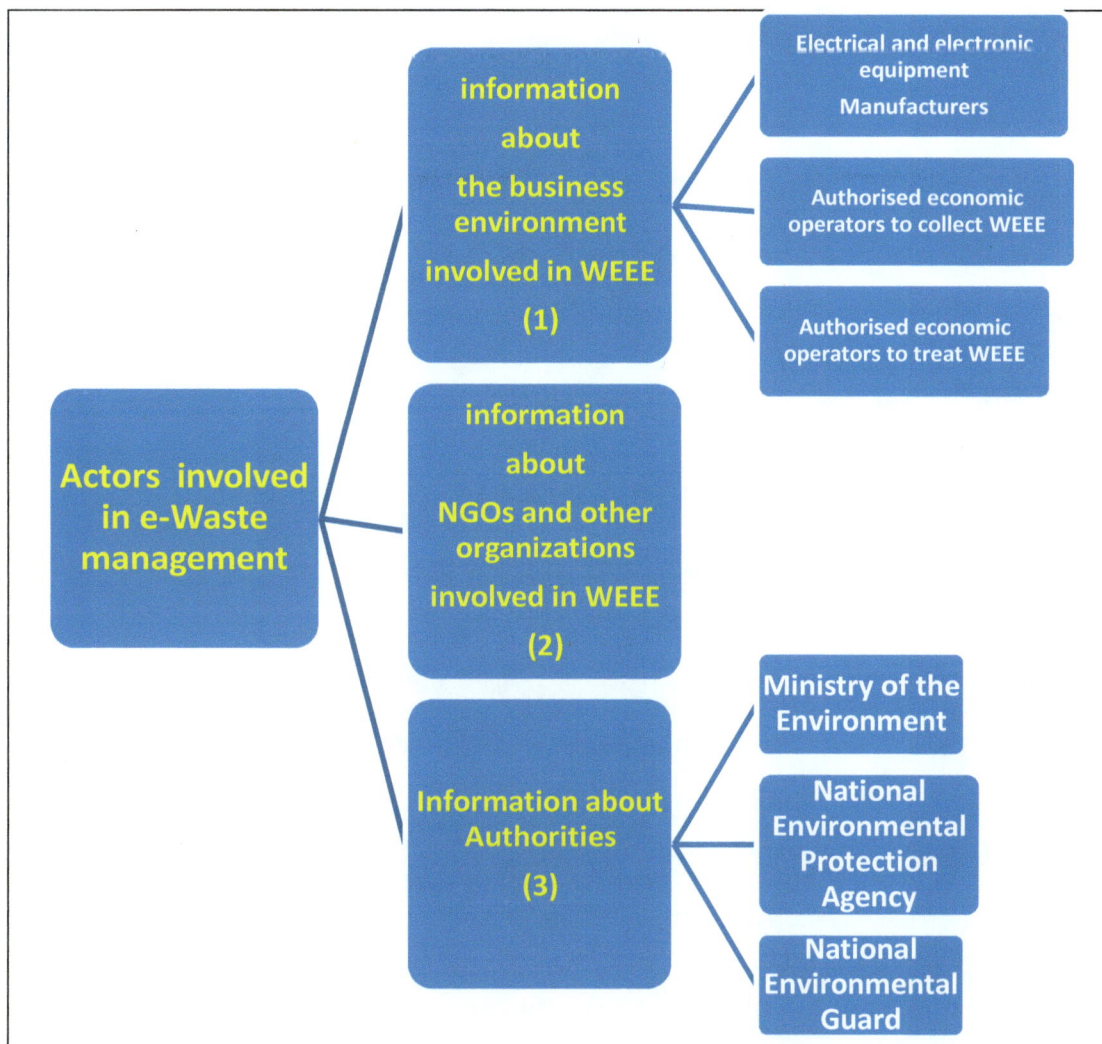

Figure 3: The module concerning the Actors involved in e-Waste management

The *Studies, research, basic knowledge of the WEEE guidelines* module offers information about Doctoral thesis and Dissertations concerning WEEE (1), Books – in this field of interest (2), Guides and specialized studies on WEEE (3) and about Papers published in Specialized Journals, in Conference volumes in the field of WEEE (4), (Figure 4).

Required information can be access in Romanian or English, by selecting the language and also, information can be saved.

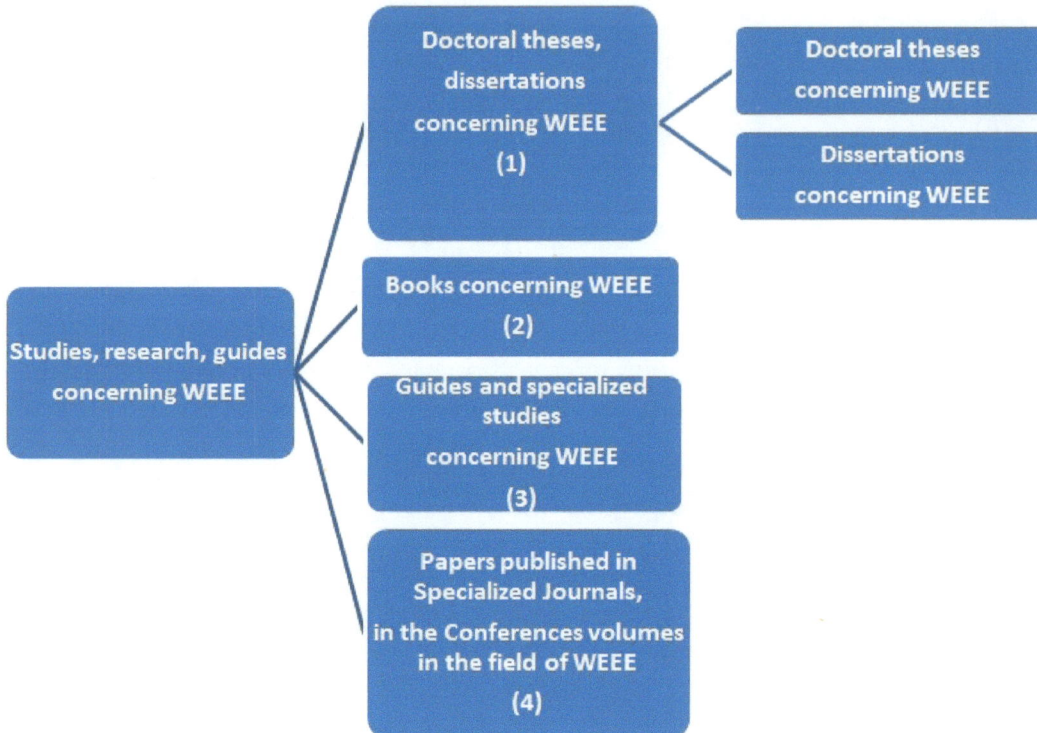

Figure 4: The module concerning the Studies, research, basic knowledge of the WEEE guidelines

If the user selects the module about *Eco – innovation technologies* to obtain information, it is possible to access the *Implementation model of eco – innovative technologies*, a way to eco – innovation for the company. In this case the user downloads a file or a computing program, which can be saved.

If the user wishes to consult the Library with *Examples of existing successful eco - innovative technologies*, he must select this sub – module. He will obtain a text file, a PDF file or a link to desired information, which can be saved, (Figure 5), [5], [6].

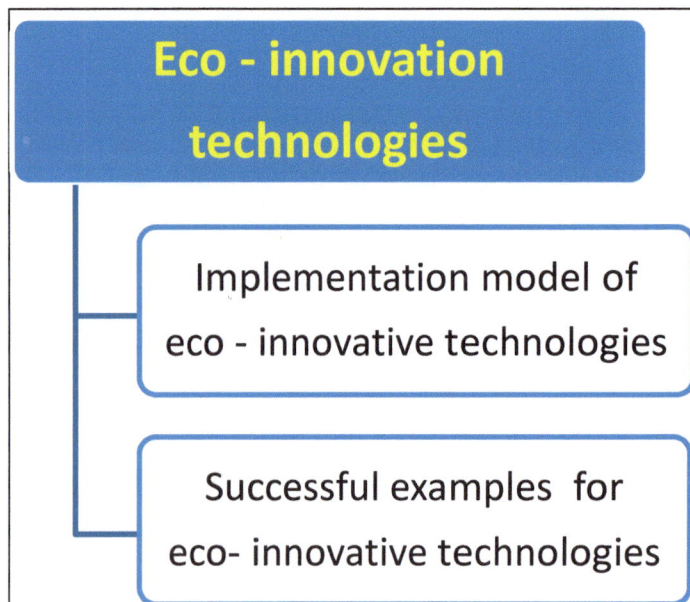

Figure 5: The module concerning the Eco- innovation technologies

If it is selected the module regarding *News from eco-innovation domain*, it is possible for the user to select: *Information about WEEE from websites, Information about trade fairs, exhibitions, research fairs concerning WEEE, Information about conferences, seminars, workshops on WEEE*, (Figure 6), [7], [8]. For each case, it may be select the search for a keyword or group of keywords, in Romanian or English, yielding a text page, a PDF file or a link to the desired information. Finally , there is the possibility of saving the required information.

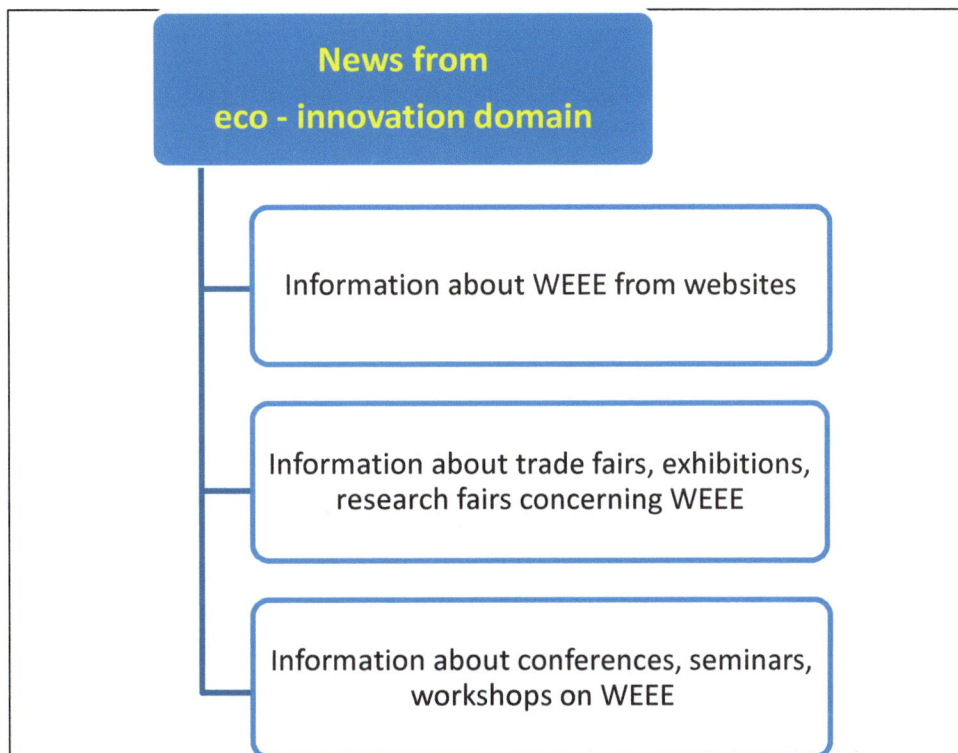

Figure 6: The module concerning the News from Eco- innovation domain

4. CONCLUSION

Developing new instrument as the virtual eco-innovation hub is a way to increase competitiveness in the field of waste electrical and electronic equipment recycling. The target of the society is to obtain, in the following years, firms benefit from a real economic development, based on technological research for reducing raw materials and energy consumption, expertise and consultancy on eco – efficient materials use, green products and services, informing systems for development of environment infrastructure. Increasing awareness level of economic actors will help as a direct contribution to sustainable economic development.

Virtual engineering eco - innovation library for the management issues of the waste of electrical and electronic equipment is a way to the stimulation to innovation and technology transfer, to consider economic and social value of environment resources. It is important to influence the business environment for a rational use and development of initiatives for decreasing economic activities' impact on environment through research-development solutions, ITC, technological transfer and automation.

ACKNOWLEDGMENT

This work was supported by MEN –UEFISCDI, Joint Applied Research Projects programme, project number **PN-II-PT-PCCA-2013-4-1400**, contract 320/2014.

REFERENCES

[1] European Commission, 2003. Directive 2002/96/EC of the European Parliament and of the Council of 27 January 2003 on waste electrical and electronic equipment (WEEE). Official Journal of the European Union, L 37/24, 13.02.2003, Brussels, BE. Online: http://eur-lex.europa.eu/resource.html?uri=cellar:ac89e64f-a4a5-4c13-8d96-1fd1d6bcaa49.0004.02/DOC_1&format=PDF.

[2] European Commission, 2015. Waste Electrical & Electronic Equipment (WEEE). Online: http://ec.europa.eu/environment/waste/weee/index_en.htm.

[3] Agenția Națională pentru Protecția Mediului (ANPM), 2015. Deșeuri de echipamente electrice și electronice. Online: http://www.anpm.ro/deseuri-de-echipamente-electrice-si-electronice

[4] Ministerul Mediului, Apelor și Pădurilor, 2015, Online: http://www.mmediu.ro/articol/strategii-planuri-studii/37

[5] Camera de Comerț și Industrie a Municipiului București, Studiu. Identificarea și evaluarea activității organizațiilor eco – inovatoare din România pentru realizarea inventarului Eco Invent ce va fi inclus în portalul EcoTehnoNet și realizarea unei rețele naționale a organizațiilor eco –inovatoare cu potențial de colaborare cu organizații similare din Norvegia, București, 2010.

[6] Camera de Comerț și Industrie a Municipiului București, Studiu. Identificarea principalelor categorii de furnizori de tehnologii de mediu (producători, distribuitori, importatori) prezenți pe piața din România pentru includerea acestora în baza de date accesibilă de pe portalul EcoTehnoNet, București, 2010.

[7] http://www.waste-management-world.com/articles

[8] http://ec.europa.eu/environment/waste/weee

6th International Conference
"Computational Mechanics and Virtual Engineering "
COMEC 2015
15-16 October 2015, Braşov, Romania

ON LIGHTING LIMITATIONS OF MARKER MOTION ANALYSIS USING COMMERCIAL SOFTWARE APPLICATIONS

Mircea Mihalcica

Transylvania University, Brasov, ROMANIA, mihalcica.mircea@unitbv.ro

Abstract: In this article we determine the limitations of non-professional marker motion analysis systems, designed using commercial software, considering the lighting conditions. We use a high speed video camera to record the circular motion of a metallic disc at different angular velocities and we are able to determine the maximum velocity range where the motion analysis software tools can efficiently analyze the trajectory of a marker installed on the disc. This velocity is important especially if the motion analysis system is designed to be portable and can be used by untrained or inexperienced personnel..

Keywords: motion analysis, inverse kinematics, software for motion analysis

1. INTRODUCTION

In our former research, we designed a motion capture and analysis system which uses video analysis of markers installed on the human body to determine trajectories of the main joints of the body or body part. Between the advantages of this system, we mention that the system is portable and easy to use even by non-expert personnel. For this to be true, we have to take into consideration any kind of perturbing factors which might influence the quality of the recording, which might lead to the inability to use the experimental data delivered by the system. One of the most important perturbing factors is the lighting - poor lighting surely leads to poor video data. Being able to set a maximum limit range at which the system can operate in practical conditions (which can include poor lighting) is an important aspect in the design of this kind of motion capture and analysis system, and our main aim in this paper.

2. MATERIALS AND METHODS

We tried to simulate a practical situation where poor lighting might occur. Inside, without external lighting, on a cloudy day, we recorded the circular motion of a disc, at different angular velocities. A white paper marker was installed on the metallic disc, and for the recordings we used the AOS-X PRI high speed video recording camera, using 120 frames per second. We started at the velocity of 80 resolutions per minute (rpm) and then gradually increased the velocity, with a step of 20, therefore video recording the disc at 80, 100, 120, 140 and 160 revolutions per minute. The video materials were then imported in Adobe After Effects and Kinovea (two software applications able to follow the motion of a marker) in order to determine the limits at where the motion of the marker can be efficiently followed by these applications.

We obtained 10 sets of video materials for each considered angular velocity. Each of these video materials were then imported in Adobe After Effects and Kinovea, and we used the built-in marker tracking tool in order to see where the marker "jumps" out of the trajectory.

Up until 120 resolutions per minute (including), the experiment went perfectly fine, both of the applications being able to efficiently follow the marker. Starting at 140 rpm's, the marker begun to leave the trajectory, outside human intervention being necessary in order to "drag" the marker back to its trajectory - this is possible and useful for some random analysis of motion parameters, but highly inefficient if we have to process massive amounts of data using a motion capture and analysis system of this kind.

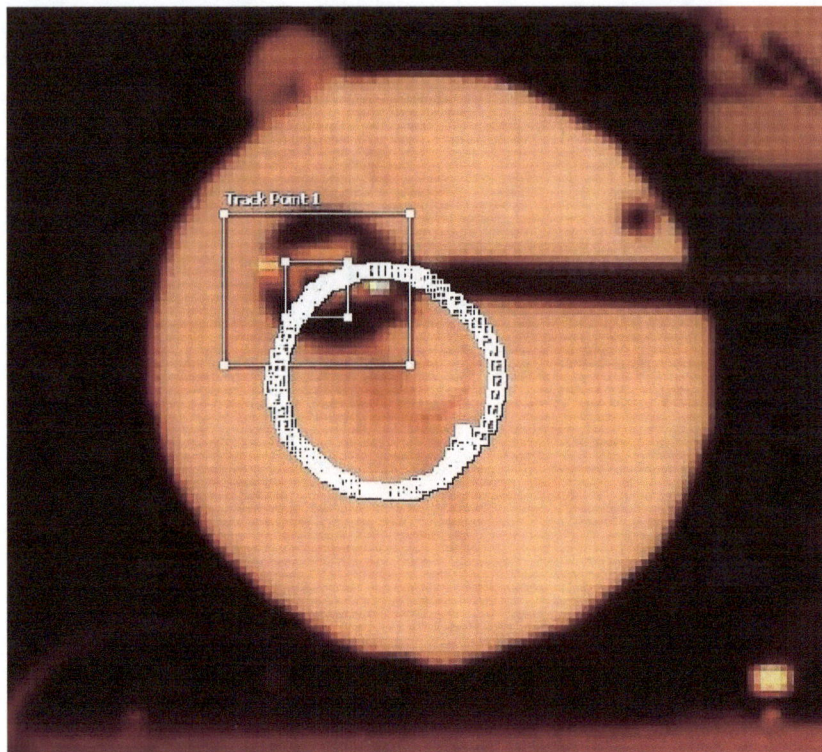

Figure 1: Marker recording at 120 resolutions per minute - the trajectory can be obtained automatically, by just using the tracking tool in the software

At 160 resolutions per minute, the automatic tracking of the marker was simply impossible, practically the trajectory of the marker was lost in less than a second, huge amounts of human intervention being necessary in order to somehow obtain some data (this data we probably cannot trust anyway.

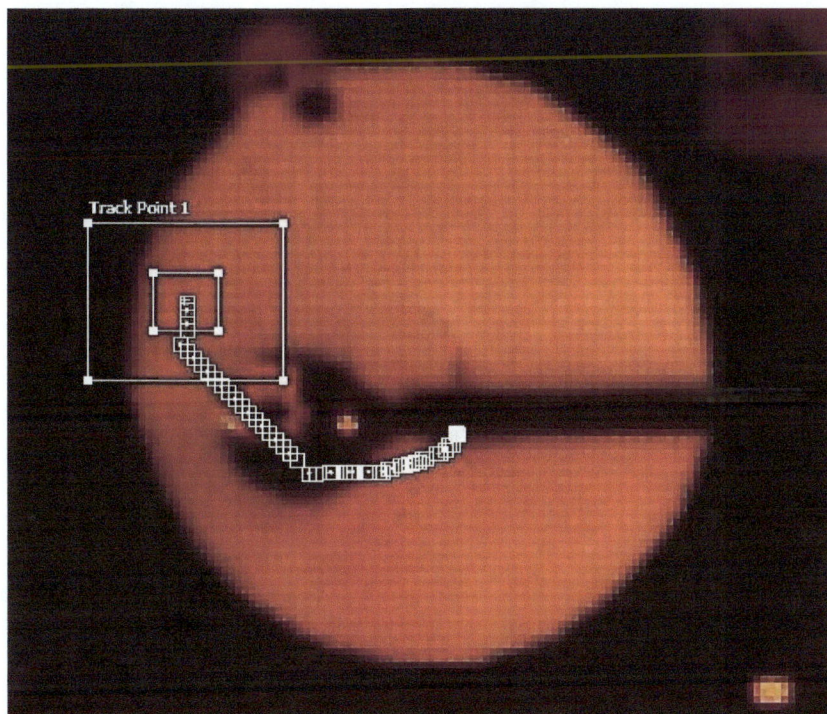

Figure2: Marker recording at 160 resolutions per minute - the trajectory cannot be obtained automatically, the marker tracking tool cannot follow the trajectory for more than a second

3. CONCLUSION

We established a point where this kind of motion capture and analysis system reaches its limits, that point being around 140 resolutions per minute, when following the circular motion of a body. When using these commercial applications above and a system of this kind, we recommend to design the experiments for a maximum circular velocity of 120 rpm.

ACKNOWLEDGEMENT

This paper is supported by the Sectoral Operational Programme Human Resources Development (SOP HRD), financed from the European Social Fund and by the Romanian Government under the project number POSDRU/159/1.5/S/134378.

REFERENCES

[1] Ambrósio, J.A.C., Kecskeméthy, A., Multibody dynamics of biomechanical models for human motion via optimization, Advances in Computational Multibody Dynamics, Springer, Dordrecht, The Netherlands, p.245-272 (2007)
[2] Higginson BK. Methods of running gait analysis, Current Sports Medicine Reports 8(3): p.136-141 (2009)
[3] Mihalcica M, Guiman V, Munteanu V. A Cheap and Portable Motion Analysis System. The 3rd International Conference „Research & Innovation in Engineering" COMAT 2014, 16-17 October 2014;2: p. 109-111
[4] Nixon M, Tan T, Chellappa R. Human Identification Based on Gait. Springer, Dordrecht, 2006
[5] Pandy MG. Computer modeling and simulation of human movement, Annual Review of Biomedical Engineering 3, 2001, p. 245–273
[6] Smith J. Adobe After Effects CS5 Digital Classroom, Wiley Publishing; 2010
[7] Safonova A, Hodgins JK, Pollard NS. Synthesizing Physically Realistic Human Motion in Low-Dimensional, Behaviour-Specific Spaces. ACM Transactions on Graphics, 2004;23(3):514-521

6th International Conference
"Computational Mechanics and Virtual Engineering "
COMEC 2015
15-16 October 2015, Braşov, Romania

NEW ASPECTS REGARDING CONTROL AND OPERATION OF THE AUTOCAR ABS SYSTEM

Eng. Marius-Gabriel Pătrășcan[1], Eng. Virgilius-Justinian Rădulescu[2],Eng. Viorel Ilie[3], Univ. Prof. Dr. Eng. Gheorghe Frățilă, Ph.D.[4]

[1]Credit Europe Insurance, Bucharest, Romania, marius.patrascan@gmail.com
[2]Metrorex, Bucharest, Romania, radulescu1961@yahoo.com
[3]S.A.-R. Astra S.A., Bucharest, Romania, viorel_ilie@yahoo.com
[4]Polytechnic University of Bucharest, Bucharest, Romania, ghe_fratila@yahoo.com

Abstract. The study embraced theoretical and experimental aspects related to the functioning of the vehicle ABS system. We evidentiated new strategies and algorithms of ABS control and a spectral analysis of the ABS operation, consisting of frequency and time-frequency analysis. Moreover, experimental data emphasizes the influence of factors on the functioning ABS by calling the sensitivity, dispersional and informational analysis.
Keywords: autocar, ABS system, electronic control, time-frequency analysis

1. INTRODUCTION

More severe requirements imposed on automobile power performance, economy and emissions were added and active safety requirements for passenger, cargo and road traffic. The most advanced active safety systems fitted to existing vehicles, such as ESC and ESP, ABS system based on the best known and most used electronic control system. For this reason, this study addresses the issue of ABS operation when driving the car on different runways and using different control algorithms.

2. MODELING OF OPERATION AND CONTROL ABS

Modeling ABS operation has two large components [6]: mathematical algorithm of vehicle dynamics during braking and control law (size control). The paper is illustrated only well-known model in the literature as "quarter-vehicle", also called *the model with a wheel*. For this model, the equations of motion of the car during braking, considering that it moves horizontally on a runway and neglecting the aerodynamic- and rolling resistance, are:

$$\begin{cases} m\dot{v} = -F_x \\ J\dot{\omega}_r = rF_x - M_f \end{cases} \tag{1}$$

where: m-mass of the vehicle, v-longitudinal speed of the car, F_x-adherence strength, J-moment of inertia of the wheel, ω_r-angular speed of the wheel, r-radius of the wheel, M_f-braking moment. From their appearance and to date, the ABS systems have benefited from control strategies and control algorithms depending on the technological level of the three major components: sensors, actuators and onboard computer. Thus, initially the control system was based on braking pressure modulation algorithm coupled/decoupled (*on/off system* so-called *bang-bang control*) following *the control the wheel deceleration*; over time the control was exquisite, relying today on artificial intelligence algorithms in order to control *the wheel slipping and the mixed control system (slip and deceleration)*. Among the most commonly used control algorithms are: control with feedback loop, bang-bang control (signum), PID control (proportional-integral-derivative), adaptive control, control based on artificial intelligence algorithms (based on fuzzy logic, based on neural network, neuro-fuzzy, based on genetic algorithms), robust control etc. For example, bang-bang control has the mathematical description [6]:

$$u(t) = \begin{cases} 1, & \text{dacă } \varepsilon(t) > 0 \\ 0, & \text{dacă } \varepsilon(t) = 0 \quad ; \quad \varepsilon(t) = \lambda_i - \lambda(t) \\ -1, & \text{dacă } \varepsilon(t) < 0 \end{cases} \qquad (2)$$

where, $u(t)$-size of the command (control law), $e(t)$ – error and the last size λ_i - represents the difference between slipping imposed to ABS system (usually, $\lambda_i=0.20$ which wheel-ground adherence is maximized) and λ real wheel slippage.

3. SIMULATION OF THE ABS SYSTEM

For simulation of ABS system is used its own software *Matlab* and *Simulink Toolbox*. This simulation allows pointing some functional features when car is moving on different runways and using some practical control algorithms.

In Figure 1 is shown the values of v -vehicle speed, v_r-wheel peripheral speed and u -command size (control law) *when traveling on wet pavement* using *bang-bang controller for ABS*, for ABS providing condition to ensure an imposed longitudinal sliding coefficient $\lambda_i=0.20$ properly corresponding to maximum adhesion coefficient φ_{max} =0.45 for wet asphalt.

Furthermore, Figure1a reveals that the initial speed of ABS system operation was v_0 =25 m/s (90 km/h). Likewise, it is observed the existence of an oscillatory feature of the wheel curve with v_r-wheel peripheral speed, leading to *the need of analyzing the ABS function frequency. In addition, as the frequency increases and amplitude decreases towards the end of braking is required a time-frequency analysis* of the functioning ABS, too. In the graph are shown the values of the braking time t_f=6.19s and the final velocity of the vehicle and v_f=1.3 m/s. Also, as noted, the wheel peripheral speed v_r is always lower than that of the vehicle v, obviously due to slippage issue also valid to the mean value v_{mr} compared to v_m.

In Figure1b reflects u -size control presented in expressions (2), which is the bang-bang control law; the chart is shown as continuous representation (commonly used, but unreal) and discrete representation (rarely used, but real). From the graph it appears that the size control has the values +1 and -1but not zero. We also notice that the frequency command increases in size over time.

Vehicle speed, wheel peripheral speed and size command, traveling on the wet asphalt, quarter model of vehicle, $\lambda_i=0.20$

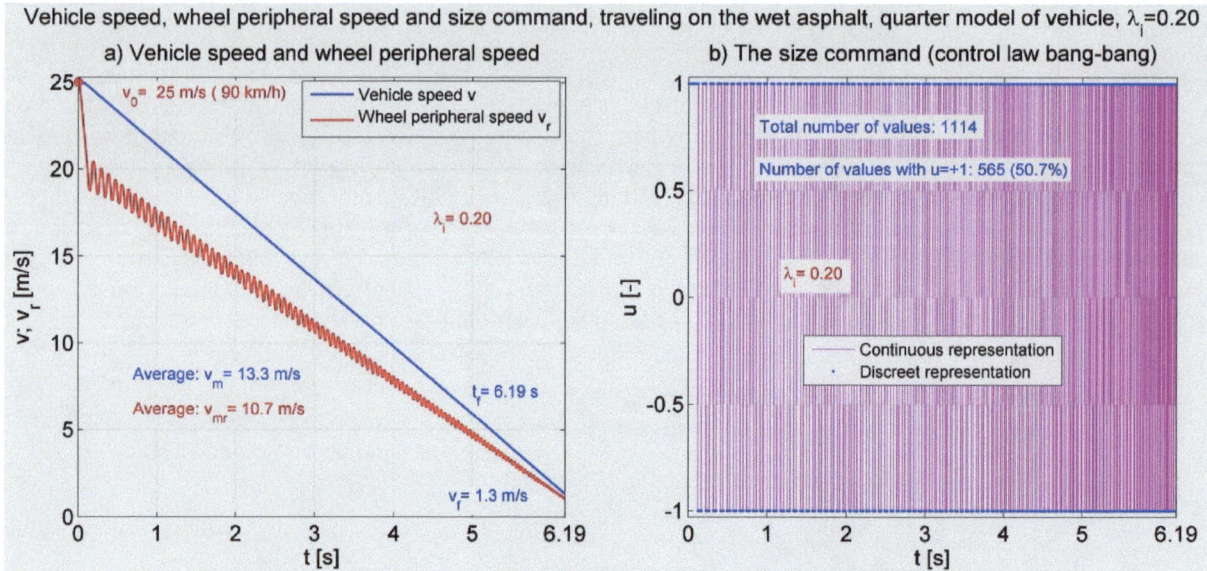

Figure 1 Representation of vehicle speed, wheel peripheral speed and size command

Figure 2 shows the adhesion coefficient values φ and braking space S_f for sliding imposed to the system $\lambda_i=0.20$. The graph from Figure 2a reveals that adherence coefficient φ is not always strictly related slip imposed λi, but most of the values are around it. Thus, the graph shows that 98% of the adherence coefficient is in the range of 0.40-0.45 lying around $\lambda i=0.20$ related slip. Also, the graph reveals that the mean value of the adhesion coefficient is 0.43.

Figure 2b develops that the value of braking space is $S_f=82.3$ m. The right-hand chart shown the value of the wheel equivalent space S_r, which is smaller than that braking space S_f, obviously because of slipping, and thereby situating the curve wheel peripheral speed v_r under the curve car speed v (Figure 1); such as, from Figure 2b can be deduced that wheel slipped 19.7% from total area covered, and 80.3% of it performed a normal running.

Adherence coefficient and braking space, traveling on the wet asphalt, quarter model of vehicle, $\lambda_i=0.20$

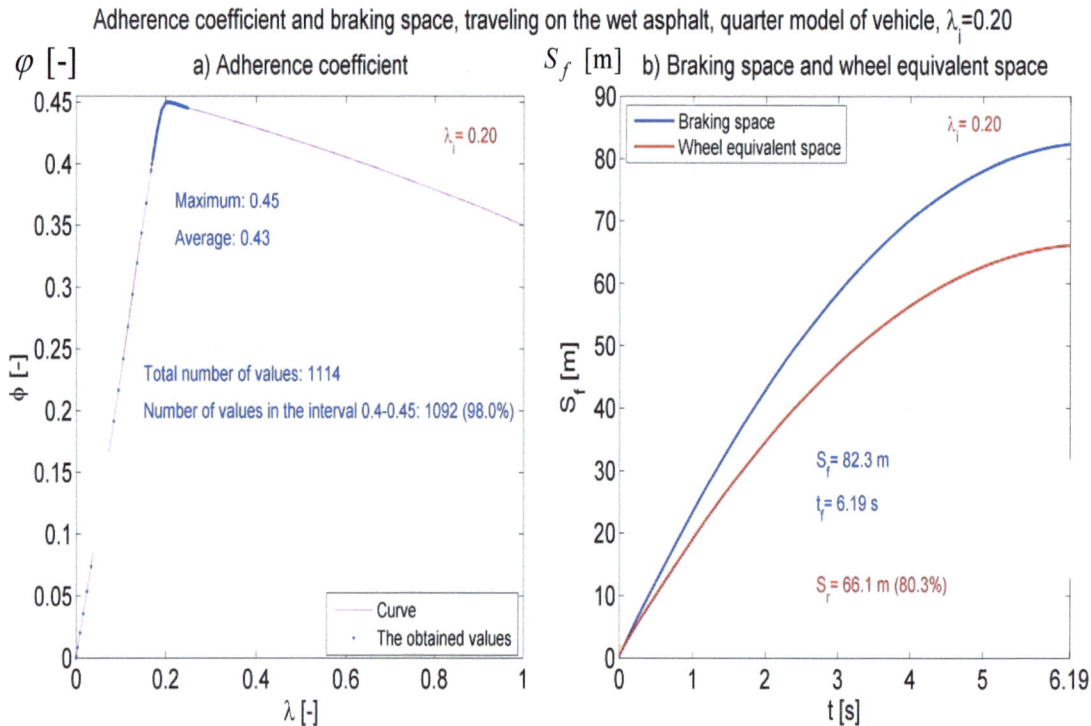

Figure 2 The adherence coefficient and braking space

Similarly can be exemplified other cases on different driveways with different control algorithms and different slip values imposed to ABS system, λ_i (e.g. λ_i=0.15 and λ_i=0.25). Studies done in this regard permit to deduce the following conclusions: with increasing slip imposed the oscillatory character of the wheel movement is change: decrease the frequency and increase the amplitude of the oscillations. Therefore, with increased of the slipping imposed rises the difference between the car mean speed v_m and wheel mean speed peripheral v_{mr}, so the two curves from Figure1a drive away. So, the minimum of the braking time and the braking space related to the slip λ_i=0.2 permits a maximum of the adhesion coefficient. It has been observed that as increasing imposed slip decreases othe wheel equivalent space, S_r. Since slipping imposed is higher, the variation rate of command size is smaller (switching from one discrete value to another beeing uncommon); braking time and - space grow on driveways with reduced adhesion, for example on ice in comparison with braking on the wet pavement. Accordingly, using PID controller or fuzzy PD controller, the braking-time increases and decreases braking-space comparing if used controller Signum (bang-bang), but the differences are reduced.

4. SPECTRUM ANALYSIS OF ABS OPERATION SYSTEM

From the above has shown that during braking process with ABS operation occur oscillatory processes with different frequencies and amplitudes of measurements such as the wheels peripheral speed (as well as the angular speed), the braking moment, the braking pressure, the friction coefficient, the coefficient of sliding etc. Therefore, it is necessary a spectral analysis of the ABS operation, as well as a frequency analysis and time-frequency analysis of it [2].
Frequency analysis of the functioning ABS system appeal to classical Fourier transform, which ensure emphasizing of the harmonic components with high energy intake among dynamic series. Instead, *time-frequency analysis* of the ABS operation enables time positioning of the various harmonic components, including those with high energy intake. This analysis uses the Class Cohen transform (Wigner-Ville, Gabor, Zak, Choi-Williams, Zao-Atlas-Mark, Born-Jordan, Page-Levin etc.), the wavelet transform, the Stockwell transform etc. [2].
Figure 3a illustrates the frequency analysis, and in Figure 3b the time-frequency analysis of the ABS operation, the target being the wheel peripheral speed when driving on wet asphalt with λi =0.2 given above in Figure 1a. As expected from these graphs, only time-frequency analysis can detect the disposition in time of the harmonic components, including those with high energy intake (Figure 3c).
Indeed, as seen in Figure 3b, the highest energy intake is bring the harmonics with frequency around 6.1 Hz acting at time moments of around 0.35 seconds.However, the graph from Figure 3b confirms that *wheel oscillations are reduced in amplitude and increase in frequency towards the end of braking.*

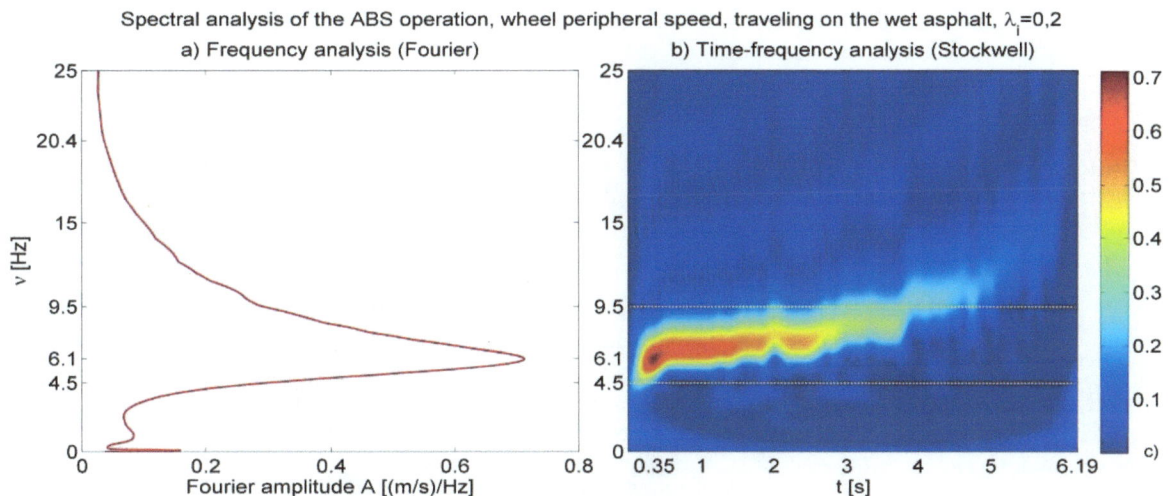

Figure 3 Spectral analysis of the ABS operation

Also, from Figure 3 is reaveling *the harmonics frequency spectrum as well having the highest energy contribution* of 4.5-9.5 Hz for the analyzed case of the wheel peripheral speed.

5. INFLUENCE OF CERTAIN FACTORS ON THE ABS OPERATION

ABS operation is influenced by many factors that affect its performance, the main including braking time and braking space of vehicles [3; 6]. To determine the influence of those functional factors on the performance of ABS operation, the main methods used in practice are sensitivity analysis, informational analysis and dispersion analysis. Those methods set quantitative influence on the ABS functioning [1; 2; 4; 5]. *The sensitivity* expresses a property of a resultative size in order to modify its values under the influence of factorial size and influence factor, respectively [4]. Among the indicators that quantify sensitivity is Sobol's index noted by S and represents the ratio between afferent dispersion of the targeted factor and the total dispersion of the outcome size. As an example, Figure 4 shows the Order I Sobol's index for braking space S_f and braking time t_f (two resultative sizes) when driving on wet asphalt with Signum controller, the six influence factors being (x-axis) 1. φ -adhesion coefficient; 2. λ -slipping coefficient; 3. u–command size ; 4. v_r–wheel peripheral speed; 5. v –vehicle speed; 6. M_f-braking moment.

Analysis of sensitivity (Sobol index), influences on braking space and braking time, wet asphalt, signum controller

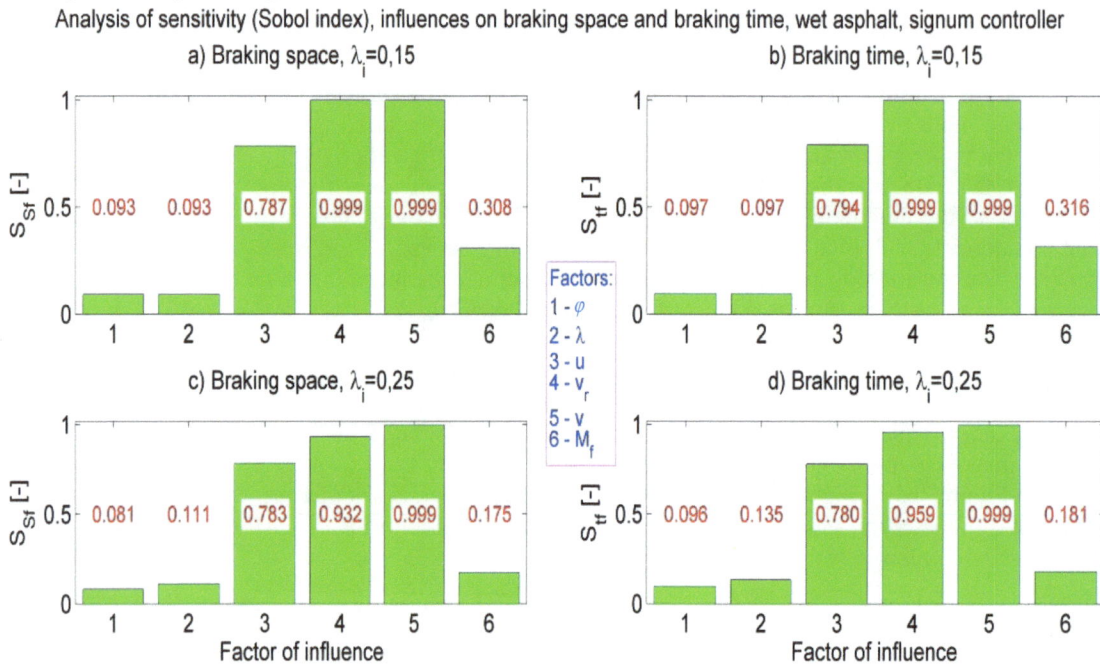

Figure 4 Analysis of Sensitivity

As expected from these graphs, braking time and braking space are the most sensitive to changes in peripheral speed of the wheel, v_r and the vehicle's speed, v (expected), followed by size u-command size and at a greater distance for braking moment, M_f. In addition, the lowest influences show φ -adhesion coefficient and λ-slipping coefficient. Also, these graphs reveal that the mentioned influences are apparent both at low values of slip coefficient imposed $\lambda_i=0.15$, at the left peak of the curve $\varphi = f(\lambda)$ with $\lambda=0.2$, and at higher values of it $\lambda i=0.25$, at the right peak of the curve $\varphi = f(\lambda)$. It should be noted that when braking moment but is more the slipping coefficient imposed $\lambda i=0.15$.

Informational analysis is based on two main concepts of informational theory: entropy and information, including mutual information [5]. Mutual information is a concept that provides a quantitative measure to minimize uncertainty, so to increase the prediction. The more the mutual information gain higher values, the smaller uncertainties will become and therefore higher predictions. For this reason, mutual information is a basic concept for the study of the dynamics of the systems and *is a measure of the interdependence of the variables*.

As an example, in the chart from Figure 5 is shown the results of the informational analysis when consider the braking space a resultative size (placed at the top) and taking into account the 6 factorial sizes (determinants) such as: the adhesion coefficient, the slipping coefficient, the size command, the wheel peripheral speed, the vehicle speed and the braking moment. The nodes of the graph are given entropy values, H. It appears that most entropy is possess by braking space of $H=4.4$ bits, and the smallest size command of $H=2.6$ bits. On the graphic curves are noted the mutual information values I between two sizes.

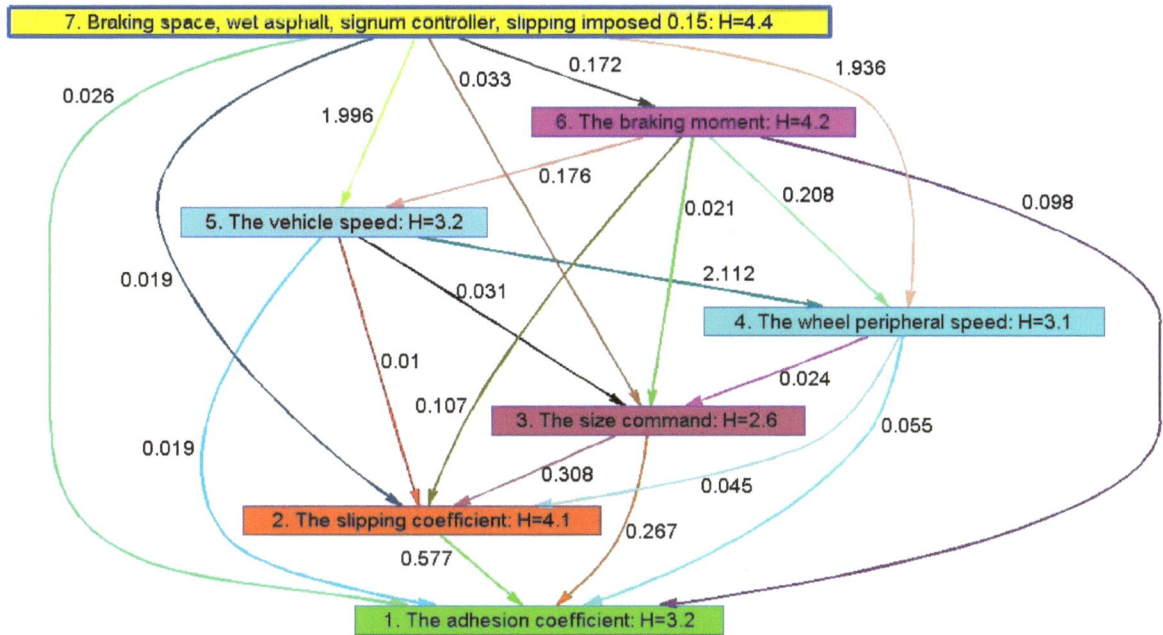

Figure 5 Informational analysis

From Figure 5 it is noted that the first two relevant variables (with the greatest influence on the braking-space) are the car speed (mutual information with braking space of 1.996 bits) and the peripheral speed of the wheel (I=1,936 bits), same as the sensitivity analysis (Figure 4a, with λi =0.15).

6. EXPERIMENTAL RESEARCH

To perform the experimental study on the ABS operation were performed some tests with recording the speed of movement and the wheel speed revving to both types of car, Skoda Octavia and BMW 523i. Acquisition and data storage was possible because of the existence of embedded sensors, a board computer, a specialized tester and an embedded computer with specialized software. The data were processed subsequently purchased a computer and appropriate software.Figure 6 shows the experimental results of the braking time t_f, the braking space S_f, initial speed V_0 and the final speed V_f when the vehicle Skoda Octavia was 15 times tested by *driving on wet asphalt* and in Figure 7 the analytical afferent expressions offering functional dependencies of the first three sizes.

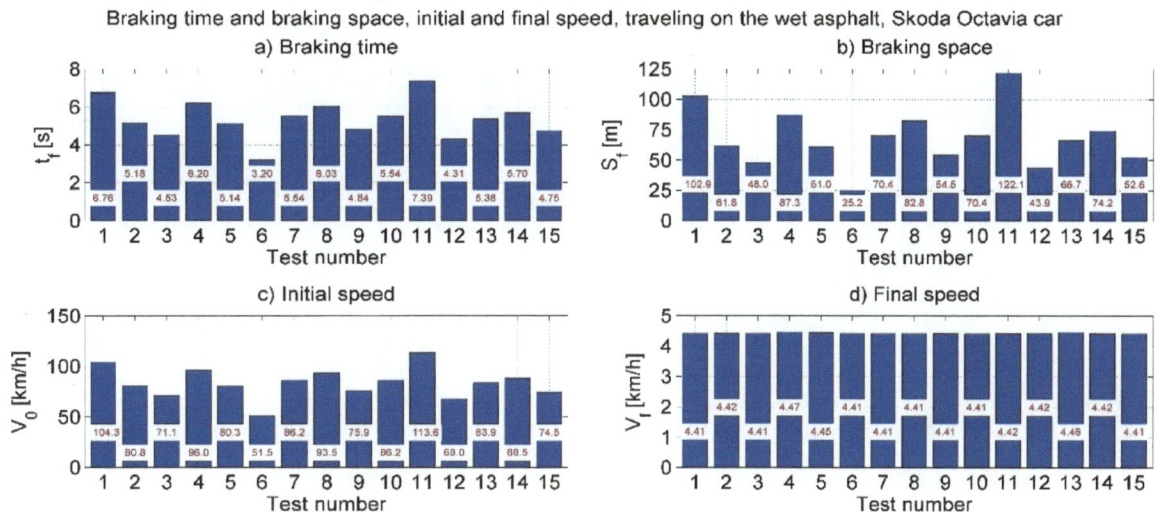

Figure 6. Sizes defining the braking process

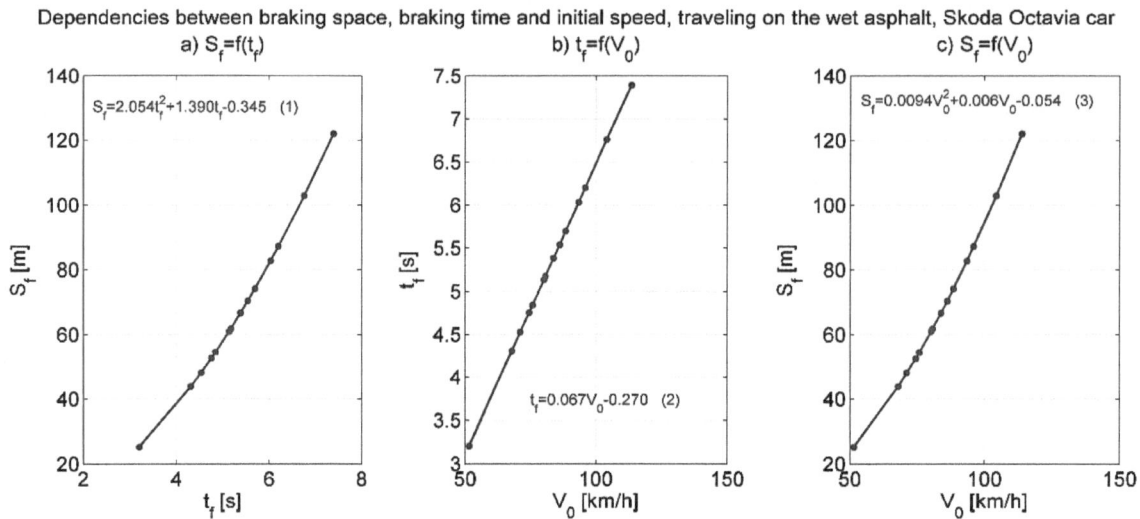

Dependencies between braking space, braking time and initial speed, traveling on the wet asphalt, Skoda Octavia car

a) $S_f=f(t_f)$ b) $t_f=f(V_0)$ c) $S_f=f(V_0)$

$S_f=2.054t_f^2+1.390t_f-0.345$ (1)

$t_f=0.067V_0-0.270$ (2)

$S_f=0.0094V_0^2+0.006V_0-0.054$ (3)

Figure 7. Functional dependencies between brake sizes

As shown in Figure 6, the braking time varied in the range from t_f=3.2-7.39 s and braking space S_f= 25.2-122.1 m, the initial speed V_0=51,5-113,6 km/h and final speed V_f=4.41-4.47 km/h.

7. CONCLUSIONS

To study the ABS operation based on experimental data allows emphasizing some specific features that resort to methods and specific algorithms for systems dynamics. It follows that the influence of various factors on the ABS operation must be consider and the interactions between factors as well as that of the factorial sizes vary simultaneously during operation, two main differences to the study of classical literature.

REFERENCES

[1] Carey G., Multivariate Analysis of Variance, MANOVA, Colorado State University, 1998.

[2] Copae I., Lespezeanu I., Cazacu C., Dinamica autovehiculelor, Editura ERICOM, Bucureşti, 2006.

[3] Frăţilă Gh., *Calculul şi construcţia automobilelor*. Ed. Didactică şi Pedagogică, Bucureşti, 1977.

[4] Glenn G., Isaacs K., Estimating Sobol sensitivity indices using correlations, Environmental Modelling & Software, 37 (2012), pp. 157-166, 2012.

[5] Gray R., Entropy and information theory, Stanford University, New York, 2007.

[6] Savaresi S., Tanelli M., Active Braking Control Systems Design for Vehicles, Springer-Verlag London, 2010

6th International Conference
"Computational Mechanics and Virtual Engineering "
COMEC 2015
15-16 October 2015, Braşov, Romania

STUDY ON THE SEPARATING PROCESSES OF IMPURITIES FROM WASTE WATER BY USING TANGENTIAL FILTERS

Daniela Zărnoianu[1], Simion Popescu[2], Carmen Brăcăcescu[3]

[1] S.C. Oltchim S.A, Rîmnicu –Vîlcea, Romania, e-mail: daniela.zarnoianu@yahoo.com
[2] Transilvania University of Brasov, Romania, e-mail: simipop38@yahoo.com
[3] National Institute of Research-Development for Agricultural Machinery (INMA), Bucharest, Romania, e-mail: carmenbraca@yahoo.com

Abstract: *The paper presents the methodology and the experimental research installation of the efficiency of impurities separation from the wastewaters by the use of the tangential filters connected in series. The experimental installation (pilot installation) used consists of two tangential filters mounted in series, with filtering elements with stainless steel sieves, having a fineness of 475 µm, and 80 µm. To establish the filtration efficiency to a given feed flow rate were determined by measurements the liquid pressures and concentrations of the suspension in clear (in mg/l) at the entrances and at the exits from the filters, after certain durations of operation (throughout of 60-minute). Finally are presented the conclusions on the efficiency of separation of mechanical impurities of the analyzed filtration system.*
Keywords: *wastewater, concentrations of suspensions, tangential filtration, experimental installation, filtration efficiency,*

1. INTRODUCTION

Filtration is the separation process of removing solid particles, microorganisms or droplets from a liquid by depositing them on a filter medium (granular layers, sieves, membranes), also called filters, which is essentially permeable to only the fluid, haze of the mixture being separated. The particles are deposited either at the outer surface of the filter medium and/or within its depth. The permeation of the fluid phase through the filter medium is connected to a pressure gradient [1; 2; 3].
The liquid more or less thoroughly separated from the solids is called the filtrate, effluent, permeate or, in case of water treatment, clean water. As in other separation processes, the separation of phases is never complete: liquid adheres to the separated solids (cake with residual moisture) and the filtrate often contains some solids (solids content in the filtrate or turbidity). The purpose of filtration may be clarification of the liquid or solids recovery or both. In clarification the liquid is typically a valuable product and the solids are of minor quantity and are often discarded without further treatment. During a dynamic filtration the collected solids on the filter media are continuously removed, mostly with a tangential flow to the filter medium (cross-flow filtration). Cross-flow filtration is a standard operation with membranes as a filter medium [4; 5; 6]. The flow parallel to the filter medium reduces the formation of a filter cake or keeps it at a low level. So it is possible to get a quasi-stationary filtrate flow for a long time. The four idealized filtration models are: cake filtration, blocking filtration, deep-bed filtration and cros-flow filtration (Fig.1).
Cross-flow filtration, called *tangential filtration*, is a form of a dynamic filtration. In cross-flow filtration the build-up of a filter cake on the surface of the filter media is hindered by a strong flow tangentially (parallel) to the filter surface. Clear liquid passes through the filter medium (mostly a membrane) and the concentrate (respectively the retentate) with higher concentrations of the rejected components is discharged from the filter. The cross-flow stream over the filter media has

often a linear velocity in the range of 1 to 6 m/s. This cross-flow is achieved in most cases by pumping the suspension through a membrane module, which e.g., contains the membrane in form of a bundle of membrane tubes. Alternatively the cross-flow is achieved by rotating inserts. The limitation of the deposition of particles and macromolecules on the membrane surface enables to filter very fine particles, which otherwise would form a cake with prohibitively high resistance. Even submicron, nonparticulate matter can be retained by membranes with a corresponding separation characteristic according to this principle. The shearing action on at the membrane surfaces limits the deposition of retained matter due to lift forces and diffusion processes back from the membrane. In normal operation the filtrate (respectively permeate) is collected and the concentrate is re-circulated until the desired concentration of retained components is achieved, or pumping cannot be longer performed due to raised viscosity of the suspension. Typical units comprise one or more membrane modules and a recirculation pump. Depending on the size of the species retained, a distinction is made between microfiltration, ultrafiltration, nanofiltration, and reverse osmosis.

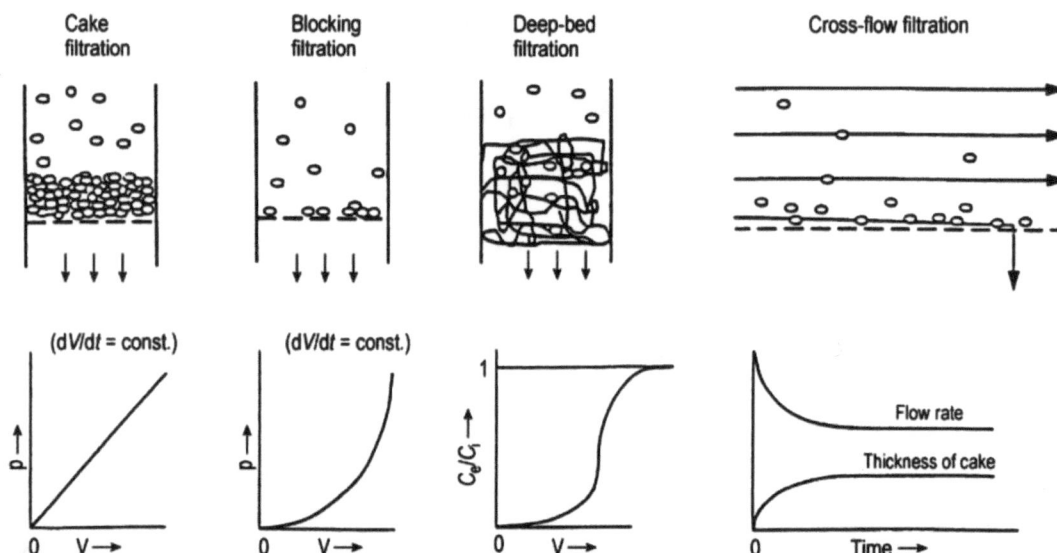

Figure 1: The four idealized filtration models: cake filtration, blocking filtration, deep-bed filtration and cros-flow filtration

Microfiltration retains particles and microorganisms down to 0.1 μm in size. In steady-state cross-flow filtration these particles are conveyed onto the membrane by convection due to the filtrate flow and transported away from it by hydrodynamic lift forces due to the parallel shear flow (forces FY and FL in Fig. 2). For particles below a certain size the lift force becomes smaller than the convection, $FL < FY$, and they are deposited on the membrane. After deposition they are retained by van der Waals' adhesion forces FA and not easily swept away even if there is no filtrate flow (irreversible cake formation). In case of a steady-state microfiltration there is an equilibrium between the lift force and the drag force due the convection by the filtrate flow and a critical particle size. This steady-state flow is often mentioned as critical flux. In most practical applications also colloidal particles are present and they are transported according to the mechanisms of ultrafiltration or they are able to permeate through the microporous membrane of microfiltration. The cut-off size of a microfiltration membrane is usually defined as the size of particles of a test suspension (generally a suspension of microorganisms) which are nearly absolute retained.

Ultrafiltration retains colloidal particles or macromolecules. The cut-off sizes are ca. < 0.1 μm. Such particles are subject to Brownian diffusion (*Br* in Fig. 2) and only weakly influenced by the hydrodynamic lift force. The smaller these particles are, the more they are transported away from the filter medium by diffusion. This is why in ultrafiltration the smaller particles are preferably swept away and the bigger ones are collected on the membrane (contrary to microfiltration). The cut-off size of an ultrafiltration membrane is usually defined as the molar mass (in g/mol) of a test solution (generally protein or dextrin molecules of defined molar mass in the range of 103–105 g/mol). In practical application the cut-off size depends not only on the pore size of the membrane, but to a large degree on a gel layer on the membrane consisting of retained colloids.

Nanofiltration combines features of ultrafiltration and reverse osmosis with a high selectivity. Its name is derived from its approximate cut-off size of some nanometers or more exactly molar masses of 200–600 g/mol. This is achieved with special nanofiltration membranes which still have pores of a defined size, but their retention depends also on the electrostatic charge of the molecules to be separated (bivalent anions are typically retained).

Reverse osmosis retains molecules or ions using selective membranes without pores. Certain molecules permeate through the membrane because they are soluble in the membrane material. Other molecules are not (or less) soluble and are retained (or concentrated) on the upstream side of the membrane. A particular feature of reverse osmosis is the high pressure required to overcome the osmotic pressure of the retained molecules. An important application is desalination of seawater. In the food industry it is applied to concentrate juices and other sugar solutions at low temperatures.

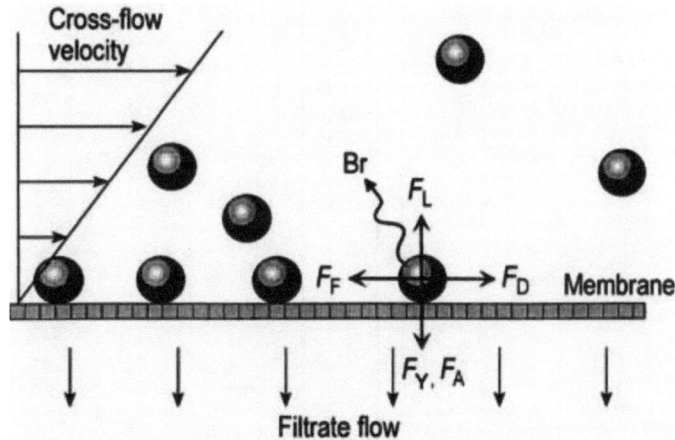

Figure 2: Forces and transport effects onto a particle in cross-flow filtration:
Br - Brownian motion; F_A -adhesion to the membrane; F_D - drag from the cross-flow; F_F- friction force; F_L-hydrodynamic lift force; F_Y - drag from the filtrate flow

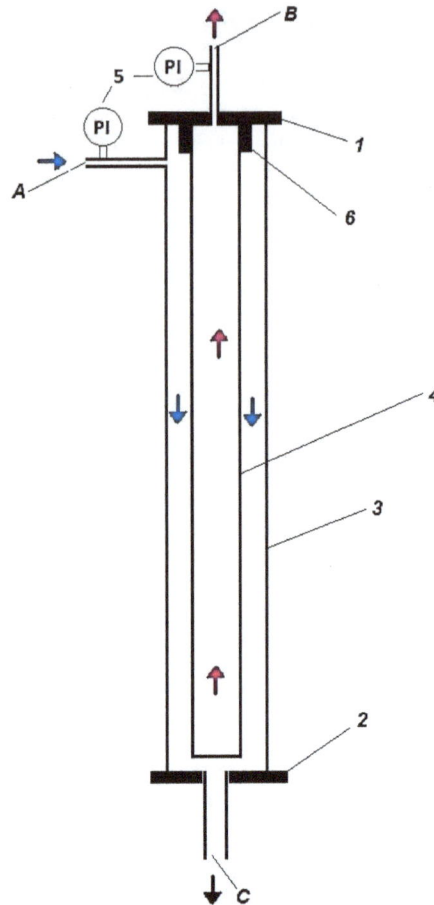

Figure 3: The constructive and functional diagram of a tangential flow filter: A - liquid inlet; B- filtered liquid outlet; C- evacuation fluid that has not passed through the membrane:
1- upper lid; 2- lower lid; 3- filter body; 4- filtering element; 5- differential pressure gauge connection fittings; 6- fixing plate of the filtering element.

Tangential filtration is a more efficient method compared with the filtration of surface and of depth, because most of the liquid to be filtered flows parallel to the surface of the filter, a much smaller part flowing through the filter membrane. Due to the effect of "sweeping" and of cleaning is prevented premature the filters clogging and the occurrence of differences in concentration of the filtered fluid. Tangential flow filtration allows the use of much higher flows than to the perpendicular flow (normal), of surface and of depth, being used increasingly in the industrial processes [3]. Regeneration of the filter permeability structure can be done by washing the filter element by directing the liquid in counter-flow [4]

Figure 3 shows the functional diagram of a tangential flow filter, at which the cylindrical casing of the filtering element 4 is closed with two caps, upper 1 and lower cap 2. After a certain operating time of the filter is possible the plugging (clogging) the pores or orifices of the filtering element, with negative effects on the efficiency of the filtration process. To eliminate this inconvenience the filter is provided with the possibility of performing a washing procedure by the inverse filtration. For determining when it is necessary the inverse filtering, the inlet and outlet connections from filter are coupled through a differential pressure gauge, which measures the pressure difference between the entrance of the wastewater and the exit of the filtrated liquid. In practice, cleaning the filters by washing performed by inverse filtering is recommended when the differential manometer indicates a pressure drop of more than 2 bar between the filter inlet and outlet.

2. MATERIALS AND METHODS

The basic objective of the undertaken researches consists in experimental determination of the optimum moment at which is imposed the washing of tangential filters through the reverse filtering operation. For this is used the method of the measurement of pressure differences between Ins and Outs of filters, whose values depend on the degree of fouling (clogging) of filter elements.

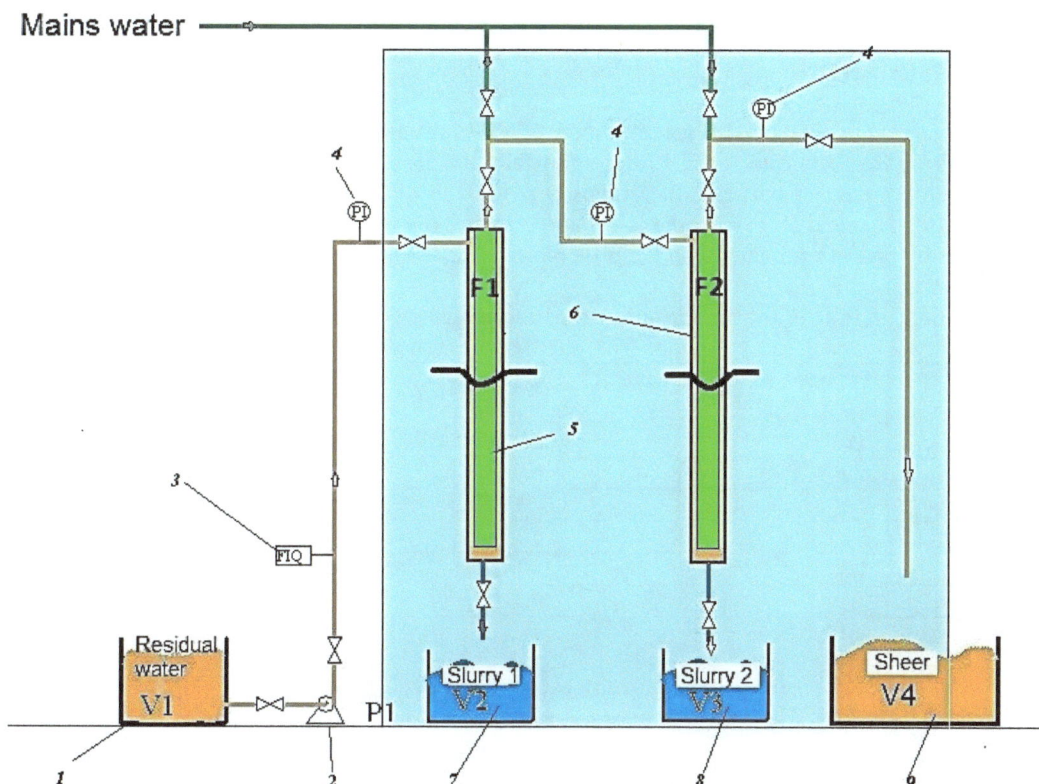

Figure 4: The scheme of the experimental plant (pilot plant) with two filtration steps coupled in series:
1-tank (V-1) with waste water that will be submitted to filtration; 2- hydraulic pump (P1) for supplying the filters ; 3-. electromagnetic flowmeter (FIQ) ; 4- manometers (PI); 5. coarse filter element (F1) with 475μm steel sieves ; 6- fine filter element (F2) with steel sieves of 80 μm ; 7- tank (V2) to collect slurry / sterile resulting from the coarse filter F1 ; 8- tank (V3) to collect slurry / sterile resulting from the fine filter F2; 9- tank (V4) to collect the filtered product (clear)

For the experimental study of the functional behaviour and of the filtration efficiency when using tangential filtration systems of the industrial waste water was designed and built an experimental installation (pilot plant) with two filtration steps (Fig. 4), which allows retaining the solid suspensions from 4000 mg / l up to a value of approx. 400 mg / l. The experimental installation consists of two tubular filtering modules serially connected: the coarse filtration module F1 equipped with filter with stainless steel sieves having a fineness of 475μm and the microfiltration module (fine filtration) F2, equipped with stainless steel filter with sieves having a fineness of 80 μm. To measure the flow discharged by the pump 1 is used the electromagnetic flowmeter FIQ (type YOKOGAWA) and for measuring the pressures are used the pressure gauges PI (type Bourdon tube type with separation membrane MS) All the elements of the installation are mounted on a supporting plate.

The waste water subjected to analysis is stored in the tank V1 (with a capacity of approx. 200 l) and is sent in the filtration system by means of the centrifugal pump P1. In the first stage the water reaches the coarse filter F1 (with the sieve of 475 μm), at the upper part being discharged the filtered liquid (who has passed through the filtering element F1) and at the lower part occurs the discharge in the tank V2 (approx. 60 l) of the liquid with impurities (the liquid that has not passed through the filter membrane F1. In the second stage, the filtered liquid into the filter F1 reaches into the fine filter F2 (with the sieve of 80 μm), out of which at the upper side is discharged the filtered fluid which is collected in the tank V4 (approx. 200 l), and at the lower part occurs the discharge in the tank V3 (approx. 50 l) of the liquid with impurities which has not passed through the filter F2 . The filtrate quality is targeted in the clear liquid accumulated in the tank V4. To clean (desilting) the filters is used the water from the current network (the circuit marked with green), with exhaust in the system slurry / sterile of the two filters. To determine the effectiveness of the filtration system tested in the tank V1 was introduced a wastewater with an initial concentration of suspensions of approx. 4.000 mg/l. After certain durations of operation of the filtering installation, in the range of from 0 to 60 min, were measured the pressures at the gauges mounted in the filter F1 entry and at the exits from the filters F1 and F2 and were determined by measuring the concentrations of the mechanical suspensions (in mg/l). The concentration of the liquid suspensions was determined by gravimetric analysis of the taken samples

3. RESULTS AND DISCUSSION

The time variations of the pressures and concentrations of the mechanical suspensions from liquid experimentally determined at the outflows from the filters of the filtering installation are presented in Table 1.
Based on data from Table 1 were constructed the graphs with the variations in time of the pressures (Fig. 5) and of the suspensions concentrations in clear (Fig. 6) at the inlets and outlets from the filters of the filtration installation, allowing an analysis of the filtering process conducted by the pilot plant used.

Table 1: The variations in time of the suspensions pressures and concentrations l at the entry into F1 filter and the goings from the filters F1 and F2

Time	Pressure before F1	Pressure at the exit of F1	Pressure at the exit of F2	Concentration suspension before the F1	Concentration suspension at the exit of F1	Concentration suspension at the exit of F2
min	bar	bar	bar	mg/l	mg/l	mg/l
0	3	2.1	0.9	4058	1059	587
20	3.2	2.2	1.1	4058	826	436
40	3.4	2.6	1.2	4058	514	325
60	4	3.1	1.3	4058	468	216

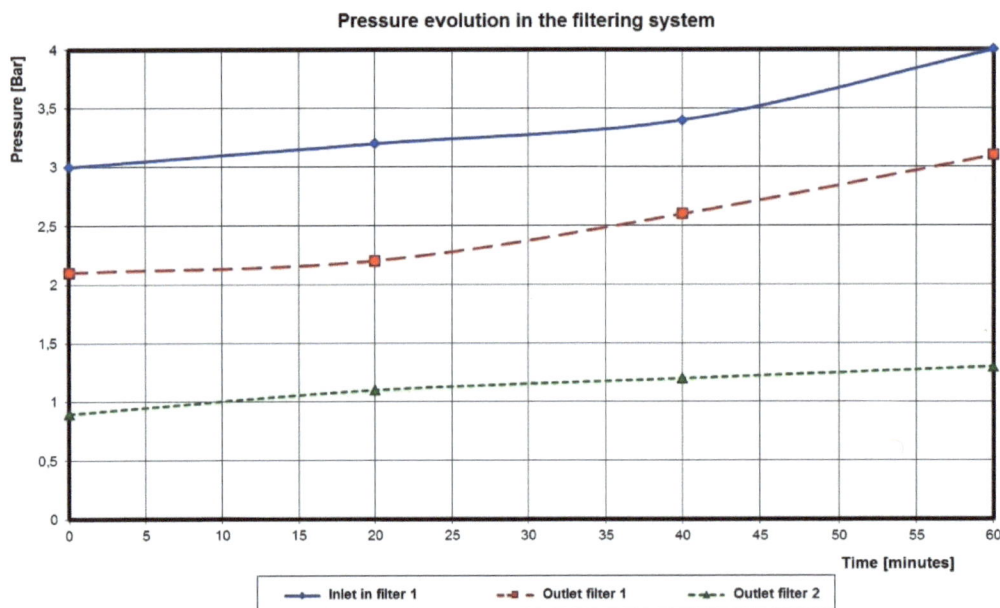

Figure 5: Evolution in time of the pressures in filtering system at the entrance into the filter F1 and the goings of the filters F1 and F2

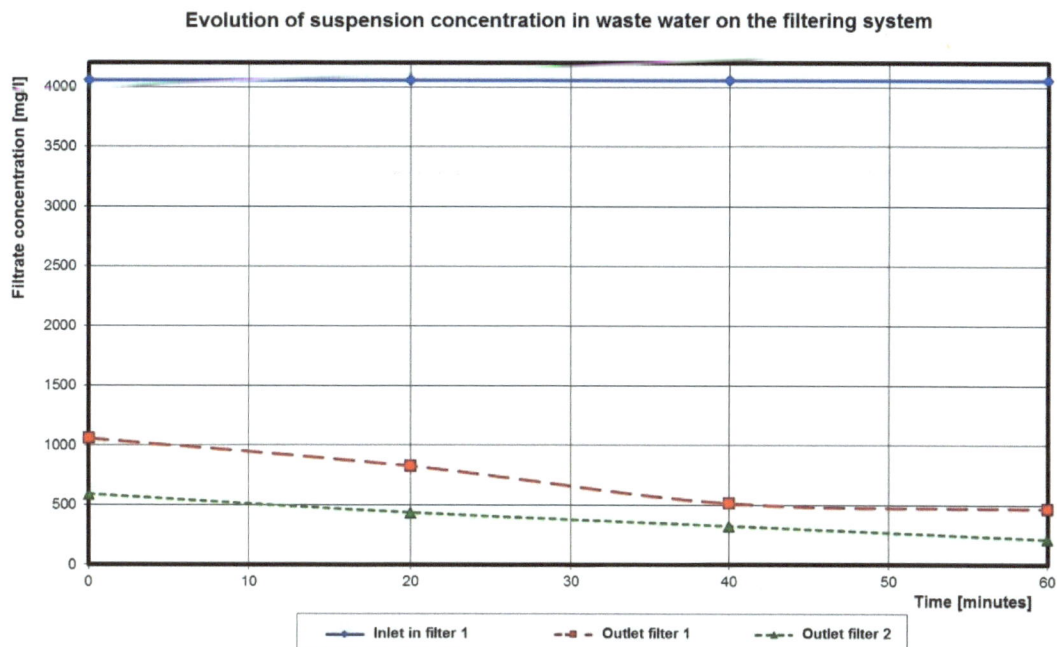

Figure 6: Evolution in time of the suspension concentration in the waste water in the filtering system at the entrance into the filter F1 and the goings of the filters F1 and F2

4. CONCLUSIONS

From the analysis of the results presented as tables and graphics revealed the following conclusions:

- the pressures at the entrance into the first filter (coarse filter) and at the outlet of the both filters increase with the duration of the filtration process;
- the first filter (coarse filter) retained around 95% of the suspensions mass contained in the waste water analyzed;
- By using the system with tangential filters coupled in series the suspension concentration from the analyzed wastewater was reduced from the original value of approx. 4000 mg/l to a final value of approx. 200 mg/l in the filtered liquid (clear), obtaining a reduction of over 90% of the initial content of mechanical suspensions from the wastewater.

ACKNOWLEDGMENTS.

This paper is supported by the Sectoral Operational Programme Human Resources Development (SOP HRD), ID134378, financed from the European Social Fund and by the Romanian Government

REFERENCES

[1]. Rivet, G., Guide de la separation liquide-solide, Editura IDEXPO, Cachan, 1985.
[2]. Robescu, Diana et al., Tehnici de epurare a apelor uzate, Editura Tehnica, Bucuresti, 2011
[3]. Rus, F. Operaţii de separare în industria alimentară. Editura Universităţii Transilvania, Braşov, 2001.
[4]. Sutherland, K., Filters and Filtration Handbook, 5th ed., Elsevier Butterworth-Heinemann, Amsterdam 2008.
[5]. Techobanoglous, G., Burton, F.L. and Stensel, H.D. Wastewater Engineering. (4th Edition). McGraw-Hill Book Company, 2003.
[6].Wakeman, K. and Tarleton, E. S., Solid/liquid Separation, 1st ed., Elsevier, Oxford 2005

6th International Conference
"Computational Mechanics and Virtual Engineering "
COMEC 2015
15-16 October 2015, Braşov, Romania

THEORETICAL AND EXPERIMENTAL RESEARCH ON THE SEPARATION PROCESS OF IMPURITIES FROM WASTE WATER THROUGH DECANTATION

Daniela Zârnoianu[1], Simion Popescu[2], Carmen Brăcăcescu[3]

[1] S.C. Oltchim S.A, Rîmnicu –Vilcea, Romania, e-mail: daniela.zarnoianu@yahoo.com
[2] Transilvania University of Brasov, Romania, e-mail: simipop38@yahoo.com
[3] National Institute of Research-Development for Agricultural Machinery, Bucharest, Romania,
e-mail: carmenbraca@yahoo.com

Abstract: *The paper presents the results obtained from the experimental research on the decantation processes for residual water (wastewater) on an experimental plant (pilot station) formed by two constructive varieties for the clarifiers with tangential inlet and spillway threshold: without a device for scraping (removing) decanted sludge (mud) and equiped with device for scraping sludge (mud). The compared analysis of sediment separating efficiency for the two types of clarifiers was conducted by measuring the time variation of the concentration of suspensions at the entrance and exit from the two types of clarifiers and developing some comparative charts for the concentration variation of the suspensions and the efficiency of clarifiers.*
Keywords: *residual water, separation, sedimentation, impurity concentration, clarifier with tangential inlet, scraping equipment, separating efficiency.*

1. INTRODUCTION

Industrially used water, also called „waste water" or „residual water" comes from different production and processing processes ad, from a physical point of view, they represent multiphase fluids (mixtures). When stopping, the phases are separated by gravity by a downward movement (gravimetric separation), due to the differences in specific mass of the particles in the suspension, thus achieving a sedimentation or decantation process. The solid particles, which fall off to the bottom of the sedimentation vessel, form a solid – liquid mixture, more or less concentrated, in the form of a sediment called precipitate, slurry or sludge. Depending on the type of sediments and their state of dispersion wastewater, the particles, or impurities, have different dimensions. Thus there are discrete granulated particles (sand, gravel), colloidal particles (groups of molecules or substances of 0.5 … 500 nm size) and molecules or macromolecules in the case of dissolved substances of less than one nanometre in size [1; 2].
The equipment used the for gravimetric separation, called decantor, is composed by a sedimentation bazin, filled with suspension water that will be decanted, a device for recovering the cleared liquid (decanted water) that, usually, surpasses the basin surface and leaks above its level. Additionally, the clarifier is equipped with a device for eliminating (extracting) the sediment (precipitate) from the bottom of the basin. In the general case several distinctive areas can be identified in a decanter: the inlet area, the sedimentation area, the accumulation area of sediments (sludge area), and the evacuation area of sediments. The mixture containing sediments enters the inlet area in turbulent flow and distributes via a uniform, piston or plunger type motion of speed v_m in the entire cross-section of the basin [5; 6].
The connection between the distribution and the decanting area can be achieved by means of a wall with calibrated holes or by means of a deflector that ensures a steady, laminar flow of the water, free of turbulence (vortices). In reality,

however, also secondary convection currents occur caused by the temperature differences and parasite flows generated by the differences in density of the various areas in the basin. These aspects evidently affect also the separation efficiency of the decanter. The sludge is evacuated swiftly and continuously from the sedimentation area without disturbing the aqueous solution; this is due to the evacuation area meant to ensure the necessary conditions such as to not disturb the flow in the sedimentation area and to collect the whole flow from the entire cross-section of the basin.

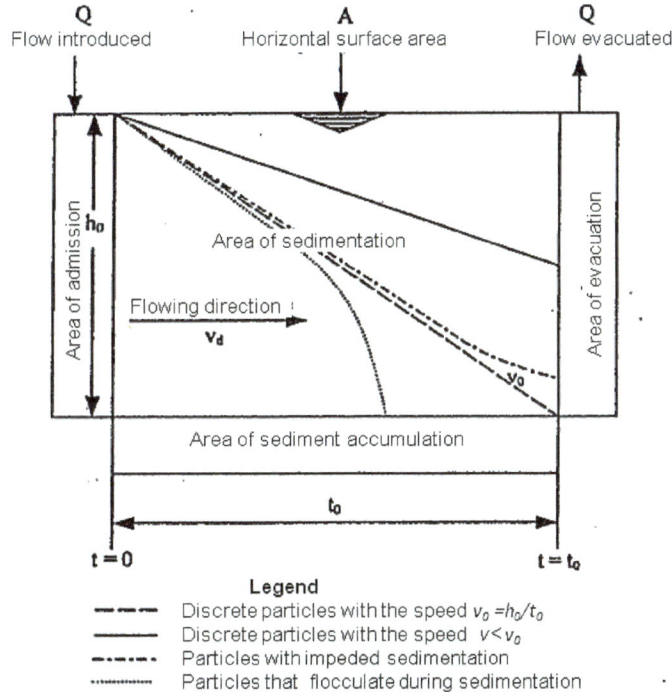

Figure 1: Diagram for defining the parameters of a conventional decanter

Consequently it can be assumed that the concentration of equal size suspended particles is the same in all points of the cross-section located at the end of the inlet area. In the sedimentation area the particles settle at the same speed v_s as the steady, static fluid. In reality, in the sedimentation basin the mixture subjected to decantation (water with sediments) moves horizontally to that a particle situated in suspension in this environment performs a movement composed with the absolute speed v_a (fig.2), resulted from summing up the two movements: the movement caused by the flowing of the mixture on a horizontal direction with the speed of transport v_m and the vertical sedimentation movement with the speed v_s caused by the gravitational field, meaning: $\vec{v}_a = \vec{v}_m + \vec{v}_s$. In a rectangular basin, where the height of the suspension water layer is H (Fig. 2), the maximum time necessary for achieving the sedimentation of a particle with a given diameter is calculated by the relation: $t_s = H/v_s$.

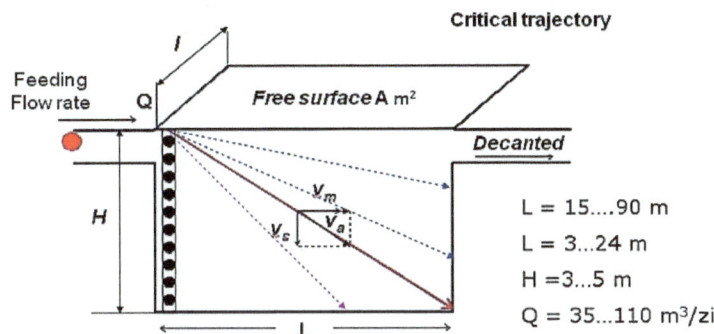

Figure 2: Schematic defining the parameters in a conventional decanter and the movement diagram for the particle in case of transversal (horizontal) circulation of the mixture (water with sediments)

If the useful length of the basin is L (s. Fig.2), the time t_0 for stationary movement of the fluid on the length of the basin with a sedimentation speed v_s is calculated by the relation: $t_0 = L/v_m$. Due to the fact that a particle is considered sediment in the moment when it reaches the bottom of the decanter, in order to achieve the sedimentation of the particles in the moving is necessary that the time t_0 of fluid stationing in the basin to be larger or at least equal to the time t_s necessary of particle sedimentation, expressed by the relation: $t_0 \geq t_s$. Therefore, the condition of achieving the sedimentation process $(t_0 \geq t_s)$ is expressed by the relations:

$$\frac{L}{v_m} \geq \frac{H}{v_s} \tag{1}$$

or

$$\frac{v_m}{v_s} \geq \frac{L}{H} \tag{2}$$

Because, generally, the dimensions L and H of the basin are known and the value of the sedimentation speed v_s for a certain type of suspension is also known, from relation (2) is obtained the relation through which is determined the limit value of the v_m speed of movement of the mixture:

$$v_m \geq v_s \cdot \frac{L}{H} \tag{3}$$

Relation (3) emphasis the fact that by reducing height H of the fluid layer in the decantation basin, v_m speed of movement of the mixture can be increased. Height H of the layer of fluid can't be reduced below a certain limit, because the situation may be reached where the separation of particles situated in suspension is done incompletely, the current of fluid also engaging a part of the particles laid on the bottom of the basin.

Considering a rectangular shaped basin (fig. 2), it result that the width l of the basin section depends on the value of the feeding flow Q, the height H of the basin and the average speed of movement (transport) v_m of the water in the basin section. From the known equation of continuity, namely $Q = H.l.v_m$, result the value of the basin width:

$$l = Q/H.v_m. \tag{4}$$

and the free surface A of the decanter (Fig. 2) is determined with the relation:

$$A = L.Q/H.v_m \tag{5}$$

The diagram of circulation of material flow in the case of a decanter (clarifier) is presented intuitively in figure 3, on the basis of which the equation of total balance and partial balance is established. The equation of total balance refers to the quantities of material that enter and exit the system and is expressed through the relation:

$$Q_0.\rho_0 = Q_1.\rho_1 + Q_2.\rho_2 \tag{6}$$

where:

Q_0, Q_1 and Q_2 are the volume flows for the water-suspensions mixture, the decanted (clear) water and respectively, of the precipitate (sludge);

ρ_0, ρ_1 and ρ_2 – densities of the water-suspensions mixture, the decanted (clear) water and respectively, of the precipitate (sludge);.

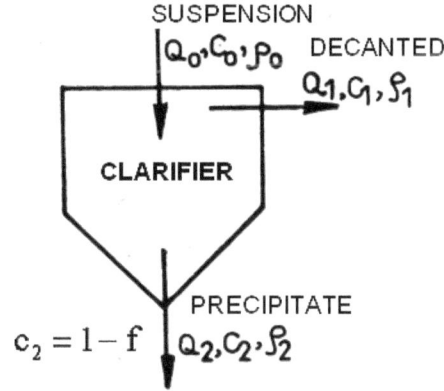

Figure 3: Diagram for the circulation of material flows in the case of a clarifier (decanter)

The partial balance equation refers to the content of solid particles situated in the material flows of the system and is express by the relation:

$$Q_0 \rho_0 c_0 = Q_1 \rho_1 c_1 + Q_2 \rho_2 c_2 \qquad (7)$$

where c_0, c_1 and c_2 are the mass concentrations in solid particles from the mixture subjected to decantation (waste water), from the clear water (decanted water) and from the precipitate (sludge). The mass concentration c represents the quantity of impurities (particles), expressed in mass units on the volume unit and, normally, is expressed in mg/l.

If is considered, hypothetically, that the cleared water does not contain any solid particles, meaning that the clarifier ensures a total yield of detaining the particles of the waste water suspension, meaning $c_1 = 0$, the total balance equation is given by the relation:

$$Q_0 \cdot c_0 = Q_2 \cdot c_2 \qquad (8)$$

and the partial balance equation obtains the from:

$$Q_0 \rho_0 c_0 = Q_2 \rho_2 c_2 \qquad (9)$$

The volume flow of the decanted water (cleared) Q_1 depending on the volume flow Q_0 of the water-suspension mixture subjected to decantation is obtained from the relation above, meaning:

$$Q_1 = Q_0 \rho_0 c_0 / \rho_2 c_2 \qquad (10)$$

Replacing relation (10) in the partial balance equation (7) is obtained the relation for calculating the volume flow of the decanted water Q_1 depending on the flow volume Q_0 of the mixture subjected to decantation, expressed in the form:

$$Q_1 = Q_0 \cdot \frac{\rho_0}{\rho_1} \cdot \left(1 - \frac{c_0}{c_2}\right) \qquad (11)$$

From relations (13) and (14) is obtained the relation between the volume flow of the decanted water Q_1 and the one of precipitations Q_2, namely:

$$Q_1 = Q_2 \cdot \frac{\rho_2}{\rho_1} \cdot \frac{c_2}{c_0} \left(1 - \frac{c_0}{c_2}\right) \qquad (12)$$

Due to the fact that the precipitate is presented in the form of sludge, characterized by a certain moisture u (expressed in %), the concentration in solid particles of the precipitate is determined with the relation:

$$c_2 = (1-u)/100 \qquad (13)$$

Moisture u of the precipitate is defined by the ratio between the mass of the fluid phase in the precipitate (sludge) and the mass of the precipitate.

The efficiency of a clarifier is assessed by the percentage coefficient E of retaining impurities from used water, depending on the initial concentration c_0 and the final concentration c_1, defined by the relation:

$$E = \frac{c_0 - c_1}{c_1} \cdot 100 = \left(1 - \frac{c_0}{c_1}\right) \cdot 100 \quad [\%] \tag{14}$$

The value of the coefficient of separation E depends on the clarifier type and on the installations that equip it and, usually, has values comprised in the limits $E = 35...65\%$

2. MATERIALS AND METHODS

The main objective of experimental researches consists in the compared analysis of functional performances of a clarifier with tangential inlet and spillway threshold built in two models: clarifier with simple construction, meaning without an additional system for scraping sludge deposits (called scraper) and modernized clarifier, additionally equipped with device for interior scraping of sludge deposits. By installing the scraper was aimed to keep the volume of decantation constant by the contiguous removal of sludge deposits from the inferior walls of the clarifier.

In order to establish the factors influencing the efficiency of the process of sedimentation in clarifiers for industrial waste water and finding the optimal constructive option for the clarifier, an experimental installation (pilot station) was built, which will allow to have a compared analysis between a model of clarifier without scraper and a model with scraper (Fig. 4). The experimental installation is built by the feeding vessel (tank) 1 (V-1), fitted with a mechanical agitator for homogenizing the composition of the residual water subjected to experiments and a centrifuge pump 2 (P-1) for feeding clarifiers D-1 (without scraper) and D2 (with scraper). The temperature of the used water when entering in the clarifiers is measured with electromagnetic flow meters 4 and 12, mounted on the feeding pipe of clarifiers D-1 and, respectively, D-2. The evacuating flow of the sludge is measured with electromagnetic flow meters 6, respectively 14, mounted at the exit of sludge pipes of clarifiers D-1 and D-2. The adjustment of the feeding flow for clarifier D-1 is done by the regulating valve 5, connected to the flow meter 4, and in the case of clarifier D-2, adjustment of the feeding flow is done by the regulating valve 13, connected to the flow meter 12. The adjustment of the evacuating flow for clarifiers D-1 and D-2 is done by the regulating valve 7, connected to the flow meter 6, respectively with the regulating valve 15, connected to the flow meter 14. The scraping organ 11 of clarifier D-2 is actuated mechanically from a driving device 9 situated in the exterior.

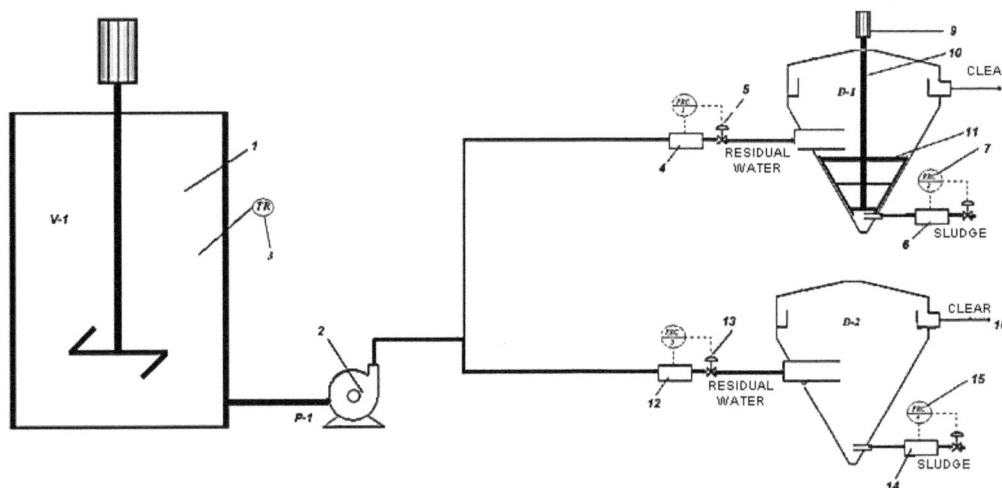

Figure 4: Diagram for the installation (pilot station) for conducting experiments on two options for clarifiers with tangential inlet and spillway threshold: D-1 - clarifier with scraper; D-2 clarifier without scraper:
1 – vessel for feeding waste water (V-1)fitted with agitator; **2** – centrifugal pump (P-1); **3** – thermometer with thermal resistance; **4** – flow meter for measuring the flow at the entrance in clarifier D-1; **5** – valve for adjusting water flow when entering clarifier D-1; 6- flow meter for measuring the sludge flow when exiting clarifier D-1; 7- valve for adjusting the sludge flow when exiting clarifier D-1; **8** – pipe for evacuating clear liquid from clarifier D-1; 9 -

equipment for driving the organ for scraping impurities (sludge); 10- drive shaft for the scraping organ; 11-organ for scraping impurities (sludge); 12- flow meter for measuring the flow at the entrance in clarifier D-2; 13- valve for adjusting water flow when entering clarifier D-2; 14- flow meter for measuring the sludge flow when exiting clarifier D-2; ; 7 valve for adjusting the sludge flow when exiting clarifier D-1; 15 – pipe for evacuating clear liquid from clarifier D-2

After filling vessel V-1 with used water subjected to testing, having a known initial concentration of impurities (determined by measuring), and pump 2 is started (P-1) for feeding clarifiers D-1, respectively D-2. In the process of filling clarifiers with used water, sludge in the inferior part is purged (eliminated) periodically in order to avoid the clogging of the system for evacuating clear water. In addition, after filling clarifier D-2, the device for mechanical evacuation of sludge with scraping organ 11 enters into operation, driven by the driving system 9 through drive shaft 10. For the tests conducted, samples of material were periodically harvested and were measured the concentration of suspensions from the used waters introduced in clarifiers, from the clear liquid evacuated and from the sludge evacuated from the clarifier.

3. RESULTS AND DISCUSSIONS

Testing the operation of each option of experimental clarifier was achieved by harvesting samples at a 24 hour interval. For each sample were determined by measures the values of concentration of impurities (in mg/l) at the entrance of waste water in the two clarifiers, from the clear water and at the exit from the clarifier and from the sludge evacuated from clarifiers. By processing the data were determined by calculation the percentage values for the efficiency of the separation process E for each periodical testing, using relation (14). The result obtained after measuring and processing the data are summarized in Table 1. On the basis of experimental data mentioned in table 1 were built graphics that represent the variation of coefficients for the efficiency of separation E (%) of the clarifier depending on the initial concentration of the solid in the residual (waste) water at the entrance in the clarifiers. These graphics allow making a compared analysis of the separation process for each type of clarifier and the evolution in time of the separation efficacy for the two types of clarifiers studied.

Table 1: Results from experimental determinations conducted for the two types of clarifiers

Time	Solids concentration residual water	Clarifier with scraper (D-1)			Clarifier without scraper (D-2)		
		Solids concentration		Separation efficiency E	Solids concentration		Separation efficiency E
		clear	mud		clear	mud	
[hours]	[mg/l]	[mg/l]	[mg/l]	%	[mg/l]	[mg/l]	%
0	9687	4016	24591	58,54	5742	23704	40,72
24	8945	6462	25264	27,76	7459	24849	16,61
48	10308	3518	25662	65,87	5356	14150	48,04
72	10260	6852	28654	33,22	9094	19302	11,36
96	5730	3015	27962	47,38	3612	12758	36,96
120	6220	3050	16734	50,96	3440	12194	44,69
144	4248	2938	16572	30,84	3108	15846	26,84
168	3286	1804	9936	45,10	2162	8416	34,21
192	4320	1838	23350	57,45	2132	11532	50,65
216	9324	3812	29105	59,12	4536	17654	51,35
240	10706	4750	25388	55,63	5656	15130	47,17
264	8852	2946	25344	66,72	4200	15356	52,55
288	5674	3354	12598	40,89	3926	10420	30,81
312	9800	5182	15982	47,12	6064	12898	38,12
336	3250	2020	15176	37,85	2206	8760	32,12

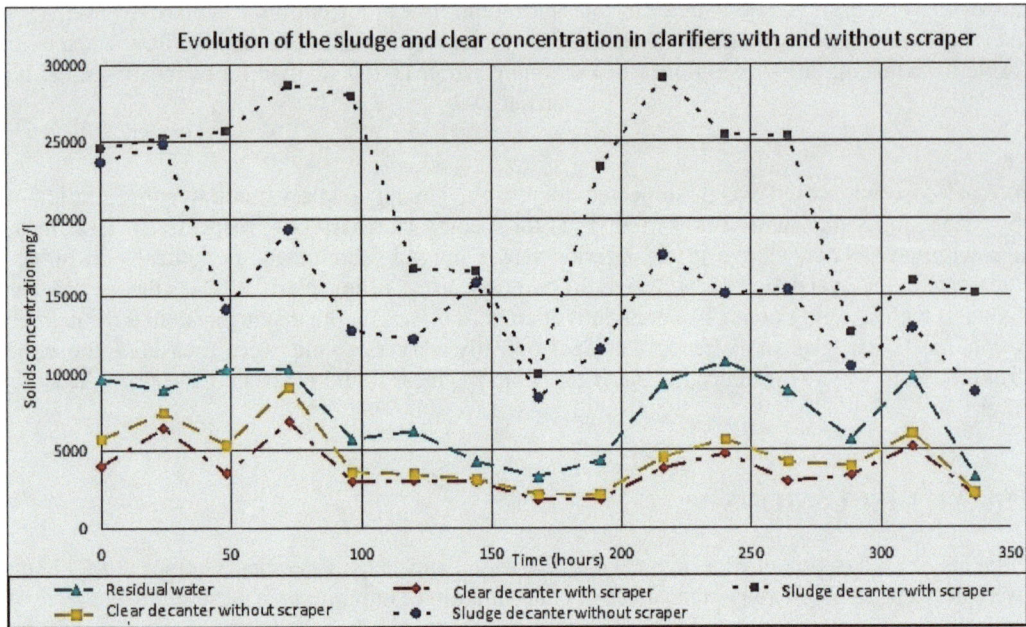

Figure 6: Variation in time of the concentration for the suspension in clear water and mud for the clarifier with scaper and without scraper

In figure 6 are presented the compared graphics for the time variation of the suspension's concentration in clear water and in mud for the two types of clarifiers and in figure 7 are presented the compared variations in time for the efficacity of separating for the two types of decantors (without scraper and with scraper).

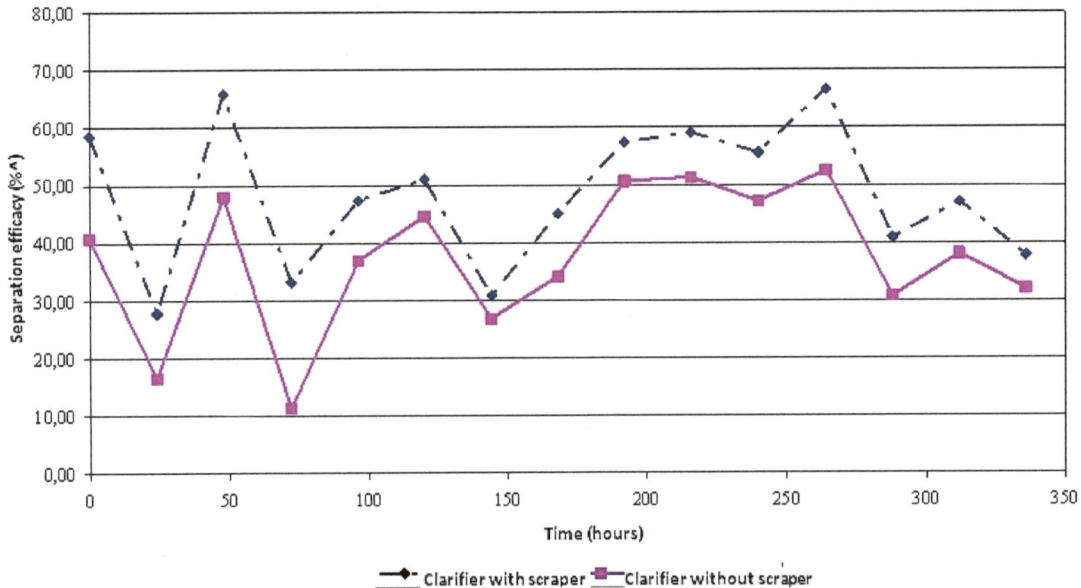

Figure 7: Variation in time of the efficacity of separating impurities in the clarifier with scraper and without scraper

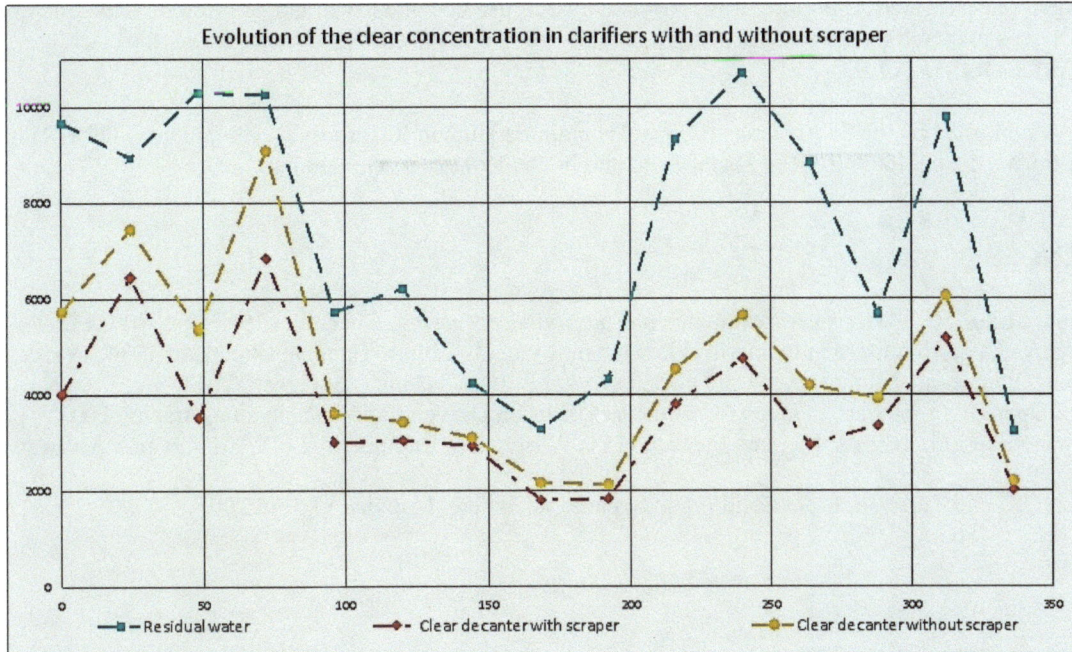

Figure 8: Variation in time of the concentration of the suspension in clear water for the two types of clarifiers

In figure 8 are presented the charts for the variation in time of the suspension's concentration of clear water and in figure 9, the variation of the suspension's concentration in mud for the two types of clarifiers (without scraper and with scraper).

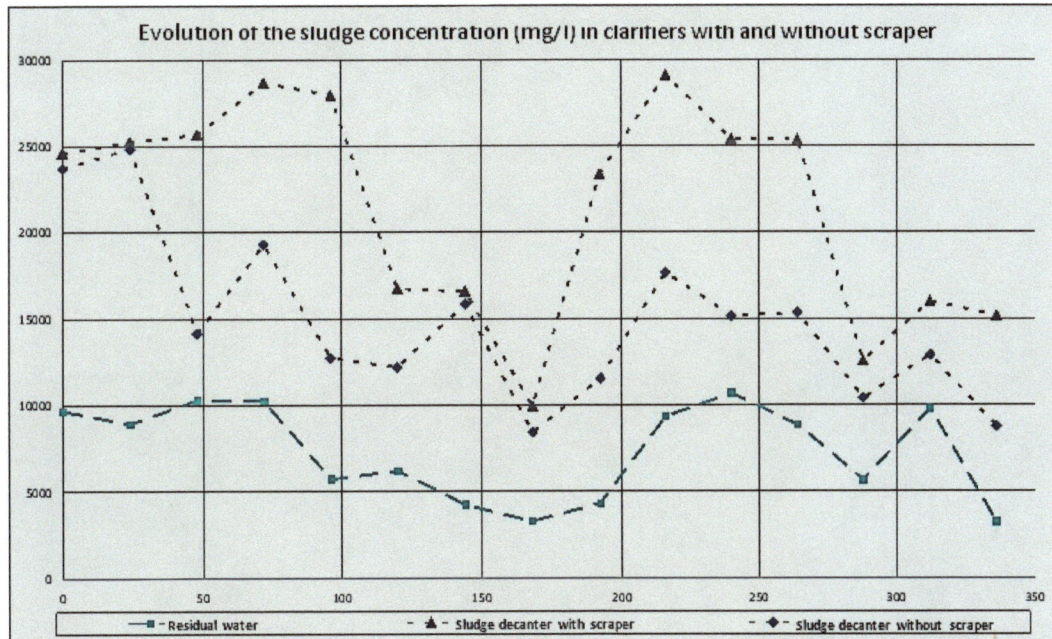

Figure 9: Variation in time of the suspension's concentration of mud for the two types of clarifiers

4. CONCLUSIONS

The efficiency of separating suspensions from residual waters is superior for the clarifiers with devices for scraping, compared to the one of the clarifiers without scraping devices, a fact leading to the reduction of solids concentration in the discharged (clear) water. Therefore, it is recommended to use clarifiers with tangential inlet and spillway threshold equipped devices for scraping the sludge, which come into operation automatically after filling the clarifier.

ACKNOWLEDGMENTS.

This paper is supported by the Sectoral Operational Programme Human Resources Development (SOP HRD), ID134378, financed from the European Social Fund and by the Romanian Government

REFERENCES

[1]. Beychok, Milton R., Wastewater treatment. *Hydrocarbon Processing*, December 1971), pp. 109–112
[2]. Bratu, E.A., Operaţii unitare în ingineria chimică, vol. I,vol. II, Editura Tehnică, Bucuresti, 1984, 1985.
[3]. Robescu, Diana et al. Tehnici de epurare a apelor uzate, Editura Tehnica, Bucuresti, 2011
[4]. Rus, F., Operaţii de separare în industria alimentară. Editura Universităţii Transilvania, Braşov, 2001
[5]. Techobanoglous, G., Burton, F.L. and Stensel, H.D., Wastewater Engineering. (4th Edition ed.). McGraw-Hill Book Company, 2003.
[6].Wakeman, K. and Tarleton, E. S., Solid/liquid Separation, 1st ed., Elsevier, Oxford 2005

**6th International Conference
"Computational Mechanics and Virtual Engineering"
COMEC 2015
15-16 October 2015, Brașov, Romania**

APPLICATIONS OF THE VIRTUAL INTELLIGENT PORTABLE VIPRO PLATFORM FOR 3D CONTACT PROBLEMS WITH FRICTION IN THE HUMANOID ROBOTS CONTROL

Ionel-Alexandru Gal, Nicolae Pop, Victor Vladareanu, Mihaiela Iliescu, Daniel Mitroi and Luige Vladareanu*

Institute of Solid Mechanics of Romanian Academy, Bucharest, ROMANIA, luigiv2007@yahoo.com.sg
* Corresponding author

Abstract: In this paper we investigate the influence of the friction force while the robot is doing different actions. By computing the limit conditions for the robot walking according to the friction force between ground and feet, we have found the maximum safe values for a walking step and other stability conditions, according to the walking slope and type of the support surface. The motion trajectory positions of the robot leg end-effector and joints with reaction forces are analyzed and the virtual projection method is adopted using the Versatile Intelligent Portable Robot Platform VIPRO. The presented simulations demonstrate through a numeric modeling of the 3D contact problems with friction, that we can detect the slip/stick phenomenon for a walking robot motion on a uneven terrain, so it can improve the real time control to predict and avoid robot overthrow. The obtained results lead to the development of new technological capabilities of the control systems.
Keywords: mobile robots, stick/slip motion, NAO robot, virtual projection

1. INTRODUCTION

Humanoid robots use bipedal walking to move from point to point [1]. This means that the dynamic motion control should be planned according to every possible perturbation that can hinder the walking process [2, 3]. One of these is the slipping conditions [4, 5] for each robot feet which can destabilize the walking process. To compensate for this, many walking robots compensate through dynamic control all the forces that are present within the robot joints. But only some robots actually take into account the slip conditions to compensate additional forces [2, 6].
In bipedal walking, robots will often encounter slip conditions [5-7]. These positions must be avoided, and unless their trajectory will keep them outside the dangerous areas of kinematic positions, the slip conditions become active. In these cases, the robot joint control will also have to compensate for the slip forces, so that the slip/stick conditions will fall back to the stick state [4, 5].
For the friction problem, a reference on analysis and numerical approximations of contact problems involving elastic materials with or without friction is given by Kikuchi and Oden [8]. This is concerned by the effects of friction at the interface accurately and of the non-penetration constraints at contact boundary of deformable bodies being in a mutual contact [9-13,19]. The non-smooth friction law and the non-penetration constrains depend on time [4].
The aim of this paper is to obtain a mathematical algorithm and a chart for the slip/stick conditions for a bipedal walking robot according to different center of mass position regarding the robot feet and on different values of the friction

119

coefficient. The robot chosen is the NAO robot, from which we have taken its measurements and added into the testing simulation of our algorithm.

The obtained results are an application of the virtual intelligent portable VIPRO platform [14, 18], which analyzes the problem of friction during robot walking. By using the VIPRO platform, we can improve the stability performances of robot motion in a virtual and real environment on unstructured and uneven terrains. This will lead in building more efficient mobile robots by adding different control methods [15-17, 18, 19] which can benefit from the robot feet friction research [4, 5].

2. STICK/SLIP CONDITIONS

The three-dimensional frictional model consists of a mass M, with three translational degrees of freedom, u_1, u_2 and u_3, with constrain $u_3 \geq 0$, that can make contact with a rigid plane surface at which the frictional coefficient is μ. The mass M is connected to a generalized linear elastic support with stiffness and damping matrices K, C, respectively, and is subjected to an externally applied forces F. When the mass makes contact with the plane ($u_3=0$), induces a reaction force R (R_1, R_2, R_3) [4].

2.1. States of the system

For any given time there are three states of the contact nodes: separation (open contact), stick or slip. But in 3D case the slip state has two components in the contact plane surface [4]. These states are defined by the conditions:

1) Separation (open contact)
For this case the mass may lose contact with plane and there is no reaction between the mass and plane:

$$u_3 \geq 0 \text{ and } R = 0 \tag{1}$$

and for $R = 0$ and $u_3=0$ the state is a limiting state of separation.

2) Stick
Stick is the state when the mass makes contact with the plane and is not moving:

$$u_3 = 0$$
$$\dot{u} = 0 \tag{2}$$

and the normal reaction be compressive:

$$R_3 > 0 \tag{3}$$

The Coulomb friction law demands that:

$$\sqrt{R_1^2 + R_2^2} < \mu R_3 \tag{4}$$

3) Slip
For slip must have:

$$u_3 = 0$$
$$R_3 > 0 \tag{5}$$

and the Coulomb friction law demands that:

$$R_k = -\frac{\mu R_3 \dot{u}_k}{\sqrt{\dot{u}_1^2 + \dot{u}_2^2}}, \qquad k = 1,2 \tag{6}$$

i.e. the result frictional force has the magnitude of and is in a direction opposing the instantaneous motion direction [4].

3. MATH MODEL AND SLIP/STICK DETECTION ALGORITHM

Before explaining the mathematic model we need to present the robot structure on which we have made the testing and simulations. The bipedal robot that we have used is the robot NAO (figure 1).

Figure 1: NAO robot

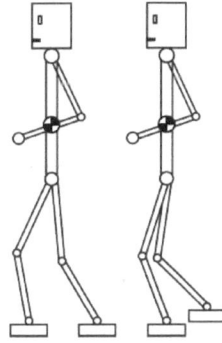

Figure 2: Walking robot summary diagram

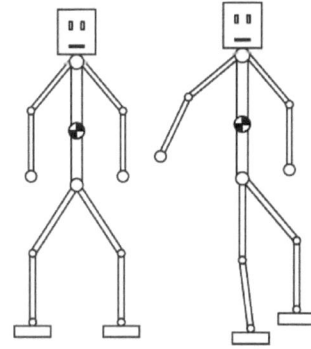

Figure 3: Other robot positions

This robot has become very popular within the scientific community and in the academic area in which students learn to control its behavior for different stimuli. This is why we have chosen this robot which is accurately presented by Goualier et. all [20], from where we have taken the robot technical data. Starting from the original kinematic diagram, we have made figures 2 and 3. These images show the simplified diagram of the robot, removing most of the links and joints to a point that helped us in presenting different positions of the robot while not removing the joints that helped the robot to reach certain positions within its kinematic space.

To test the slip/stick conditions, we didn't have to use the whole system of links and joints. For that we have made a simplified model of the legs and center of gravity to allow us to easily compute and demonstrate the slip/stick conditions.

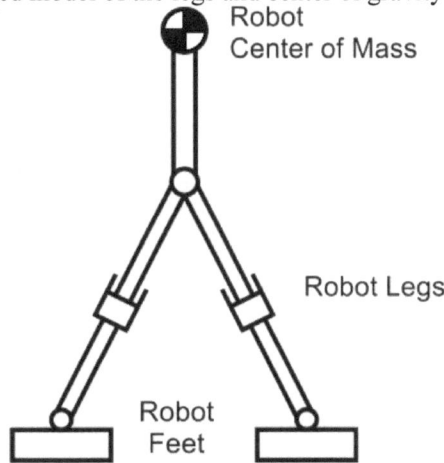

Figure 4: Over simplified robot diagram to show only the points of interest

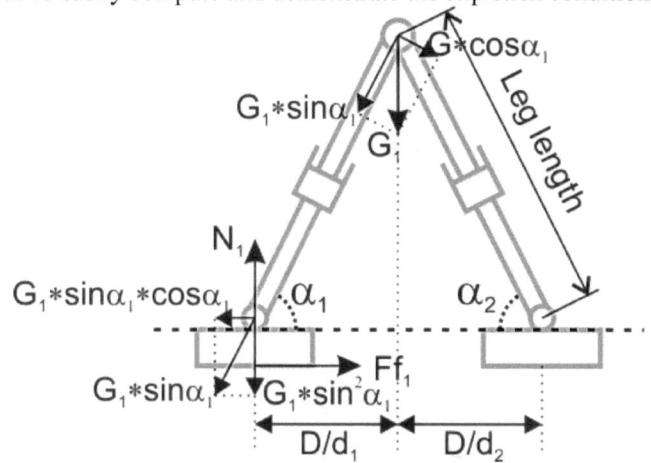

Figure 5: Weight force decomposition for leg 1

Figure 4 presents the walking robot diagram as needed for the slip/stick detection algorithm. The detection algorithm needs the position of the Center of Mass (COM) of the entire robot, which for our simulation we'll take only above the hip and change the hip position to achieve all the possible kinematic positions. For this we have another simplification. The legs have rotation joints within the robot knees but for our simulation we can replace these with translation joints because we'll only need the leg extension which is equal to the distance between the robot feet to its hip.

Taking this into account, we have made the diagram from figure 5, which presents the force decomposition for one leg of the robot, where G is the weight of the entire robot, G_1 and G_2 are the distributed weights of the robot on each foot, α_1 and α_2 are the angles for each leg with the horizontal, D is the distance between the two feet, and d_1 and d_2 are the distances between each foot and the projection of the center of mass on the horizontal plane.

$$d_1 + d_2 = D \tag{7}$$

Before computing the friction force and the force acting on the foot, we need to compute the weight ration which is distributed on each foot.

$$G_{1,2} = G \frac{d_{1,2}}{D} \tag{8}$$

Knowing the weight value for each foot from equation (8), we then compute the force acting on the foot and the one that adds torque to the ankle, which are presented in equations (9) and (10), respectively.

$$G_x = G \sin(\alpha) \tag{9}$$

$$G_y = G \cos(\alpha) \tag{10}$$

Equation (10) will give us the static torque which will act on the robot foot after multiplying it with the distance from the force to the joint. But for our research we'll not use it. What we use is the force that is transmitted along the leg to the foot (G_x). This force will give us the friction force as in equation (11). The other force remaining is the one that pulls on the robot foot which is presented in equation (12).

$$F_f = \mu N = \mu G_x \sin(\alpha) = \mu G \sin^2(\alpha) \tag{11}$$

$$F = G_x \cos(\alpha) = G \sin(\alpha)\cos(\alpha) \tag{12}$$

Using equation 11 and 12, in which μ is the friction coefficient and α is the angle between leg and horizontal, along with the stick condition in (4), we achieve the stick condition for our legs:

$$F < F_f \Leftrightarrow G \sin(\alpha)\cos(\alpha) < \mu G \sin^2(\alpha) \tag{13}$$

Which is simplified to:

$$\cos(\alpha) < \mu \sin(\alpha) \tag{14}$$

This results in the next equation which we have used in our simulations:

$$\arctan\left(\frac{1}{\mu}\right) < \alpha \tag{15}$$

So it all reduces to a condition between the angle that the horizontal is making with the line that goes through the robot ankle joint and its hip joint and the friction coefficient between the robot's feet and the walking surface. In addition to this, we can add the angle that the surface has with the ground horizontal, when the robot is walking on a slope and achieve a different result. But for our test cases we have used a horizontal walking surface so we can detect and easily interpret the results.

Algorithm 1: used for determining the stick/slip condition:
   Step 1: Generate the distance between the two feet.
   Step 2: Generate a new position of COG (center of gravity) of the robot.
   Step 3: Validate COG position in respect to each foot.
   Step 4: Compute mass distribution for each foot using equation (8).
   Step 5: Compute angle between legs and horizontal, while knowing the position of COG and each foot.
   Step 6: Check the slip/stick conditions for each leg using Equations (4), (11) and (12).
   Step 7: Go Back to Step 2, until no new position can be found.
   Step 8: Go Back to Step 1, until no new valid distance can be found.

Algorithm 1 is the actual algorithm we have used in our simulations, and the results are presented in the next section.

5. RESULTS AND CONCLUSIONS

Using the mathematical model and algorithm presented in this paper, we have implemented a Matlab simulation to prove our research. In the conducted simulations the initial conditions which we have used are presented in table 1.

Table 1: Simulation data

Parameter	Value
Minimum leg length	55 [mm]
Maximum leg length	200 [mm]
Robot weight	5,12 [Kg]
Distance between feet	5 [mm] – 140 [mm]

One initial condition we have imposed in the simulation is that the two feet are in constant contact with the support surface. If we'll have only one foot in contact with the ground then we'll have two cases the robot could be in. The first one is the support case in which the robot will support its entire weight on one leg and the center of mass will have to be within the support surface that for now is the footprint. The other case is when the center of mass is outside the support surface, in which case the robot will start falling and at some point slipping. But the slipping condition will not be the main problem at that point.

Having our first initial condition so that the feet are in constant contact with the support surface, we then add another one. The second condition is that the center of mass will be within the bounded area of the support surface, which is given by the two feet in contact with the ground. This will allow us to have a stable robot which will not be in critical overthrowing condition that is not related to the stick/slip effect.

Having the initial conditions, we then change the distance between the two feet and test in each case for different friction coefficients, the kinematic area in which the robot hip can be positioned and the stick/slip condition on each foot. If the condition on one of the legs for a certain point in space of the robot hip will meet the slip condition then that point will be considered dangerous and be marked on the diagram appropriately.

In figures 6 we have presented the results after simulation for different inputs, in which:

- The light grey areas are areas in which the center of mass can't be positioned due to the conditions in leg length.
- The darker grey areas are the positions for the center of mass where both feet will not slip and the stick condition is fulfilled.
- The black areas are the position for the center of mass in which the stick condition is not fulfilled and one of the feet can slip.

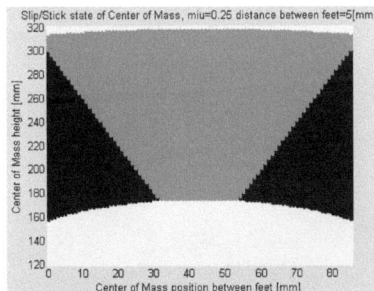

a1) μ=0.25; D=5mm

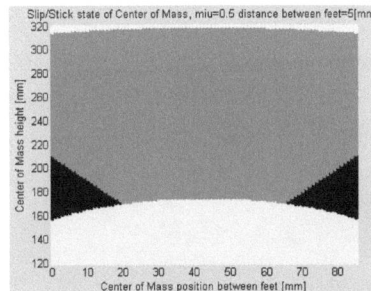

b1) μ=0.5; D=5mm

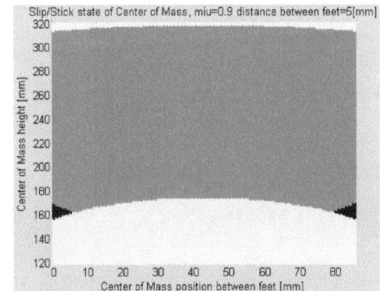

c1) μ=0.9; D=5mm

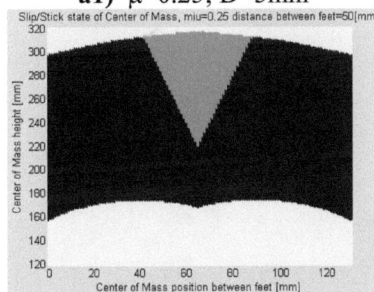

a2) μ=0.25; D=50mm

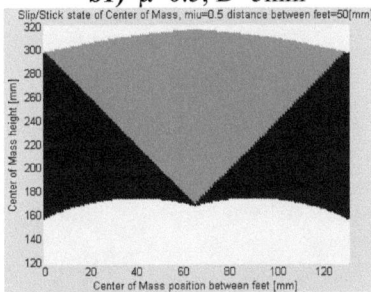

b2) μ=0.5; D=50mm

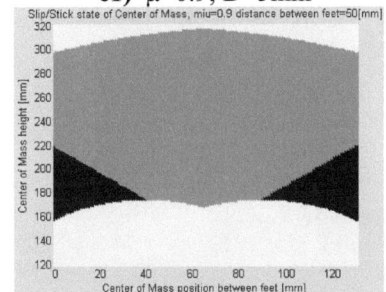

c2) μ=0.9; D=50mm

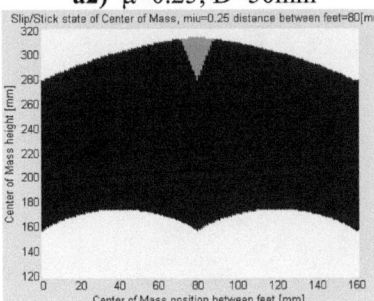

a3) μ=0.25; D=80mm

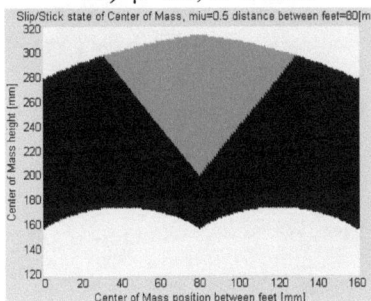

b3) μ=0.5; D=80mm

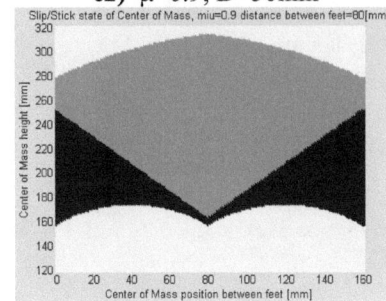

c3) μ=0.9; D=80mm

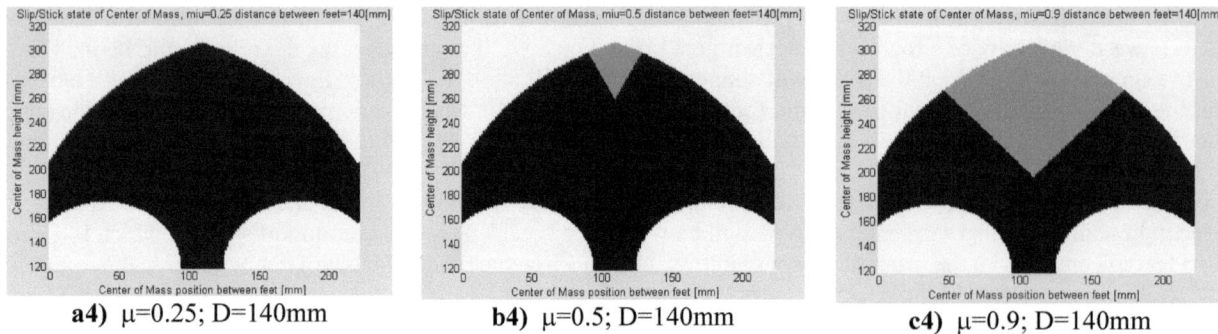

a4) μ=0.25; D=140mm   **b4)** μ=0.5; D=140mm   **c4)** μ=0.9; D=140mm

Figures 6: Simulation results for different values of D (distance between feet) and the friction coefficient

For the simulation, we have changed the friction coefficient to vary between 0.25 and 0.9, but we have shown only 3 values of it which are 0.25, 0.5 and 0.9. While varying the friction coefficient we have also change the distance between the two feet from 5mm up to 140mm.

In the conducted simulations for which we have presented figures 6 with the results, we can clearly see as expected that for smaller friction coefficients the robot can enter the slip area faster. Also, while increasing the distance between feet we can see that the stick condition is not fulfilled.

Another important observation is that for the robot to be in the stick area and its legs not slip on the support surface, it must have its center of mass as high as possible while the step length should be shorter, when the friction coefficient is smaller. This means that the control laws for robot trajectory and planning should take into account for the stick/slip conditions. In this way, the control law that acts on the robot joints torque can easily compensate for other dynamic forces and not worry about the slipping effect. It results that the torque needed for each joint to reach its target will not need to compensate for the torque value that can be found using equation (10).

ACKNOLEDGEMENT

This work was accomplished through the Partnerships Program in priority fields - PN II, developed with the support of MEN-UEFISCDI, PN-II-PT-PCCA-2013-4, ID2009, VIPRO project no. 009/2014, Romanian Academy and FP7-PEOPLE-2012-IRSES RABOT project no. 318902.

REFERENCES

[1] Hong, Young-Dae, Chang-Soo Park, and Jong-Hwan Kim, Stable bipedal walking with a vertical center-of-mass motion by an evolutionary optimized central pattern generator, Industrial Electronics, IEEE Transactions on 61.5: 2346-2355, 2014.

[2] Hereid, A., Kolathaya, S., Jones, M. S., Van Why, J., Hurst, J. W., & Ames, A. D., Dynamic multi-domain bipedal walking with ATRIAS through SLIP based human-inspired control, In Proceedings of the 17th international conference on Hybrid systems: computation and control (pp. 263-272). ACM, 2014.

[3] Hill, Joshua, and Farbod Fahimi, Active disturbance rejection for walking bipedal robots using the acceleration of the upper limbs, Robotica 33.02: 264-281, 2015.

[4] N. Pop, L. Vladareanu, I. N. Popescu, C. Ghita, I.A. Gal, Shuang Cang, Hongnian Yu, Vasile Bratu, Mingcong Deng, "A numerical dynamic behaviour model for 3D contact problems with friction", Computational Materials Science, Volume 94, November 2014, Pages 285-291, ISSN 0927-0256

[5] L. Vladareanu, N. Pop, I.A. Gal, M. Deng, The 3D elastic quasi-static contact applied to robots control – ICAMECHS 2013, pp. 517-523, ISBN 978-1-4799-2519-3, IEEE, 2013.

[6] H. Takemura, M. Deguchi, J. Ueda, Y. Matsumoto, T. Ogasawara, Slip-adaptive walk of quadruped robot. Robotics and Autonomous Systems, 53(2), 124-141, 2005.

[7] C.C. Ward, K. Iagnemma, A Dynamic-Model-Based Wheel Slip Detector for Mobile Robots on Outdoor Terrain, IEEE Transactions on Robotics, VOL. 24, NO. 4, AUGUST 2008.

[8] N. Kikuchi & J. T. Oden, Contact problems in elasticity: a study of variational inequalities and finite element methods. Philadelphia, PA: SIAM., 1988.

[9] M. F. David, Gh. Voicu, L. David, C. O. Rusănescu, Experimental Analysis Considered the Dynamics of Mobiles Agricultural Aggregate, Bulletin of University of Agricultural Sciences and Veterinary Medicine Cluj-Napoca. Agriculture, 66(1) 51-58, 2009.

[10] Alan D. Berman, William A. Ducker, Jacob N. Israelachvili, Origin and Characterization of Different Stick–Slip Friction Mechanisms, Langmuir, 1996, 12 (19), pp 4559–4563, DOI: 10.1021/la950896z

[11] K.L. Kuttler, Dynamic friction contact problems for general normal and friction laws, Nonlinear Anal.TMA 28 (3) (1997) 559–575.

[12] Cho, H. and Barber, J.R., Dynamic behavior and stability of simple frictional systems. Math. Comput. Modeling vol. 28, pp. 37-53. 1998.

[13] Cho, H. and Barber, J.R., Stability of the three-dimensional Coulomb friction law, Proc. R. Soc. Lond. A, vol. 455, pp. 839-861, 1999.

[14] L. Vlădăreanu, "Versatile Intelligent Portable Rescue Robot Platform through the Adaptive Networked Control", Proceedings of 5th European Conference of Mechanical Engineering, 5th European Conference of Mechanical Engineering,Florence, Italy, 22-24 11.2014, ISSN 2227 – 4596.

[15] I.A. Gal, L. Vladareanu, F. Smarandache, H. Yu, M. Deng, Neutrosophic Logic Approaches Applied to" RABOT" Real Time Control, , Vol. 1, pp. 55-60, EuropaNova, Bruxelles, 2014,

[16] L. Vladareanu, I.A. Gal, H. Yu, M. Deng, Robot control intelligent interfaces using the DSmT and the neutrosophic logic. International Journal of Advanced Mechatronic Systems, 6(2-3), 128-135, print ISSN: 1756-8412, eweb ISSN: 1756-8420, 2015.

[17] I.A. Gal, L. Vladareanu, R.I. Munteanu, "Sliding Motion Control with Bond Graph Modeling Applied on a Robot Leg", Rev. Roum. des Sc. Tech.-Serie El. et En., 60(2), 215-224, ISSN 0035-4066, 2015.

[18] M. Iliescu, C. Spirleanu, A. Pătraşcu and L.Vladareanu, "Distributed Control System for Machining Process Optimization in Drilling Mineral Composites Reinforced by 3% Glass Fibers", TEHNOMUS Journal, no. 22/2015, pg. 426 – 431, ISSN-1224-029X

[19] Vladareanu, V; Schiopu, P; Vladareanu, L, "Theory and application of extension hybrid force-position control in robotics", UPB Sci Bull -Series A, Volume: 76 Issue: 3 43-54, 2014, ISSN: 1223-7027

[20] D. Gouaillier, V. Hugel, P. Blazevic, C. Kilner, J. Monceaux, P. Lafourcade, B. Maisonnier, The nao humanoid: a combination of performance and affordability. CoRR abs/0807.3223, 2008.

6th International Conference
"Computational Mechanics and Virtual Engineering "
COMEC 2015
15-16 October 2015, Braşov, Romania

ON THE NUMERICAL SOLUTION OF NON-LINEAR ROTOR

Dumitru Nicoara
Transylvania University of Brasov, Brasov, ROMANIA

Abstract: *The present paper deals with the dynamical analysis and numerical solutions of non-linear rotor-bearing systems. The numerical solution is calculated using the Wilson-θ method in conjunction with an iteration procedure. The model of rotor-bearing system comprises of a continuous elastic shaft mounted on several non-linear bearing. One or several disks are mounted on the shaft. Timoshenko beam model is adopted and gyroscopic effect is taken into account.*
Keywords: *non-linear rotor-bearings systems, numerical solutions*

1. INTRODUCTION

The behavior of dynamical systems undergoing time dependent changes (transients). Transient state analysis has been an active research area in many engineering problems.
For a linear mechanical system the equations of motion are

$$[M]\{\ddot{x}\} + [C]\{\dot{x}\} + [K]\{x\} = \{F\} \qquad (1)$$

where $\{x\}$, $\{\dot{x}\}$, and $\{\ddot{x}\}$ refer to displacement, velocity, and acceleration vectors, respectively, and $[M]$, $[C]$, and $[K]$ are mass, damping and stiffness matrices.
In practical situation the systems are usually nonlinear and time-dependent. For example, the equation of motion for rotor-bearing systems with nonlinear bearings, systems that will be studied in this paper, is nonlinear

$$[M]\{\ddot{x}\} + [C]\{\dot{x}\} + [K(\{x\})]^{nl}\{x\} = \{F\} \, q \qquad (2)$$

The analytical solutions for such problems are difficult to obtain. The numerical methods have to be used.
Many numerical integration methods are available for approximate solutions of equations of motions. Direct numerical integrations methods is based on two ideas. First, instead of trying to satisfy dynamic governing equations at any time t, it is aimed to satisfy the equations only at discrete time interval Δt. The second idea is that, for each time interval Δt is assumed a specific type of variation of the displacement, $\{x\}$ velocity $\{\dot{x}\}$, and acceleration $\{\ddot{x}\}$.

Numerical integration methods are usually divided into two categories, implicit and explicit.
Consider the ODEs

$$\dot{\mathbf{x}} = \mathbf{f}(\mathbf{x}, t) \qquad (3)$$

In an explicit numerical scheme, the ODEs are represented in terms of known values at a priori time step

$$\mathbf{x}_{i+1} = \mathbf{x}_i + \Delta t \, f(\mathbf{x}_i, t_i) \qquad (4)$$

while in an implicit scheme

$$\mathbf{x}_{i+1} = \mathbf{x}_i + \Delta t \, f(\mathbf{x}_{i+1}, t_i) \qquad (5)$$

Explicit numerical schemes are conditionally stable. Implicit numerical schemes are unconditionally stable, i.e. do not impose a restriction on the size of time step Δt.

Thus several numerical integration schemes are available depending on the type of variations assumed for $\{x\}$, $\{\dot{x}\}$, and $\{\ddot{x}\}$ within each time interval Δt.

In this paper, the numerical time response solution for the non-linear rotor-bearing system is calculated using the Wilson-θ method in conjunction with an iteration procedure.

The Wilson-θ method is an extension of the linear-acceleration method. That is, within a time step the acceleration vector is proportional to time [1].

2. MODEL OF NON-LINEAR ROTOR-BEARING SYSTEMS

The model consists of a rotor treated as a continuous elastic shaft with several rigid disks, supported on the bearings with a non-linear behaviour. Consider that the dynamic equilibrium configuration of the rotor-bearing system the undeformed shaft is along the x- direction of an inertial x, y, z coordinate system. In the study of the lateral motion of the rotor, the displacement of any point is defined by two translations (v, w) and two rotations (φ_y, φ_z). The model use the beam C^1 finite element type based on Timoshenko beam model. The beam finite element has two nodes. In the case of the dynamic analysis four degrees of freedom per node are considered: two displacements and two slopes measured in two perpendicular planes containing the beam, [4], [5].

It is well known that the behaviour of both lubricated journal bearings and rolling element bearings is strongly non-linear and can cause rotors to behave in a non-linear way. In this paper the nonlinearities are involved only in the "elastic" part of the system. The non-linear bearings have a cubic non-linear term [3], [6], where the force–displacement relation for a non-linear spring element can be written as a function of the complex displacement z by the law

$$f(z) = k\,(1+|z|^2)\,z \tag{6}$$

This law is particularly well suited for modeling some rolling element bearings, in particular preloaded angular contact bearings [3], but has a more general application. Equation (6) has been widely used in non-linear dynamics, starting on the work of Duffing [2].

The global mass matrix and the damping matrix are the same as in the linearized model of the bearings, but the global stiffness matrix contains non-linear terms $k + \hat{k}u^2$ in nodal displacements, due to the stiffness matrix of the bearings

$$[k]^{nl} = \begin{bmatrix} \left[k_{yy}^{nl}\right] & | & [0] \\ --- & -- & --- \\ [0] & | & \left[k_{zz}^{nl}\right] \end{bmatrix} \tag{7}$$

$$\left[k_{yy}^{nl}\right] = \begin{bmatrix} k_{yy} + \hat{k}v^2 & 0 \\ 0 & 0 \end{bmatrix} = \begin{bmatrix} k_{yy} & 0 \\ 0 & 0 \end{bmatrix} + \begin{bmatrix} \hat{k}v^2 & 0 \\ 0 & 0 \end{bmatrix} \tag{8}$$

$$\left[k_{zz}^{nl}\right] = \begin{bmatrix} k_{zz} + \hat{k}w^2 & 0 \\ 0 & 0 \end{bmatrix} = \begin{bmatrix} k_{zz} & 0 \\ 0 & 0 \end{bmatrix} + \begin{bmatrix} \hat{k}w^2 & 0 \\ 0 & 0 \end{bmatrix}$$

Thus, the global stiffness matrix can be written as the sum of two matrices by isolating the non-linear part

$$[K(\{x\})]^{nl} = [K] + [\hat{K}(\{x\})]^{nl} \tag{9}$$

where the first matrix is the stiffness matrix of the structure which refers to the shaft and to the constant terms of the bearing stiffness and the second matrix appears due to the non-linearity of the bearings.

The equations of motions of anisotropic rotor-bearing systems which consist of a flexible non-uniform axisymmetric shaft, rigid disk and anisotropic bearings are obtained in second order form, by assembling the element matrices and may be written as

$$[M]\{\ddot{x}\} + [C]\{\dot{x}\} + [K(\{x\})]^{nl}\{x\} = \{F\} \tag{10}$$

3. NUMERICAL SOLUTION. THE WILSON – Θ METHOD

The numerical time response solution for the non-linear system is calculated using the Wilson-θ method in conjunction with an iteration procedure. From the Wilson-θ method the resulting equations for acceleration and velocity vector can be expressed as

$$\{\ddot{x}\}_{t+\theta\Delta t} = \frac{6}{\theta^2\Delta t^2}\left(\{x\}_{t+\theta\Delta t} - \{x\}_t\right) - \frac{6}{\theta\Delta t}\{\dot{x}\}_t - 2\{\ddot{x}\}_t \tag{11}$$

$$\{\dot{x}\}_{t+\theta\Delta t} = \frac{3}{\theta\Delta t}\left(\{x\}_{t+\theta\Delta t} - \{x\}_t\right) - 2\{\dot{x}\}_t - \frac{\theta\Delta t}{2}\{\ddot{x}\}_t \tag{12}$$

From the equation of motion (10) written at the time $t + \theta\Delta t$ by using the equations (11) and (12) we obtain the non-linear algebraic system

$$\left[\tilde{K}\right]^{nl}\{x\}_{t+\theta\Delta t} = \{\tilde{F}\} \tag{13}$$

with $\{x\}_{t+\theta\Delta t}$ as the unknown. In the equation (13) we used the following notations

$$\left[\tilde{K}\right] = [K] + \frac{6}{(\theta\Delta t)^2}[M] + \frac{3}{\theta\Delta t}[C] \tag{14}$$

$$\{\tilde{F}\} = \{F\}_t + \theta\left(\{F\}_{t+\Delta t} - \{F\}_t\right) + [M]\left(\frac{6}{(\theta\Delta t)^2}\{x\}_t + \frac{6}{\theta\Delta t}\{\dot{x}\}_t + 2\{\ddot{x}\}_t\right) +$$
$$+ [C]\left(\frac{3}{\theta\Delta t}\{x\}_t + 2\{\dot{x}\}_t + \frac{\theta\Delta t}{2}\{\ddot{x}\}_t\right) \tag{15}$$

In the non-linear rotors case, the stiffness matrix $\left[\tilde{K}\right]$ in Eq. (13) has non-linear terms, which depend on the values of the elements in vector $\{x\}_{t+\theta\Delta t}$. Equation (13) is a set of non-linear algebraic equations now. Therefore an iteration procedure is utilized in conjunction with the Wilson-θ method to find $\{x\}_{t+\theta\Delta t}$ and then the displacements $\{x\}_{t+\Delta t}$. The following steps describe the numerical procedure:

1. At $t = 0$ specify initial conditions $\{x\}_0, \{\dot{x}\}_0$.

2. From the Eq. (13), using the known initial conditions from step 1, $\{\ddot{x}\}_0$ is calculated as

$$\{\ddot{x}\}_0 = [M]^{-1}\left(\{F\}_{t=0} - [C]\{\dot{x}\}_0 - [K]^{nl}\{x\}_0\right). \tag{16}$$

3. $i = i+1$

4. For the time step i at the moment of time t, assume a displacement vector $\{x\}_t \equiv \{x\}_i = \{x\}_i^*$.

5. Calculate the non-linear terms of the stiffness matrix $\left[\tilde{K}\right]$ using values from the assumed displacement vector.

6. Calculate displacement vector $\{x\}_{t+\Delta t} \equiv \{x\}_{i+1}$ using

$$\{x\}_{t+\Delta t} = \{x\}_t + \Delta t\{\dot{x}\}_t + \frac{\Delta t^2}{6}\left(\{\ddot{x}\}_{t+\Delta t} + 2\{\ddot{x}\}_t\right) \tag{17}$$

and the stiffness matrix from Step (5).

7. The vector $\{x\}_i^* := \left(\{x\}_i^* + \{x\}_{i+1}\right)/2$ is modified.

8. Compare $\{x\}_{i+1}$ with assumed displacement vector

$$\sqrt{\left(\{x\}_{i+1} - \{x\}_i^*\right)^T\left(\{x\}_{i+1} - \{x\}_i^*\right)} < tol \tag{18}$$

9. If the difference is not within specified tolerances (tol) use an average value of the assumed displacement vector from Step (7) and the calculated displacement vector from Step (6) for the new assumed vector and return to Step (4). If the difference is within tolerance then update the assumed vector for t_{i+1} and go to Step (3). Continue for $i = N$ iterations to obtain the steady state solution.

Numerical example.
The model comprises of a continuous elastic shaft mounted on tree nonlinear bearing, Figure 1. One disk is mounted on the shaft. Timoshenko beam model is adopted and gyroscopic effect is taken into account [4].

Table 1: Rotor data – numerical example

Shaft :	Disk :	Bearings:
$L = 1.2$ m	$M = 75$ Kg	$k_{yy} = 5 \times 10^8$ N/m
$a = 0.3$ m	$J_T = .190$ Kg m^2	$k_{zz} = 3 \times 10^8$ N/m
$d_1 = d_2 = d_3 = 0 .08$ m	$J_P = 0.368$ Kg m^2	$k_{yz} = k_{zy} = 0$
$E = 2.068e11$ N/m^2	$e = 0.01$ m	$c_{yy} = c_{zz} = 1 \times 10^4$ Ns/m
$\rho = 7.833$ Kg/m^3		$c_{zy} = c_{yz} = 0$

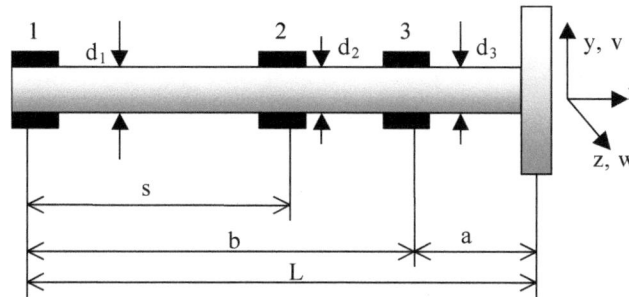

Figure 1: Rotor configuration

where: J_T, J_P are transverse, respectively, polar mass moment of inertia; k_{ij}, c_{ij} (i, j = y, z) are stiffness and damping coefficients; e is the eccentricity of the disk. In this numerical example in the equations (7) and (8) we deal with the values: $\hat{k} = 10^{14}\ N / m^3$, $s = 0.25$ m, *nrot* = 60, *nstep* = 4.096; N/m.

In an implicit schema the difference equations are combined with the equation of motion and the displacement are calculated directly by solving the equations. The graph response, vertical displacements of the flywheel node is shown in Figure 2.

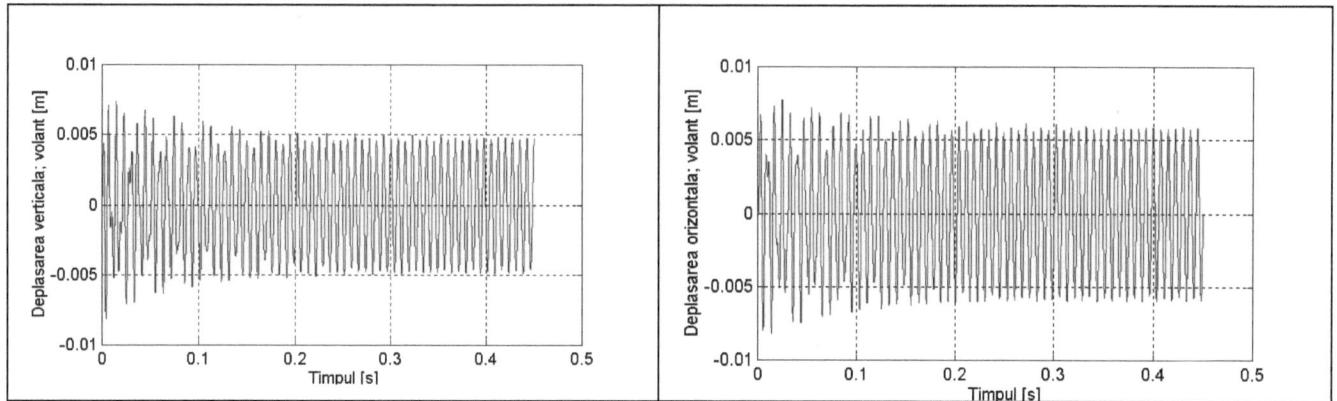

Figure 2: Vertical movement of the flywheel node

4. CONCLUSION

There is a basic approach to numerically evaluate the dynamic response of non-linear rotors. The implicit scheme Wilson-Theta method in conjunction with an iteration procedure is useful to a multipledegrees-of-freedom (MDOF) non-linear system with non-proportional damping.

REFERENCES

[1] Bathe, K-J., Wilson E.L., Numerical Methods in Finite Analysis, Englewood Cliffs, New Jersey: Prentice-Hall, Inc,1976.

[2] Duffing, G., Erzwungene Schwingungen bei Veränderlicher Eigenfrequenz., F. Vieweg u. Sohn, Braunschweig, 1918.

[3] Genta, G., Vibration of Structures and Machines: Practical Aspects, Springer-Verlag, New-York Berlin Heidelberg, 1993.

[4] Hughes, J.R.T., The Finite Element Method, Prentice –Hall Inc., 1987.

[5] Krämer, E., Dynamics of Rotors and Foundations, Springer Verlag, Berlin, 1993.

[6] D.D. Nicoara, D.D., Munteanu, M. Gh., Contributions on the Optimization of Rotor –Bearing Systems. Anspruch und Tendenzen in der experimentellen Strukturmechanik, VDI-Gesellschaft Mess-und Automatisierungstechnik, 369-375, VDI Verlag., Düsseldorf, 1999.

6th International Conference
"Computational Mechanics and Virtual Engineering "
COMEC 2015
15-16 October 2015, Braşov, Romania

TIME-FREQUENCY ANALYSIS OF VRANCEA FAULT EARTHQUAKES

Luciana Majercsik[1], Ion Simulescu[1]

[1] Technical University of Civil Engineering, Bucharest, ROMANIA, luciana.majercsik@gmail.com

Abstract: *As the classical time or frequency Fourier analysis alone are insufficient to provide comprehensive information about the seismic signals who are known to be nonstationary, of ondulatory type and with time-frequency dependent energy power distribution, the more complex bi-dimensional time-frequency analysis methods are necessary to be employed. The studies of the seismic signal using time-frequency analysis are scarce. The authors' main goal is to evaluate the effectiveness of those methods relatively to Vrancea Fault generated earthquakes. The paper, for editorial space reasons, provides just a very brief description of the theoretical and numerical results of the time-frequency analysis conducted on 13 recorded strong ground motions generated by the Vrancea Fault.*
Keywords: Time-frequency analysis, seismic signals, energy density, S-Transform, Zhang-Sato distribution

1. INTRODUCTION

Let's consider a seismic signal $x(t)$, defined as a continuous real functions of time $t \in [0,T]$, extended for mathematical reasons to $t \in \mathrm{R}$. The classical *time* or *frequency analysis,* employing the dual set of complex Fourier transforms, represented by the *Fourier transform (FT)*

$$X(\omega) = \frac{1}{\sqrt{2\pi}} \int_{-\infty}^{\infty} x(t)e^{-i\omega t} dt \qquad (1)$$

and the *inverse Fourier transform (IFT)*

$$x(t) = \frac{1}{\sqrt{2\pi}} \int_{-\infty}^{\infty} X(\omega)e^{i\omega t} d\omega \qquad (2)$$

alone, are insufficient to provide comprehensive information about this kind of *nonstationary signals,* of *ondulatory type* and with *time-frequency dependent energy power distribution*. In the above formulae, the variable $\omega \in \mathrm{R}$ represents physically the frequency. Must be mentioned that the complex Fourier transforms pair, extensively used by engineers in their analysis of the seismic signal, can still provide reliable information about the frequency range of the signal, (1), and can be used to generate ondulatory functions, (2), but is unable to offer any information about its time-dependent power spectrum. If information about the changes of the frequency content over time is required, the more sophisticated *bi-dimensional time-frequency analysis methods (TFAM)* are absolutely necessary to be employed.

The basic idea of time-frequency analysis is to devise a joint function of time t and frequency ω, named $P(t,\omega)$, that describes *the signal energy density simultaneously in time and frequency*, playing a similar role to that of probability density used in the probability theory. The description of the energy distribution of a signal in the time-frequency plane is obstructed by the lack of a basis of signals perfectly confined to a point of the time frequency plane, with respect to whom any signal could be represented as a superposition integral [6]. In the circumstances of missing such a basis, several time-frequency distributions have been proposed, in an attempt to have the least amount of energy spread in the time-frequency plane. Each one of these distributions has its advantages and disadvantages, [2], [3]. Therefore a considerable difference between a time-frequency energy distribution *(TFD)* of the signals and the probability densities is due to the fact that for a given signal the *(TFD) is not unique.*

Employing the modern, more complex, *bi-dimensional time-frequency analysis methods* require a more sophisticated mathematical formulation. The pure mathematical characterization of the seismic signal $x(t)$ provided by any *TFAM* must be verified to correctly reflect the physical properties of the phenomenon under scrutiny. Consequently, the mathematical formulae are additionally constrained by a number of relations generated by the physical characterization of the seismic signal. For the particular case of the seismic signal, the *energy distribution over the time-frequency plane* is the most important physical characteristic with effective repercussion in the dynamic behavior of civil and industrial structures.

Under the circumstances of missing unicity, the general approach consists in a two steps procedure. In step (1) the bi-dimensional energy distribution $P(t, \omega)$ is obtained either by: *(a)* atomic decomposition methods or *(b)* Cohen class representations. The first type methods employ a complex transformation of the signal $x(t)$ to express the $P(t, \omega)$, while the second type methods are proposing an analytical expression for $P(t, \omega)$. The physical truthfulness of the previously obtained $P(t, \omega)$, by either method, is evaluated in step (2) by imposing a *set of mathematical and physical constraints* relevant for the class of the seismic signals. The most important constraints ensuring that a time-frequency distribution $P(t, \omega)$ can be interpreted as a joint energy density, are:

    a.   Non-negativity:

$$P(t, \omega) \geq 0 \tag{3}$$

This property allows also the interpretation of $P(t, \omega)$ as a true probability density function (if properly normalized).

    b.   Correctness of the energy marginal densities in time $|x(t)|^2$ and frequency $|X(\omega)|^2$, respectively:

    1.
$$P_t(t) = \int_{-\infty}^{\infty} P(t, \omega) d\omega = |x(t)|^2 \tag{4}$$

    2.
$$P_\omega(\omega) = \int_{-\infty}^{\infty} P(t, \omega) dt = |X(\omega)|^2 \tag{5}$$

    c.   Conservation of the total energy of the signal $x(t)$:

$$E_{total} = \int_{-\infty}^{\infty} \int_{-\infty}^{\infty} P(t, \omega) dt d\omega < \infty \tag{6}$$

Flandrin [5] and Cohen [2] had described several additional constraints for the distribution $P(t, \omega)$ evaluation, but those do not have such an important physical significance for the case of the seismic signal.

Remark 1: Relation *(a)* is in general violated by the classical Cohen class distributions $P(t, \omega)$ and, for practical applications, it is replaced by a more relaxed constrained. It is imposed that the distribution must be at least a real function.
Remark 2: Relation *(c)* indicates the existence of a limited total energy of the seismic signal.
Remark 3: It must also be mentioned that the energy is conserved if *(b1)* and *(b2)* are satisfied [1].

Theoretically, according to Rao [9], the one-dimensional marginal densities, in this case either the *time energy density* $|x(t)|^2$, $t \in \mathbb{R}$ or the *frequency energy density* $|X(\omega)|^2$, $\omega \in \mathbb{R}$, are uniquely defined by the corresponding moments up to order *n*. For practical applications, only the first four moments of the energy marginal densities or combination of them are used. The first moment of time energy density, the *mean time*:

$$t_m = \int_{-\infty}^{\infty} t \cdot |x(t)|^2 \cdot dt \tag{7}$$

indicates the time value where the time energy density is concentrated. The concentration of the density around the mean time is measured by the standard deviation defined as:

$$\sigma_t = \left[\int_{-\infty}^{\infty} (t - t_m)^2 \cdot |x(t)|^2 \cdot dt \right]^{1/2} \tag{8}$$

Similarly, the first moment of the frequency energy density, the *mean frequency:*

$$\omega_m = \int_{-\infty}^{\infty} \omega \cdot |X(\omega)|^2 \cdot d\omega \tag{9}$$

indicates where the frequency value energy density is concentrated. The spread of the spectral density around the mean frequency, known as *bandwidth*, is measured by the standard deviation defined as:

$$\sigma_\omega = \left[\int_{-\infty}^{\infty} (\omega - \omega_m)^2 \cdot |X(\omega)|^2 \cdot d\omega \right]^{1/2} \tag{10}$$

The third and fourth moments are used to calculate the *skewness* and *kurtosis coefficients* The skewness coefficient indicates the degree of asymmetry of the energy distribution around its mean, while the kurtosis coefficient measures the relative flatness of a distribution relative to a normal distribution. In time, these moment are defined as:

$$skew_t = \frac{\int_{-\infty}^{\infty} (t - t_m)^3 \cdot |x(t)|^2 \cdot dt}{\sigma_t^3}; \quad kurt_t = \frac{\int_{-\infty}^{\infty} (t - t_m)^4 \cdot |x(t)|^2 \cdot dt}{\sigma_t^4} \tag{11}$$

and, in a similar manner, in frequency:

$$skew_\omega = \frac{\int_{-\infty}^{\infty} (\omega - \omega_m)^3 \cdot |X(\omega)|^2 \, d\omega}{\sigma_\omega^3}; \quad kurt_\omega = \frac{\int_{-\infty}^{\infty} (\omega - \omega_m)^4 \cdot |X(\omega)|^2 \, d\omega}{\sigma_\omega^4} \tag{12}$$

Remark 4: It must be emphasized that when dealing with seismic signals, the four moments described above, which are obtained directly from the original signal or its frequency energy density, can be compared with the similar results obtained from the time-frequency approach.

2. TIME-FREQUENCY DISTRIBUTIONS

For better understanding of the bi-dimensional energy distributions obtained, the investigation is conducted using two types of time-frequency analyses: *(a)* atomic decompositions (*Short Time Fourier Transform* and the *S-transform*) and *(b)* the classic Cohen class representations (*Wigner-Ville, Zhang-Sato* and *Choi-Williams* distributions).

2.1. Atomic Decomposition Methods

The first approach, known as the *Atomic Decomposition Methods*, is to write the signal as the superposition of time-frequency functions (atoms) derived from translating, modulating and scaling a basis function, having a definite time and frequency localization. For a real signal $x(t)$, this kind of time-frequency representation is given by:

$$x(t) = \int_{-\infty}^{\infty} \int_{-\infty}^{\infty} TF(\tau, \omega) \cdot g_{\tau, \omega}(t) \cdot d\tau d\omega \tag{13}$$

where

$$TF(t, \omega) = \int_{-\infty}^{\infty} x(\tau) \cdot \overline{g_{t, \omega}(\tau)} \cdot d\tau \tag{14}$$

and $g_{t, \omega}$ are the time-frequency atoms, assumed to have finite energy. A normal transition between those atomic decompositions, which are linear transformations and their corresponding time-frequency energy distributions, is made by taking $P(t,\omega) = |TF(t,\omega)|^2$. The two methods of atomic decomposition used in this paper are the classical *Short Time Fourier Transform (STFT)* which leads to the distribution known as the *spectrogram (SP)* and the more recent *S-Transform (ST)* [10] and its corresponding distribution known as *ST-spectrogram (ST-SP)*. The form of the time-frequency atoms used in both methods is presented in Table 1.

Table 1: Form of time-frequency atoms

Decomposition Method	STFT	S-Transform
$g_{t,\omega}(\tau)$	$g(\tau-t)e^{-i\tau\omega}$	$g(\tau-t;\omega)e^{-i\tau\omega}$

It must be mentioned that while for the STFT, the function $g(t)$ can be any complex window with small time support, for the S-Transform, it is a Gaussian window $g(t;\omega) = \dfrac{1}{\sigma(\omega)\sqrt{2\pi}}\exp\left(-\dfrac{t^2}{2\sigma(\omega)^2}\right)$ with $\sigma(\omega) = \dfrac{2\pi}{|\omega|}$ a scale factor, meant to change the width of the window.

2.2. Cohen Class Representations

In 1966 Cohen, [4], provided and demonstrated the existence of a general formula to create a time-frequency distribution with desirable physical properties:

$$P(t,\omega) = \frac{1}{4\pi^2}\int_{-\infty}^{\infty}\int_{-\infty}^{\infty}\int_{-\infty}^{\infty}\overline{x\left(u-\frac{\tau}{2}\right)}\cdot x\left(u+\frac{\tau}{2}\right)\cdot\varphi(\theta,\tau;x)\cdot e^{-i\theta t-i\tau\omega+i\theta u}\cdot du\,d\theta\,d\tau \qquad (15)$$

where $\varphi(\theta,\tau;x)$ is the *kernel of the distribution,* a function playing *a determinant role in the development of the distribution* properties. If the kernel is independent of the signal, $\varphi(\theta,\tau)$ then the distribution is called *bilinear distribution*. The problem of those distributions (15) is that, in general, the constraint (a) is violated, [2], and this can cause problems with their physical interpretation. Therefore, it is agreed that if the constraint (a) is not satisfied, at least they should be real valued.

In the present paper only three classic distributions from the Cohen's class (the *Wigner-Ville, Choi-Williams* and *Zhang-Sato* distributions) are selected and presented, based on their popularity in the technical literature and generality. From absence of publishing space, herein, for these three distributions the kernels are presented in *Table 2*.

Table 2: Kernels of the selected time-frequency distributions

Distribution	Wigner-Ville	Choi-Williams	Zhang-Sato
Kernel $\varphi(\theta,\tau)$	1	$\exp\left(-\dfrac{\theta^2\tau^2}{\alpha}\right)$	$\exp\left(-\dfrac{\theta^2\tau^2}{\alpha}\right)\cos(2\pi\beta\tau)$

The verification of the general constraints *(a) - (c)* for all five time-frequency distributions discussed above are summarized in Table 3, see [2], [7], [8], [11].

Table 3: Properties of the time-frequency distributions

Property	Constraint	WVD	CWD	ZSD	SP	ST-SP
Non-negativity	(a)	-	-	-	√	√
Time Marginal	(b1)	√	√	√	-	-
Frequency Marginal	(b2)	√	√	-	-	-
Energy	(c)	√	√	√	√	-
Reality		√	√	√	√	√

3. TIME-FREQUENCY ANALYSIS OF VRANCEA FAULT EARTHQUAKE RECORDS

All time-frequency methods mentioned before were used in the analyses of the 13 recorded strong ground motions, generated by the Vrancea Fault and reported on an extended study [12]. Herein, *for editorial space reasons only a very brief description of the theoretical and numerical results of this study is presented*. Precisely, only the *TDF* analyses of NS component of Incerc 1977 earthquake accelerogram is offered. The numerical computation was carried employing the Matlab code [13] and a number of modified functions from [14].

In *Figure 1a* the NS component of the accelerogram registered at Incerc in 1977, used as numerical investigation base, is shown. The 3D representation of the corresponding Spectrogram is pictured in *Figure 1b*. The five *TFDs* of NS horizontal component are shown in *Figures 2a* through *2e*. On the left side of each bi-dimensional representation the frequency marginal superimposed with the normalized cumulative frequency energy are plotted. Below , the normalized cumulative energy variation in time, together with the time marginal are shown.

The numerical global characteristics of those distributions considered are summarized in Table 4. From the frequency domain representations, the right subplot of *Figures 2a – 2e*, the energy concentration is evident at 0.42, 0.62, 0.73, 0.81 and 1.75 Hz. Of these frequency components, the dominant is located at 0.62 Hz. Beyond 3 Hz the spectral amplitudes become insignificant. For the same component, the time domain representation indicates that the maximum amplitude is located at around 15 s.

Table 4: Numerical characteristics of the time-frequency distributions for NS component of Incerc 1977 earthquake accelerogram

Time-frequency distribution	E_{total}	t_m	σ_t	$skew_t$	$kurt_t$	ω_m	σ_ω	$skew_\omega$	$kurt_\omega$
SP	1096.8	17.7317	5.6767	2.5507	11.3080	1.0608	2.9907	7.0734	65.1241
ST - SP	2029	19.0379	6.5305	1.6134	6.8097	4.4990	5.4043	1.6335	4.7482
WVD	1098.0	17.7575	5.4403	3.0756	13.8424	1.3013	2.4518	6.9548	59.9822
CWD	1098.0	17.7575	5.4403	3.0756	13.8424	1.3013	2.4518	6.9548	59.9822
ZSD	1186.0	17.65	5.3796	3.2559	14.94	2.4552	3.0245	2.0139	7.482

As it can be seen from Table 4, the numerical results obtained by our analysis are similar for all distributions, in accordance with their theoretical characterization. A notable exception is the ST-spectrogram, for whom the numerical characteristics are quite different from the others. This can restricts its use for the characterization of the energy distribution.

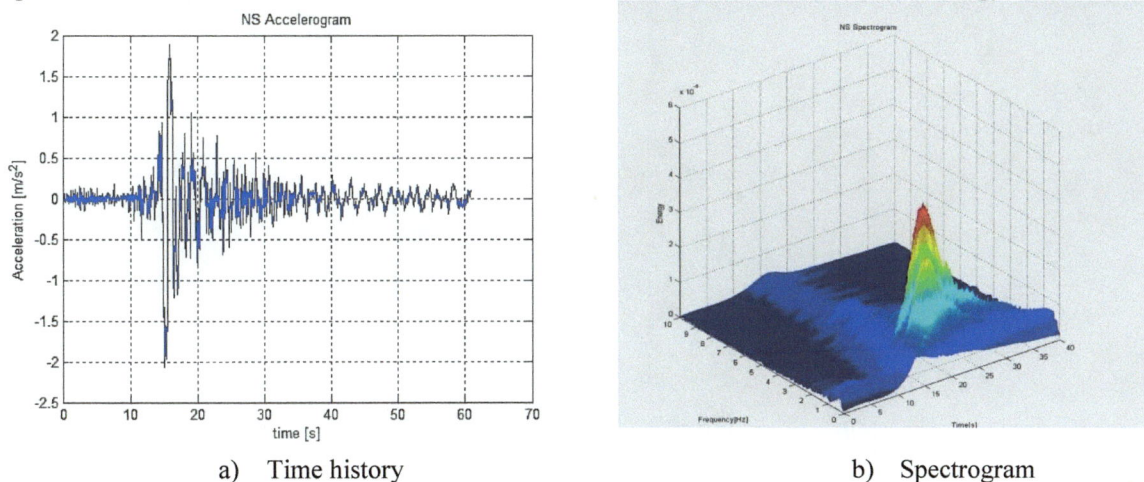

a)   Time history   b)   Spectrogram

Figure 1: NS accelerogram at INCERC 1977

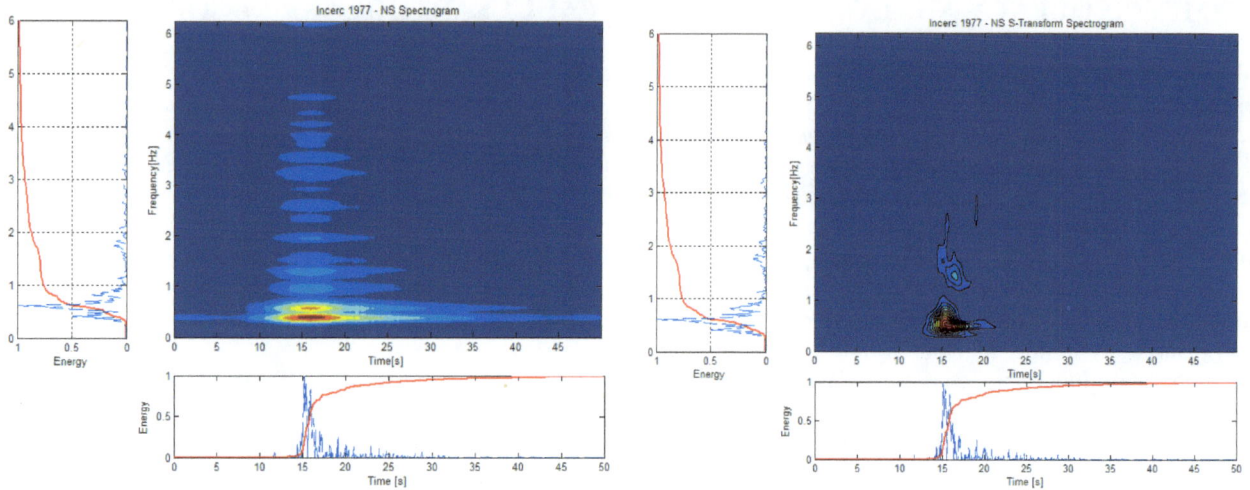

a) Spectrogram

b) ST - Spectrogram

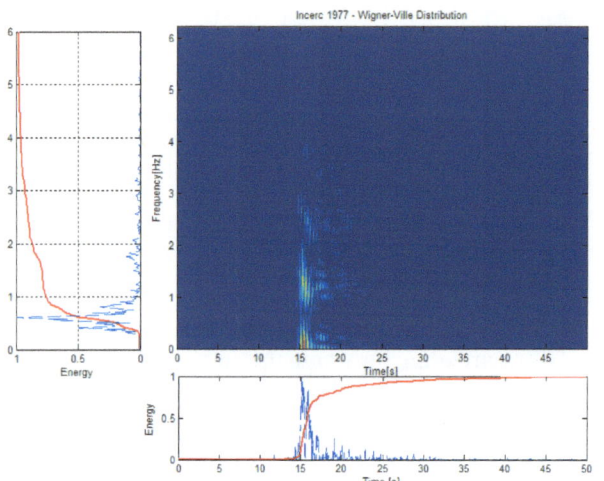

c) Wigner – Ville Distribution

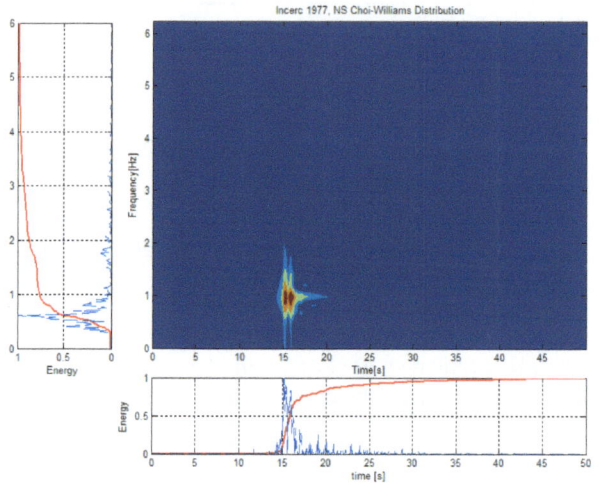

d) Choi – Williams Distribution

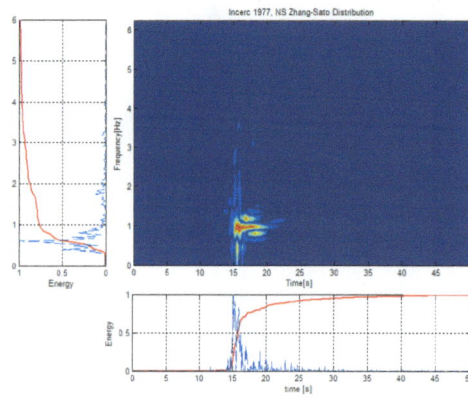

e) Zhang – Sato Distribution

Figure 2: Time-frequency energy densities for NS component of Incerc 1977 earthquake

4. CONCLUSION

A number of general conclusions can be drawn from the above numerical results presented:
- For the global characterization of the frequency content and energy distribution, the *spectrogram* is a well suited method. It emphasizes the dominant frequencies contained in the seismic signals, but the time resolution of these frequencies it not that fine.
- The time resolution of the *ST-spectrogram* is better for low frequencies, but for higher ones is limited, as we could saw from the study of other earthquake records [12]. Also, the fact that this distribution doesn't preserve the energy and the marginals can restricts its use in the description of time-frequency energy density.
- Even though the *WVD* has significant better time-frequency resolution, the information that this *TFD* carries is blurred by the existence of cross-terms. This makes the realistic interpretation of the time-frequency characteristics of the earthquake signals difficult. Therefore *WVD* is not suited for the characterization of the seismic signal energy density.
- The *CWD* suppresses much of the cross-term interference. However some of the details present in the *WVD* are lost.
- The *ZSD* facilitates a better identification of time-frequency components of the seismic signal then the other *TFDs*, even though it can suffer from small spurious concentrations of energy.

Further research is necessary and a number of adaptive and optimized methods for *TFDs* will be considered. Also, the physical interpretation of the others *TFDs* characteristics will be further studied, in order to have a characterization as complete as possible of the seismic accelerograms.

REFERENCES

[1] Cohen, L., *Time - Frequency Distributions - A review*, Proc. of the IEEE, vol. 77(7), 941-981, 1989.
[2] Cohen, L., *Time - Frequency Analysis*, Prentice Hall, Signal Processing Seris, New Jersey, 1995.
[3] Cohen, L., *The Uncertainty Principle For The Short - Time Fourier Transform*, Proc. Int. Soc. Opt. Eng., 22563, 80-90, 1995.
[4] Cohen, L., *Generalized Phase-Space Distribution Functions*, J. Math. Phys., vol. 7, no. 5, 781-786, 1966.
[5] Flandrin, P., *Temps-Frequence,* Hermes, Paris, 1993. Trait des NouvellesTechnologies, serie Traitement du Signal
[6] Janssen, A.J.A., *Positivity and Spread of Bilinear Time - Frequency Distributions,* Amsterdam, The Wigner Distribution, Elsevier, 1-58, 1997.
[7] Moukadem, A., Schmidt, S., and Dieterlen, A., *High Order Statistics and Time-Frequency Domain to Classify Heart Sounds for Subjects under Cardiac Stress Test*, Computational and Mathematical Methods in Medicine, vol. 2015, Article ID 157825, 15 pages, 2015.
[8] Moukadem, A., Ould Abdeslam, D., Dieterlen, A., *Time-Frequency Domain for Segmentation and Classification of Non-stationary Signals: The Stockwell Transform Applied on Bio-signals and Electric Signals,* John Wiley & Sons, FOCUS: Digital Signal and Image Processing Series, 2014.
[9] Rao, C.R., *Linear Statistical Inference and its Applications,* John Wiley & Sons, 2009
[10] Stockwell, R. G., Mansinha, L. and Lowe, R. P., *Localization of the complex spectrum: The S- transform*, IEEE Trans. on Signal Processing, vol. 44, no. 4, pp. 998-1001, 1996.
[11] Ventosa, S., Simon, C., Schimmel, M., Dañobeitia, J. J. and Mànuel, A., *The S-Transform From a Wavelet Point of View*, IEEE Transactions on Signal Processing, vol. 56, no. 7, 2771-2780, 2008.
[12] Majercsik, L., *Research report: Time –frequency analysis of seismic signals*, University of Pitesti, 2015
[13] The Mathworks, Inc. (2012), Matlab [computer software], Natick, MA, USA.
[14] Time-frequency Toolbox, http://tftb.nongnu.org/

6th International Conference
"Computational Mechanics and Virtual Engineering "
COMEC 2015
15-16 October 2015, Brașov, Romania

APPLICATION OF NON-EQUILIBRIUM THERMODYNAMICS TO THE ESTIMATION OF PHENOMENOLOGICAL TRANSFER COEFFICIENTS IN CONVECTIVE WOOD DRYING

Daniela Șova [1], Olivia A. Florea [2]

[1] Transilvania University, Faculty of Mechanical Engineering, Brașov, ROMANIA, sova.d@unitbv.ro

[2] Transilvania University, Faculty of Mathematics and Computer Science, Brașov, ROMANIA, olivia.florea@unitbv.ro

Abstract: Convective drying is an important process within the wood industry since it improves the physical, mechanical and technological properties of wood. Starting from the critical moisture content, the evaporation surface gradually becomes deeper inside the wood board and thus the intensity of evaporation is different as compared to that one that occurs on the wood surface. The present study aimed to determine the heat and mass fluxes occurring during the falling-rate drying period in the boundary layer, based on the non-equilibrium thermodynamics, according to which the forms of molecular transfer (momentum, energy and mass) are coupled. The variation of the fluxes along the wood board was investigated and the influence of each flux component on the heat and mass fluxes was assessed. The phenomenological coefficients, as variables in the fluxes equations were also calculated.
Keywords: non-equilibrium thermodynamics, convective wood drying, phenomenological coefficients

1. INTRODUCTION

A method for analyzing the process of convective wood drying process, where the coupled heat and mass transfer is influenced by the moisture evaporation that occurs between the surface of the wet wood material and the surrounding air is the non-equilibrium thermodynamics. It consists of phenomenological theories regarding simultaneous molecular transport of heat, mass and momentum in fluids and it is based on the laws of mass and energy conservation and Onsager's theorem of entropy increase [1], [2], [3]. According to Prigogine [4], the corresponding fluxes are linear relations of all thermodynamic forces involved in the process and the mutual and self phenomenological coefficients. There were a few attempts to determine the phenomenological coefficients specific to different irreversible heat and mass transfer processes. In his paper [5], Verros has determined the self diffusion and mutual diffusion coefficients by using the principles of non-equilibrium thermodynamics applied to diffusion (solvent evaporation from polymer solutions). Lewis and Malan [6] expressed liquid, vapor and energy fluxes based on phenomenological transfer coefficients and thermodynamic forces for the coupled heat and mass transfer processes in drying of non-hygroscopic and hygroscopic capillary particulate materials. They analyzed the mass transfer (molecular diffusion, thermal diffusion and capillary flow) and the heat transfer that occurred due to mass transfer, phase change and thermal conductivity. These transfer mechanisms are strongly coupled and non-linear functions of the thermodynamic state (temperature and moisture content). They cited Perrin and Javelas [7] and Fortes and Okos [8], according to which the phenomenological coefficients are typically highly non-linear functions of the thermodynamic state. They must be determined for each specific material by means of experimental and regression techniques. Stenholm [9] mentioned about the linear coefficient matrix from Onsager's formulation of irreversible thermodynamics that they can be measured experimentally or calculated by expanding around the thermal states. He named the fluxes "currents" and defined the forces as positive or negative deviations from equilibrium. The eigenvalues of the coefficient matrix are expressed in the paper. Based on the theory of non-equilibrium thermodynamics, transport equations for coupled mass and heat transport were derived by Kuhn et al. [10] for water pervaporation through a zeolite membrane. To model the paper drying process, Zvolinschi et al. [11] have used balance equations and common flux equations for heat and mass transfer in the paper drying process and comparative, flux equations based on the theory of non-equilibrium thermodynamics. They mentioned in the paper that the transfer coefficients for the interface were evaluated by Bedeaux and Kjelstrup [12] in the absence of a parallel flow at the evaporation surface. The latter ones used non-equilibrium thermodynamics to obtain the appropriate boundary conditions at the evaporating surface of different liquids. Interfacial transfer coefficients, appearing in these boundary conditions, were determined from the experiments. They all were found as positive values.

Another method presented in previous works uses the theory of transport phenomena in porous materials developed by Luikov in order to analyze the conjugate problem of heat and moisture transport during thermal processes. In this regard, Kadem et al. [13] solved the three-dimensional Navier–Stokes equations along with the energy and concentration equations for the fluid coupled with the energy and mass conservation equations for the solid (wood) to study the transient heat and mass transfer during high thermal treatment of wood. A three-dimensional, unsteady-state model solving simultaneous heat and mass transfer in porous media was developed by Kocaefe et al. [14] The model took into account the variation of properties with temperature and/or moisture content and used Luikov's formulation. It was applied to the thermal treatment of wood at high temperature. Coles and Murio [15] introduced a stable numerical method for the determination of the temperature and moisture distribution in a Luikov system with space dependent diffusion coefficients during thermal drying in a porous medium. The mathematical model based on the set of coupled heat and mass transfer equations proposed by Luikov was applied to wood drying by several authors, for instance Irudayaraj et al. [16] and Younsi et al. [17], [18].

The study presented in this paper was focused on applying Onsager's coupled heat and mass transfer equations to convective wood drying. Heat and mass fluxes, thermodynamic forces and phenomenological coefficients were evaluated considering the retraction of the evaporation surface in wood.

2. MATERIALS AND METHODS

The convective heat and mass transfer process along a horizontal flat plate of moist wood located in a clear fluid, in this case heated air as drying agent, was considered in this study. The temperature and vapor concentration of the ambient medium are denoted T_∞ and $C_{1\infty}$, respectively. The flowing fluid has an incoming velocity u_∞. The flow regime is laminar, but this is monitored during the numerical runs, by computing the Reynolds number at each location of x. The x-coordinate is measured along the plate surface and the y-coordinate normal to it.

From the drying theory it is known that starting from the critical moisture content (which is about 30%) the evaporation surface gradually becomes deeper inside the wood [1]. Therefore, the evaporation occurs at a depth ξ below the wood surface (Fig. 1). This is also mentioned by Salin [19] who defines dry shell the wood thickness between the evaporation front and the wood surface, where there is no free water. Within the dry shell there is a moisture gradient which is the driving force for moisture migration from the evaporation front and a temperature gradient which is the driving force for the heat conduction to the evaporation front. Thus, an additional thermal resistance is developed in wood.

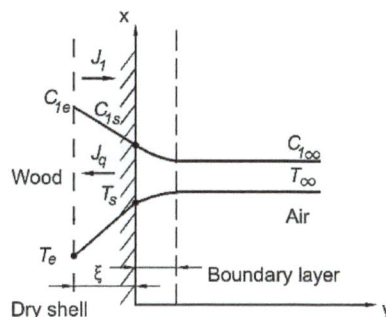

Figure 1: Temperature and moisture profiles across wood-air interface

Transfer of energy and mass is accompanied by an increase of entropy in a macroscopic system. According to the basic relation of irreversible processes thermodynamics, the entropy source is proportional to the sum of the products of fluxes and thermodynamic forces. Thus, the entropy production per unit time and unit fluid volume [20] can be written as

$$\sigma = \sum_i J_i X_i \tag{1}$$

where J_i are fluxes and X_i are thermodynamic forces.

Based on the phenomenological theories of irreversible processes, the irreversible fluxes are linear functions of the thermodynamic forces [20]

$$J_i = \sum_k L_{ik} X_k \tag{2}$$

The quantities L_{ik} are called phenomenological coefficients. Therefore, for a fluid volume unit located in the boundary layer, where molecular heat and mass transfer occurs, considering [3] and [20], the coupled heat and mass fluxes per unit of area can be written as follows:

$$J_q = -L_{TT}\frac{DT}{T^2} + L_{T1}\left[\frac{DT}{T^2}(r_0 + c_{p1}T - c_{p2}T) - \frac{R_1}{c_1 c_2}DC_1\right] - \frac{L_{TE_1}}{T}D\left(\frac{\rho_1 u^2}{2}\right) - \frac{L_{TE_2}}{T}D\left(\frac{\rho_2 u^2}{2}\right) \tag{3}$$

$$J_1 = L_{11}\left[\frac{DT}{T^2}(r_0 + c_{p1}T - c_{p2}T) - \frac{R_1}{c_1 c_2}DC_1\right] - L_{1T}\frac{DT}{T^2} - \frac{L_{1E_1}}{T}D\left(\frac{\rho_1 u^2}{2}\right) - \frac{L_{1E_2}}{T}D\left(\frac{\rho_2 u^2}{2}\right) \tag{4}$$

where L_{TT}, L_{T1}, L_{TE_1}, L_{TE_2}, L_{11}, L_{1T}, L_{1E_1}, L_{1E_2} are phenomenological transfer coefficients, T is the average air temperature, r_0 is the latent heat of vaporization of water at 273.15 K, c_{p1} and c_{p2} are constant-pressure specific heats of the air components - water vapor (index 1) and the dry air (index 2), R_1 is the gas constant of water vapor, c_1 and c_2 are mass fractions of the air components, ρ_1 and ρ_2 are the densities of the air components, u is the air velocity along the wood board, DT is the temperature gradient, DC_1 is the vapor concentration gradient, $D\left(\frac{\rho_1 u^2}{2}\right)$ and $D\left(\frac{\rho_2 u^2}{2}\right)$ are the kinetic energy gradients for the two air components, D stands for d/dy.

The phenomenological coefficients are in equations (3) and (4) unknown quantities. They can be determined if the fluxes and thermodynamic forces are known. Accordingly, they must determined by other methods. The thermodynamic forces can be expressed from the mediation theorem as

$$D\left(\frac{1}{2}\rho u^2\right) = \frac{\rho u_\infty^2}{2\delta_u(x)} \tag{5}$$

$$DT = \frac{T_\infty - T_s}{\delta_T(x)} \tag{6}$$

$$DC_1 = \frac{C_{1s} - C_{1\infty}}{\delta_C(x)} \tag{7}$$

where ρ is the density of air, u_∞, T_∞ and $C_{1\infty}$ are the air velocity, air temperature and water vapor concentration at the boundary layer outer edge, T_s and C_{1s} are the temperature and vapor concentration on the wood surface, δ_u, δ_T and δ_C are the thicknesses of the dynamic, thermal and concentration boundary layer.

According to Eckert and Drake [21] the dynamic boundary layer thickness and the thermal boundary layer thickness are

$$\delta_{u(x)} = 4.64\sqrt{\frac{v \cdot x}{u_\infty}} \tag{8}$$

$$\delta_{T(x)} = \delta_{u(x)}\frac{1}{1.026 \cdot \sqrt[3]{Pr}} \tag{9}$$

where v is the coefficient of kinematic viscosity, Pr is Prandtl number.

By using the analogy between heat and mass transfer, the concentration boundary layer can be written as

$$\delta_{C(x)} = \delta_{u(x)}\frac{1}{1.026 \cdot \sqrt[3]{Sc}} \tag{10}$$

where Sc is Schmidt number.

If the heat and mass transfer includes both boundary layer and dry shell, the heat and mass flux equations are

$$J_q = h'(T_\infty - T_e) \tag{11}$$

$$J_1 = h_m'(\rho_{1e} - \rho_{1\infty}) \tag{12}$$

where h' and h_m' are the heat and mass transfer coefficients, ρ_{1e} and $\rho_{1\infty}$ are the vapor densities on the evaporation front and at the boundary layer outer edge. The heat transfer coefficient from the heated air to the evaporation surface is [1]

$$h' = \left(\frac{1}{h} + \frac{\xi}{k_s}\right)^{-1} \tag{13}$$

and the mass transfer coefficient from the evaporation surface to the heated air [19]

$$h_m' = \left(\frac{1}{h_m} + \frac{\xi}{D_s}\right)^{-1} \tag{14}$$

where h is the heat transfer coefficient for the corresponding case with a wet surface, ξ is the effective thickness of the dry shell, k_s is the thermal conductivity within the dry shell, h_m is the mass transfer coefficient for the case with a wet surface, D_s is the diffusion coefficient within the dry shell.

From the conservation conditions of the heat and mass fluxes, the following balance equations can be written

$$h(T_\infty - T_s) = \frac{T_s - T_e}{\frac{\xi}{k_s}} \tag{15}$$

$$h_m(\rho_{1s} - \rho_{1\infty}) = \frac{\rho_{1e} - \rho_{1s}}{\frac{\xi}{D_s}} \tag{16}$$

where T_s and ρ_{1s} are the temperature and vapor density on the wood surface, T_e and ρ_{1e} are the temperature and vapor density on the evaporation surface.

By substituting equations (13) and (14) in (15) and (16), the next equalities between the two kinds of heat and mass transfer coefficients can be expressed

$$h' = h\frac{T_\infty - T_s}{T_\infty - T_e} \tag{17}$$

$$h_m' = h_m \frac{\rho_{1s} - \rho_{1\infty}}{\rho_{1e} - \rho_{1\infty}} \tag{18}$$

In order to determine h' and h_m', the heat and mass transfer coefficients h and h_m must be evaluated before. They can be obtained from criterial equations specific to the drying process. Nesterenko, cited by [1] developed the following relations for the heat and mass transfer in evaporation processes

$$Nu_x = 2 + A \cdot Pr^{0.33} Re_x^n Gu^m \tag{19}$$

$$Sh_x = 2 + A' \cdot Sc^{0.33} Re_x^{n'} Gu^{m'} \tag{20}$$

where $Gu = (T_\infty - T_s)/T_\infty$ is Gukhman number and

$$A = 0.51, \, m = 0.175, \, n = 0.61; \, A' = 0.49, \, m' = 0.135, \, n' = 0.61$$

are experimental coefficients, valid in the range $(3.15 \times 10^3, \, ..., \, 2.2 \times 10^4)$ of Reynolds number. We mention at this point that this range was encountered in the present analysis. The local Nusselt and Sherwood numbers are

$$Nu_x = hx/k_f \tag{21}$$

$$Sh_x = h_m x/D_f \tag{22}$$

where subscript f refers to the fluid (air).

The physical quantities were calculated by using data from thermodynamic tables [22] at the average temperature of air. The input data for numerical analysis were $T_\infty = 333.15$ K, $u_\infty = 1.5$ m/s and the relative humidity of air at the boundary layer outer edge was $\varphi_\infty = 45\%$. The pressure was constant $p = 10^5$ Pa.

3. RESULTS AND DISCUSSIONS

The phenomenological coefficients were calculated for a range of x-values between 0.1 and 0.25 m, the latter corresponding to the length of a wood board dried in a laboratory drier (wind tunnel). They are indicated in Table 1. According to Onsager's reciprocal relations [20], $L_{ik} = L_{ki}$. In the present paper the phenomenological coefficients were calculated based on the assumption that $L_{TI} = L_{IT}$. The phenomenological coefficients are physical constants, which are always positive. There is a condition mentioned by Prigogine [4] that the self coefficients L_{TT} and L_{II} are positive and the mutual coefficients L_{TI}, L_{TEI}, L_{TE2}, L_{IEI}, L_{IE2} can be either positive or negative. The negative values indicate that the gradients have opposite directions than those initially considered, since the heat and mass fluxes are also positive quantities. It also can be observed that L_{TEI} and L_{IEI} have opposite signs with respect to L_{TE2} and L_{IE2}, meaning that the air components, water vapor and dry air move in opposite directions during the drying process. Accordingly, a component separation occurs in the boundary layer due to the influence of the kinetic energy of each component.

Table 1: Phenomenological coefficients

L_{TT} [WK/m]	$L_{TI} = L_{IT}$ [kgK/ms]	L_{TEI} [m²K/s]	L_{TE2} [m²K/s]	L_{II} [kgKs/m³]	L_{IEI} [Ks]	L_{IE2} [Ks]
1.238×10^7	4.412	5.876×10^{11}	-3.59×10^{10}	1.561×10^{-6}	2.755×10^6	-1.683×10^5

The heat and mass flux equations (3) and (4) can be compared with those obtained from Fourier's law $J_q = -k_f DT$ and Fick's law $J_1 = -D_f \rho DC_1$ expressed for the boundary layer if the other gradients become zero. There can be written therefore relationships between the phenomenological coefficients determined before and the thermal conductivity and coefficient of diffusion, respectively. From these considerations we obtained for the thermal conductivity of the fluid, $k_f = 0.37$ W/mK (the value $k_f = 0.027$ W/mK was used in equation (20)) and for the mass diffusion coefficient of the fluid, $D_f = 1.87 \times 10^{-2}$ m²/s (the value $D_f = 3.2 \times 10^{-5}$ W/mK was used in equation (21)).

The heat and mass fluxes per unit of area as functions of the x-coordinate are shown in Figures 2 and 3. There is a sensible decrease in the intensity of heat and mass transfer along the wood board.

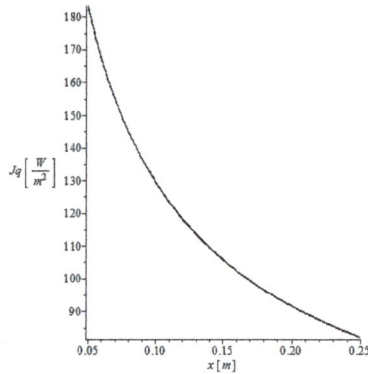

Figure 2: Heat flux variation along the wood board

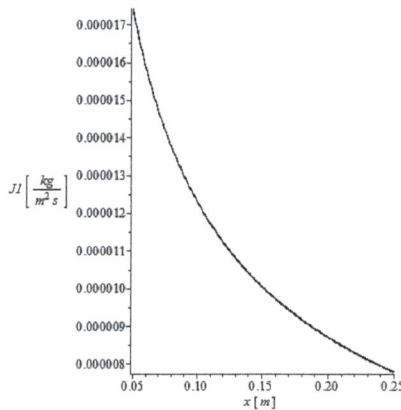

Figure 3: Mass flux variation along the wood board

142

Figure 4 shows the change of the dry shell thickness along the wood board. As can be seen, the thickness also decreases along the board being influenced by the heat and mass flux magnitudes.

4. CONCLUSION

Phenomenological equations for the coupled heat and mass transfer during wood drying that relate fluxes to thermodynamic forces were applied in the paper for the determination of the phenomenological coefficients. The fluxes were calculated from criterial equations and the thermodynamic forces from the mediation theorem. There was a decrease of both fluxes along the wood board, meaning that there is a non-uniformity of drying on the board length. The retraction of the evaporation front introduced an additional thermal resistance in wood that reduced the magnitude of the heat and mass transfer coefficients on the board length with respect to the same transfer coefficients if the evaporation front would have been on the wood surface. By using the method of non-equilibrium thermodynamics the influence of each form of transfer (heat, mass and momentum) is pointed out in the heat and mass fluxes, showing that they are strongly coupled.

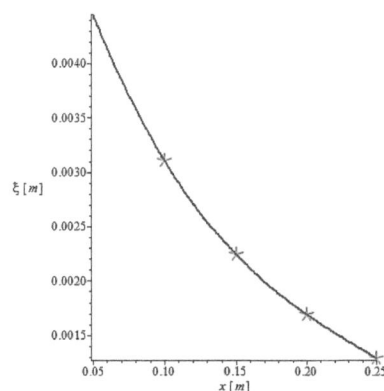

Figure 4: Dry shell thickness variation along the wood board

REFERENCES

[1] Luikov A.V., Heat and mass transfer in capillary-porous bodies, Pergamon Press, Oxford, 1966.
[2] Wang J., Modern thermodynamics – New concepts based on the second law of thermodynamics, Progress in Natural Science 19:125–135, 2009.
[3] Sova V., Sursa de entropie in procesele de evaporare de pe suprafete libere, Studii si cercetari de fizica 23(2):133-144, Bucuresti, 1971.
[4] Prigogine I., Introduction a la thermodynamique des processus irréversible, Dunod, Paris, 1968.
[5] Verros G.D., Application of non-equilibrium thermodynamics and computer aided analysis to the estimation of diffusion coefficients in polymer solutions: The solvent evaporation method, Journal of Membrane Science, 328:31–57, 2009.
[6] Lewis R.W., Malan A.G., Continuum thermodynamic modeling of drying capillary particulate materials via an edge-based algorithm, Comput. Methods Appl. Mech. Energ., 194:2043–2057, 2005.
[7] Perrin B., Javelas R., Simultaneous heat and mass-transfer in consolidated materials used in civil engineering, Int. J. Heat Mass Transfer 30(2):297–309, 1987.
[8] Fortes M., Okos M.R., Heat and mass transfer in hygroscopic capillary extruded products, AIChE J. 27(2): 255–262, 1981.
[9] Stenholm S., On entropy production, Annals of Physics 323:2892–2904, 2008.
[10] Kuhn J., Stemmer R., Kapteijn F., Kjelstrup S., Gross J., A non-equilibrium thermodynamics approach to model mass and heat transport for water pervaporation through a zeolite membrane, Journal of Membrane Science 330:388–398, 2009.
[11] Zvolinschi A., Johannessen E., Kjelstrup S., The second-law optimal operation of a paper drying machine, Chemical Engineering Science 61:3653 – 3662, 2006.
[12] Bedeaux D., Kjelstrup S., Transfer coefficients for evaporation, Physica A 270:413-426, 1999.
[13] Kadem S., Lachemet A., Younsi R., Kocaefe D., 3d-transient modeling of heat and mass transfer during heat

treatment of wood, International Communications in Heat and Mass Transfer 38:717–722, 2011.

[14] Kocaefe D., Younsi R., Chaudry B., Kocaefe Y., Modeling of heat and mass transfer during high temperature treatment of aspen, Wood Sci Technol 40: 371–391, 2006.

[15] Coles C., Murio D., Parameter estimation for a drying system in a porous medium, Computer and mathematics with applications 51:1519-1528, 2006.

[16] Irudayaraj J., Wu Y., Ghazanfari A., Yang W., Application of simultaneous heat, mass, and pressure transfer equations to timber drying, Numerical Heat Transfer, Part A: Applications, 30:233 – 247, 1996.

[17] Younsi R., Kocaefe D., Poncsak S., Kocaefe, Y., Thermal modeling of the high temperature treatment of wood based on Luikov's approach, Int. Journal of Energy Research 30: 699–711, 2006.

[18] Younsi, R.; Kocaefe, D.; Kocaefe, Y., Three-dimensional simulation of heat and moisture transfer in wood, Applied Thermal Engineering 26, 1274–1285, 2006.

[19] Salin J.G. in Perré P., Fundamentals of wood drying, COST & A.R.BO.LOR., 2007.

[20] De Groot S.R., Mazur P., Non-equilibrium thermodynamics, Dover Publications, New York, 1984.

[21] Eckert E.R.G., Drake R.M., Heat and mass transfer, McGraw-Hill Book Company, New York, 1959.

[22] Ražnevič, K., Tabele si diagrame termodinamice, Editura Tehnica, Bucuresti, 1978.

6th International Conference
"Computational Mechanics and Virtual Engineering"
COMET 2015
15-16 October 2015, Braşov, Romania

DYNAMICS OF A SYSTEM OF RIGID BODIES WITH GENERAL CONSTRAINTS BY A MULTIBODY APPROACH

Nicolae–Doru Stănescu

University of Piteşti, Piteşti, ROMANIA, e-mail s_doru@yahoo.com

Abstract: *This paper generalizes the equations of motion for a single rigid body, published in a previous work, for the case of a system containing an arbitrary number of rigid bodies. It is not important the way in which the rigid bodies are linked one to another. We also present an application to highlight the theory.*

Keywords): *constraint, multibody, equations of motion, reactions*

1. INTRODUCTION

The study of the multibody systems is a great task of nowadays researchers. The general case of the rigid body with arbitrary constraints is published in [1]. In this paper we will generalize the equation of motion published there for the case of a mechanical system with an arbitrary number of rigid bodies linked one to another.

The matrix equation of motion may be obtained in two ways. The first approach is to use the general theorems (the theorem of momentum and the theorem of moment of momentum). In this approach we make the following notations

$$[\mathbf{M_q}] = \begin{bmatrix} [\mathbf{m}] & [\mathbf{A}][\mathbf{S}]^T[\mathbf{Q}] \\ [\mathbf{Q}]^T[\mathbf{S}][\mathbf{A}]^T & [\mathbf{Q}]^T[\mathbf{J}_O][\mathbf{Q}] \end{bmatrix}, \quad \{\mathbf{F_q}\} = \begin{bmatrix} \{\mathbf{F_s}\} \\ \{\mathbf{F_\beta}\} \end{bmatrix}, \quad \{\widetilde{\mathbf{F}}_\mathbf{q}\} = \begin{bmatrix} \{\widetilde{\mathbf{F}}_s\} \\ \{\widetilde{\mathbf{F}}_\beta\} \end{bmatrix}, \quad \{\widetilde{\mathbf{F}}_s\} = -\left[[\mathbf{A}][\mathbf{S}]^T[\dot{\mathbf{Q}}] + [\dot{\mathbf{A}}][\mathbf{S}]^T[\mathbf{Q}]\right]\{\dot{\beta}\},$$

$$\{\widetilde{\mathbf{F}}_\beta\} = -\left[[\mathbf{Q}]^T[\mathbf{J}_O][\dot{\mathbf{Q}}] + [\mathbf{Q}]^T[\omega][\mathbf{J}_O][\mathbf{Q}]\right]\{\dot{\beta}\}, \qquad (1)$$

and the matrix equation of motion takes the form

$$[\mathbf{M_q}]\{\ddot{\mathbf{q}}\} = \{\mathbf{F_q}\} + \{\widetilde{\mathbf{F}}_\mathbf{q}\} \qquad (2)$$

where

$$[\mathbf{A}] = [\psi][\theta][\varphi], \quad [\mathbf{A}_\psi] = [\mathbf{U}_\psi][\mathbf{A}], \quad [\mathbf{A}_\theta] = [\psi][\mathbf{U}_\theta][\varphi]^T[\mathbf{A}], \quad [\mathbf{A}_\varphi] = [\mathbf{A}][\mathbf{U}_\varphi], \quad [\dot{\mathbf{A}}] = \dot{\psi}[\mathbf{A}_\psi] + \dot{\theta}[\mathbf{A}_\theta] + \dot{\varphi}[\mathbf{A}_\varphi],$$

$$[\omega] = [\mathbf{A}]^T[\dot{\mathbf{A}}], \quad [\theta_Q] = [[\theta]^T\{\mathbf{u}_\psi\} \ \{\mathbf{u}_\theta\} \ \{\mathbf{u}_\varphi\}], \quad [\mathbf{Q}] = [\varphi]^T[\theta_Q], \quad \{\omega\} = [\mathbf{Q}]\begin{bmatrix} \dot{\psi} \\ \dot{\theta} \\ \dot{\varphi} \end{bmatrix}, \quad [\mathbf{S}] = \begin{bmatrix} 0 & -mz_C & my_C \\ mz_C & 0 & -mx_C \\ -my_C & mx_C & 0 \end{bmatrix},$$

$$[\dot{\mathbf{Q}}] = -\dot{\varphi}[[\mathbf{U}_\varphi][\varphi]^T\{\mathbf{u}_\psi\} \ [\mathbf{U}_\varphi][\varphi]^T\{\mathbf{u}_\theta\} \ \{0\}] - \dot{\theta}[[\varphi]^T[\theta]^T[\mathbf{U}_\theta]\{\mathbf{u}_\psi\} \ \{0\} \ \{0\}]. \qquad (3)$$

If we use the Lagrange equations, then we denote by q_k, and F_{q_k}, $k = \overline{1, n}$, the generalized coordinates, and forces, respectively, and the Lagrange equations read

$$\frac{\mathrm{d}}{\mathrm{d}t}\left(\frac{\partial E_c}{\partial \dot{q}_k}\right) - \frac{\partial E_c}{\partial q_k} = F_{q_k}, \quad k = \overline{1, n}, \qquad (4)$$

where E_c is the kinetic energy, and the generalized forces contain both the given and the constraint forces. Using the matrix notations

$$\left\{\frac{dE_c}{\partial \dot{\mathbf{q}}}\right\} \equiv \left[\frac{\partial E_c}{\partial \dot{q}_1} \frac{\partial E_c}{\partial \dot{q}_2} \cdots \frac{\partial E_c}{\partial \dot{q}_n}\right]^T, \left\{\frac{dE_c}{\partial \mathbf{q}}\right\} \equiv \left[\frac{\partial E_c}{\partial q_1} \frac{\partial E_c}{\partial q_2} \cdots \frac{\partial E_c}{\partial q_n}\right]^T, \{\mathbf{F_q}\} \equiv \left[F_{q_1} \ F_{q_2} \ \cdots \ F_{q_n}\right]^T, \tag{5}$$

one obtains the matrix form of the Lagrange equations

$$\frac{d}{dt}\left\{\frac{\partial E_c}{\partial \dot{\mathbf{q}}}\right\} - \left\{\frac{\partial E_c}{\partial \mathbf{q}}\right\} = \{\mathbf{F_q}\}. \tag{6}$$

Denoting

$$\{\mathbf{s}\} = \left[X_O \ Y_O \ Z_O\right]^T, \ \{\boldsymbol{\beta}\} = \left[\psi \ \theta \ \varphi\right]^T, \ \{\mathbf{q}\} = \left[X_O \ Y_O \ Z_O \ \psi \ \theta \ \varphi\right]^T, \tag{7}$$

the kinetic energy reads

$$E_c = \frac{1}{2}\{\dot{\mathbf{q}}\}^T [\mathbf{M_q}]\{\dot{\mathbf{q}}\} \tag{8}$$

wherefrom we get

$$\left\{\frac{\partial E_c}{\partial \dot{\mathbf{q}}}\right\} = [\mathbf{M_q}]\{\dot{\mathbf{q}}\}, \left\{\frac{\partial E_c}{\partial \mathbf{q}}\right\} = \frac{1}{2}\left[\{\dot{\mathbf{q}}\}^T \frac{\partial[\mathbf{M_q}]}{\partial X_O}\{\dot{\mathbf{q}}\} \cdots \{\dot{\mathbf{q}}\}^T \frac{\partial[\mathbf{M_q}]}{\partial \varphi}\{\dot{\mathbf{q}}\}\right]^T \tag{9}$$

and the equation (4) becomes

$$[\mathbf{M_q}]\{\ddot{\mathbf{q}}\} + [\dot{\mathbf{M}}_\mathbf{q}]\{\dot{\mathbf{q}}\} - \left\{\frac{\partial E_c}{\partial \mathbf{q}}\right\} = \{\mathbf{F_q}\}. \tag{10}$$

Denoting

$$\left\{\tilde{\tilde{\mathbf{F}}}_\mathbf{q}\right\} = -[\dot{\mathbf{M}}_\mathbf{q}]\{\dot{\mathbf{q}}\} + \left\{\frac{\partial E_c}{\partial \mathbf{q}}\right\}, \tag{11}$$

it results

$$[\mathbf{M_q}]\{\ddot{\mathbf{q}}\} = \{\mathbf{F_q}\} + \left\{\tilde{\tilde{\mathbf{F}}}_\mathbf{q}\right\}. \tag{12}$$

2. THE MATRIX EQUATION OF MOTION FOR A RIGID BODY WITH CONSTRAINTS

One may prove [1] that the equations (2) and (12) are equivalent, that is, the matrices $\{\tilde{\mathbf{F}}_\mathbf{q}\}$, $\{\tilde{\tilde{\mathbf{F}}}_\mathbf{q}\}$ are one and the same . In this way, the generalized and the constraint forces can be replace by the sum between the matrix $\{\mathbf{F_q}\}$ of the generalized given forces and the matrix $[\mathbf{B}]^T\{\boldsymbol{\lambda}\}$, where $[\mathbf{B}]$ is the matrix of constraints, and $\{\boldsymbol{\lambda}\}$ is the matrix of the Lagrange multipliers. It results the differential equation of motion

$$\begin{bmatrix} [\mathbf{M}] & -[\mathbf{B}]^T \\ [\mathbf{B}] & [\mathbf{0}] \end{bmatrix}\begin{bmatrix} \{\ddot{\mathbf{q}}\} \\ \{\boldsymbol{\lambda}\} \end{bmatrix} = \begin{bmatrix} \{\mathbf{F_q}\} + \{\tilde{\mathbf{F}}_\mathbf{q}\} \\ \{\dot{\mathbf{C}}\} - [\dot{\mathbf{B}}]\{\dot{\mathbf{q}}\} \end{bmatrix}, \tag{13}$$

where
$$[\mathbf{B}]\{\dot{\mathbf{q}}\} = \{\mathbf{C}\} \tag{14}$$
is the equation of the constraints, and

$$\{\mathbf{s}\} = \left[X_O \ Y_O \ Z_O\right]^T, \qquad \{\boldsymbol{\beta}\} = \left[\psi \ \theta \ \varphi\right]^T, \qquad \{\mathbf{q}\} = \left[X_O \ Y_O \ Z_O \ \psi \ \theta \ \varphi\right]^T, \qquad [\mathbf{M}] = \begin{bmatrix} [\mathbf{m}] & [\mathbf{A}][\mathbf{S}]^T[\mathbf{Q}] \\ [\mathbf{Q}]^T[\mathbf{S}][\mathbf{A}]^T & [\mathbf{Q}]^T[\mathbf{J}_O][\mathbf{Q}] \end{bmatrix},$$

$$\{\tilde{\mathbf{F}}_\mathbf{q}\} = \begin{bmatrix} \{\tilde{\mathbf{F}}_\mathbf{s}\} \\ \{\tilde{\mathbf{F}}_\boldsymbol{\beta}\} \end{bmatrix}, \ \{\tilde{\mathbf{F}}_\mathbf{s}\} = -\left([\mathbf{A}][\mathbf{S}]^T[\dot{\mathbf{Q}}] + [\dot{\mathbf{A}}][\mathbf{S}]^T[\mathbf{Q}]\right)\{\dot{\boldsymbol{\beta}}\}, \ \{\tilde{\mathbf{F}}_\boldsymbol{\beta}\} = -\left([\mathbf{Q}]^T[\mathbf{J}_O][\dot{\mathbf{Q}}] + [\mathbf{Q}]^T[\boldsymbol{\omega}][\mathbf{J}_O][\mathbf{Q}]\right)\{\dot{\boldsymbol{\beta}}\}, \tag{14}$$

The matrix of constraints $[\mathbf{B}]$ has the general form

$$[\mathbf{B}] = \begin{bmatrix} B_{11} & B_{12} & \dots & B_{16} \\ B_{21} & B_{22} & \dots & B_{26} \\ \dots & \dots & \dots & \dots \\ B_{p1} & B_{p2} & \dots & B_{p6} \end{bmatrix} ; \tag{15}$$

hence, the generalized force that corresponds to the constraint of index i will have the components

$$\{\mathbf{F}_{Gs}^{(i)}\} = \lambda_i \begin{bmatrix} B_{i1} \\ B_{i2} \\ B_{i3} \end{bmatrix}, \ \{\mathbf{F}_{G\beta}^{(i)}\} = \lambda_i \begin{bmatrix} B_{i4} \\ B_{i5} \\ B_{i6} \end{bmatrix}, \tag{16}$$

in which $\lambda_i B_{i1}$, $\lambda_i B_{i2}$ and $\lambda_i B_{i3}$ represent the projections of the force of constraint onto the axes of the fixed reference frame, while the projections of the moments about the point O, onto the mobile axes are obtained from the expression $\left[[\mathbf{Q}]^{-1}\right]^{\mathrm{T}}\{\mathbf{F}_{G\beta}^{(i)}\}$.

3. THE MOVING EQUATION FOR A SYSTEM OF RIGID BODIES

The moving equation (13) may be easily generalized for the case of n rigid bodies with constraints. We make the following modifications:
– the matrix of constraints $[\mathbf{B}]$ reads

$$[\mathbf{B}]^{\mathrm{T}} = \begin{bmatrix} \dfrac{\partial f_1}{\partial X_{O_1}} & \dots & \dfrac{\partial f_i}{\partial X_{O_1}} & \dots & \dfrac{\partial f_p}{\partial X_{O_1}} \\[2mm] \dfrac{\partial f_1}{\partial Y_{O_1}} & \dots & \dfrac{\partial f_i}{\partial Y_{O_1}} & \dots & \dfrac{\partial f_p}{\partial Y_{O_1}} \\[2mm] \dfrac{\partial f_1}{\partial Z_{O_1}} & \dots & \dfrac{\partial f_i}{\partial Z_{O_1}} & \dots & \dfrac{\partial f_p}{\partial Z_{O_1}} \\[2mm] \dfrac{\partial f_1}{\partial \psi_1} & \dots & \dfrac{\partial f_i}{\partial \psi_1} & \dots & \dfrac{\partial f_p}{\partial \psi_1} \\[2mm] \dfrac{\partial f_1}{\partial \theta_1} & \dots & \dfrac{\partial f_i}{\partial \theta_1} & \dots & \dfrac{\partial f_p}{\partial \theta_1} \\[2mm] \dfrac{\partial f_1}{\partial \varphi_1} & \dots & \dfrac{\partial f_i}{\partial \varphi_1} & \dots & \dfrac{\partial f_p}{\partial \varphi_1} \\[2mm] \vdots & \vdots & \vdots & \vdots & \vdots \\[2mm] \dfrac{\partial f_1}{\partial X_{O_n}} & \dots & \dfrac{\partial f_i}{\partial X_{O_n}} & \dots & \dfrac{\partial f_p}{\partial X_{O_n}} \\[2mm] \dfrac{\partial f_1}{\partial Y_{O_n}} & \dots & \dfrac{\partial f_i}{\partial Y_{O_n}} & \dots & \dfrac{\partial f_p}{\partial Y_{O_n}} \\[2mm] \dfrac{\partial f_1}{\partial Z_{On}} & \dots & \dfrac{\partial f_i}{\partial Z_{On}} & \dots & \dfrac{\partial f_p}{\partial Z_{On}} \\[2mm] \dfrac{\partial f_1}{\partial \psi_n} & \dots & \dfrac{\partial f_i}{\partial \psi_n} & \dots & \dfrac{\partial f_p}{\partial \psi_n} \\[2mm] \dfrac{\partial f_1}{\partial \theta_n} & \dots & \dfrac{\partial f_i}{\partial \theta_n} & \dots & \dfrac{\partial f_p}{\partial \theta_n} \\[2mm] \dfrac{\partial f_1}{\partial \varphi_n} & \dots & \dfrac{\partial f_i}{\partial \varphi_n} & \dots & \dfrac{\partial f_p}{\partial \varphi_n} \end{bmatrix}, \tag{17}$$

in which $f_i(\mathbf{q}) = 0$, $i = \overline{1, p}$, are the constraints;
– the matrix $[\mathbf{M}]$ has the expression

$$[\mathbf{M}] = \begin{bmatrix} [\mathbf{M}_1] & [0] & [0] & \cdots & [0] \\ [0] & [\mathbf{M}_2] & [0] & \cdots & [0] \\ [0] & [0] & [\mathbf{M}_3] & \cdots & [0] \\ \cdots & \cdots & \cdots & \cdots & \cdots \\ [0] & [0] & [0] & \cdots & [\mathbf{M}_n] \end{bmatrix}, \tag{18}$$

where

$$[\mathbf{M}_k] = \begin{bmatrix} [m_k] & [\mathbf{A}_k][\mathbf{S}_k]^T[\mathbf{Q}_k] \\ [\mathbf{Q}_k]^T[\mathbf{S}_k][\mathbf{A}_k]^T & [\mathbf{Q}_k]^T[\mathbf{J}_{O_k}][\mathbf{Q}_k] \end{bmatrix}, \tag{19}$$

O_k being the origin of the mobile reference system linked to the rigid body k, $k = \overline{1,n}$;

– the matrices $\{s\}$ and $\{\beta\}$ take the form

$$\{\mathbf{s}\} = \begin{bmatrix} X_{O_1} & Y_{O_1} & Z_{O_1} & \cdots & X_{O_n} & Y_{O_n} & Z_{O_n} \end{bmatrix}^T, \quad \{\boldsymbol{\beta}\} = \begin{bmatrix} \psi_1 & \theta_1 & \varphi_1 & \cdots & \psi_n & \theta_n & \varphi_n \end{bmatrix}^T, \tag{20}$$

that is,

$$\{\mathbf{s}\} = \begin{bmatrix} \{\mathbf{s}_1\}^T & \{\mathbf{s}_2\}^T & \cdots & \{\mathbf{s}_n\}^T \end{bmatrix}^T, \quad \{\boldsymbol{\beta}\} = \begin{bmatrix} \{\boldsymbol{\beta}_1\}^T & \{\boldsymbol{\beta}_2\}^T & \cdots & \{\boldsymbol{\beta}_n\}^T \end{bmatrix}^T; \tag{21}$$

– the vector $\{\mathbf{q}\}$ writes as

$$\{\mathbf{q}\} = \begin{bmatrix} \{\mathbf{s}_1\}^T & \{\boldsymbol{\beta}_1\}^T & \cdots & \{\mathbf{s}_n\}^T & \{\boldsymbol{\beta}_n\}^T \end{bmatrix}^T \tag{22}$$

and it has $6n$ components;

– the vector $\{\widetilde{\mathbf{F}}_\mathbf{q}\}$ becomes

$$\{\widetilde{\mathbf{F}}_\mathbf{q}\} = \begin{bmatrix} \{\widetilde{\mathbf{F}}_{\mathbf{s}_1}\}^T & \{\widetilde{\mathbf{F}}_{\boldsymbol{\beta}_1}\}^T & \cdots & \{\widetilde{\mathbf{F}}_{\mathbf{s}_n}\}^T & \{\widetilde{\mathbf{F}}_{\boldsymbol{\beta}_n}\}^T \end{bmatrix}^T, \tag{23}$$

where $\{\widetilde{\mathbf{F}}_{\mathbf{s}_k}\}$ and $\{\widetilde{\mathbf{F}}_{\boldsymbol{\beta}_k}\}$, $k = \overline{1,n}$, have the same known expressions;

Let us remark that if there is no link between the rigid bodies, then the matrix of constraints takes the form

$$[\mathbf{B}] = \begin{bmatrix} \left[\dfrac{\partial \mathbf{f}_1}{\partial \mathbf{q}_1}\right] & [0] & \cdots & [0] \\ [0] & \left[\dfrac{\partial \mathbf{f}_2}{\partial \mathbf{q}_2}\right] & \cdots & [0] \\ \cdots & \cdots & \cdots & \cdots \\ [0] & [0] & \cdots & \left[\dfrac{\partial \mathbf{f}_n}{\partial \mathbf{q}_n}\right] \end{bmatrix}, \tag{24}$$

where $\mathbf{f}_k = \begin{bmatrix} f_{p_1} & f_{p_2} & \cdots & f_{p_k} \end{bmatrix}^T$ are the constraints of the rigid solid k, $k = \overline{1,n}$, each constraint of this kind containing only the parameters $\{\mathbf{q}_k\}$. In this way, one obtains n independent equations of type (13).

4. EXAMPLE

We consider two bars AC, and CB of lengths l_1, and l_2, masses m_1, and m_2, respectively, linked one to another by a spherical joint at the point C. The mobile reference systems $O_1 x_1 y_1 z_1$, $O_2 x_2 y_2 z_2$ are principal central system of inertia for which one knows the values J_{x_1}, J_{y_1}, J_{z_1}, J_{x_2}, J_{y_2}, J_{z_2}. For the fixed reference system $O_0 XYZ$ the axis $O_0 Z$ is vertical ascendant.

Choosing as rotational parameters the Bryan angles for each body, we have

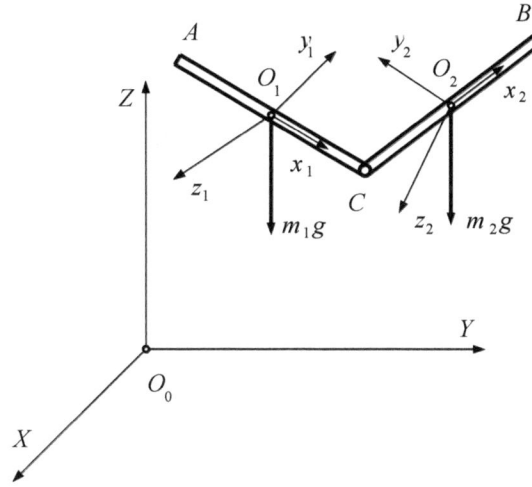

Figure 1: Example

$$[\psi_i] = \begin{bmatrix} 1 & 0 & 0 \\ 0 & \cos\psi_i & -\sin\psi_i \\ 0 & \sin\psi_i & \cos\psi_i \end{bmatrix}, \quad [\theta_i] = \begin{bmatrix} \cos\theta_i & 0 & \sin\theta_i \\ 0 & 1 & 0 \\ -\sin\theta_i & 0 & \cos\theta_i \end{bmatrix}, \quad [\varphi_i] = \begin{bmatrix} \cos\varphi_i & -\sin\varphi_i & 0 \\ \sin\varphi_i & \cos\varphi_i & 0 \\ 0 & 0 & 1 \end{bmatrix}, \tag{25}$$

$$[A_i] = [\psi_i][\theta_i][\varphi_i] = \begin{bmatrix} c\theta_i c\varphi_i & -c\theta_i s\varphi_i & s\theta_i \\ s\psi_i s\theta_i c\varphi_i + c\psi_i s\varphi_i & -s\psi_i s\theta_i s\varphi_i + c\psi_i c\varphi_i & -s\psi_i c\theta_i \\ c\psi_i s\theta_i c\varphi_i + s\psi_i s\varphi_i & c\psi_i s\theta_i s\varphi_i + s\psi_i c\varphi_i & c\psi_i c\theta_i \end{bmatrix}, \tag{26}$$

$$[S_i] = \begin{bmatrix} 0 & 0 & 0 \\ 0 & 0 & 0 \\ 0 & 0 & 0 \end{bmatrix}, \quad [U_{\psi_i}] = \begin{bmatrix} 0 & 0 & 0 \\ 0 & 0 & -1 \\ 0 & 1 & 0 \end{bmatrix}, \quad [U_{\theta_i}] = \begin{bmatrix} 0 & 0 & 1 \\ 0 & 0 & 0 \\ -1 & 0 & 0 \end{bmatrix}, \quad [U_{\varphi_i}] = \begin{bmatrix} 0 & -1 & 0 \\ 1 & 0 & 0 \\ 0 & 0 & 0 \end{bmatrix},$$
(27)

$$[\psi_{p_i}] = [U_{\psi_i}][\psi_i] = \begin{bmatrix} 0 & 0 & 0 \\ 0 & -\sin\psi_i & \cos\psi_i \\ 0 & \cos\psi_i & -\sin\psi_i \end{bmatrix}, \qquad [\theta_{p_i}] = [U_{\theta_i}][\theta_i] = \begin{bmatrix} -\sin\theta_i & 0 & \cos\theta_i \\ 0 & 0 & 0 \\ \cos\theta_i & 0 & -\sin\theta_i \end{bmatrix},$$

$$[\varphi_{p_i}] = [U_{\varphi_i}][\varphi_i] = \begin{bmatrix} -\sin\varphi_i & -\cos\varphi_i & 0 \\ \cos\varphi_i & -\sin\varphi_i & 0 \\ 0 & 0 & 0 \end{bmatrix}, \tag{28}$$

$$[A_{\psi_i}] = [\psi_{p_i}][\theta_i][\varphi_i] = \begin{bmatrix} 0 & 0 & 0 \\ c\psi_i s\theta_i c\varphi_i - s\psi_i s\varphi_i & -c\psi_i s\theta_i s\varphi_i - s\psi_i c\varphi_i & -c\psi_i c\theta_i \\ s\psi_i s\theta_i c\varphi_i + c\psi_i s\varphi_i & -s\psi_i s\theta_i s\varphi_i + c\psi_i c\varphi_i & -s\psi_i c\theta_i \end{bmatrix},$$

$$[A_{\theta_i}] = [\psi_i][\theta_{p_i}][\varphi_i] = \begin{bmatrix} -s\theta_i c\varphi_i & s\theta_i s\varphi_i & c\theta_i \\ s\psi_i c\theta_i c\varphi_i & -s\psi_i c\theta_i s\varphi_i & s\psi_i s\theta_i \\ -c\varphi_i c\theta_i c\varphi_i & s\psi_i c\theta_i s\varphi_i & -c\psi_i s\theta_i \end{bmatrix},$$

$$[A_{\varphi_i}] = [\psi_i][\theta_i][\varphi_{p_i}] = \begin{bmatrix} -c\theta_i s\varphi_i & -c\theta_i c\varphi_i & 0 \\ -s\psi_i s\theta_i s\varphi_i + c\psi_i c\varphi_i & -s\psi_i s\theta_i c\varphi_i - c\psi_i s\varphi_i & 0 \\ c\psi_i s\theta_i s\varphi_i + s\psi_i c\varphi_i & c\psi_i s\theta_i c\varphi_i - s\psi_i s\varphi_i & 0 \end{bmatrix}, \tag{29}$$

$$[\dot{\mathbf{A}}_i] = \dot{\psi}_i[\mathbf{A}_{\psi_i}] + \dot{\theta}_i[\mathbf{A}_{\theta_i}] + \dot{\varphi}_i[\mathbf{A}_{\varphi_i}], \quad \{\mathbf{u}_{\psi_i}\} = \begin{bmatrix} 1 \\ 0 \\ 0 \end{bmatrix}, \quad \{\mathbf{u}_{\theta_i}\} = \begin{bmatrix} 0 \\ 1 \\ 0 \end{bmatrix}, \quad \{\mathbf{u}_{\varphi_i}\} = \begin{bmatrix} 0 \\ 0 \\ 1 \end{bmatrix}, \tag{30}$$

$$[\mathbf{Q}_i] = [\boldsymbol{\varphi}_i]^T [\boldsymbol{\theta}_i]^T \{\mathbf{u}_{\psi_i}\} \{\mathbf{u}_{\theta_i}\} \{\mathbf{u}_{\varphi_i}\}] = \begin{bmatrix} \cos\varphi_i \cos\theta_i & \sin\varphi_i & 0 \\ -\sin\varphi_i \cos\theta_i & \cos\varphi_i & 0 \\ \sin\theta_i & 0 & 1 \end{bmatrix}, \qquad [\mathbf{Q}_{\varphi_i}] = \begin{bmatrix} -\sin\varphi_i \cos\theta_i & \cos\varphi_i & 0 \\ -\cos\varphi_i \cos\theta_i & -\sin\varphi_i & 0 \\ 0 & 0 & 0 \end{bmatrix},$$

$$[\mathbf{Q}_{\theta_i}] = \begin{bmatrix} -\cos\varphi_i \sin\theta_i & 0 & 0 \\ \sin\varphi_i \sin\theta_i & 0 & 0 \\ \cos\theta_i & 0 & 0 \end{bmatrix}, \tag{31}$$

$$[\dot{\mathbf{Q}}_i] = \dot{\varphi}_i[\mathbf{Q}_{\varphi_i}] + \dot{\theta}[\mathbf{Q}_{\theta_i}] = \begin{bmatrix} -\dot{\varphi}_i \sin\varphi_i \cos\theta_i - \dot{\theta}_i \cos\varphi_i \sin\theta_i & \dot{\varphi}_i \cos\varphi_i & 0 \\ -\dot{\varphi}_i \cos\varphi_i \cos\theta_i + \dot{\theta}_i \sin\varphi_i \sin\theta_i & -\dot{\varphi}_i \sin\varphi_i & 0 \\ \dot{\theta}_i \cos\theta_i & 0 & 0 \end{bmatrix}, \tag{32}$$

$$\{\boldsymbol{\omega}_i\} = [Q_i]\begin{bmatrix} \dot{\psi}_i \\ \dot{\theta}_i \\ \dot{\varphi}_i \end{bmatrix} = \begin{bmatrix} \dot{\psi}_i \cos\varphi_i \cos\theta_i + \dot{\theta}_i \sin\varphi_i \\ -\dot{\psi} \sin\varphi_i \cos\theta_i + \dot{\theta}_i \cos\varphi_i \\ -\dot{\psi}_i \sin\theta_i - \dot{\varphi}_i \end{bmatrix}, \tag{33}$$

$$[\boldsymbol{\omega}_i] = \begin{bmatrix} 0 & -\dot{\psi}\sin\theta_i - \dot{\varphi}_i & -\dot{\psi}\sin\varphi_i \cos\theta_i + \dot{\theta}_i \cos\varphi_i \\ \dot{\psi}\sin\varphi_i + \dot{\varphi}_i & 0 & -\dot{\psi}_i \cos\varphi_i \cos\theta_i - \dot{\theta}_i \sin\varphi_i \\ \dot{\psi}\sin\varphi_i \cos\theta_i - \dot{\theta}_i \cos\varphi_i & \dot{\psi}_i \cos\varphi_i \cos\theta_i + \dot{\theta}_i \sin\varphi_i & 0 \end{bmatrix}, \tag{34}$$

$$[\mathbf{m}_i] = \begin{bmatrix} m_i & 0 & 0 \\ 0 & m_i & 0 \\ 0 & 0 & m_i \end{bmatrix}, \quad [\mathbf{J}_{O_i}] = \begin{bmatrix} J_{x_i} & 0 & 0 \\ 0 & J_{y_i} & 0 \\ 0 & 0 & J_{z_i} \end{bmatrix}, \tag{35}$$

$$[\mathbf{A}_i][\mathbf{S}_i]^T[\mathbf{Q}_i] = \begin{bmatrix} 0 & 0 & 0 \\ 0 & 0 & 0 \\ 0 & 0 & 0 \end{bmatrix}, \quad [\mathbf{Q}_i]^T[\mathbf{S}_i][\mathbf{A}_i]^T = \begin{bmatrix} 0 & 0 & 0 \\ 0 & 0 & 0 \\ 0 & 0 & 0 \end{bmatrix}, \tag{36}$$

$$[\mathbf{Q}_i]^T[\mathbf{J}_{O_i}][\dot{\mathbf{Q}}_i]^T =$$

$$= \begin{bmatrix} J_{x_i} c\varphi_i c\theta_i (\dot{\varphi}_i s\varphi_i c\theta_i - \dot{\theta}_i c\varphi_i s\theta_i) - \\ - J_{y_i} s\varphi_i c\theta_i(-\dot{\varphi}_i c\varphi_i c\theta_i + \dot{\theta}_i s\varphi_i s\theta_i) + J_{z_i} s\theta_i c\theta_i \dot{\theta}_i & J_{x_i} c^2\varphi_i c\theta_i \dot{\varphi}_i + -J_{y_i} s^2\varphi_i c\theta_i \dot{\varphi}_i & 0 \\ J_{x_i} s\varphi_i(\dot{\varphi}_i s\varphi_i c\theta_i - \dot{\theta}_i c\varphi_i s\theta_i) + \\ + J_{y_i} c\varphi_i(-\dot{\varphi}_i c\varphi_i c\theta_i + \dot{\theta}_i s\varphi_i s\theta_i) & J_{x_i} s\varphi_i c\varphi_i \dot{\varphi}_i - -J_{y_i} s\varphi_i c\varphi_i \dot{\varphi}_i & 0 \\ J_{z_i} c\theta_i \dot{\theta}_i & 0 & 0 \end{bmatrix}, \tag{37}$$

$$[\mathbf{Q}_i]^{\mathrm{T}}[\omega_i][\mathbf{J}_{O_i}][\mathbf{Q}_i] =$$

$$= \begin{bmatrix} \begin{array}{l} -J_{x_i}c\varphi_i c\theta_i\left(\dot{\varphi}_i s\varphi_i c\theta_i + \dot{\theta}_i c\varphi_i s\theta_i\right) - \\ -J_{y_i}s\varphi_i c\theta_i\left(-\dot{\varphi}_i c\varphi_i c\theta_i + \dot{\theta}_i s\varphi_i s\theta_i\right) + \\ +J_{z_i}s\theta_i c\theta_i\dot{\theta}_i \end{array} & \begin{array}{l} -J_{x_i}s\varphi_i\left(\dot{\varphi}_i s\varphi_i c\theta_i + \dot{\theta}_i c\varphi_i s\theta_i\right) + \\ +J_{y_i}c\varphi_i\left(-\dot{\varphi}_i c\varphi_i c\theta_i + \dot{\theta}_i s\varphi_i s\theta_i\right) \end{array} & J_{z_i}c\theta_i\dot{\theta}_i \\[4mm] \begin{array}{l} J_{x_i}c^2\varphi_i c\theta_i\left(\dot{\psi}_i s\theta_i + \dot{\varphi}_i\right) - \\ -J_{y_i}s^2\varphi_i c\theta_i\left(-\dot{\psi}_i s\theta_i - \dot{\varphi}_i\right) - \\ -J_{z_i}s\theta_i c\varphi_i\dot{\psi}_i \end{array} & \begin{array}{l} J_{x_i}s\varphi_i c\varphi_i\left(\dot{\psi}_i s\theta_i + \dot{\varphi}_i\right) - \\ -J_{y_i}s\varphi_i c\varphi_i\left(\dot{\psi}_i s\theta_i + \dot{\varphi}_i\right) \end{array} & -J_{z_i}c\theta_i\dot{\theta}_i \\[4mm] \begin{array}{l} J_{x_i}c\varphi_i c\theta_i\left(\dot{\psi}_i s\varphi_i c\theta_i - \dot{\theta}_i c\varphi_i\right) - \\ -J_{y_i}s\varphi_i c\theta_i\left(\dot{\psi}_i c\varphi_i c\theta_i + \dot{\theta}_i s\varphi_i\right) \end{array} & \begin{array}{l} J_{x_i}s\varphi_i\left(\dot{\psi}_i s\varphi_i c\theta_i - \dot{\theta}_i c\varphi_i\right) + \\ +J_{y_i}c\varphi_i\left(\dot{\psi}_i c\varphi_i c\theta_i + \dot{\theta}_i s\varphi_i\right) \end{array} & 0 \end{bmatrix}. \qquad (38)$$

We denote by $\beta_j^{(i)}$, $j = \overline{1,3}$, $i = \overline{1,2}$, the components of the vector

$$\left\{\widetilde{\mathbf{F}}_{\beta_i}\right\} = \begin{bmatrix} \beta_1^{(i)} \\ \beta_2^{(i)} \\ \beta_3^{(i)} \end{bmatrix} = -\left[[\mathbf{Q}]^{\mathrm{T}}[\mathbf{J}_O][\dot{\mathbf{Q}}] + [\mathbf{Q}]^{\mathrm{T}}[\omega][\mathbf{J}_O][\mathbf{Q}]\right]\begin{bmatrix} \dot{\psi}_i \\ \dot{\theta}_i \\ \dot{\varphi}_i \end{bmatrix}. \qquad (39)$$

We also have

$$[\mathbf{Q}_i]^{\mathrm{T}}[\mathbf{J}_{O_i}][\mathbf{Q}_i] = \begin{bmatrix} J_{x_i}c^2\varphi_i c^2\theta_i + J_{y_i}s^2\varphi_i c^2\theta_i + J_{z_i}s^2\theta_i & J_{x_i}s\varphi_i c\varphi_i c\theta_i - J_{y_i}s\varphi_i c\varphi_i c\theta_i & J_{z_i}s^2\theta_i \\ J_{x_i}s\varphi_i c\varphi_i c\theta_i - J_{y_i}s\varphi_i c\varphi_i c\theta_i & J_{x_i}s^2\varphi_i + J_{y_i}c^2\varphi_i & 0 \\ J_{z_i}s\theta_i & 0 & J_{z_i} \end{bmatrix} \qquad (40)$$

and let us denote by $\gamma_{jk}^{(i)}$, $j, k = \overline{1,3}$, $i = \overline{1,2}$, the components of these matrices.

We will consider the order of parameters X_{O_1}, Y_{O_1}, Z_{O_1}, ψ_1, θ_1, φ_1, X_{O_2}, Y_{O_2}, Z_{O_2}, ψ_2, θ_2, φ_2.

We have only one constraints given by the belonging of the point C to the two bars. For the bar AC we may write $x_C = l_1/2$, $y_C = 0$, $z_C = 0$, while for the bar BC we have $x_C = -l_2/2$, $y_C = 0$, $z_C = 0$. From the relations

$$\begin{bmatrix} X_C \\ Y_C \\ Z_C \end{bmatrix} = \begin{bmatrix} X_{O_1} \\ Y_{O_1} \\ Z_{O_1} \end{bmatrix} + [\mathbf{A}]\begin{bmatrix} \dfrac{l_1}{2} \\ 0 \\ 0 \end{bmatrix}, \quad \begin{bmatrix} X_C \\ Y_C \\ Z_C \end{bmatrix} = \begin{bmatrix} X_{O_2} \\ Y_{O_2} \\ Z_{O_2} \end{bmatrix} + [\mathbf{A}]\begin{bmatrix} -\dfrac{l_2}{2} \\ 0 \\ 0 \end{bmatrix} \qquad (41)$$

it results the expressions

$$X_{O_1} + \frac{l_1}{2}\cos\theta_1\cos\varphi_1 = X_{O_2} - \frac{l_2}{2}\cos\theta_2\cos\varphi_2,$$

$$Y_{O_1} + \frac{l_1}{2}\left(\sin\psi_1\sin\theta_1\cos\varphi_1 + \cos\psi_1\sin\varphi_1\right) = Y_{O_2} - \frac{l_2}{2}\left(\sin\psi_2\sin\theta_2\cos\varphi_2 + \cos\psi_2\sin\varphi_2\right),$$

$$Z_{O_1} + \frac{l_1}{2}\left(-\cos\psi_1\sin\theta_1\cos\varphi_1 + \sin\psi_1\sin\varphi_1\right) = Z_{O_2} - \frac{l_2}{2}\left(-\cos\psi_2\sin\theta_2\cos\varphi_2 + \sin\psi_2\sin\varphi_2\right) \qquad (42)$$

and the constraint function read

$$f_1\left(X_{O_1}, ..., \varphi_2\right) = X_{O_1} - X_{O_2} + \frac{l_1}{2}\cos\theta_1\cos\varphi_1 + \frac{l_2}{2}\cos\theta_2\cos\varphi_2 = 0 \, ,$$

$$f_2\left(X_{O_1}, ..., \varphi_2\right) = Y_{O_1} - Y_{O_2} + \frac{l_1}{2}\left(\sin\psi_1\sin\theta_1\cos\varphi_1 + \cos\psi_1\sin\varphi_1\right) +$$

$$+ \frac{l_2}{2}\left(\sin\psi_2\sin\theta_2\cos\varphi_2 + \cos\psi_2\sin\varphi_2\right) = 0 \, ,$$

$$f_3\left(X_{O_1}, ..., \varphi_2\right) = Z_{O_1} - Z_{O_2} + \frac{l_1}{2}\left(-\cos\psi_1\sin\theta_1\cos\varphi_1 + \sin\psi_1\sin\varphi_1\right) +$$

$$\qquad\qquad(43)$$

$$+ \frac{l_2}{2}\left(-\cos\psi_2\sin\theta_2\cos\varphi_2 + \cos\psi_2\sin\varphi_2\right) = 0 \, .$$

The components of the matrix of constraints are

$$B_{11} = \frac{\partial f_1}{\partial X_{O_1}} = 1 \, , \quad B_{12} = \frac{\partial f_1}{\partial Y_{O_1}} = 0 \, , \quad B_{13} = \frac{\partial f_1}{\partial Z_{O_1}} = 0 \, , \quad B_{14} = \frac{\partial f_1}{\partial \psi_1} = 0 \, , \quad B_{15} = \frac{\partial f_1}{\partial \theta_1} = -\frac{l_1}{2}\sin\theta_1\cos\varphi_1 \, ,$$

$$B_{16} = \frac{\partial f_1}{\partial \varphi_1} = -\frac{l_1}{2}\cos\theta_1\sin\varphi_1 \, , \quad B_{17} = \frac{\partial f_1}{\partial X_{O_2}} = -1 \, , \quad B_{18} = \frac{\partial f_1}{\partial Y_{O_2}} = 0 \, , \quad B_{19} = \frac{\partial f_1}{\partial Z_{O_2}} = 0 \, , \quad B_{110} = \frac{\partial f_1}{\partial \psi_2} = 0 \, ,$$

$$B_{111} = \frac{\partial f_1}{\partial \theta_2} = -\frac{l_2}{2}\sin\theta_2\cos\varphi_2 \, , \quad B_{112} = \frac{\partial f_1}{\partial \varphi_2} = -\frac{l_2}{2}\cos\theta_2\sin\varphi_2 \qquad\qquad (44)$$

$$B_{21} = \frac{\partial f_2}{\partial X_{O_1}} = 0 \, , \quad B_{22} = \frac{\partial f_2}{\partial Y_{O_1}} = 1 \, , \quad B_{23} = \frac{\partial f_2}{\partial Z_{O_1}} = 0 \, , \quad B_{24} = \frac{\partial f_2}{\partial \psi_1} = \frac{l_1}{2}\left(\cos\psi_1\sin\theta_1 - \sin\psi_1\sin\varphi_1\right),$$

$$B_{25} = \frac{\partial f_2}{\partial \theta_1} = \frac{l_1}{2}\sin\psi_1\cos\theta_1\cos\varphi_1 \, , \quad B_{26} = \frac{\partial f_2}{\partial \varphi_1} = \frac{l_1}{2}\left(-\sin\psi_1\sin\theta_1\sin\varphi_1 + \cos\psi_1\cos\varphi_1\right), \quad B_{27} = \frac{\partial f_2}{\partial X_{O_2}} = 0 \, ,$$

$$B_{28} = \frac{\partial f_2}{\partial Y_{O_2}} = -1 \, , \qquad B_{29} = \frac{\partial f_2}{\partial Z_{O_2}} = 0 \, , \qquad B_{210} = \frac{\partial f_2}{\partial \psi_2} = \frac{l_2}{2}\left(\cos\psi_2\sin\theta_2\cos\varphi_2 - \sin\psi_2\sin\varphi_2\right),$$

$$B_{211} = \frac{\partial f_2}{\partial \theta_2} = \frac{l_2}{2}\sin\psi_2\cos\theta_2\cos\varphi_2 \, , \quad B_{212} = \frac{\partial f_2}{\partial \varphi_2} = \frac{l_2}{2}\left(-\sin\psi_2\sin\theta_2\sin\varphi_2 + \cos\psi_2\cos\varphi_2\right),$$

$$(45)$$

$$B_{31} = \frac{\partial f_3}{\partial X_{O_1}} = 0 \, , \quad B_{32} = \frac{\partial f_3}{\partial Y_{O_1}} = 0 \, , \quad B_{33} = \frac{\partial f_3}{\partial Z_{O_1}} = 1 \, , \quad B_{34} = \frac{\partial f_3}{\partial \psi_1} = \frac{l_1}{2}\left(\sin\psi_1\sin\theta_1\cos\varphi_1 + \cos\psi_1\sin\varphi_1\right),$$

$$B_{35} = \frac{\partial f_2}{\partial \theta_1} = -\frac{l_1}{2}\cos\psi_1\cos\theta_1\cos\varphi_1 \, , \quad B_{36} = \frac{\partial f_3}{\partial \varphi_1} = \frac{l_1}{2}\left(\cos\psi_1\sin\theta_1\sin\varphi_1 + \sin\psi_1\cos\varphi_1\right), \quad B_{37} = \frac{\partial f_3}{\partial X_{O_2}} = 0 \, ,$$

$$B_{38} = \frac{\partial f_3}{\partial Y_{O_2}} = 0 \, , \qquad B_{39} = \frac{\partial f_3}{\partial Z_{O_2}} = -1 \, , \qquad B_{310} = \frac{\partial f_3}{\partial \psi_2} = \frac{l_2}{2}\left(\sin\psi_2\sin\theta_2\cos\varphi_2 + \cos\psi_2\sin\varphi_2\right),$$

$$B_{311} = \frac{\partial f_3}{\partial \theta_2} = -\frac{l_2}{2}\cos\psi_2\cos\theta_2\cos\varphi_2 \, , \quad B_{312} = \frac{\partial f_3}{\partial \varphi_2} = \frac{l_2}{2}\left(\cos\psi_2\sin\theta_2\sin\varphi_2 + \sin\psi_2\cos\varphi_2\right). \qquad (46)$$

We may also write

$$\{\mathbf{F}_{s_1}\} = \begin{bmatrix} 0 & 0 & -m_1 g \end{bmatrix}^T, \; \{\mathbf{F}_{s_2}\} = \begin{bmatrix} 0 & 0 & -m_2 g \end{bmatrix}^T, \; \{\mathbf{F}_{\beta_1}\} = \begin{bmatrix} 0 & 0 & 0 \end{bmatrix}^T, \; \{\mathbf{F}_{\beta_2}\} = \begin{bmatrix} 0 & 0 & 0 \end{bmatrix}^T, \; \{\widetilde{\mathbf{F}}_{s_1}\} = \begin{bmatrix} 0 & 0 & 0 \end{bmatrix}^T, \; \{\widetilde{\mathbf{F}}_{s_2}\} = \begin{bmatrix} 0 & 0 & 0 \end{bmatrix}^T$$

$$, \; \{\widetilde{\mathbf{F}}_{\beta_1}\} = \begin{bmatrix} \beta_1^{(1)} & \beta_2^{(1)} & \beta_3^{(1)} \end{bmatrix}^T, \; \{\widetilde{\mathbf{F}}_{\beta_2}\} = \begin{bmatrix} \beta_1^{(2)} & \beta_2^{(2)} & \beta_3^{(2)} \end{bmatrix}^T, \qquad\qquad (47)$$

$$[\mathbf{M}_i] = \begin{bmatrix} m_i & 0 & 0 & 0 & 0 & 0 \\ 0 & m_i & 0 & 0 & 0 & 0 \\ 0 & 0 & m_i & 0 & 0 & 0 \\ 0 & 0 & 0 & \gamma_{11}^{(i)} & \gamma_{12}^{(i)} & \gamma_{13}^{(i)} \\ 0 & 0 & 0 & \gamma_{21}^{(i)} & \gamma_{22}^{(i)} & \gamma_{23}^{(i)} \\ 0 & 0 & 0 & \gamma_{31}^{(i)} & \gamma_{32}^{(i)} & \gamma_{33}^{(i)} \end{bmatrix}, \qquad \{\mathbf{F_{q_1}}\} = \begin{bmatrix} 0 & 0 & -mg_1 & 0 & 0 & 0 \end{bmatrix}^T, \qquad \{\mathbf{F_{q_2}}\} = \begin{bmatrix} 0 & 0 & -mg_2 & 0 & 0 & 0 \end{bmatrix}^T,$$

$$\{\widetilde{\mathbf{F}}_{q_1}\} = \begin{bmatrix} 0 & 0 & 0 & \beta_1^{(1)} & \beta_2^{(1)} & \beta_3^{(1)} \end{bmatrix}^T, \qquad \{\widetilde{\mathbf{F}}_{q_2}\} = \begin{bmatrix} 0 & 0 & 0 & \beta_1^{(2)} & \beta_2^{(2)} & \beta_3^{(2)} \end{bmatrix}^T, \qquad \{\mathbf{F_q}\} = \begin{bmatrix} 0 & 0 & -m_1 g & 0 & 0 & 0 & 0 & 0 & -m_2 g & 0 & 0 & 0 \end{bmatrix}^T,$$

$$\{\widetilde{\mathbf{F}}_q\} = \begin{bmatrix} 0 & 0 & 0 & \beta_1^{(1)} & \beta_2^{(1)} & \beta_3^{(1)} & 0 & 0 & 0 & \beta_1^{(2)} & \beta_2^{(2)} & \beta_3^{(2)} \end{bmatrix}^T,$$

$$\{\mathbf{F_q}\} + \{\widetilde{\mathbf{F}}_q\} = \begin{bmatrix} 0 & 0 & -m_1 g & \beta_1^{(1)} & \beta_2^{(1)} & \beta_3^{(1)} & 0 & 0 & -m_2 g & \beta_1^{(2)} & \beta_2^{(2)} & \beta_3^{(2)} \end{bmatrix}^T, \qquad \{\mathbf{C}\} = \begin{bmatrix} 0 & 0 & 0 \end{bmatrix}^T, \qquad \{\dot{\mathbf{C}}\} = \begin{bmatrix} 0 & 0 & 0 \end{bmatrix}^T,$$

$$\{\dot{\mathbf{B}}\} = \begin{bmatrix} \dot{B}_{11} & \dot{B}_{12} & \cdots & \dot{B}_{112} \\ \dot{B}_{21} & \dot{B}_{22} & \cdots & \dot{B}_{212} \\ \dot{B}_{31} & \dot{B}_{32} & \cdots & \dot{B}_{312} \end{bmatrix}^T, \qquad [\dot{\mathbf{B}}]\{\dot{\mathbf{q}}\} = \begin{bmatrix} \dot{B}_{15}\dot{\theta}_1 + \dot{B}_{15}\dot{\varphi}_1 + \dot{B}_{111}\dot{\theta}_2 + \dot{B}_{112}\dot{\varphi}_2 \\ \dot{B}_{24}\dot{\psi}_1 + \dot{B}_{25}\dot{\theta}_1 + \dot{B}_{26}\dot{\varphi}_1 + \dot{B}_{210}\dot{\psi}_2 + \dot{B}_{211}\dot{\theta}_2 + \dot{B}_{212}\dot{\varphi}_2 \\ \dot{B}_{34}\dot{\psi}_1 + \dot{B}_{35}\dot{\theta}_1 + \dot{B}_{36}\dot{\varphi}_1 + \dot{B}_{310}\dot{\psi}_2 + \dot{B}_{311}\dot{\theta}_2 + \dot{B}_{312}\dot{\varphi}_2 \end{bmatrix},$$

$$\{\mathbf{q}\} = \begin{bmatrix} X_{O_1} & Y_{O_1} & Z_{O_1} & \psi_1 & \theta_1 & \varphi_1 & X_{O_2} & Y_{O_2} & Z_{O_2} & \psi_2 & \theta_2 & \varphi_2 \end{bmatrix}^T, \quad \{\boldsymbol{\lambda}\} = \begin{bmatrix} \lambda_1 & \lambda_2 & \lambda_2 \end{bmatrix}^T. \tag{48}$$

We denote by $[\mathbf{M}]$ the fifteenth order squared matrix

$$[\mathbf{M}] = \begin{bmatrix} m_1 & 0 & 0 & 0 & 0 & 0 & 0 & 0 & 0 & 0 & 0 & 0 & -B_{11} & -B_{21} & -B_{31} \\ 0 & m_1 & 0 & 0 & 0 & 0 & 0 & 0 & 0 & 0 & 0 & 0 & -B_{12} & -B_{22} & -B_{32} \\ 0 & 0 & m_1 & 0 & 0 & 0 & 0 & 0 & 0 & 0 & 0 & 0 & -B_{13} & -B_{23} & -B_{33} \\ 0 & 0 & 0 & \gamma_{11}^{(1)} & \gamma_{12}^{(1)} & \gamma_{13}^{(1)} & 0 & 0 & 0 & 0 & 0 & 0 & -B_{14} & -B_{24} & -B_{34} \\ 0 & 0 & 0 & \gamma_{21}^{(1)} & \gamma_{22}^{(1)} & \gamma_{23}^{(1)} & 0 & 0 & 0 & 0 & 0 & 0 & -B_{15} & -B_{25} & -B_{35} \\ 0 & 0 & 0 & \gamma_{31}^{(1)} & \gamma_{32}^{(1)} & \gamma_{33}^{(1)} & 0 & 0 & 0 & 0 & 0 & 0 & -B_{16} & -B_{26} & -B_{36} \\ 0 & 0 & 0 & 0 & 0 & 0 & m_2 & 0 & 0 & 0 & 0 & 0 & -B_{17} & -B_{27} & -B_{37} \\ 0 & 0 & 0 & 0 & 0 & 0 & 0 & m_2 & 0 & 0 & 0 & 0 & -B_{18} & -B_{28} & -B_{38} \\ 0 & 0 & 0 & 0 & 0 & 0 & 0 & 0 & m_2 & 0 & 0 & 0 & -B_{19} & -B_{29} & -B_{39} \\ 0 & 0 & 0 & 0 & 0 & 0 & 0 & 0 & 0 & \gamma_{11}^{(2)} & \gamma_{12}^{(2)} & \gamma_{13}^{(2)} & -B_{110} & -B_{210} & -B_{310} \\ 0 & 0 & 0 & 0 & 0 & 0 & 0 & 0 & 0 & \gamma_{21}^{(2)} & \gamma_{22}^{(2)} & \gamma_{23}^{(2)} & -B_{111} & -B_{211} & -B_{311} \\ 0 & 0 & 0 & 0 & 0 & 0 & 0 & 0 & 0 & \gamma_{31}^{(2)} & \gamma_{31}^{(2)} & \gamma_{33}^{(2)} & -B_{112} & -B_{212} & -B_{312} \\ B_{11} & B_{12} & B_{13} & B_{14} & B_{15} & B_{16} & B_{17} & B_{18} & B_{19} & B_{110} & B_{111} & B_{112} & 0 & 0 & 0 \\ B_{21} & B_{22} & B_{23} & B_{24} & B_{25} & B_{26} & B_{27} & B_{28} & B_{29} & B_{210} & B_{211} & B_{212} & 0 & 0 & 0 \\ B_{31} & B_{32} & B_{33} & B_{34} & B_{35} & B_{36} & B_{37} & B_{38} & B_{39} & B_{310} & B_{311} & B_{312} & 0 & 0 & 0 \end{bmatrix} \tag{49}$$

and it results the equation of motion

$$[\overline{\mathbf{M}}] \begin{bmatrix} \{\ddot{\mathbf{q}}\} \\ \{\boldsymbol{\lambda}\} \end{bmatrix} = \begin{bmatrix} \{\mathbf{F_q}\} + \{\widetilde{\mathbf{F}}_q\} \\ -[\dot{\mathbf{B}}]\{\dot{\mathbf{q}}\} \end{bmatrix}. \tag{50}$$

The reaction at the point C has the components

$$\lambda_1 \begin{bmatrix} B_{11} \\ B_{12} \\ B_{13} \end{bmatrix} = \begin{bmatrix} \lambda_1 \\ 0 \\ 0 \end{bmatrix}, \quad \lambda_2 \begin{bmatrix} B_{21} \\ B_{22} \\ B_{23} \end{bmatrix} = \begin{bmatrix} 0 \\ \lambda_2 \\ 0 \end{bmatrix}, \quad \lambda_3 \begin{bmatrix} B_{31} \\ B_{32} \\ B_{33} \end{bmatrix} = \begin{bmatrix} 0 \\ 0 \\ \lambda_3 \end{bmatrix} \tag{51}$$

or, equivalently,

$$\lambda_1 \begin{bmatrix} B_{17} \\ B_{18} \\ B_{19} \end{bmatrix} = \begin{bmatrix} -\lambda_1 \\ 0 \\ 0 \end{bmatrix}, \quad \lambda_2 \begin{bmatrix} B_{27} \\ B_{28} \\ B_{29} \end{bmatrix} = \begin{bmatrix} 0 \\ -\lambda_2 \\ 0 \end{bmatrix}, \quad \lambda_3 \begin{bmatrix} B_{37} \\ B_{38} \\ B_{39} \end{bmatrix} = \begin{bmatrix} 0 \\ 0 \\ -\lambda_3 \end{bmatrix}, \tag{52}$$

verifying the principle of action and reaction.

On the other hand,

$$\left[\left[Q_i\right]^{-1}\right]^{\mathrm{T}} = \frac{1}{\cos\theta_i}\begin{bmatrix} \cos\varphi_i & \sin\varphi_i\cos\theta_i & -\cos\varphi_i\sin\theta_i \\ -\sin\varphi_i & \cos\varphi_i\cos\theta_i & \sin\varphi_i\sin\theta_i \\ 0 & 0 & \cos\theta_i \end{bmatrix}, \tag{53}$$

$$\left\{\mathbf{F}_{G\beta_1}^{(1)}\right\} = \lambda_1\begin{bmatrix} B_{14} \\ B_{15} \\ B_{16} \end{bmatrix} = -\lambda_1\frac{l_1}{2}\begin{bmatrix} 0 \\ \sin\theta_1\cos\varphi_1 \\ \cos\theta_1\sin\varphi_1 \end{bmatrix}, \qquad \left\{\mathbf{F}_{G\beta_2}^{(1)}\right\} = \lambda_2\begin{bmatrix} B_{24} \\ B_{25} \\ B_{26} \end{bmatrix} = \lambda_2\frac{l_1}{2}\begin{bmatrix} \cos\psi_1\sin\theta_1\cos\varphi_1 - \sin\psi_1\sin\varphi_1 \\ \sin\psi_1\cos\theta_1\cos\varphi_1 \\ -\sin\psi_1\sin\theta_1\sin\varphi_1 + \cos\psi_1\cos\varphi_1 \end{bmatrix},$$

$$\left\{\mathbf{F}_{G\beta_3}^{(1)}\right\} = \lambda_3\begin{bmatrix} B_{34} \\ B_{35} \\ B_{36} \end{bmatrix} = \lambda_3\frac{l_1}{2}\begin{bmatrix} \sin\psi_1\sin\theta_1\cos\varphi_1 + \cos\psi_1\sin\varphi_1 \\ -\cos\psi_1\cos\theta_1\cos\varphi_1 \\ \cos\psi_1\sin\theta_1\sin\varphi_1 + \sin\psi_1\cos\varphi_1 \end{bmatrix} \tag{54}$$

and it results the projections of the components of the moment of the reaction that acts upon the first body at the point C, onto the axes of the system $O_1 x_1 y_1 z_1$

$$\begin{bmatrix} M_{x_1} \\ M_{y_1} \\ M_{z_1} \end{bmatrix} = \left[\left[Q_1\right]^{-1}\right]^{\mathrm{T}}\left\{\left\{\mathbf{F}_{G\beta_1}^{(1)}\right\} + \left\{\mathbf{F}_{G\beta_2}^{(1)}\right\} + \left\{\mathbf{F}_{G\beta_3}^{(1)}\right\}\right\} = \frac{l_1}{2}\begin{bmatrix} E_{11}\cos\varphi_1 + E_{21}\sin\varphi_1\cos\theta_1 - E_{31}\cos\varphi_1\sin\theta_1 \\ -E_{11}\sin\varphi_1 + E_{21}\cos\varphi_1\cos\theta_1 + E_{31}\sin\varphi_1\sin\theta_1 \\ E_{31}\cos\theta_1 \end{bmatrix}, \tag{55}$$

where

$$E_{11} = \lambda_2\left(\cos\psi_1\sin\theta_1\cos\varphi_1 - \sin\psi_1\sin\varphi_1\right) + \lambda_3\left(\sin\psi_1\sin\theta_1\cos\varphi_1 + \cos\psi_1\sin\varphi_1\right),$$

$$E_{21} = -\lambda_1\sin\theta_1\cos\varphi_1 + \lambda_2\sin\psi_1\cos\theta_1\cos\varphi_1 - \lambda_3\cos\psi_1\cos\theta_1\cos\varphi_1,$$

$$E_{31} = -\lambda_1\cos\theta_1\sin\varphi_1 + \lambda_2\left(-\sin\psi_1\sin\theta_1\sin\varphi_1 + \cos\psi_1\cos\varphi_1\right) +$$
$$+ \lambda_3\left(\cos\psi_1\sin\theta_1\sin\varphi_1 + \sin\psi_1\cos\varphi_1\right). \tag{56}$$

We also have

$$\left\{\mathbf{F}_{G\beta_1}^{(2)}\right\} = \lambda_1\begin{bmatrix} B_{110} \\ B_{111} \\ B_{112} \end{bmatrix} = -\lambda_1\frac{l_2}{2}\begin{bmatrix} 0 \\ -\sin\theta_2\cos\varphi_2 \\ \cos\theta_2\sin\varphi_2 \end{bmatrix}, \quad \left\{\mathbf{F}_{G\beta_2}^{(2)}\right\} = \lambda_2\begin{bmatrix} B_{210} \\ B_{211} \\ B_{212} \end{bmatrix} = \lambda_2\frac{l_2}{2}\begin{bmatrix} \cos\psi_2\sin\theta_2\cos\varphi_2 - \sin\psi_2\sin\varphi_2 \\ \sin\psi_2\cos\theta_2\cos\varphi_2 \\ -\sin\psi_2\sin\theta_2\sin\varphi_2 + \cos\psi_2\cos\varphi_2 \end{bmatrix},$$

$$\left\{\mathbf{F}_{G\beta_3}^{(2)}\right\} = \lambda_3\begin{bmatrix} B_{310} \\ B_{311} \\ B_{312} \end{bmatrix} = \lambda_3\frac{l_2}{2}\begin{bmatrix} \sin\psi_2\sin\theta_2\cos\varphi_2 + \cos\psi_2\sin\varphi_2 \\ -\cos\psi_2\cos\theta_2\cos\varphi_2 \\ \cos\psi_2\sin\theta_2\sin\varphi_2 + \sin\psi_2\cos\varphi_2 \end{bmatrix} \tag{57}$$

and it results the projections of the components of the moment of the reaction that acts upon the second body at the point C, onto the axes of the system $O_2 x_2 y_2 z_2$

$$\begin{bmatrix} M_{x_2} \\ M_{y_2} \\ M_{z_2} \end{bmatrix} = \left[\left[Q_2\right]^{-1}\right]^{\mathrm{T}}\left\{\left\{\mathbf{F}_{G\beta_1}^{(2)}\right\} + \left\{\mathbf{F}_{G\beta_2}^{(2)}\right\} + \left\{\mathbf{F}_{G\beta_3}^{(2)}\right\}\right\} = \frac{l_2}{2}\begin{bmatrix} E_{12}\cos\varphi_2 + E_{22}\sin\varphi_2\cos\theta_2 - E_{32}\cos\varphi_2\sin\theta_2 \\ -E_{12}\sin\varphi_2 + E_{22}\cos\varphi_2\cos\theta_2 + E_{32}\sin\varphi_2\sin\theta_2 \\ E_{32}\cos\theta_2 \end{bmatrix}$$
$$\tag{58}$$

where

$$E_{12} = \lambda_2\left(\cos\psi_2\sin\theta_2\cos\varphi_2 - \sin\psi_2\sin\theta_2\right) + \lambda_3\left(\sin\psi_2\sin\theta_2\cos\varphi_2 + \cos\psi_2\sin\varphi_2\right),$$

$$E_{22} = -\lambda_1\sin\theta_2\cos\varphi_2 + \lambda_2\sin\psi_2\cos\theta_2\cos\varphi_2 - \lambda_3\cos\psi_2\cos\theta_2\cos\varphi_2,$$

$$E_{32} = \lambda_2\left(-\sin\psi_2\sin\theta_2\sin\varphi_2 + \cos\psi_2\cos\varphi_2\right) +$$
$$-\lambda_1\cos\theta_2\sin\varphi_2 + \lambda_3\left(\cos\psi_2\sin\theta_2\sin\varphi_2 + \sin\psi_2\cos\varphi_2\right). \tag{59}$$

3. CONCLUSION

In this paper we presented the generalization of the equation of motion for a mechanical system of an arbitrary number of rigid bodies. The reader may easily observe the procedure which extends the matrix of inertia, the matrix of constraints, and the matrix of generalized forces. The absence of the link between two bodies leads to a simpler form for the matrix equation of motion.

If the constraints are independent, then the left-hand matrix in the equation of motion is invertible (not necessary in a classical way). Discussions about this property of inversion could be found in [2], where the authors deal with the Moore-Penrose inverse for the matrix of constraints.

For the planar cases one may consider a simplified version of the method (the translation along the axis Oz, and the rotations about the axes Ox and Oy vanish). This is equivalent to consider particular initial conditions for this parameters, and no motion corresponding to them. The paper also includes a complete solved example.

REFERENCES

[1] Pandrea N., Stănescu N.–D., Dynamics of the Rigid Solid with General Constraints by a Multibody Approach, John Wiley & Sons, Chichester, 2015.

[2] Udwadia, F. E., Kalaba, R. E., Analytical Dynamics. A New Approach, Cambridge University Press, Cambridge, 1996.

6ᵗʰ International Conference
"Computational Mechanics and Virtual Engineering "
COMEC 2015
15-16 October 2015, Braşov, Romania

DETECTING HUMAN EMOTIONS WITH AN ADAPTIVE NEURO-FUZZY INFERENCE SYSTEM

Phd. Gheorghe Gîlcă[1], Prof Nicu-George Bîzdoacă[1]

[1] University of Craiova, Craiova, ROMANIA, gigi@robotics.ucv.ro, nicu@robotics.ucv.ro

Abstract: This article deals with an emotion recognition system based on fuzzy sets. Human faces are detected in images with the Viola - Jones algorithm and for their tracking in video sequences we used the Camshift algorithm. The detected human faces are transferred to the sugeno type decisional fuzzy system, which is based on the variable fuzzyfication measurements of the face: eyebrow, eyelid and mouth. The system can easily determine the emotional state of a person.
Keywords: emotion, face detection, face tracking, fuzzy rules, training

1. INTRODUCTION

The verbal and nonverbal communication is the most common in relationships between people, respectively between machines and people. Mehrabian [1] explained in his work in what percentage each form of communication contributes to the overall message delivered between two individuals: the linguistic message is only 7%, the vocal intonation is 38% while the facial expression is 55% of the sent message .

The recognition and classification of the human facial expressions is an important aspect to develop in an automatic recognition system in order to facilitate the communication between machines and people.

Facial emotions are used in areas such as telemedicine, distance learning, robotics and automotive. There are six main types of basic emotions: happiness, anger, fear, sadness, disgust and surprised. All other emotions are variations of these basic emotions. Each emotion is characterized by psychological and behavioral qualities, including movement, posture, voice, facial expression and heart rate fluctuation. Although there are many different types of emotions, they all have some common characteristics.

Charles Darwin was the first to bring into discussion the universality of emotions, using this idea in his book [2], to support the theory of evolution, arguing that emotions are mental reaction patterns imprinted in the nervous system.

The American psychologist Paul Ekman from the University of California (expert in the study of emotions, their manifestations and the study of the mechanism of lies), based on the results of a long research on all the continents, finds that there are six facial expressions which can be recognized by any person belonging to any culture on the planet: fear, anger, sadness, surprise, disgust and joy [3]. The universality of these events may be considered to be a strong indication that these emotions are six basic emotions related to human nature.

Adaptive Neuro-Fuzzy Inference System (ANFIS) classifies the expressions of a supplied face into seven basic categories like: surprised, neutral, sad, disgusted, fear, happy and angry.

2. RELATED WORKS

Fatemeh Shahrabi Farahani et al., present a new method based on fuzzy logic for recognizing emotion based on the eyes and mouth features. Their approach shows an emotion recognition system which has 3 levels: face detection, feature extraction and classification. The fuzzy logic used by the authors in the third stage is performed using the Mamdani type inference relations with 94 rules of the trapezoidal membership functions to encode facial attributes and their mapping of the emotion space [4].

Joseph W. Matiko et al. present in their paper, an algorithm designed to classify emotions into two categories: positive and negative. Their proposed algorithm is based on fuzzy sets making the fuzzyfication of signal from the EEG. Their method is advantageous compared to the literature, because the classification includes the type of emotions, but also its power [5].

In their paper, Yong-Hwan Lee et al. proposed a method of extraction and recognition of facial expression and emotions on mobile cameras. They formulated a classification model using 65 landmarks in order to estimate facial expressions [6]. With their method we can recognize three types of emotions: neutral, happy and angry. Khandait S.P. et al. perform in their work a comparative study of two methods: The Fuzzy Inference Adaptive Neuro-System (ANFIS) and The Back Propagation Neural Network (BPNN) to classify facial expressions [7].

3. METODOLOGY

We have proposed an emotions recognition system with a simple structure as in Figure 1.

The system consists of three subsystems: the first is to detect the human face in images, the second is to track the human face, and the third is to decide what emotion is shown on the human face.

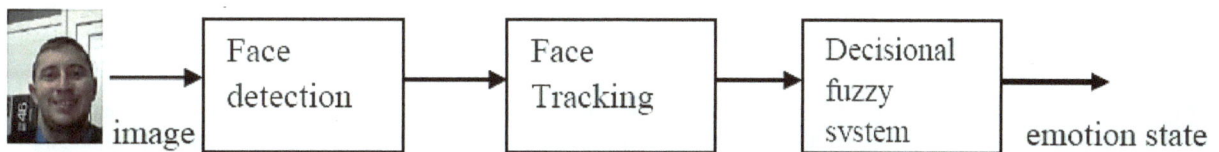

Figure 1: Structure of proposed system

3.1. Face detection using the Viola-Jones algorithm

Figure 2 [8] illustrates the Viola and Jones object detection algorithm in two steps: the detecting of the face in an image region and then applying on it a cascade of boosted classifiers. The first step detection window is realized by scanning the same image many times, each time with a new size. During the second stage each window is passed through a cascade of classifiers which in its turn is divided in various steps, each step uses a set of weak learners.

Each stage is trained to select only the wanted images using a technique called boosting. Boosting has the advantage of training a very accurate classifier to choose a weighted average of the decisions taken by the weak learners. Each level of the labeled region belonging to the classifier is defined as being the current location of the sliding window, being positive or negative.

The detector reports a found face in the current location when the final level classifies the region as being positive. If we have the k classifiers in a cascade, the result of detection rate, D, and false positive rate, F, is given by the product rates on each stage classifier [9]:

$$D = \prod_{i=1}^{k} d_i , \ F = \prod_{i=1}^{k} f_i \tag{1}$$

,where d_i is the detection rate of the i the classifier in the examples that get through to it and f_i is the false detection rate of the i the classifier in the examples that get through to it.

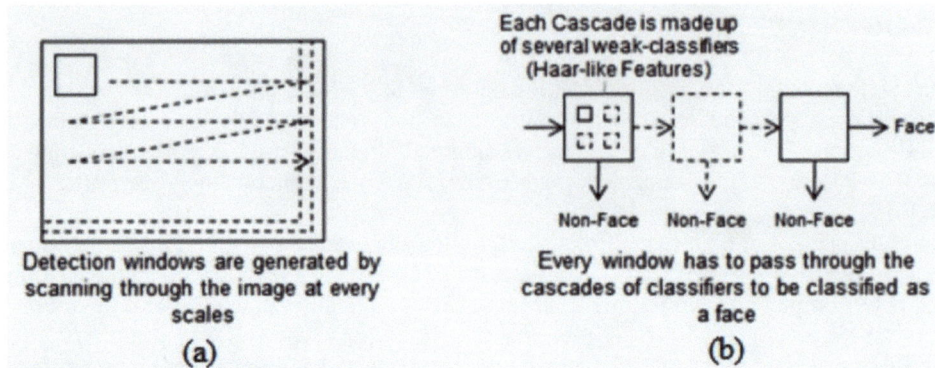

Figure 2: The Viola-Jones Object Detection Algorithm: a) Detection windows ; b) Classified windows

3.2. Face tracking using the Camshift algorithm

For face tracking in a video sequence, we used the camshift algorithm proposed by the authors in their paper [10]:

a) Set the calculation region of the probability distribution for images.
b) Choose an initial location of the Mean Shift search window; it will be the tracked target.
c) Calculate a color probability distribution of the region centered on the Mean Shift search window, where the target is slightly larger than Mean Shift window size.
d) Run Mean Shift algorithm to find the centroid of the probability image. Store the zero moment (the area) and the centroid location.
e) For the next frame, put the search window at the mean location found in Step d) and set the window size to a function of the zeroth moment. Go to Step c).

3.3. The decisional fuzzy system

After the operations of detection and tracking of the human face, it goes through a fuzzy system modeled using the Fuzzy Logic Toolbox of Matlab. The system takes into account in making the decision three face variables that change for each emotional state. Our system is modeled to recognize seven states: happiness, anger, fear, sadness, disgust, surprise and neutral. More details of this system are given in the next section.

4. THE DESIGN OF THE ADAPTIVE NEURO-FUZZY LOGIC SYSTEM IN DETERMINING EMOTION RECOGNITION

The fuzzy model projected by us looks like Figure 3:

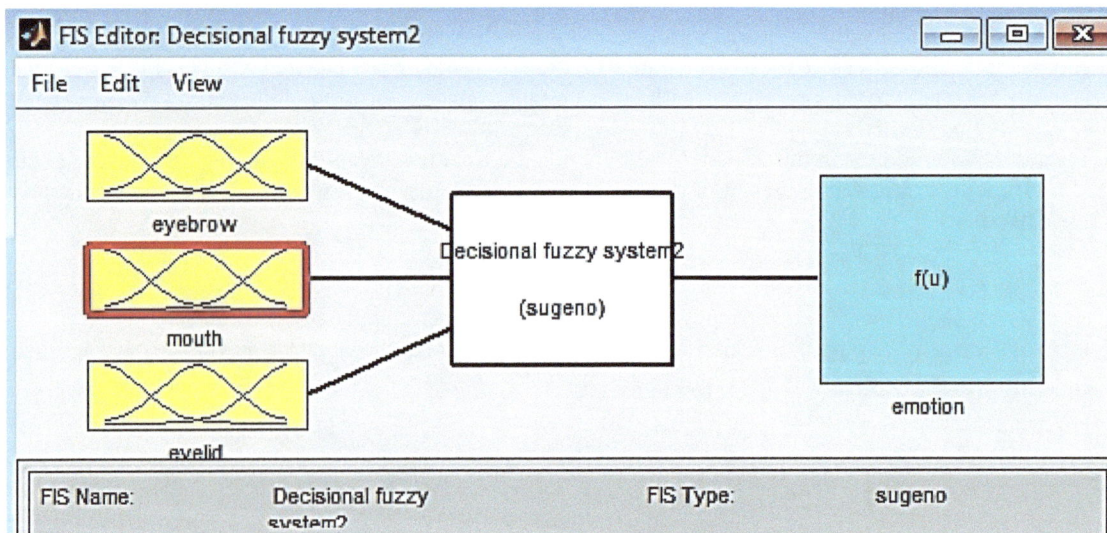

Figure 3: The fuzzy system model

The three yellow rectangles represent the three inputs of the fuzzy logic system: eyelid, eyebrow and mouth. In these three blocks the fuzzyfication takes place. The light gray rectangle represents the process of inference and the turquoise rectangle represents the output fuzzyfication block (in these rectangle the defuzzyfication takes place).

4.1. Inputs

In this subsection we define universes of discussion for the 3 inputs variables thus:
- Eyelid: [40; -100];
- Eyebrow: [100;-100]:
- Mouth: [10;-100].

The associated fuzzy set of input variables are: small, moderate and large.

4.2. Outputs

For the output variable emotion, the universe of discussion is [0;1.4] and the associated fuzzy sets are: happy, sad, surprised, fear, anger, disgusted and neutral, each with the following universes of discussion, in order: [0;0.2], [0.2;0.4], [0.4;0.6], [0.6;0.8], [0.8;1], [1;1.2], [1.2;1.4].

The Sugeno linear output model has the general form:

$$a_0 X_0 + a_1 X_1 + ... + a_i X_i + z \tag{2}$$

Where a_i is a constant parameter and X_i is input variables. For constant functions, the values of a_i is equal to zero and, hence Y = z (constant value). For the linear function model, the a_i values are entered through the ANFIS editor, while X_i are the values of the input factors. The correlation between the input and the output variables is done through a set of fuzzy rules. Each rule uses AND/OR connectors to connect various input factors with a particular output emotion.

4.3. The defining of a set of rules for the decisional fuzzy system

The system works with fuzzy rules type if ..., then ..., that condition of if must be complied with in order to deduce the desired output, which makes the system to be decisional. We have created a knowledge base for the system with the seven following rules:
- If (the mouth is moderate) and (Eyelid is small) and (Eyebrow is moderate) then (Emotion is happy);
- If (mouth is small) and (Eyelid is large) and (Eyebrow is moderate) then (Emotion is sad);
- If (Mouth is moderate) and (Eyelid is large) and (Eyebrow is small) then (Emotion is anger);
- If (Mouth is large) and (Eyelid is large) and (Eyebrow is large) then (Emotion is surprised);
- If (Mouth is moderate) and (Eyelid is large) and (Eyebrow is large) then (Emotion is fear);
- If (Mouth is small) and (Eyelid is moderate) and (Eyebrow is small) then (Emotion is disgust);
- If (Mouth is small) and (Eyelid is moderate) and (Eyebrow is moderate) then (Emotion is neutral);

E.g. if we have the following values of the input variables: Mouth is -45, Eyelid is -100, and Eyebrow is 0 then the output will be happy.

4.4. The results design of the fuzzy system decisional system

Figure 4 shows the operation of the fuzzy system designed by us for the example above. The results are found by the system as being valid only for the first rule, where the values of the three variables belong to the fuzzy sets in the fuzzification process. The system output is that the variable has the value equal with 0.2, so it was classified as happy.

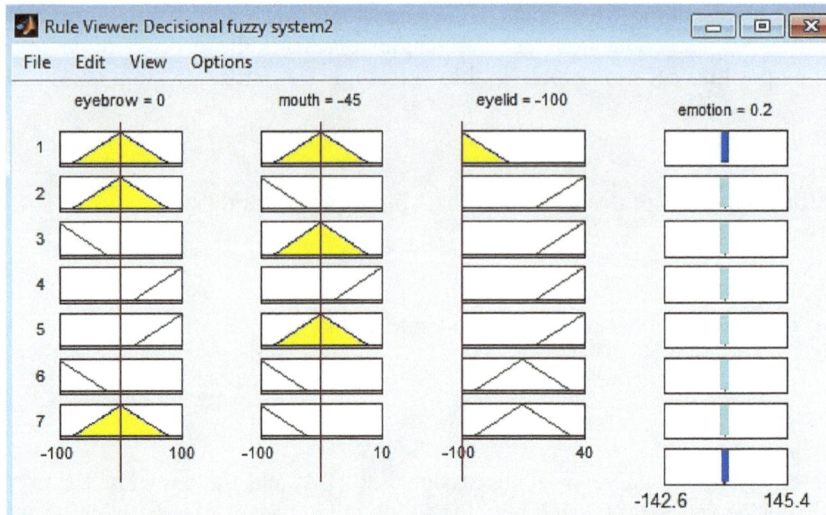

Figure 4: Happy emotional state classification by the decisional fuzzy system

4.5. The ANFIS Model Structure

The neuro-fuzzy model structure consists of five layers: the first layer is the input variables (mouth, eyebrow, eyelid), the second layer is the membership functions of inputs, which are the type gaussfm, gauss2mf and psigfm; the third layer is the rules system (seven in total); the fourth layer is the emotional state (also seven), and the fifth layer is the output system.

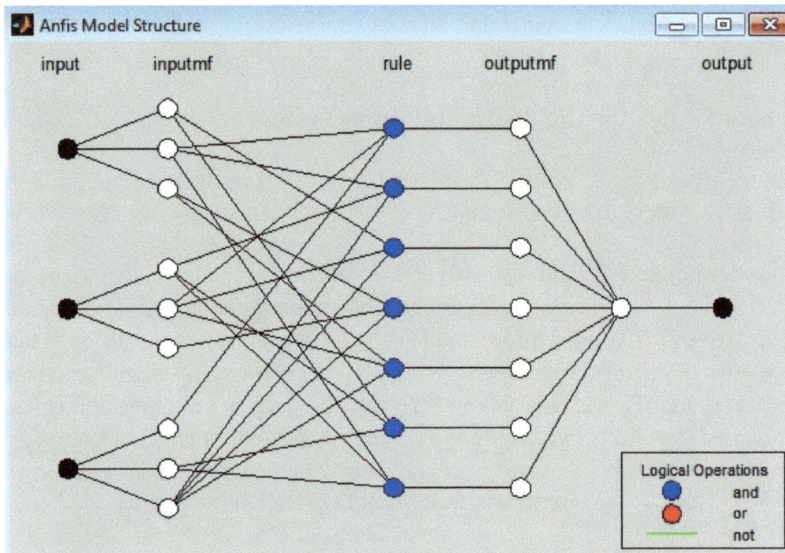

Figure 6: The Neuro-Fuzzy Structure for the proposed system

We have achieved the best results for the training of neuro-fuzzy inference system with the membership function psigmf. The results are displayed for each input in Figure 7.

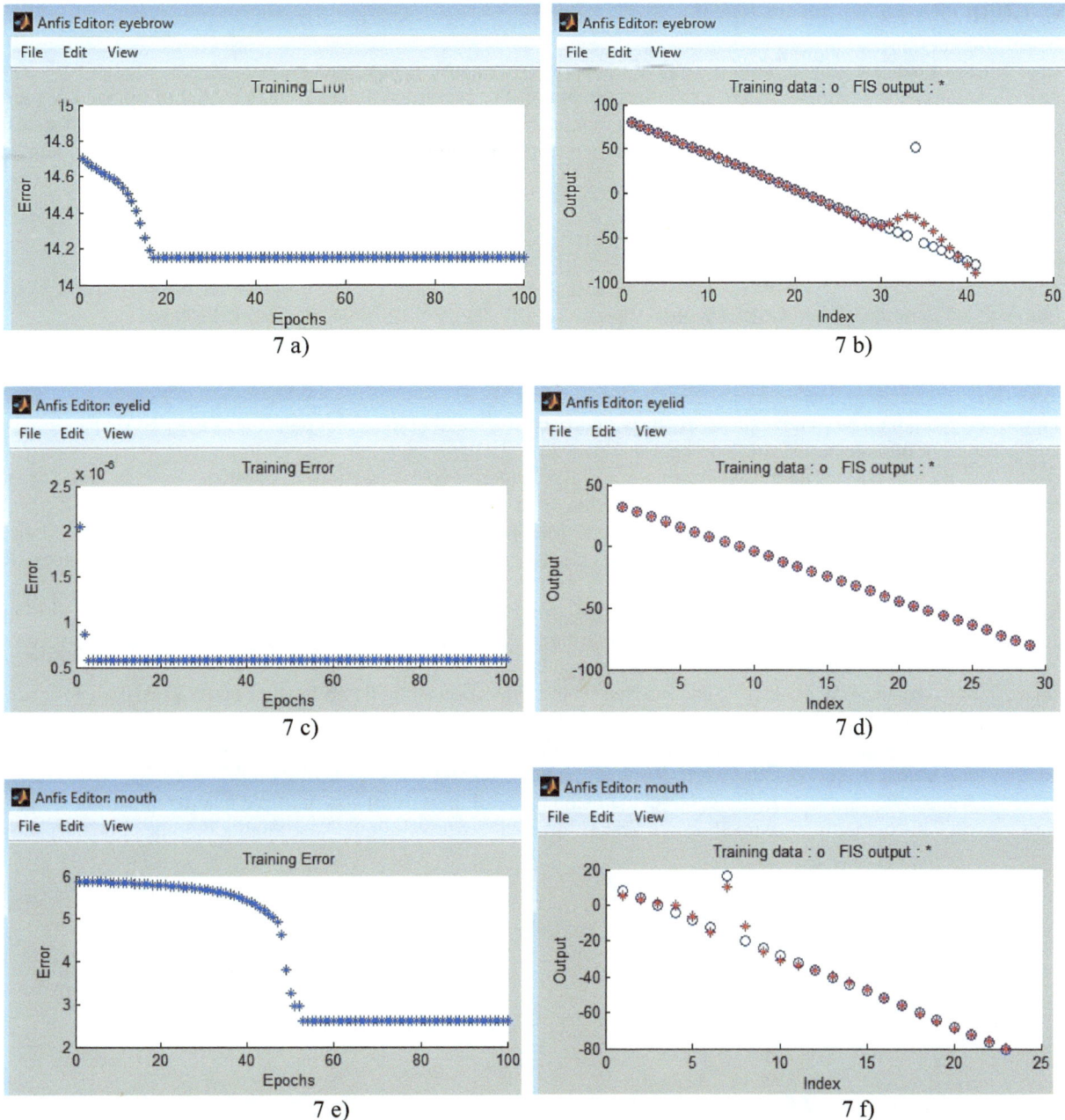

Figure 7: The experimental results for the training of neuro-fuzzy inference system

In Figure 7 a) the graph of training error of the eyebrow input variable is presented, and 7 b) shows the graph of points of its (blue circles) and the fuzzy system output (red asterisk). In Figure 7 c) the graph of training error of the eyelid input variable is presented, and 7 b) shows the graph of points of its (blue circles) and the fuzzy system output (red asterisk). In Figure 7 e) the graph of training error of the mouth input variable is presented, and 7 f) shows the graph of points of its (blue circles) and the fuzzy system output (red asterisk).

From Figure 7 we can deduce that the system has the smallest error at the eyelid input variable, which is equal with $\varepsilon=0.5 \times 10^{-6}$ and the output data system perfectly follows the input data. For the other two variables the output system is acceptable and the training error is a little higher than the the eyebrow variable.

6. CONCLUSION

The proposed system can recognize emotions quickly and easily, because it has strict rules in its knowledge base and membership functions of the output variable are not overlapping. We presented a neuro fuzzy model of human emotion detection using three factors of the human face that can make a big emphasis in the classification of the emotional states. As future work proposed system will be expanded to recognize human emotion in a real time, which will be implemented SociBot robot.

REFERENCES

[1] Mehrabian. A, Communication without Words , Psychology Today, Vol. 2, No.4, pp 53-56, 1968.

[2] Darwin, C., The expression of the emotions in man and animals, London, John Murray, 1872.

[3] Ekman, P., Friesen, W., Facial Action Coding System: Investigator's Guide, Palo Alto, CA: Consulting Psychologists Press, 1978.

[4] Farahani, F.S., Sheikhan, M., Farrokhi A., A fuzzy approach for facial emotion recognition, Fuzzy Systems (IFSC), 2013 13th Iranian Conference on, vol., no., pp.1,4, 27-29 Aug. 2013.

[5] Matiko, J.W., Beeby, S.P., Tudor, J., Fuzzy logic based emotion classification, Acoustics, Speech and Signal Processing (ICASSP), 2014 IEEE International Conference on , vol., no., pp.4389,4393, 4-9 May 2014.

[6] Lee, Y-H., Han, W., Kim, Y., Kim, B, Facial Feature Extraction Using an Active Appearance Model on the iPhone, Innovative Mobile and Internet Services in Ubiquitous Computing (IMIS), 2014 Eighth International Conference on , vol., no., pp.196,201, 2-4 July 2014.

[7] S.P.Khandait, R.C.Thool, P.D.Khandait, Comparative Analysis of ANFIS and NN Approach for Expression Recognition using Geometry Method, International Journal of Advanced Research in Computer Science and Software Engineering, Volume 2, Issue 3, ISSN: 2277 128X, March 2012.

[8] Mustafah, Y.M., Azman, A.W., Out-of-Plane Rotated Object Detection using Patch Feature based Classifier, Procedia Engineering, Volume 41, 2012, Pages 170-174, ISSN 1877-7058.

[9] Castrillón, M., Déniz, O., Hernández, D., Lorenzo, J., A comparison of face and facial feature detectors based on the Viola–Jones general object detection framework, Volume 22, Issue 3, Pages 481-494, ISSN 0932-8092, May 2011.

[10] Allen, G.J., Xu, R.Y.D., Jesse, S.J., Object tracking using camshift algorithm and multiple quantized feature spaces, Proceedings of the Pan-Sydney area workshop on Visual information processing, Australian Computer Society, Inc., 2004.

6th International Conference
"Computational Mechanics and Virtual Engineering "
COMEC 2015
15-16 October 2015, Braşov, Romania

PRESSURE WAVE GENERATOR FOR DEVELOPING RESONANCE IN A CLOSED PIPE, USED FOR AN ALTERNANT-HYDRAULIC TRANSMISSION

Sebastian Radu [1], Horia Abăităncei [2], Gheorghe Alexandru Radu [3], Marton Iakab-Peter

[1] Transilvania University, Braşov, ROMANIA, s.radu@unitbv.ro,
[2] Transilvania University, Braşov, ROMANIA, h.abaitancei@unitbv.ro,
[3] Transilvania University, Braşov, ROMANIA, radugal@unitbv.ro,
[4] Transilvania University, Braşov, ROMANIA, iakab.peter.marton@gmail.com,

Abstract : *To meet the current requirements imposed by the vehicle manufacturers to reduce significantly the fuel consumption and associated CO_2 emissions, it is necessary the research and development of new technical emission and consumption quantity reducing systems. A direction that has emerged in recent years is to develop hybrid-hydraulic propulsion systems which permit to reconsider the research objectives of the engines for an optimal functioning. The scientific approach is based on the fundamental research regarding the pressure waves energy generation - propagation and conversion process, research based on 2D simulations of the flow and on experimental results obtained by waves' piezoelectric measuring. The conclusions from the fundamental study are further used for developing mechano-hydraulic resonant systems designed to be integrated into the vehicle's propulsion system.*
Keywords : *fluid power, pressure wave, CFD, hybrid propulsion system*

1. INTRODUCTION

This paper presents and analyses the physical phenomenon that fundaments the structure of the mechano-hydraulic systems which use the resonance phenomenon and which can be integrated in the propulsion system of auto vehicles, in order to highlight the complexity of the factors which influence the development of the dynamic working mode of the system. The idea of using dynamic phenomenon theory comes from the following reasoning: identifying a point characterized by maximum potential energy is followed by searching for the trajectories of which this point belongs to, and then decide, using technical criteria, the optimal trajectory. After that, the system has to support the transformations suggested by the mathematical trajectory, for the environment to be capable to support the dynamic phenomenon, and to have the necessary parametric sensitivity for the transition between the initial conditions and maximum energy point. Technically, this transition needs to be done with energetic benefit, and to identify the potential transition trajectories through the analysis of the dynamic events
A dynamic oscillatory system has more chance for developing dynamic events if the dissipation in the system is minimal. This feature is favorable for liquids where oscillation damping is very low [1].

2. SYSTEM DESCRIPTION.

Structurally, the technical solution "Hydraulic system for road vehicle propulsion", is a driving system, as it has a hydraulic energy source, transmission which contains the command system and the actuators that converts hydraulic energy into mechanical energy [2]. (**Error! Reference source not found.**1. – upper side).

The hydraulic energy generator is composed of a cam mechanism driving a piston, exerting a force F of speed **v**, and generating the Q_1 flow. The systems continue with a pipe capable of sustaining wave propagation, phenomenon which at the other end of the pipe is characterized by Q_2 flow. The phenomenon is unsteady, and depending on load opposing the piston at the end of the pipe, will generate p_1 and p_2 pressures on both ends of the pipe, due to the system load. Using a piston – lever – one- way clutch system, the hydraulic power is converted into rotational mechanical power described by the torque **M** and angular speed ω

Dynamically, the system is made of two mechanical components and a hydraulic component in between them, resulting in a mechano-hydraulic coupled system. The assembly is characterized as a dynamic system with m_1 mass and elastic constant k_1 of the generator, resonant angular frequency ω_{1p}, inertia **L**, capacity C, and the resonant pulsation of the liquid column ω_{Lp}, while representative for the execution element are the mass m_2, the elastic constant **k2**, with the resonant angular frequency of $\omega2p.$

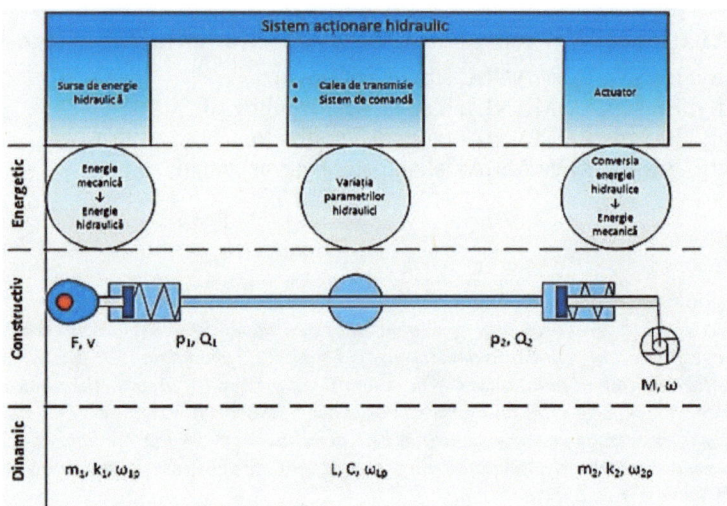

Fig. 1. The conceptual correlations between the propulsion system and the resonant phenomenon.

Dynamically, the hydraulic energy generator is a mass suspended by an elastic element, subjected to periodic excitation. Its movement is damped by wall friction. The mass-elastic element assembly has a resonant frequency.

The liquid column, in contact with the hydraulic generator, is characterized by mass, (as an inertial characteristic), and elastic properties which allow accumulation and dissipation effect due to the friction between the fluid and the walls and other local resistances [3]. As a dynamic effect, the liquid column is an oscillatory system with resonant frequency, with external excitation, and, depending on inertial properties, cumulative and dissipative, and also on correlations existent at frequency level with the oscillations source, the output parameters are resulted.

In this paper was analyzed a pressure wave generator, mechanically actuated by a translating piston, observing the formation, propagation, reflection and interference of the pressure waves.

The analysis was made theoretically, through simulations using dedicated software, and also experimentally.

The main objective was to identify the potential of the analyzed solution to be used for automotive propulsion.

3. SCOPE OF THE RESEARCH

The scope of the research is to determine the conditions in which the resonance is reached in a pipe with a mechanical check valve comprised of a sealing element and a spring. It was followed:
- Analysis of the pressure waves generated in a resonant regime, establishment of the oscillation parameters and the average pressure in the pipe;

- Analysis of the slowly generated pressure wave, using a sinusoidal or pseudo sinusoidal law;
- Analysis of the propagation of the wave;
- Analysis of the wave reflection at the closed end of the pipe.

It was also intended to determine the aspect of the wave, the pressure rise and fall periods, and also the duration of the high pressure period.

4. DESCRIPTION OF THE TECHNICAL SOLUTION

The studied system is presented in the principle scheme in Fig.2. Comprising: a pipe (1) closed at one end by a mechanical check valve (2), and at the other end by a piston (8) actuated by a harmonic cam (9). On the pipe, there are piezo-resistive transducers (4) at both ends, manometers (5) and flowmeter (6). Experimental data are recorded on the computer (7). The pipe is supplied with liquid from a stationary steady-pressure source. The perturbation in induced by rotating the cam using an electric motor.

For the maximum speed of the system, 11.000 l/min, the optimum pipe length was calculated so that when the wave runs through the end of the pipe and back, the reflected wave confronts a new impulse. A series of taps (3) was mounted to compensate the eventual unidentified perturbation sources, allowing for the pipe to be lengthened or shortened by 100mm.

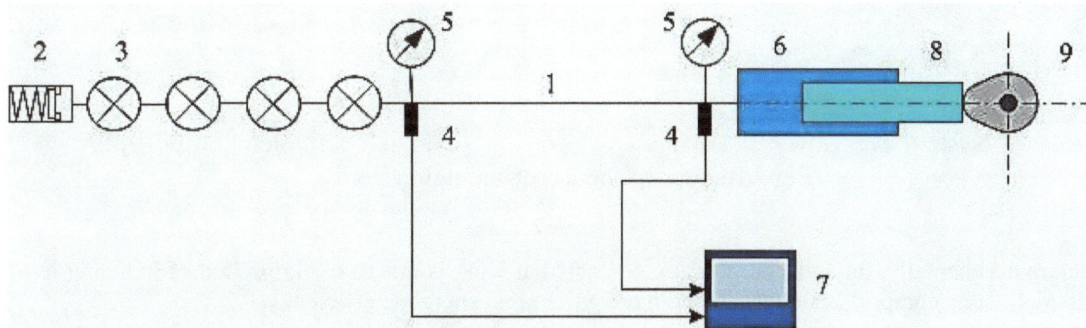

Fig.2. The system used for the slow generation of the wave

5. THE MATHEMATICAL MODEL USED

In order to analyze the mechanism which forms the pressure wave, the mode in which the wave propagates, and its stability analysis, CFDesign, finite element method software was used.

Like all the numerical methods applied on all problems with no analytical solution, it was necessary to reduce the problem of a number of infinite freedom degrees (DOF) to a finite number of DOF.

The finite volume method allows solving complex, three-dimensional, aninsentrope, and unsteady equations. Equations systems do not admit exact solutions, thus, it is necessary to apply numerical methods. Applying the numerical methods requires special calculation resources [4].

Applying the method also involves integration of the equations characteristic to finite space called finite volume or control volume. Each finite volume is associated with one or more points (nodes) in which the unknown variables are calculated. The geometric property of the finite volume is the fact that they are in contact through nodes which are placed at the boundaries of the finite volume. It is useful that the finite volume limits to exist at the boundary of the calculation domain, and in the areas where sudden property variations appear. Thus, for applying the method, it is necessary to replace the geometrical model with a set of finite volumes which geometrically approximates the studied domain. The finite volumes are tied through nodes, and these form the mesh network. The eulerian formulation of the flow equations is applied to these geometrical conditions. The solution for the problem is calculated at node level, while for the boundaries, the interpolation is used.

The base of the numerical method consists of converting the generalized equation with partial derivatives into an algebraic equation which relates the value of Φ variable in a random point P, belonging to the meshed domain, with the values in its neighborhood. This is done through integration of the general equation on a typical control volume, approximating different terms of the integral so that they will be expressed according to the values of the mesh nodes.

6. VALIDATION OF THE SIMULATION MODELS

The experimental research objective was to validate the simulation models.

The experimental study is done on functional models using piezo-electric transducers to measure instantaneous pressure, and with conventional manometers to measure the average pressure. The pressure variations are measured near the end of the pipe to observe the influence of the reflection at the closed end, and the way the waves propagate in time, in the straight, and curved pipe respectively.

The experimental results consisted of generating the wave at the speeds indicated in Table 1, for the 12.5mm diameter pipe. For this condition, the instantaneous pressures at the ends of the pipe have been recorded, according to Fig. 2.

System overview   Piston actuation components

Fig 3 Images of the experimental system

The fact that, experimentally, the induced pressure stagnated at 3 bar is due to the limitation of inducing oscillations in permanently high value compressed environment, having the same energetic efficiency.

In Fig. 4 are shown the instantaneous pressure evolution corresponding to 1000 l/min and 11000 l/min respectively. It shows that for high frequencies, the wave front is strongly distorted, the liquid properties being covered, influence-wise, by the externally induced perturbation. At low frequencies, the oscillation is freely manifested according to the properties and the flow conditions.

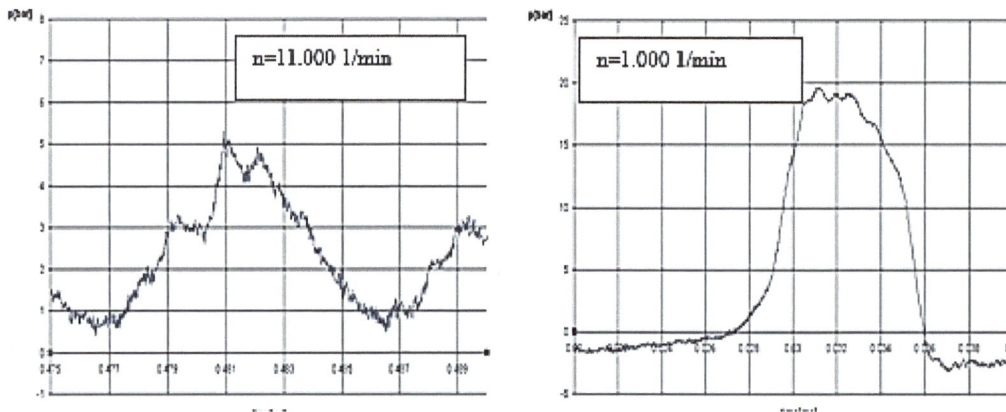

Fig. 4 The comparison of the elementary waves for different excitation frequencies

7. PRESENTING THE SIMULATION RESULTS

In order achieve the determinations; the electric motor spins the camshaft, which transforms the rotational movement into translational movement of a piston which pushes the liquid in the pipe. The studied versions and their characteristic dimensions are identical with the ones in Table 1.

Table 1. The constructive parameters of the studied system

Parameter	Measuring unit	Value
Pipe length	[m]	0,5; 1,0; 3.8
Pipe diameter	[mm]	2,0; 9,0; 12.5
Generation speed	[1/min]	1000; 2500; 11000
Cam number	[-]	1

The goal of the simulations was to find the way the waves form in pipes of different diameters, different lengths and using different frequencies. In the analysis, it was considered the amplification mechanism in a resonant regime by sinusoidal function.

For the simulation analysis of the resonant regime, standard conditions are considered to be able to compare solutions. A 1 meter long pipe, 1 m/s speed oscillation amplitude have been considered and the pulsations are induced without reflections at the mobile end (at each return, the oscillation is subjected to a new impulse).

The simulation process is done for the liquid that is in a straight pipe, having the diameters from Table 1. The simulation results from CFDesign will be presented only for the 12.5mm pipe diameter. The main parameters of the simulation are synthesized in Table 2.

In all the parameter representations got by simulation analysis, on the horizontal axis is represented the time for a certain number of oscillations (as requested by the simulation program).

Table 2 The main simulation parameters

Parameter	Measuring unit	Value	
No. of nodes	[-]	4334	
No. of elements	[-]	3930	
Initial speed	[m/s]	0	
Oscillation pulsation	[1/s]	4398	
Time step	[s]	10^{-4}	
No. of time steps Corresponding to a number of oscillations:	[-]	100 7	500 35

The velocity is calculated in 7 points along the pipe and is illustrated in Fig. 5.

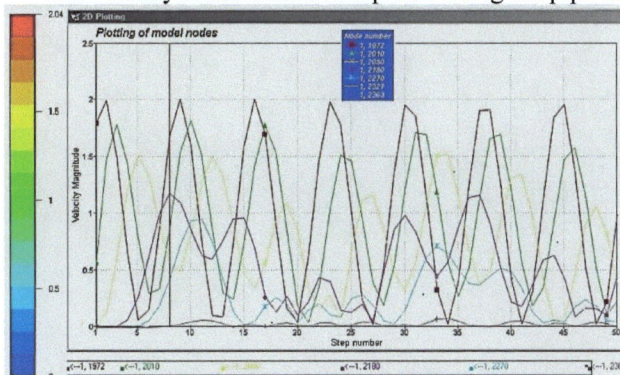

Fig. 5. The time evolution of the velocity induced by a sinus law in a resonant regime

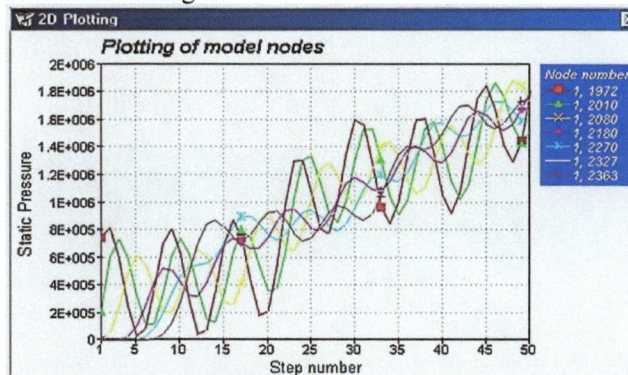

Fig. 6 Pressure evolution induced by a sinusoidal law (1 time step = 2 x 10-4 s)

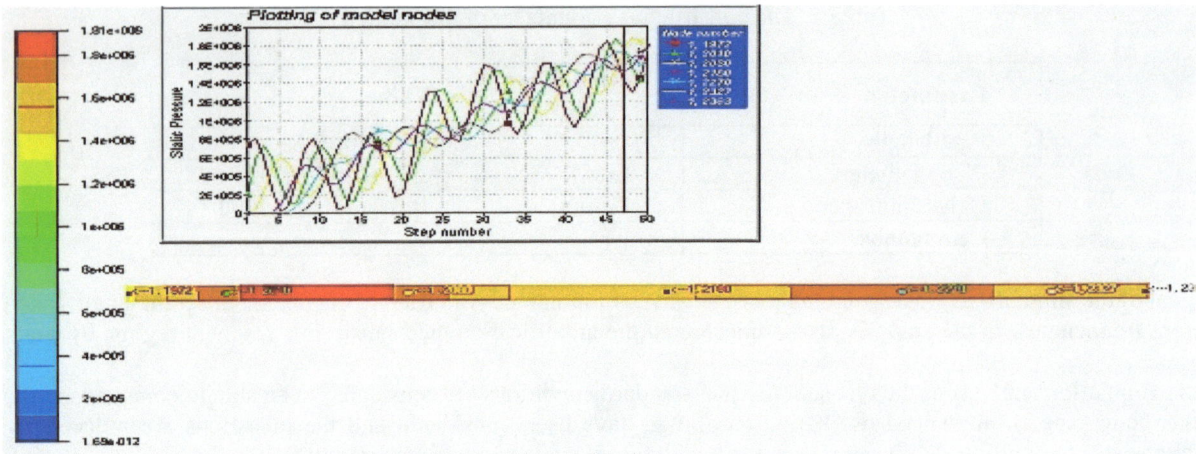

Fig. 7 The degree of uniformity of the pressure in the pipe

The pressure is induced by a sinusoidal law in resonant regime, and its temporal evolution is represented in Fig. 6.

It is noticeable that on seven sinusoidal oscillations induced in a resonant regime for 10^{-2} seconds, for an velocity amplitude of 2m/s, a 20 bar pressure is reached. The influence of the reflections on the amplification mechanism, through the fact that it has an oscillating evolution around a general linear evolution, marked separately on the chart, is also noticed. Through the pressure rise in the resonant regime, there is a unification of the total pressure variations, which leads to an improvement on the pipe pressure uniformity. (Fig. 7)

The described mechanism illustrates an 18 bar / (2×10^{-4}) s amplification. Extrapolating this result, the 360 bar target pressures could be reached in 4×10^{-2} s. The simulated amplification mechanism considers the liquid compressibility, and the propagation conditions are considered if other perturbation sources don't exist. In Fig. 8, the wave fronts deform spatially, being unstable under the interference action.

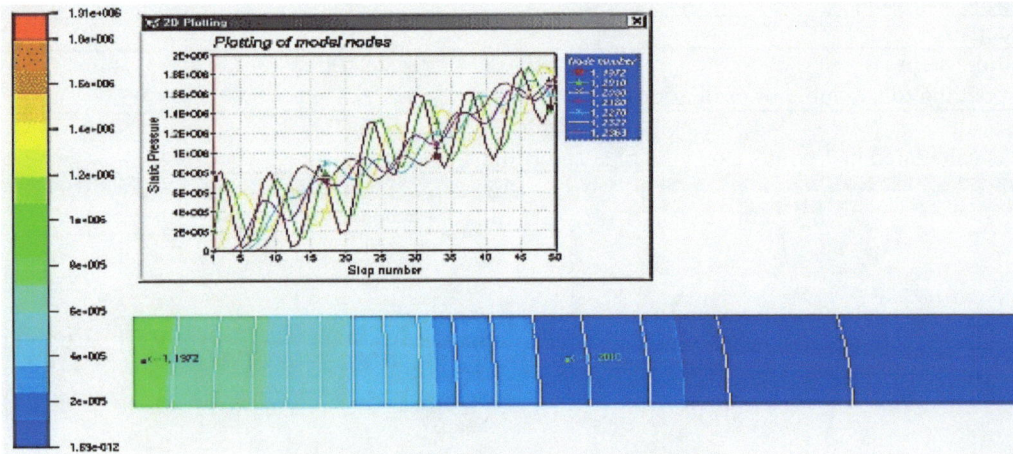

Fig. 8 Wave front evolution for pressure

8. CONCLUSIONS

• The analysis of the simulation results shows that the pressure generator for resonant regime in a closed pipe, with a mechanically coupled oscillating system generates a linear rise of the average pressure in the system;

• The studies regarding the possibility of inducing resonance illustrated the wave formation mechanism in the permanent perturbation regime.

• It was noted that although the oscillations are more distorted, when the frequencies are high, the sustained oscillations amplitude is higher and it induces a remnant pressure in the system, three times higher than the initial pressure. The process is revealed both theoretically and experimentally. The extrapolation of the results, demonstrated by simulation, shows a slight rise in the remnant pressure, induced by resonance.

- Experimentally, the amplification mechanism was identified to be less efficient, with local perturbation sources in the transducers zones and in the variable section area between the piston and the pipe, and also 4 reflections on the closed wall between successive impulse generations.

- The resonance generation mechanism is a continuous process if the generation conditions are maintained. Through impulse generation, the theoretical potential has an exponential evolution in regard to the energy induced to the system by impulse. Numerical simulation showed a linear evolution, while the experiment showed a fast evolution in the resonant regime, followed by stagnation.

- The resulted conclusion shows that it is necessary to have a new system design for the oscillatory system so that it is functionally closer to the impulse perturbed oscillator.

ACKNOWLEDGEMENT

This paper is supported by the Sectoral Operational Programme Human Resources Development (SOP HRD), financed from the European Social Fund and by the Romanian Government under the project number POSDRU/159/1.5/S/134378.

REFERENCES
[1]. Carlsson, E.: Modeling Hydrostatic Transmission in Forest Vehicle, LITH - ISY - EX - -06/3801 - - SE, 2006
[2]. Abăităncei, H.; Irimia, C.; Cârciumaru, G.; Soare, I.T.; **Radu, I. S.**: The free piston Stirling principle as prime mover for alternant hydraulic propulsion systems,. FISITA 2012 WORLD Automotive Congress, Beijing
[3]. Wang, Q.F.; Zhang, Y.T.; Xiao, Q.: Evaluation for energy saving effect and simulation research on energy saving of hydraulic system in hybrid construction machinery, Chinese Journal of Mechanical Engineering, 2005.
[4]. CFDesign solver technical reference finite element fluid flow and heat transfer solver version 3.0, Blue ridge numerics

6th **International Conference**
"Computational Mechanics and Virtual Engineering "
COMEC 2015
15-16 October 2015, Braşov, Romania

AN APPROACH TO SOME NON-CLASSICAL EIGENVALUE PROBLEMS OF STRUCTURAL DYNAMICS

H. Sandi[1], I.S. Borcia (†)[2]
[1] Academy of Technical Sciences of ROMANIA. horeasandi@yahoo.com.
[2] Dr., senior researcher (†), URBAN – INCERC, ROMANIA.

Abstract: *Two main shortcomings of usual formulations, encountered in literature concerning the linear problems of structural dynamics are revealed: the implicit, not discussed, postulation, of the use of Kelvin – Voigt constitutive laws (which is often infirmed by experience) and the calculation difficulties involved by the attempts to use other constitutive laws. In order to surpass these two categories of shortcomings, the use of the bilateral Laplace – Carson transformation is adopted. Instead of the dependence on time, t, of a certain function f (t), the dependence of its image f# (p) on the complex parameter $p = \chi + i\omega$ (ω: circular frequency) will occur. This leads to the formulation of associated non-classical eigenvalue problems. The basic relations satisfied by the eigenvalues $\lambda_r{}^{\#}(p)$ and the eigenvectors $v_r{}^{\#}(p)$ of dynamic systems are examined (among other, the property of orthogonality of eigenvectors is replaced by the property of pseudo-orthogonality). The case of points p = p', where multiple eigenvalues occur and where, as a rule, chains of principal vectors are to be considered, is discussed. An illustrative case, concerning a non-classical eigenvalue problem, is presented. Plots of variation along the ω axis, for the real and imaginary components of eigenvalues and eigenvectors, are presented. A brief final discussion closes the paper.*
Keywords: *Non-classical eigenvalue problems, Laplace – Carson transformation, pseudo-orthogonality, multiple eigenvalues, singular eigenvectors.*

1. INTRODUCTION

The main object of the paper is represented by non-classical eigenvalue problems encountered in the linear dynamics of structures (having, formally, a finite number of degrees of freedom). Matrix formulations are used (vectors: lower case, bold, matrices: upper case bold, characters).
The main tool for calculations, used in the paper, is represented by the bilateral Laplace – Carson transform [9], according to which the relations between the *original functions* g(t) and their corresponding *image functions* $g^{\#}(p)$ are

$$g^*(p) = p \int_{-\infty}^{\infty} e^{-pt} g(t)\, \mathrm{d}t \qquad \text{where } \chi = \operatorname{Re} p \in (\alpha, \beta) \qquad (1.1a)$$

$$g(t) = (1 / 2\pi i) \int_{c-i\infty}^{c+i\infty} [e^{pt} g^*(p) / p)]\, \mathrm{d}p \qquad \text{where } c \in (\alpha, \beta) \qquad (1.1b)$$

The starting point of the developments of the paper is represented by a view at the linear equation of motion for the original functions,

$$\boldsymbol{M} \, \mathrm{d}^2\boldsymbol{u} / \mathrm{d}t^2 + \boldsymbol{C} \, \mathrm{d}\boldsymbol{u} / \mathrm{d}t + \boldsymbol{K}\,\boldsymbol{u} = \boldsymbol{f}(t) \qquad (1.2a)$$

which becomes for the image functions, determined by the bilateral Laplace – Carson transform,

$$(p^2\,\boldsymbol{M} + p\,\boldsymbol{C} + \boldsymbol{K})\,\boldsymbol{u}^*(p) = \boldsymbol{f}^*(p) \qquad (1.2b)$$

The option for a critical reconsideration of the equation (1.2a) is determined by the fact that the constitutive laws implicitly postulated in the formulation of this equation, which are of Kelvin – Voigt type, lead often to results that are not confirmed by physical experience, while an attempt to adopt a different type of constitutive laws leads to considerable difficulties for the calculation techniques in case one tries to deal with the original functions.

A (rather simplistic) frequently encountered approach in structural dynamics is represented by postulating for the beginning the existence of ideally elastic constitutive law, which lead for the equations (1.2) to a classical eigenvalue problem,

$$(-\mu^2 M + K)\, v = 0 \tag{1.3}$$

for which the solution is represented by a system of constant and real eigenvalues μ_r and eigenvectors v_r. Thereafter, a correction is introduced just for the eigenvalues, in a way that is similar to the one usable in case of single degree of freedom systems. This approach keeps the eigenvectors real, while the eigenvalues become complex functions of the variable p.

A more correct procedure would lead to the need to consider together all three matrices occurring in the equations (1.2). It is shown [5] that the eigenvalue problem is in general irreducible to an eigenvalue problem for the equation

$$(-\mu^2 M + i\,\mu\, C + K)\, v = 0 \tag{1.3'}$$

It is reducible to that type of equation only in case the matrix C can be represented as a linear combination of terms $[K\, (M^{(-1)}\, K)^{\,j}]$, where the values of j are integer, arbitrary. In case the eigenvalue problem corresponding to the equation (1.3') is no longer reducible to a classical problem, a correct way, dealt with in literature, is as follows: the non-linear n - dimensional problem corresponding to the equation (1.3'), is replaced by a linear, $2n$ – dimensional one. In this latter case the matrices used become usually non-symmetrical, while the solutions become complex. Note that this latter approach is usable only in case Kelvin – Voigt constitutive laws are admitted.

A different approach [7], [8], is adopted in the paper. An image equation

$$[p^2\, M + K^{\#}(p)]\, u^*(p) = f^*(p) \tag{1.4}$$

where, the case a Kelvin – Voigt constitutive law is admitted, is referred to. The matrix $K^{\#}(p)$ would become in this case equal to the matrix $p\, C + K$ of equation (1.2b), dealt with previously. In order to set up the constitutive laws of structural components, the use of generalized Maxwell laws [6] is proposed [7], [8]. The generalized Maxwell law is as follows: an ideally elastic (Hooke) component is connected in parallel with several Maxwell type components. The solution adopted in this way benefits from two main advantages: on one hand, experimentally determined characteristics can be approximated upon a desired interval of the ω-axis; on the other hand, there exist methodological advantages raised by the analytical properties of the laws postulated, characterized by the existence of poles (the matrix $K^{\#}(p)$ is meromorphic)

2. ANALYTICAL DEVELOPMENTS

2.1. Properties of some constitutive laws

To start, a discussion on some alternative constitutive laws is dealt with. These laws should allow defining the most appropriate kind of equations of motion for the dynamic systems investigated. Keeping in view the fact that the main tool for analysis is represented by the use of the bilateral Laplace – Carson transform (1.1), this approach is based on the use of the transforms of original functions depending on time. Two basic entities are considered, for illustration of the use of the transforms referred to: the normal stress, $\sigma^{\#}(p)$ and the homologous local deformation $\varepsilon^{\#}(p)$. An explicit extension to tensorial functions is not necessary in this respect.

Two reference models, in which the elasticity (or elastic stiffness) modulus, E, and the viscous stiffness modulus, η, intervene, are used:

- the ideally elastic model (called Hooke's model), $\quad \sigma^{\#}(p) = E\, \varepsilon^{\#}(p) \tag{2.1a}$
- the ideally viscous model (called Newton's model), $\quad \sigma^{\#}(p) = p\, \eta\, \varepsilon^{\#}(p) \tag{2.1b}$

The parameter E is the elasticity modulus, or the modulus of elastic stiffness. The parameter η is the viscous stiffness modulus. These models are to be dealt with in adequate ways for performing specific analyses. The parameters E and η are first used for connections in parallel or in series respectively and are to be combined in an appropriate way to correspond to various goals of analysis.

The model of solid body with retardation (called the *Kelvin – Voigt* model):

$$\sigma^{\#}(p) = E\, \varepsilon^{\#}(p) + p\, \eta_{ret}\, \varepsilon^{\#}(p) = (E + p_{ret}\, \eta)\, \varepsilon^{\#}(p) = E\, (1 + p\, \eta_{ret} / E)\, \varepsilon^{\#}(p) = E\, (1 + p\, \mathrm{T_{ret}})\, \varepsilon^{\#}(p) \tag{2.2a}$$

The model of viscous fluid body with relaxation (called the *Maxwell* model):

$$\varepsilon^{\#}(p) = \sigma^{\#}(p) / E + p \, \sigma^{\#}(p) / \eta_{rel} , \quad \sigma^{\#}(p) = [p \, E \, \eta_{rel} / (E + p \, \eta_{rel})] \, \varepsilon^{\#}(p) = [E / (1 + p \, \mathrm{T_{rel}})] \, \varepsilon^{\#}(p) \qquad (2.2b)$$

The two latter models include two parameters of physical dimension time: the retardation time, $\mathrm{T_{ret}}$, and the relaxation time, $\mathrm{T_{rel}}$, respectively.

A first combination (in parallel) of these two models is the *Poynting* model

$$\sigma^{\#}(p) = [E_0 + E_1 / (1 + p \, \mathrm{T_{ret}})] \, \varepsilon^{\#}(p) \qquad (2.2c)$$

while a generalization of it is the *generalized Maxwell* model,

$$\sigma^{\#}(p) = [E_0 + \Sigma_k E_k / (1 + p \, \mathrm{T_{ret.k}})] \, \varepsilon^{\#}(p) \qquad (2.2d)$$

The scalar models (2.1) and (2.2) may be extended to pluri – dimensional models, in order to derive specific laws to structural models. The use of the model corresponding to relation (2.2d) is subsequently preferred. This is because this option makes it possible to approximate the rheological properties of materials and, at the same time, offer controllable singularities (poles of the theory of functions of a complex variable).

NOTE:

1. The fact that the denominators of the terms of expressions (2.2d), as well as of the denominator of expression (2.2c), have real, positive, values leads to the fact that the poles intervening in the expressions of terms of index k are placed on the negative half-axis, Re $p < 0$, Im $p = 0$, at abscissae of $(-1 / \mathrm{T_{ret.k}})$.

2. The coefficients E_k of expression (2.2d) have the same physical dimension as that of the coefficient E_0, while their physical sense is to be specified for each of them.

2.2. Relations of structural dynamics for structures with various constitutive laws

Returning to the equations (2.2), a system of real eigenvectors to simultaneously diagonalize the matrix triad (M, C, K) of the equation of motion does not exist in the general case for systems having Kelvin – Voigt constitutive laws. The consequence of this fact is that, in the general case when the eigenvectors become complex, a transfer of energy between the free vibrations corresponding to different eigenmodes occurs. More generally, for a pair of matrices $[M, K^{\#}(p)]$, where the second matrix is variable, there does not exist a system of constant, real eigenvectors in case the analysis is performed for bilateral Laplace – Carson transforms (where p is the complex, independent, variable, replacing the time argument t, specific to analysis in the field of original functions).

This situation has important consequences, considered subsequently. On one hand, the property of orthogonality of eigenvectors corresponding to different eigenmodes is no longer satisfied. It is consequently appropriate to introduce some new concepts, namely the concepts of *pseudo-orthogonality* and of *pseudo-normalization*, which generalize the classical concepts of orthogonality and normalization.

Given the advantages of use of the solutions of eigenvalue problems derived for the equation of motion, which are illustrated in literature for various cases, a generalization to the non-classical case was looked for. Two orientations may be distinguished in these studies:

a) approaches which are aiming at the direct determination of singularities (more precisely, the zeroes of the determinant for the image impedance matrix $Z^{\#\wedge}(p)$ [1], [5]);

$$Z^{\#\wedge}(p) = p^2 M + K^{\#\wedge}(p) \qquad (2.3)$$

these approaches are usable, practically, in case of adoption of Kelvin – Voigt constitutive laws;

b) approaches aimed at deriving the inverse matrix $Z^{(-1)\#}(p)$ of the matrix $Z^{\#}(p)$, which is a function of the p variable [7]; this way is the only one dealt with in this paper, due to its more general usability.

Following developments are starting from the equation (1.4), where both matrices M and $K^{\#}(p)$ are symmetrical and lead to the eigenvalues $\lambda_r^{\#}(p)$ depending on the p parameter, for the homogeneous equation

$$[-\lambda^{\#}(p) M + K^{\#}(p)] \, v^*(p) = 0 \qquad (2.4)$$

for which the non-trivial solutions (in case the value of variable p to which the solution refers is not affected by singularities), is represented by the eigenvalues $\lambda_r^{\#}(p)$ and the corresponding eigenvectors $v_r^*(p)$. The existence of non-trivial solutions implies for the equation zeroes for the determinant of $[-\lambda^{\#}(p) M + K^{\#}(p)]$,

$$\mathrm{Det} \{-\lambda^{\#} M + K^{\#}(p)\} = 0 \qquad (2.5)$$

2.3. The case of points p where the eigenvalues are different (i. e. simple)

Due to the symmetry properties of matrices, in case of two different eigenvalues $\lambda_r^{\#}(p)$ and $\lambda_s^{\#}(p)$, the corresponding eigenvectors $v_r^{\#}(p)$ and $v_s^{\#}(p)$ are pseudo-orthogonal with respect to both matrices,

$$v_r^{\#}(p)^T M v_s^{\#}(p) = 0 \qquad (r \neq s) \tag{2.6a}$$

$$v_r^{\#}(p)^T K^{\#}(p) v_s^{\#}(p) = 0 \qquad (r \neq s) \tag{2.6b}$$

(note: orthogonality would have involved that one of the factor vectors should be replaced by its complex conjugate). In a similar way, the *pseudo-norm* (with respect to matrix M) of a vector $v^{\#}(p)$, $m^{\#}(v)$, is defined by the relation

$$m^{\#}(v)^2 = v^{\#}(p)^T M v^{\#}(p) \tag{2.7}$$

while the *pseudo-normalized* with respect to matrix M homologous vector $v^{\prime}(p)$, $v^{\#(M)}(p)$, is given by the relation

$$v^{\#(M)}(p) = v^{\#}(p) / m^{\#}(v) \tag{2.8}$$

In order to formulate some condensed relations, it is appropriate to define the matrix of eigenvectors, $V^{\#}(p)$. Its columns are represented by the eigenvectors $v_r^{\#}(p)$ (arranged in the order of eigenvalues of rank (r)). In the same way, it is possible to define a matrix of pseudo-normal eigenvectors, $V^{\#(M)}(p)$. The condition of pseudo-normalization may be rewritten as

$$v_r^{\#\,(M)T}(p) M v_r^{\#\,(M)}(p) = \delta_{rs} \qquad (\delta_{rs}: \text{Kronecker's symbol)} \tag{2.9}$$

while a homologous relation for the matrix $K^{\#}(p)$ is

$$v_r^{\#\,(M)T}(p) K^{\#}(p) v_s^{\#\,(M)}(p) = [\lambda_r^{\#}(p) s_r^{\#}(p)]^{1/2} \delta_{rs} \tag{2.10}$$

The vectors $v_r^{\#(M)}(p)$ span mono-dimensional subspaces which are invariant with respect to the pair of matrices $(M, K^{\#}(p))$. The relations

$$V^{(M)\#T}(p) M V^{(M)\#}(p) = I_n \qquad (I_n: \text{unit matrix of dimension } n) \tag{2.11a}$$

$$V^{(M)\#T}(p) K^{\#}(p) V^{(M)\#}(p) = \text{Diag } \{\lambda_r^{\#}(p)\} \tag{2.11b}$$

$$V^{(M)\#T}(p) M K^{(-1)\#}(p) M V^{(M)\#}(p) = \text{Diag } \{1 / \lambda_r^{\#}(p)\} \tag{2.11c}$$

$$V^{(M)\#T}(p) Z^{\#}(p) V^{(M)\#}(p) = \text{Diag } \{p^2 + \lambda_r^{\#}(p)\} \tag{2.11d}$$

$$V^{(M)\#T}(p) M Z^{(-1)\#}(p) M V^{(M)\#}(p) = \text{Diag } \{1 / [p^2 + \lambda_r^{\#}(p)]\} \tag{2.11e}$$

(where the symbol **Diag** means a diagonal matrix of dimension n, while the impedance matrix, $Z^{\#}(p)$, is defined by the relation (2.3).

Conversely, the relations (having the sense of spectral expansions for the matrices of the dynamic system dealt with) are

$$K^{\#}(p) = M V^{(M)\#}(p) \text{ Diag } \{\lambda_r^{\#}(p)\} V^{(M)\#T}(p) M \tag{2.12a}$$

$$K^{\#(-1)}(p) = V^{(M)\#}(p) \text{ Diag } \{1 / \lambda_r^{\#}(p)\} V^{(M)\#T}(p) \tag{2.12b}$$

$$Z^{\#}(p) = M V^{(M)\#}(p) \text{ Diag } \{ p^2 + \lambda_r^{\#}(p)\} V^{(M)\#T}(p) M \tag{2.12c}$$

$$Z^{\#(-1)}(p) = V^{(M)\#}(p) \text{ Diag } \{1 / [p^2 + \lambda_r^{\#}(p)]\} V^{(M)\#T}(p) \tag{2.12d}$$

It may be shown that the properties accepted for the matrix $K^{\#}(p)$ lead to the conclusion that the real and imaginary parts of the eigenvalue $\lambda_r^{\#}(p)$ satisfy the conditions

Re $\lambda_r^{\#}(p) > 0$ for Re $p \geq 0$ and

Im $\lambda_r^{\#}(p) / $ Im $p \geq 0$ for the whole plane $\{p\}$,

while the poles of the eigenvalues $\lambda_r^{\#}(p)$ can be placed only along the half-axis (Im $p = 0$, Re $p < 0$) in case the scalar constitutive laws of types (2.2) can be directly applied as constitutive laws between the specific vectors of internal forces and the specific deformation components. It may be shown also that the eigenvalues $\lambda_r^{\#}(p)$ are stationary at the point $v = v_r^{\#}(p)$, in case one considers the kind of variation of the expression $\lambda_r^{\#}(p) = v^T(p) K^{\#}(p) v^{\#}(p)$ along the *hyper-pseudosphere* $v^T M v = 1$.

The matrix $K^{\#}(p)$ and the eigenvalues $\lambda^{\#}_r(p)$ may be expanded into integer series of powers of $(p - p_0)$ [4]. The eigenvectors $v^{\#}_r(p)$ are uniform functions, which may be expanded in a similar way into series of powers. Due to the condition (2.3), and to the fact that a complete basis exists, the eigenvectors do not have, at such points p, zeroes or poles.

2.4. The case of points $p = p'$, where multiple eigenvalues exist

The case of points $p = p'$, where multiple eigenvalues $\lambda^{\#}_r(p)$ exist, raises special problems, which impose a special kind of analysis, needing a revision of the calculation techniques usually adopted for points p, where all eigenvalues are different. Some features of the variation of the eigenvectors in the neighborhood of points $p = p'$ where a multiple eigenvalue $\lambda^{\#}_{r'}(p)$ exists, may be mentioned: the variation of the system of eigenvectors in the subspace spanned by the system of eigenvectors corresponding to a multiple eigenvalue $\lambda^{\#}_{r'}(p)$ must be replaced by a chain of principal vectors [10]. The diagonal submatrix corresponding to the chain referred to will no longer be a diagonal one, but is to be replaced by a submatrix of Jordan's canonic type [10]. The relations developed previously in subsection 2.3 of the paper must be correspondingly adapted. As mentioned in the note of subsection 2.1, item 1, in the problems dealt with in the paper this is not expected to happen.

3. ILLUSTRATIVE APPLICATION

The physical problem dealt with is related to the examination of the vibration of a dynamic system S consisting of a mass connected by means of a perfectly elastic spring to a (vertical) axi-symmetrical foundation block that is connected at its turn to the elastic half-space. Note here that the connection to the half-space involves dissipative properties even in case of an ideally elastic half-space. This is due to the fact that during the vibration process the energy is radiated from the foundation block, without returning to the contact zone. In order to keep calculations as simple as possible, the dissipative properties of the contact zone are assumed to correspond to a Kelvin – Voigt constitutive model. In agreement with the modelling and the approximate relations given in [2], the contact of an axi-symmetrical block with the half-space is equivalent to a dynamic single degree of freedom system, for which following input data were adopted:
-   the mass of the rigid foundation block, including the equivalent mass pertaining to the half space material: $m_1 = 15$ t;
-   the viscous stiffness of the system of contact with the half-space: $c_1 = 4500$ t/s = 4500 kNs/m;
-   the elastic stiffness of the same contact system: $k_1 = 3\ 000\ 000$ t/s^2 = 3 000 000 kN/m;
-   the mass of the upper body: 5 t;
-   the viscous stiffness of the contact system between the two bodies: $c_2 = 0$;
-   the elastic stiffness of the same: $k_2 = 100\ 000$ t / s^2 = 100 000 kN/m

The condition of zero value of the determinant corresponding to the equation of motion is

$$m_1\, m_2\, \lambda^{\#2} - [\, m_1\ k_2 + m_2\, (\, cp + k_1 + k_2)]\, \lambda^{\#} + k_1\, k_2 = 0 \qquad (3.1)$$

with the solutions for the eigenvalues

$$\lambda_{I,II}^{\#}(p) = \ <[m_1\, k_2 + m_2\, (cp + k_1 + k_2)] \ -/+ \ \{[m_1\, k_2 + m_2\, (cp + k_1 + k_2)]^2 - 4\, m_1\, m_2\, k_1\, k_2\}^{1/2}> / (2\ m_1\, m_2);$$

and for the eigenvectors respectively

$$v_{1r}^{\#} = (k_2 - \lambda_r^{\#}\, m_2\ ) / n_r^{\#} \qquad (3.3a)$$

$$v_{2r}^{\#} = k_2 / n_r^{\#} \qquad (3.3b)$$

where the denominator $n_r^{\#}$ has the expression

$$n_r^{\#} = [m_1\, (k_2 - \lambda_r^{\#}\, m_2\ )^2 + m_2\, k_2^2]^{1/2} \qquad (3.4)$$

The solutions (eigenvalues and eigenvectors) as functions of the non-dimensional parameter $c\omega / k_1$, assuming $p = i\ \omega$, are presented in Figures 3.1 and 3.2 respectively for the non-dimensional interval (0., 2.0) of $c\omega / k_1$.

To note that the colors blue (for the real parts) and red (for the imaginary parts) respectively were used. It may be remarked that a singularity occurs for the eigenvector $v_{1}^{\#}$ at a value of about 0.15 of the non-dimensional ratio $c\omega / k_1$. Examining the plots presented, it turns out that for the system dealt with a strong dependence on the non-dimensional argument exists. Of course, one must take into account the fact that the results presented concern directly the Laplace – Carson images and that a use of them for practical purposes involves in principle a conversion to the field of originals for the functions of interest.

4. FINAL CONSIDERATIONS

The paper presented is dealing with a problem of wide interest, namely that of contributing to the adoption of an instrument of analysis of the performance of dynamic systems having components characterized by linear constitutive laws of a quite high complexity. This may lead to analyses to be more realistic than the practically exclusive use of Kelvin – Voigt constitutive laws, which are so frequently encountered in the literature, without the required comments.

The use of the bilateral Laplace – Carson transform represents a highly efficient tool of analysis. Learning this procedure is recommended to those engaged in the linear analyses of various problems of structural dynamics.

Besides the direct transform, expressed by the relation (1.1a), which consists of usual integration, the use of the inverse transform (1.1b) based on the residue theorem of the theory of complex functions is highly recommendable.

The appropriate consideration of the convergence band (α, β) that is specific to every functions dealt with, should be carefully carried out.

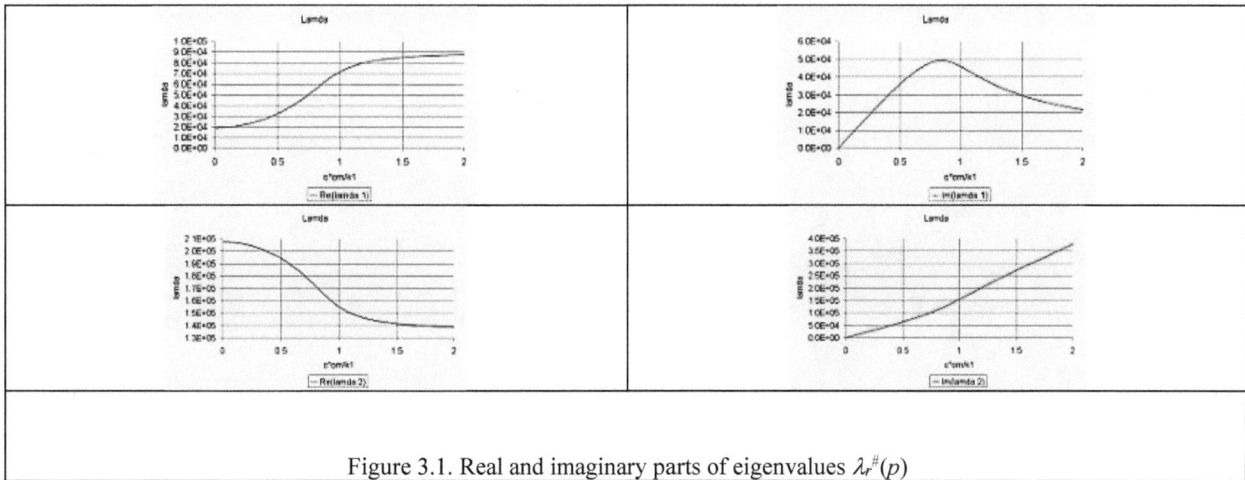

Figure 3.1. Real and imaginary parts of eigenvalues $\lambda_r^{\#}(p)$

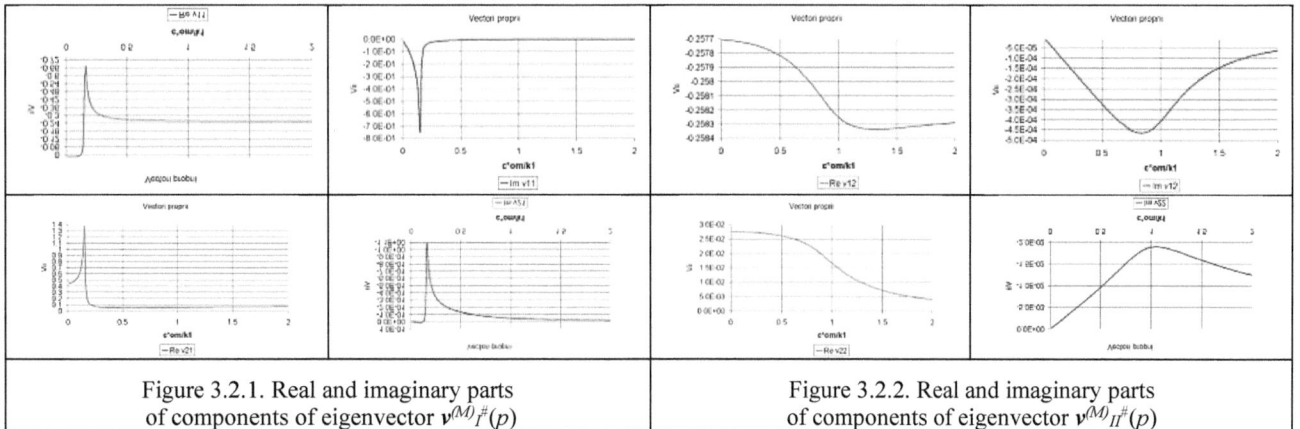

Figure 3.2.1. Real and imaginary parts of components of eigenvector $v^{(M)}{}_I^{\#}(p)$

Figure 3.2.2. Real and imaginary parts of components of eigenvector $v^{(M)}{}_{II}^{\#}(p)$

REFERENCES

1. T. K. Caughey, M. E. J. O'Kelly: "General theory of vibration of damped linear dynamic systems", California Institute of Technology, 1963.
2. R. Clough, J. Penzien "Dynamics of structures". Prentice Hall, Englewood Cliffs, 1975.
3. D. K. Faddeev, V. N. Faddeeva: "Vychisliltelnye metodî lineinoy alghebrî". GIFML, 1960.
4. K. Knopp: "Funktionentheorie". Goeschen, Heidelberg, 1930.
5. M. E. J. O'Kelly: "Vibration of viscously damped linear dynamic systems". Thesis, California Institute of Technology, 1964.
6. M. Reiner: Reologhia, GIFML, Moscow, 1964.
7. H. Sandi: "Eigenwertaufgaben und Übertragungsmatrizen für nichtkonservative mechanische Systeme". ZAMM, 5, 1970.
8. H. Sandi: "Elemente de dinamica structurilor" (Elements of structural dynamics). Editura Tehnică, Bucharest, 1983.
9. B. Van der Pol. H. Bremmer : "Operational calculus based on the two – sided Laplace transform". Oxford, 1950.
10. J. H. Wilkinson: "The algebraic eigenvalue problem". Oxford Univ. Press, 1965.

6th International Conference
"Computational Mechanics and Virtual Engineering "
COMEC 2015
15-16 October 2015, Braşov, Romania

STRUCTURAL EXAMINATIONS AND MECHANICAL TESTS TO CHARACTERIZE THE METALLURGICAL INTERFACE BASE MATERIAL AND FILLER THERMAL SPRAYED SAMPLES

Camelia BOBOC(ENACHE)[1], Polixenia Iuliana SIMION (DORCEA)[2]

[1] Department of Materials Technology and Welding, University" POLITEHNICA"of Bucharest, ROMANIA
E-mail: camelia.boboc2000upb@yahoo.com
[2] Department of Materials Technology and Welding, University" POLITEHNICA"of Bucharest, ROMANIA
dorcea_polixenia@yahoo.com

Abstract : *Assessment characteristics and properties Thermal spray coating layers is achieved by performing measurements of the characteristics and properties depend on the nature and properties of base materials and coating layers but also how to prepare base material, the selection regime parameters metal coating and the type of metallization process.*
Keywords Metal spraying, arc spraying, coatings

1. INTRODUCTION

The machines and the parts due to the operating conditions, in particular due to the environment are subject to corrosion. The methods of reconditioning machines and parts are made in dimensions of technological documentation execution [1].
A wide range of applications it has thermal metal spraying. Examples of areas where heat is used for metal spraying are:
- ✓ civil, industrial and agricultural buildings;
- ✓ Special industrial construction (in the chemical and petrochemical industries, conventional energy, wind and nuclear, food industry, water supply and sewerage;
- ✓ aviation industry;
- ✓ medicine (prosthetics orthopedics, optics, instrumentation);
- ✓ telecommunications, electronics;
- ✓ road and rail transport and marine transport and construction; (Fig. 1 Road bridge in Bucharest);
- ✓ Household Household (packaging, containers, furniture) .

Technological reconditioning process has a number of benefits: thermal metal spraying costs are low due to low cost of metallization equipment. It can save the material do not need a new piece, the one with imperfections can be reconditioned [2].
Imperfections of metal spraying are non thermal adhesive, coating layers is porous structure, voids, inclusions, oxide interface presence of material - filler. in the substrate, the outside coating layers [3].

The mandrel bending mechanical tests can determine the ability of plastic deformation by bending workpieces which was deposited by metal spraying arc with two wires non-ferrous materials, alumiu, copper and zinc. Following the inspection adherence deposited layers is observed.

177

A sample was analyzed by electron microscopy work that was submitted copper and zinc, which had a good grip. Selection of technological parameters depending on the material base and filler and how to prepare surfaces greatly affect the adhesion of the deposited layer and the appearance of imperfections [8].

Fig. 1: Road bridge in Bucharest

2. TECHNICAL REQUIREMENTS

2.1. Selection by thermal metal spraying process.

For the layers deposited by thermal spraying the following bending test is used :
Workpieces with metallization layers deposited by thermal spray as shown in the following selected research;
thermal metal spraying of electric arc (EA) was performed on sets of samples (I, II, III) using fillers diameter 1.62 mm from the first and second group (aluminum and copper) and the three group (aluminum and zinc). Base metal are steel S235JR (OL 37- Romanian standard) The samples are flat dimension 100 x 60 mm, thickness 3 mm, see fig. 2.

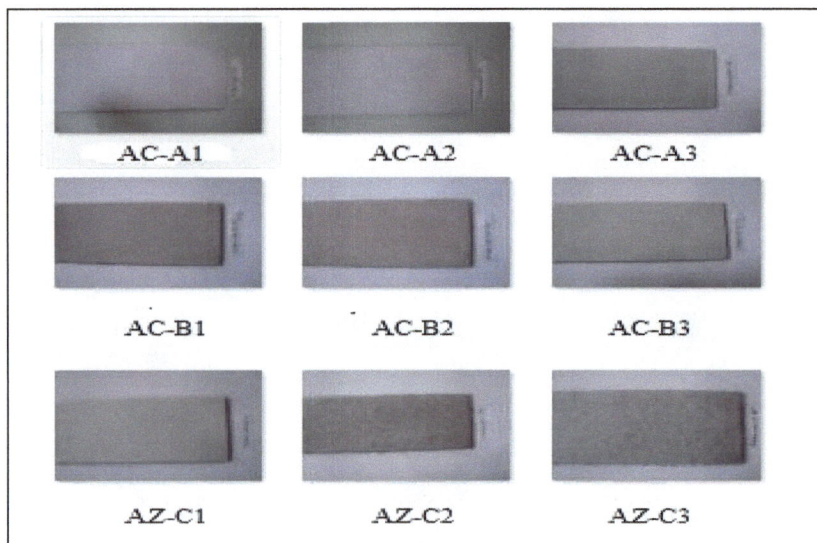

Fig. 2: Thermal metal sprayed samples

2.2. Working parametres

2.2.1 The parameters from the thermal metal spray electric arc see table 1, table 2, table 3.

Table 1: working parametres process

Parametres	Values recommended	Values used
Arc voltage, U_a [V]	30 V	30V
Electricity I [A]	30 A 50 A 60 A	30 A 50 A 60 A
Filler material	Al (99.5% Al) Cu (99.8% Cu	Al (99.5% Al) Cu (99.8% Cu
Diameter filler [mm]	1.6 1.6	1.62 1.62
Arc temperature [K]	5900	5900
Pressure compresed air [MPa]	0,5-0,7	0,6
Spray distance [mm]	160	160
Working atmosphere	Air	Air

Table 2: working parametres process

Parametres	Values recommended	Values used
Arc voltage, U_a [V]	30 V	30V
Electricity I [A]	50A	50A
Filler material	Al (99.5% Al) Cu (99.8% Cu)	Al (99.5% Al) Cu (99.8% Cu)
Diameter filler [mm]	1.6	1.62
Arc temperature [K]	5900	5900
Pressure compresed air [MPa]	0,5 0.6 0.7	0,5 0.6 0.7
Spray distance [mm]	160	160
Working atmosphere	air	Air

Table 3: working parametres process

Parametres	Values recommended	Values used
Arc voltage, U_a [V]	30 V	30V
Electricity I [A]	50A	50A
Filler material	Al (99.5% Al) Zn (99.8% Zn)	Al (99.5% Al) Zn (99.8% Zn)
Diameter filler [mm]	1.6	1.62
Arc temperature [K]	5900	5900
Pressure compresed air [MPa]	0,5	0,5
Spray distance [mm]	100 160 200	100 160 200
Working atmosphere	air	Air

2.2.2 Mark samples:
-samples set I : AC-A1, AC -A2, AC-A3,
-samples Set II :AC- B1, AC-B2, AC-B3,
-samples set III AZ-C1, AZ-C2, AZ-C3.

Arc metal spraying consists of melting the wire-electrode filler material with electrical arc which is formed between the two electrode wires Al , Cu and Zn. This melted material is then sprayed using the compressed air on the reconditioned piece surface

Is filed three layers with ARC Metal Equipment 140 S350-CL.

From plates with metallic materials deposited by thermal spraying were processed mechanical test : bending mandrel.

Mechanical test: bending: SR EN ISO 5173: 2010[6].

Test conditions: The tests were conducted at ambient temperature.

• Equipment: 400KN universal machine, traction machine type EFDZ 400,test atmosphere.

2.2.3Bending

The results of the bending test are shown in Table 4: bending test results according to SR EN ISO 5173: 2010 using a 12 mm diameter mandrel are detailed in Table 4.

Table 4: The results of the bending test

Marking samples	Thickness samples a[mm]	Width samples b[mm]	Distance of the rollsl [mm]	The diameter mandrel d[mm]	The angle of bending α[°]	The result test	Aspect bending
0	1	2	3	4	5	6	7
AC-A1-1	3,3	20,0			52	cracked	Fig. 3
AC-A1-2	3,3	19,7			46	cracked	Fig. 4
AC-A2-1	3,7	19,8	21	12	15	cracked	Fig. 5
AC-A2-2	3,6	19,1			21	cracked	Fig. 6
AC-A3-1	3,3	19,0			40	cracked	Fig. 7
AC-A3-2	3,3	19,7			22	cracked	Fig. 8
AC-B1-1	3,4	18,2			34	cracked	Fig. 9
AC-B1-2	3,4	20,6			26	cracked	Fig. 10
AC-B2-1	3,4	19,4			39	cracked	Fig. 11
AC-B2-2	3,2	18,8			70	cracked	Fig. 12
AC-B3-1	3,1	20,7	21	12	64	cracked	Fig. 13
AC-B3-2	3,2	19,0			79	cracked	Fig. 14
AZ-C1-1	3,4	20,0			34	cracked	Fig. 15
AZ-C1-2	3,3	20,1			28	cracked	Fig. 16
AZ-C2-1	3,3	19,6			46	cracked	Fig. 17
AZ-C2-2	3,3	20,5			41	cracked	Fig. 18
AZ-C3-1	3,3	20,0			43	cracked	Fig. 19
AZ-C3-2	3,3	19,7			63	cracked	Fig. 20

Fig. 3Samples AC-A1-1

Fig. 4 Samples AC-A1-2

Fig. 5 Samples AC-A2-1

Fig.6 Samples AC-A2-2

Fig. 7 Samples AC-A3-1

Fig. 8 Samples AC-A3-2

Fig. 9 Samples AC-B1-1

Fig. 10 Samples AC-B1-2

Fig. 11 Samples AC-B2-1

Fig. 12 Samples AC-B2-2

Fig. 13 Samples AC-B3-1

Fig. 14 Samples AC-B3-2

Fig. 15 Samples AZ-C1-1

Fig. 16 Samples AZ-C1-2

Fig. 17 Samples AZ-C2-1

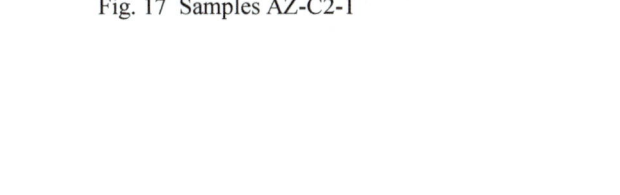

Fig. 18 Samples AZ-C2-2

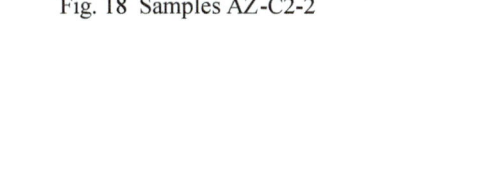

Fig. 19 Samples AZ-C3-1

Fig. 20 Samples AZ-C3-2

Results of the tests bending.

The variation of the bending angle α specimens submitted from the AC groups and the group A- Z, according to the samples of test specimens are shown in Fig. 21 and 22.

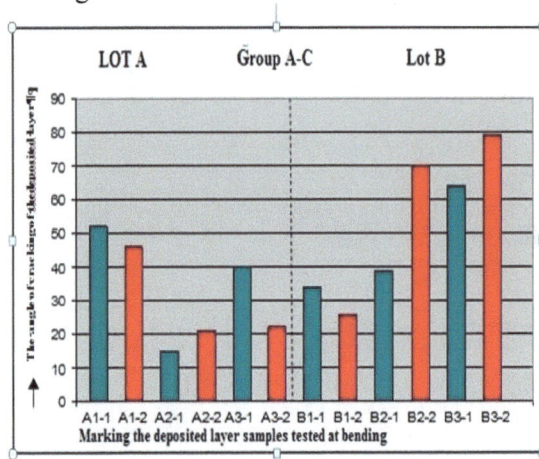

Fig. 21: Variation angle α = f (Group AC)

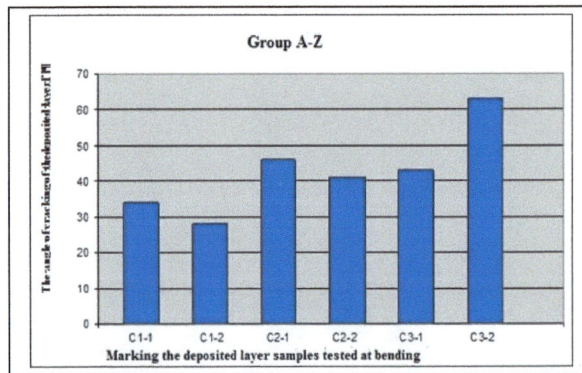

Fig. 22: Variation angle α = f (group AZ)

The AC group (Figure 21) bending angle α cracking the distinct values 15° and 79° between bending effect that made Al-Cu layers have a different behavior regarding adherence to non-alloyed iron-carbon alloys; stands of specimens with high adhesion layers deposited during the cracking angles (greater than 50°) in case specimens AC B1-1, AC B2-2, AC B3-1 and AC B3-2.

•   The group AZ (Figure 22), cracking bending angle varies between 28 ° and 63°, with higher values corresponding adherence behavior specimens meet AZ C2-1, AZC3-1 and AZC3-2.

2.2.4 Metallographic analysis of the layers deposited by thermal spraying

It was pointed phases and constituents, using a reagent - was used NITAL 2%[4,5].

The equipment used for optical microscopy analysis is FEI Inspect SEM in the laboratory LAMET faculty TMS Dept. IMST . The sample is analyzed using SEM microscopy sample AC A3 . The result is shown in Fig. 22 for the base metal ferrite-pearlite structure.

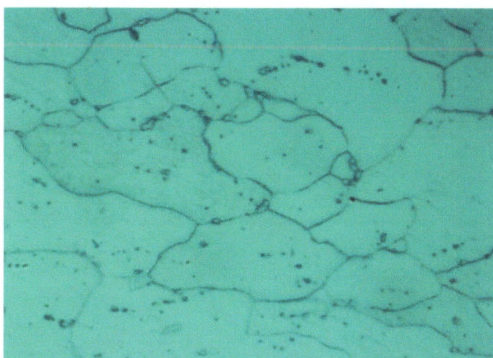

Fig.23: AC-A3, BM [atac Nital 2%, 500X]

For the deposited material result was shown in Fig. 24.

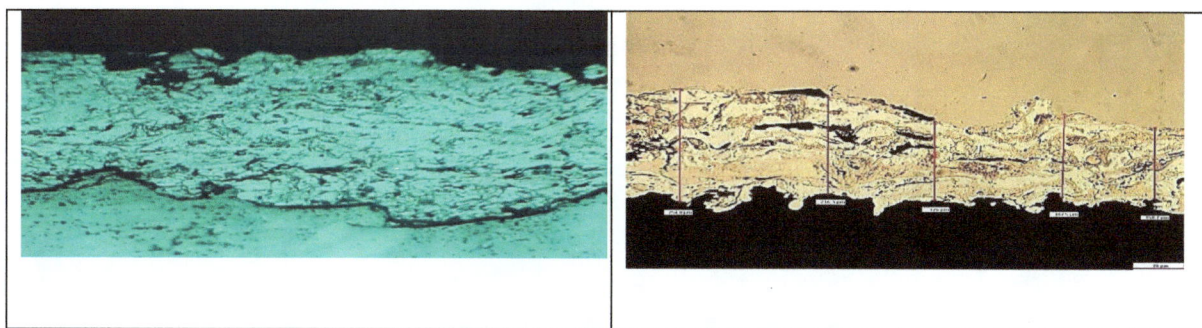

Fig.24: AC-A3, DM [atac F2, 100X]

Fig. 24 shows the joint appearance of the area between the two materials (X200). The structure is dendritic and can see perlite and ferrite grains. After microscopic analysis we can see the degree of adhesion that is good.

CONCLUSION

Changing electric current samples workpieces Group I, the pressure drop across the samples of workpieces group II and in the distance the spray gun metallization workpieces group III, caused differences on inspection adhesion layer deposited by electric-arc metallization. It was examined by electron microscopy SEM arc deposited layers and the base material and showed that a good adhesion is proven AC-A3. For the other workpieces is poor adhesion.

REFERENCES

[1] .M. Dumitru Reconditioning and repair products, Bucharest Printech Publishers(2010)
[2] Pawlowski L., The science and engineering of thermal spray coatings, J. Wiley & Sons, NewYork, 2008;
[3] Standards: SR EN 657: 2005 Thermal spraying - Terminology, classification;
[4] Standards: STAS 4203-74 Metalography. Sampling and preparation of metallographic samples;
[5] The collection of standards: STAS 5000-73 Metalography. Constituents and structures. Terminology;
[6] EN ISO 5173: 2010 Bending test;
[7] Marcu V Metal spraying, Ed Tehnică, Bucharest,1975;
[8] Information on http://www.thermalspay.org

6th International Conference
"Computational Mechanics and Virtual Engineering "
COMEC 2015
15-16 October 2015, Braşov, Romania

THE ANALYSIS OF AN ANALOGOUS HUYGENS PENDULUM CONNECTED WITH I.T.C. USING FLEXIBLE MULTIBODY DYNAMIC SIMULATIONS

Shalaby Karim[1], Lache Simona[2], Radu Florin[3]

[1] Transilvania University of Brasov, ROMANIA, karim.shalaby@unitbv.ro
[2] Transilvania University of Brasov, ROMANIA, slache@unitbv.ro
[3]Schaeffler S.R.L., Brasov, ROMANIA, raduflo@schaeffler.com

Abstract: This paper discusses the use of the Analogous Huygens Pendulum model used to investigate the dynamic behavior of the Inverted Tooth Chain (I.T.C.), where the I.T.C.is extensively applied in automotive and conveyor industries. The analysis is based on the flexible multibody dynamics approach and is expected to give a whole new perspective in evaluating the Analogous Huygens Pendulum system in order to find the different modes and natural frequencies of the system.
The relevance of using flexible multibody dynamics analysis is that it would simply show the most affected zones subjected to wear and deterioration of the I.T.C. in the Analogous Huygens Pendulum system due to different types of stresses whether on the sprocket, or on the plates, or on the contacts between them. The model developed and presented in the paper is made of flexible bodies and allows investigating the system behavior from the point of view of contacts and stresses. The model developed and presented in the paper is made of flexible bodies and allows investigating the system behavior from the point of view of contacts and stresses.
Keywords: Flexible Bodies, Contacts, Kinetic Energy, Damping Form

1. INTRODUCTION

Some common problems in chain drive systems, not related to the chain architecture (bush, roller or I.T. Chain) appear due to the contact between the chain elements and the sprocket teeth. Therefore, high contact forces are created that produce vibrations and noise.

In order to understand more profoundly the contacts between the plates of the I.T. Chain and the sprocket, a multibody dynamics simulation software MSC ADAMS and Nastran will be used for simulating the flexible plates dynamic behaviour using the multi-body theory. The general purpose is to understand the behaviour in different situations of an I.T. Chain by applying different conditions such as changing the geometries of the plates, which can be symmetric or asymmetric, or aligning the plates to the sprocket. This study is based on the Analogue Huygens Pendulum effect [1], which is caused by a fixed centroid around which the chain oscilates against the sprocket. A mono-involute motion is being created. In general, the tangent to an involute motion is normal to the evolution of the intersections of the curves made by the motion of the chain [2].

Most of the analysis will be made on flexible bodies which are in contact to make the simulations as real as possible. After knowing what forms will the contact have different positions would be analized. It becomes easier through flexible body analysis to know what are the general mass, general stiffness and the damping ratio of the system. Also to understand the different distribution of the stresses that occur during free oscillation and contacts.

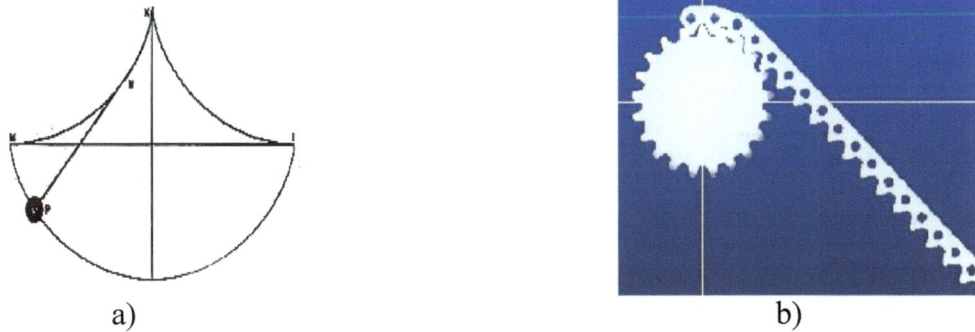

a)   b)

Figure1: Huygens Pendulum [2] (a); the model of the studied analogue Huygens Pendulum (b)

2. THEORETICAL BACKGROUND

In order to understand the flexible multibody methodology used it is important to define the flexible bodies. There are many approaches to define flexible bodies and maybe one of the most important definition from the point of view of multibody dynamics is that of Shabana [3].

2.1 Flexible Body Definition

Shabana defined flexible bodies from the moving frame approach within the setting provided by the reference point coordinates, in 1989. The natural coordinates of the body is uniquivocally used to define the large overall rigid bodies motion to which the elastic deformations variables are referred. Overall one can conclude that flexible bodies can be defined as rigid bodies multiplied by the coefficient of the elasticity of the body to express the deformation caused on the body [3].

The natural coordinates of a body do not include relative translation or rotations and are subjected to the corresponding rigid body constraints. The constraints in flexible bodies are different than that of rigid bodies as points and vectors can't be shared at the joints because the elastic deformations should be included.

Floating frame reference formulation suggested by Shabana [4] is presented below.

In general the global position coordinates of an arbitrary point "p" on a body "i" can be donated by q^i, which can be

represented as $q^i = \begin{bmatrix} R^i \\ \theta^i \\ q_f^i \end{bmatrix}$, where Ri and θ^i are reference coordinates and q_f^i represents the elasticcoordinates. One can

write the general equation of a point displacement of a body as shown in the equation below [4]:

$$\bar{r}_i = R^i + A(\bar{U}^i),$$ (1)

where:
A is the transformation matrix
R^i is the reference coordinate
$\bar{U}^i$ is general displacement of the body "i"

If one will add the element of deformability of the body, then the equation can be written by adding the vector of elasticity of the flexible body, knowing in general that the modal matrix for a point "i" has translational (S_T^i) and rotational (S_R^i) modes meaning that [4]:

$$S^i = [S_T^i \quad S_R^i] ,$$ (2)

where $\bar{U}^i$ can be (written in) expressed by the following equation [4]:

$$\bar{U}^i = S^i q_f^i .$$ (3)

$S^i q_f^i$ is the vector of elastic generalized displacement of body "i". By replacing equation 3 in equation, the general displacement of deformed bodies is obtained, as shown in the equation below [4]:

$$\bar{r}_i = R^i + A^i(\bar{U}_0^i + S^i q_f^i) .$$ (4)

The Kinetic energy can be written in the following form [4]:

185

$$T^i = \frac{1}{2}\dot{q}^{iT}M^i\dot{q}^i \quad , \tag{5}$$

where M^i is the symmetric mass matrix of the body "i" in the Multibody System.

2.2 Theoretical approach of contact modeling

Regarding contact modeling, there are many theories about how to obtain reasonable results of modeling contacts between bodies. This is mainly because more complicated systems have a huge number of DOFs due to many bodies that can fall in contact with one another and are in motion relatively with each other. Literature has assumed three main types of contacts according to the rigidity or flexibility of the parts in contacts such as: rigid-rigid bodies, flexible-rigid bodies and flexible-flexible bodies. One main and important reason to study contacts is to see the propagation of stresses and strains affecting the parts during impacts or contacts. This can only appear when one of the bodies is considered flexible or deformable.

A contact between rigid bodies can only approximate the values and shape of a contact adequate to the system. Since parts are deformable, the contacts of rigid bodies are clearly not completely accurate to represent the deformation occurred during contacts and can't find the propagation of impact stresses in the bodies.

Many theoretical approached for modeling contacts can be found in the literature. Perhaps the most commonly used is the Hertezian theory for elastic contacts [5]; it is often the basic law for direct and central impacts between rigid bodies having locally a contact surface which can be described by a quadratic function. The Hertezian law represents the contact force magnitude as a nonlinear function of the normal penetration "l" with a contact stiffness "k_p" where the contact force "f" can be represented as following [5]:

$$f = k_p l^n, \tag{6}$$

where "n" is the exponent of penetration between materials of different penalization having elliptical contacts.

The contact stiffness parameter depends on material properties as for the shape of the contact surface. For example if we take two spheres in contact, the contact stiffness can be given as [5]:

$$K_p = \frac{4}{3(\sigma_1 + \sigma_2)}\sqrt{\frac{R_1 R_2}{R_1 + R_2}} \quad , \tag{7}$$

where R_1 and R_2 are radii of the spheres, the material properties σ_i are computed from Young's modulus "E_i" and the Poisson's ratio "υ_i" of the body materials through the equation below [5]:

$$\sigma_i = \frac{1-\gamma^2}{E_i}. \tag{8}$$

Hertz's laws are limited to isotropic elastic bodies and do not account for kinetic energy losses. In order to include the energy dissipation during the contact process, the Kelvin Voigt model was created, stating that the model consists of a linear spring to represent the elastic force and a linear damper to model the energy loss, [5, 6]:

$$f = k_p l + c\dot{l} \ , \tag{9}$$

where

c is the damping coefficient or hysteresis coefficient,

$\dot{l}$ is the relative normal contact velocity.

The major drawback in the Kelvin Voigt contact model is the linearity of the contact force with respect to the load indentation; therefore the nonlinear nature of deformation is not correctly represented.

Hunt and Crossley [6] proposed to combine Hertz's laws and the Kelvin Voigt [5, 6] model in order to extend the nonlinear Hertz's law to account for the kinetic energy losses during contacts. Also in order to avoid a jump at the beginning of the impact and tension forces at the end they multiplied the classical viscous damping term $c\dot{l}$ with l^n. The internal damping contribution depends on the penetration velocity as that of the penetration length. The final model appeared in the form of the following equation [6]:

$$f = k_p l^n + c\dot{l}l^n. \tag{10}$$

They proposed to express the damping coefficient as a function of restitution e in the following form [6]:

$$C = \frac{3(1-e)}{2}\frac{k_p}{\dot{l}_s} \ , \tag{11}$$

where $\dot{l}_s$ is the initial relative normal velocity between impacting bodies.

The dependency of the damping coefficient with respect to the restitution coefficient allows controlling the amount of energy dissipated by each impact. Many have adjusted the Hunt and Crossley impact laws maybe the two most important contributions were Lankarini, Nikraveshi [7] and Flores, Machado [8]. They have developed the coefficients for the damping coefficient calculation.

3. NUMERICAL MODELING IN MSC ADAMS

All flexible bodies need to have a finite element model structure for simulating their behavior. Due to this structure the DOFs which are infinite become finite, yet with very large number of DOFs. Each part has its own local reference frame (coordinate system) that is defined by a position vector of a global reference frame.

3.1 Flexible Bodies in MSC ADAMS

ADAMS Flex-Bodies considers a small linear deformation at the local form reference frame [9]. The small linear deformations can be approximated as superposition of a number of shape vectors. The shape vectors can be determined with a modal frequency analysis called modal superposition. This can be done by ADAMS/FLEX or any FEA program. In this study *Patran* and *Nastran* were used to perform modal analysis [10]. The natural frequencies and their corresponding mode shapes are determined . The results of the analysis are stored as binary files or modal neutral files (.mnf) that ADAMS can import and represent the flexible bodies.

3.2 The Modal Superposition Theory

The location of a node "P" is defined by the vector from the ground origin to the origin of the local body reference frame "B". This is quite similar to that of Shabana's Floating Frame Theory, where $\vec{S}_p$ is the position vector from the load body reference frame of B to the un-deformed position of point "P" and $\vec{U}_p$ is the translational deformation vector of point "P". The translational deformation vector $\vec{U}_p$ is a modal superposition and can also be written in matrix form [9]

$$[U_p] = [\phi_p] * q,$$
where (12)

ϕ_p is a 3*M matrix and q is the amplitude of each shape.
The three rows are for the corresponding translational DOFs of node "P". The same will be for the rotational DOFs. The point of derivations for the flexible body generalized coordinates is introduced as [9]:

$$\xi = \{X \quad Y \quad Z \quad \psi \quad \theta \quad \phi \quad i = 1 - N\}, \tag{13}$$
where {X, Y, Z} are the generalized coordinates, $\{\psi, \theta, \phi\}$ are the Euler angles of the flexible bodies and the modal coordinates are represented by q_i (i=1-N).
A simplified equation of motion in MSC ADAMS is represented below [9]:

$$[M]\ddot{\xi} + K\xi = Q, \tag{14}$$
where Q is the generalized force applied.
The simplified equation of Kinetic Energy "T" [9]:

$$T = \frac{1}{2}\xi^T * M(\xi) * \xi . \tag{15}$$

This conveys an understanding of the kinematics created in MSC ADAMS when equations for flexible bodies are to be resolved. The importance of such data is that it illustrates how the chain plate structure is affected during contacts. This leads to the next topic which is: how does MSC ADAMS calculate contacts?

3.3 Contacts in MSC ADAMS

ADAMS does not totally base the calculations on the Hertz theory for calculating contact forces and stiffness. The geometrical shape of the plate and the geometrical shape of the sprocket are also important so to give ADAMS the ability to detect the bodies entering contacts. The stiffness of the bodies that are subjected to contacts is also important. The stiffness is calculated by the equation:

$$K = F_N / r_{p_c}^i , \tag{16}$$

where F_N is the normal force acting at the contact point and $r_{p_c}^i$ is the displacement of the point during contact.
An assumption has been created that only the plate links of the chain are flexible and subjected to deformations and the sprocket is considered as rigid body. This is in order to simplify the calculations and to find out how the contact affects the chains at low speeds and frequencies.
An important factor to calculate the contacts is the penetration depth [9]. The penetration depth is a value that defines at what amount of penetration the solver will use full damping. For example at 0 penetration 0 damping occurs. MSC

ADAMS suggests 0.01 mm is a reasonable value but the value also depends on the complexity of the model one is analyzing. In this case the penetration depth is estimated at 0.05 mm for calculation reduction.

The next important factor is the damping values [9] that define the properties of the material with which bodies are in contact. MSC ADAMS suggests a damping coefficient that is about 1 % of the stiffness coefficient.

One of the important factors is the force exponent which is a parameter of the spring force of the impact function which has recommended value >1. For example in this case hard metals, such as steel to steel, the force exponent is 2.2.

4. RESULTS AND DISCUSSIONS

The importance of studying the model from the flexible multibody dynamics point of view is to see the effect of the plates colliding with the sprocket from the kinetic energy, deformation, frequency response point of views. As mentioned before for reducing the magnitude of simulation only the plates are considered flexible bodies. It can be said that the motion of the dead body weight in time can represent the total amount of kinetic energy lost as shown in figure 2 during motion. The kinetic energy lost is due to contacts and friction of the plates with the sprocket and the friction of the plates with the pins represented in the model as revolute joints. According to the factors mentioned above it is safe to say that the (I.T.C.) looses energy in an exponential form giving an understanding that the (I.T.C.) has a healthy (good) damping behavior in its natural form and to multiple shocks without adding external excitation forces. The exponential damping of the system gives the opportunity for one to understand why (I.T.C.) can also be called "silent chains". Also the total angular displacement of the dead weight of the body shows how the system is simply damped having an exponential curve. The cause is the friction between the joints connecting the plates of the chain and the contacts between the chain plates and the sprocket.

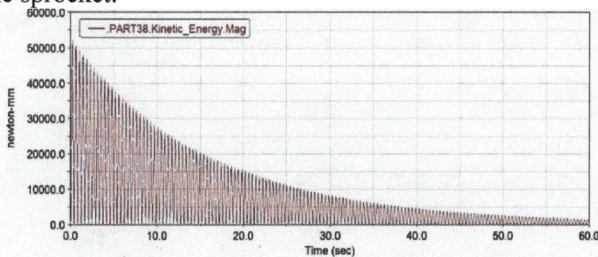

Figure 2: The kinetic energy of the deadweight

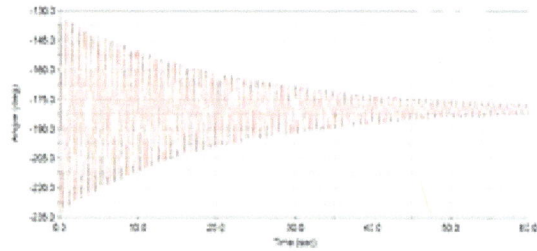

Figure 3: Angular displacement of the deadweight

We can simply see how the clean accelerations help in understanding the increase or decrease of the amount of vibrations induced due to the contacts of the moving plates with the fixed sprocket. This can be seen in the figures below by following the development of body accelerations and their corresponding frequencies responses. Figure 4 illustrates the acceleration of the last body in contact during oscillation and how the frequencies dissipate when the system decelerates.

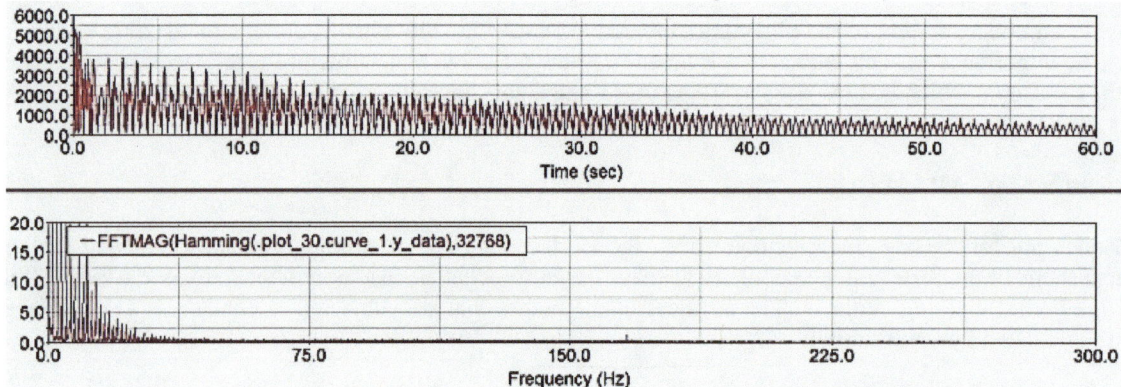

Figure 4: The acceleration of the last plate in contact in time and frequency domains

The highest contacts of pair plates of the chain can be simply subdivided to main intervals. One can notice the engagement, friction and disengagement impacts to form a total contact of the body as shown in figure 5. The second interval illustrates the creation of the impulsive impacts due to the total loss of the kinetic energy in the chain during motion causing the chain to simply reach the rest position. The contact will simply dissipate and will take the form of the last body in contact with the sprocket in the form of the second interval of a normal plate position in contact, meaning the impulsive contacts as shown in figure 6.

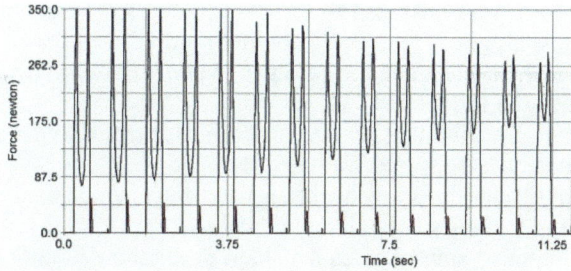

Figure 5: Contact Forces during the first interval

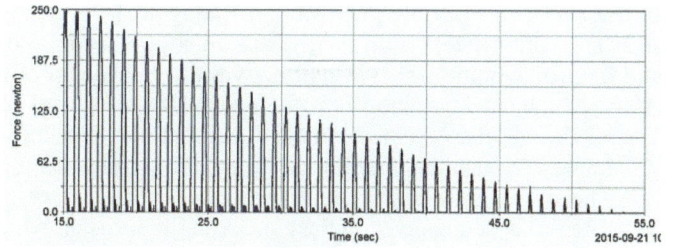

Figure 6: Contact forces dissipating till reaching impulsive contacts or the second interval of contacts

As the kinetic energy decreases the contact's usual form disappears from the point of view of engagement, friction and disengagement, giving the contacts of small friction interval and producing impulsive contacts. The birth of the impulsive contacts simply illustrates that the oscillation of the pendulum is damped and that the oscillation motion simply dissipates and disappears. Figure 7 indicates the impulsive contacts with very low bandwidth and the magnitude of the contact forces that simply disappears till no impulsive contact is observed meaning that the oscillation does not have the sufficient power to raise the plate to contact the sprocket.

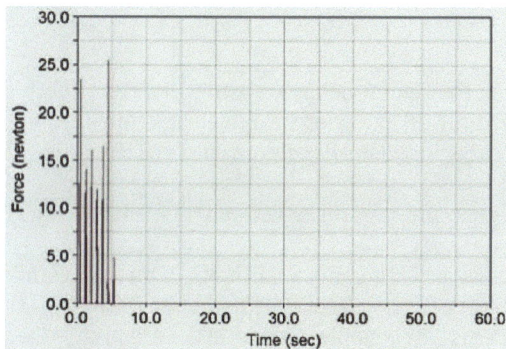

Figure 7: Impulsive contact forces

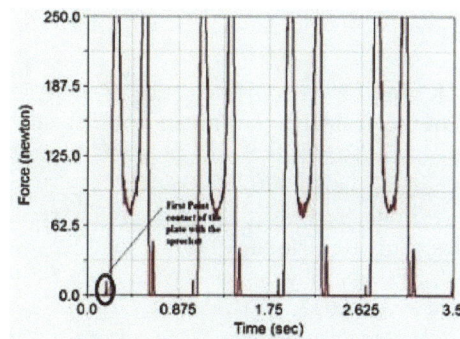

Figure 8: First contact location of a plate with the sprocket

One can approximately indicate the period and location of the first contact of the plate with the sprocket and the small spikes of pre total contacts. The impulsive spike contact as shown in figure 8 simply decreases the magnitude force of the total contact of the plate. The orientation of the plate towards the sprocket for a first impulsive sprocket is indicated in figure 9. One can deduce approximately the location of the first impulsive contact as shown in figure 10.

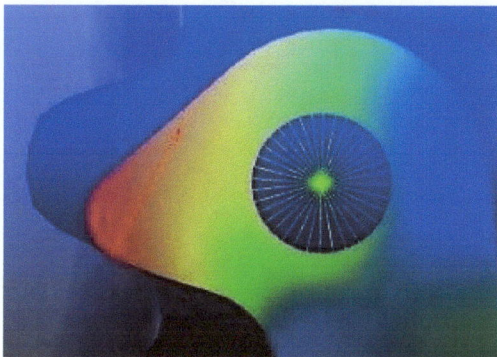

Figure 9: First contact partial view with deformation spectra due to contact

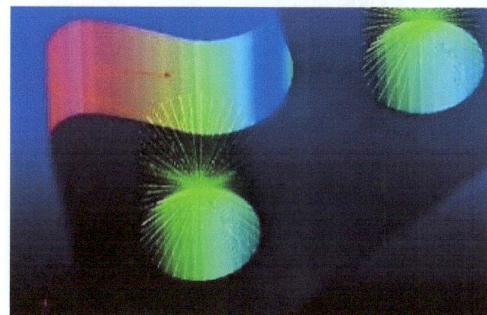

Figure 10: Estimated location of the first line of impulsive (spike) pre-contact

5. CONCLUSIONS

In this paper the flexible plates in an Analogous Huygens Pendulum were analyzed in Flexible Multi-Body Dynamic software. This type of analysis gives us a better perception of the Analogous Huygens Pendulum motion from the time and frequency points of view. It also gives the prospect of how does the system decelerates in an exponential form.
Along all the results of the simulations, one can realize and conclude the following:
1. As the angular gap between the plates and the sprocket is larger, the vibration produced becomes smaller giving an impression that the length is an inverse proportion to the frequency.
2. The single connecting plate suffers much more contact than the 2 plates in a link as the surface area is much bigger.
3. There is a first contact on the plate with the sprocket which simply gives a reduction of contact acting like a first aid damping contact even if the values are small. The first point contact simply gives a sudden deceleration so not to have a full contact at a single step.
4. One can realize if there is no first contact at the inner side of the plates during oscillation and contacts, the magnitude of the contacts is much higher and gives a higher distortion or vibration of the system. To have a first contact this also depends on the angle of the plates coming in contact with the sprocket. The double contact of the I.T. Chains gives an exponential deceleration as shown clearly at the displacements of the system and the deceleration curve.
5. The first contact assures a better positioning and assures a better engagement to reduce the total deformation of the chain plates during contacts.
6. The system in itself suffers during low frequencies and as the frequency increases the system gives a more stable behavior

REFERENCES

[1] Shalaby K. and Lache S., Kinematics of the Analogous Huygens Pendulum behavior using Inverted Tooth Chain, Bulletin of the Transilvania University of Brasov, Volume 8 (57) no.1 2015 Series I Engineering Science.
[2] Alan Emerson, Things are seldom what they seem Christian Huygens, The Pendulum and the Cycloid. NAWC Bulletin No362, June 2006, PP 295-312.
[3] Ahmed A Shabana, Flexible Multibody Dynamics: Review of past and present developments. Multibody System Dynamics, Volume I No. 2, 1997.
[4] Ahmed A Shabana, Dynamics o Multibody Systems, Fourth edition, University of Illinois, Chicago, Cambridge, University Press, 2013.
[5] Geoffrey Virlez, Modeling of Mechanical Transmission Systems in Vehicle Dynamics, University De Liege Aerospace and mechanical department. Phd Thesis, June 2014.
[6] K. Hunt and F. Crossley, Coefficient of restitution interpreted as damping in vibroimpact, Journal of Applied Mechanics, 42(2):440-445, 1975.
[7] H. Lankarini, Cononical Equation of motion and estimation of parameters in the analysis of impact problems, Phd Thesis, University of Arizona, USA, 1988.
[8] P. Flores, M. Machado, M. T. Silva and J. Martins, On the Continuous Contact Force Models For Soft Materials in Mulitbody Dynamics, Multibody System Dynamics. 25:357-375, 2011.
[9] MSC ADAMS Help 2013.
[10] MSC NASTRAN Help 2004.

6th International Conference
"Computational Mechanics and Virtual Engineering "
COMEC 2015
15-16 October 2015, Braşov, Romania

A MATHEMATICAL FORMULATION TO ROBOTICS MANIPULATION. FINITE ROTATION IN 3D

Enescu Ioan

Transilvania University, Braşov, ROMÂNIA, enescu@unitbv.ro

Abstract: The work present a mathrmatical resolv of rigid mouvement in D-3 described by $\vec{u}$ axis, rotation about this axis $(\varphi, \vec{q} = tg\,\dfrac{\varphi}{2}\vec{u}\,)$ and slip $a\vec{u}$ along the axis. Also resolv the problem to calculus of the coordinates of screw generator of the finite displacement of three points. It is also applied to the velocity firld and the field of moments generate by the reduce screw of system of forces applied.

Keywords: rotation, slip, velocity,displacement, axis

1.USING THE RODRIGUES TO COMPARE ROTATIONS

Supose we are rotating a point, **p**, in space by angle, φ, about an axis through the origin represented by the unit vector, $\vec{u}$ (fig.1).

First, we create the matrix $\tilde{u}$ which is the liniear transformation that computes the cross product of the vector a which any other vector $\vec{v}$.

$$\bar{u} \times \bar{v} = \begin{Bmatrix} u_y v_z - u_z v_y \\ u_z v_x - u_x v_z \\ u_x v_y - u_y v_x \end{Bmatrix} = \begin{bmatrix} 0 & -u_z & u_y \\ u_z & 0 & -u_x \\ -u_y & u_x & 0 \end{bmatrix} \begin{Bmatrix} v_x \\ v_y \\ v_z \end{Bmatrix} = \tilde{u}\bar{v} \tag{1}$$

with

$$\tilde{u} = \begin{bmatrix} 0 & -u_z & u_y \\ u_z & 0 & u_x \\ -u_y & u_x & 0 \end{bmatrix}$$

Now, the ro, inspace by an angle, φtation matrix can be written in terms of $\tilde{u}$ as

$$\bar{e}_1 = \tilde{u}\bar{r}_1 \qquad \bar{e}_2 = \tilde{u}\tilde{u}\bar{r}_1 \qquad Q = e^{\tilde{u}\varphi} = E\,I + \tilde{u}\sin\varphi + \tilde{u}\tilde{u}[1 - \cos(\varphi)] \tag{2}$$

2. AN GEOMETRICAL EXPLANATION ROTATION AS VECTOR COMPONENTS IN A 2D SUBSPACE

Suppose we are rotating a point p. φ, about an axis through the origin, represent by the unit vector, $\bar{u}$. The component of $\bar{r}$ (position vector of point **p)**, parallel to $\bar{u}$, $\bar{r}_{paru}$, will not change during the transformation. The component of $\bar{r}$ perpendicular to $\bar{u}$, $\bar{r}_{peru}$, the axis in the same as in 2D. The vector $\bar{r}_{peru}$ and $\bar{r}_{biperu}$ ar of the correct length and orientation to act as the x and y vectors in the 2D rotation.

$$\bar{r}_2 = \bar{r}_{paru} + \cos(\varphi)\bar{r}_{peru} + \sin(\varphi)\bar{r}_{biper}$$
$$\bar{r}_2 = (\bar{r}_1 + (\bar{u} \times (\bar{u} \times \bar{r}_1))) + \cos(\varphi)(-\bar{u} \times (\bar{u} \times \bar{r}_1))) + \sin(\varphi)(\bar{u} \times \bar{r}_1)$$
$$\bar{r}_2 = (\bar{r}_1 + \sin(\varphi)(\bar{u} \times \bar{r}_1) + (1 - \cos(\varphi))(\bar{u} \times (\bar{u} \times \bar{r}_1))$$
$$\bar{r}_2 = (I + \sin(\varphi)\tilde{u} + (1 - \cos(\varphi))\tilde{u}\tilde{u})\bar{r}_1 \tag{3}$$

3. AN ALGEBRAIC EXPLANATION ROTATION AS A DIFFERENTIAL EQUATION

Suppose we are rotating a point, **p**, in space by an angle φ about axis through the origin, represented by the unit vector, u. We will from a differential equation describing the motion of the point from time t=0 to time t= φ . Let p(t) be the position of the point at time t. The velocity of the point at any time is

$$\vec{r}_2(t) = \vec{u} \times \vec{r}_1(t) \tag{4}$$

Now, if we use the matrix formula for cross products in our differential equation, we have a first order, linear system of differential equations, $\vec{r}_2 = \tilde{u}\vec{r}_1$. The solution to that system is known to be $\vec{r}_1(t) = e^{\tilde{u}t}\vec{r}_1(0)$ so the location of the rotation of the rotated point will be $\vec{r}_1(\varphi) = e^{\tilde{u}\varphi}\vec{r}_1(0)$.

4. TAYLOR EXPANSION

Now we need to evaluate $e^{\tilde{r}\varphi}$, so we examine its Taylor expansion.
If $\tilde{r}$ =A, we have

$$e^{A\theta} = I + A\theta + \frac{(A\theta)^2}{2!} + \frac{(A\theta)^3}{3!} + \frac{(A\theta)^4}{4!} + \frac{(A\theta)^5}{5!} + \ldots\ldots \tag{5}$$

Considering how constructed A, it is easy to verify that $A^3 = -A$. Every additional application of A turns the plane $\vec{r}_{paru}$ A^2 in the appropriate places, we gest

$$e^{A\theta} = I + [.] + \left[\frac{A^2\theta^2}{2!} + \frac{A^2\theta^4}{4!} + \frac{A^2\theta^6}{6!} + \ldots\right] = I + A\left[\theta - \frac{\theta^3}{3!} + \frac{\theta^5}{5!} + \ldots\right] + A^2\left[\frac{\theta^2}{2!} - \frac{\theta^4}{4!} + \frac{\theta^6}{6!} + \ldots\right] \tag{6}$$

Now, we recognize the Taylor expansion for sin(φ) and cos(φ) in the above expression and find that

$$e^{A\varphi} = I + A\sin(\varphi) + A^2[1 - \cos(\varphi)] \tag{7}$$

Gives us the rotation matrix. This formula is known as Rodrigues'Formula

5.THE FINITE ROTATION OPERATOR

The geometrical transformation realize by the rotation of frame about an axis, with an directory vector $\vec{u}$, with the angle φ (fig.1) is described in the original base

$$\underline{B} = [\vec{e}_1, \vec{e}_2, \vec{u}] = [\tilde{u}\vec{r}_1, \tilde{u}\vec{r}_2\tilde{u}]$$

Where $\tilde{u}$ is the antireflection vector produce, side to the finale positional vector $\vec{r}_2$ is

$$\vec{r}_2 = \vec{r}_1 + (1 - \cos\varphi)\vec{e}_2 + \sin\varphi\vec{e}_1 = \left[\underline{E} + 2\sin^2\frac{\varphi}{2}\tilde{u}\tilde{u} + 2\sin\frac{\varphi}{2}\cos\frac{\varphi}{2}\tilde{u}\right]\vec{r}_1 \qquad (8)$$

Where $\underline{E}$ is the identical transformation. The operator is the isometric transformation $\underline{R}$

$$\vec{r}_2 = \underline{R}\,\vec{r}_1 \qquad (9)$$

$$\underline{R}^T = \underline{E} + 2\sin^2\frac{\varphi}{2}\tilde{u}\tilde{u} - 2\sin\frac{\varphi}{2}\cos\frac{\varphi}{2}\tilde{u}\,, \quad \underline{R}^T\underline{R} = \underline{E},$$

Theorem Rodrigues result direct by

$$\vec{r}_2 - \vec{r}_1 = 2\sin\frac{\varphi}{2}\tilde{u}\left\{\sin\frac{\varphi}{2}\tilde{u} + \cos\frac{\varphi}{2}\underline{E}\right\}\vec{r}_1$$

$$\vec{r}_2 + \vec{r}_1 = 2\left\{\underline{E} + \sin^2\frac{\varphi}{2}\tilde{u}\tilde{u} + \sin\frac{\varphi}{2}\cos\frac{\varphi}{2}\tilde{u}\right\}\vec{r}_1 \qquad (10)$$

$$\tilde{u}(\vec{r}_1 + \vec{r}_2) = 2\cos\frac{\varphi}{2}\tilde{u}\left\{\sin\frac{\varphi}{2}\tilde{u} + \cos\frac{\varphi}{2}\tilde{u}\right\}\vec{r}_1$$

With $(\tilde{u}\tilde{u}\tilde{u}\vec{r}_1 = -\tilde{u}\vec{r}_1)$

And an definite

$$\vec{r}_2 - \vec{r}_1 = tg\frac{\varphi}{2}\tilde{u}(\vec{r}_1 + \vec{r}_2) = \tilde{q}(\vec{r}_1 + \vec{r}_2)$$

Where $\tilde{q} = tg\frac{\varphi}{2}\tilde{u}$

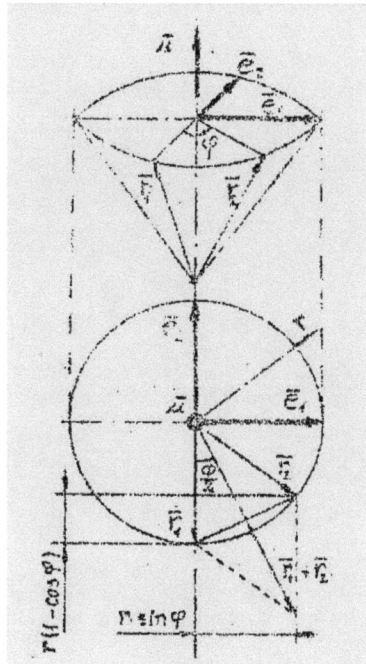

Figure 1

Result than the displacement $(\vec{r}_2 - \vec{r}_{1)}$ in rotation with the angle φ around the $\vec{u}$ axis is the vector produce of the

rotation $\vec{q} = tg \dfrac{\varphi}{2} \vec{u}$ slid on the rotation axis $\vec{u}$ applied on double median position $(\vec{r}_1 + \vec{r}_2)$.

The finite displacement of the rigid is singular display by a rotation φ around the axis along the same axis.

Front face an pole O we have

$$\vec{r}_2' - \vec{r}_1 = \vec{r}_2 - \vec{r}_1 + a\vec{u} = tg \frac{\varphi}{2} \tilde{u}(\vec{r}_2' + \vec{r}_2) + a\tilde{u}$$

and about the constant location O_1 we have

$$\vec{P}_i = \vec{P}_o + \vec{r}_i$$

$$\vec{P}_2' - \vec{P}_1 = tg \frac{\varphi}{2} \tilde{u}(\vec{P}_2' + \vec{P}_1 - 2\vec{P}_o) + a\tilde{u} \tag{11}$$

6. THE STATEMENT OF THE PROBLEM

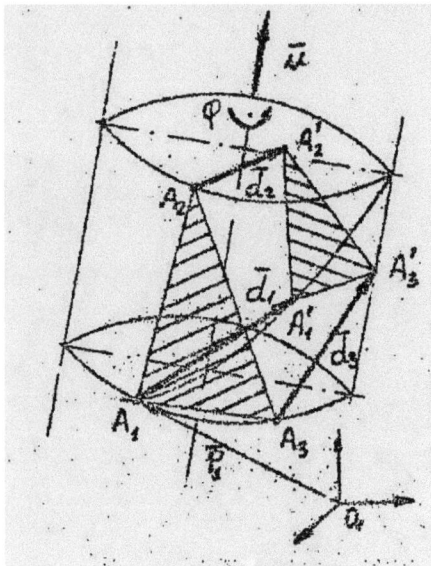

Figure.2

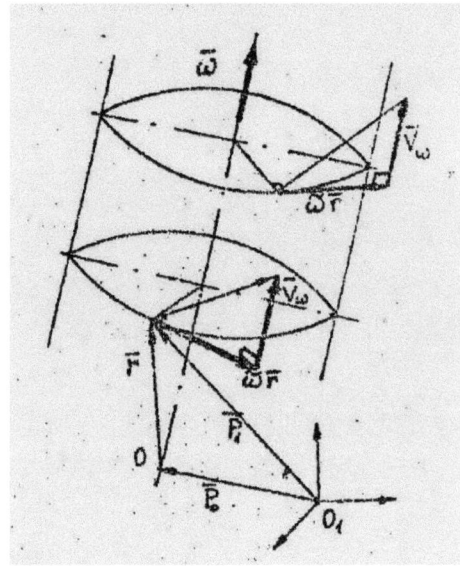

Figure.3

$$\vec{q} = tg \frac{\varphi}{2} \tilde{u} \qquad\qquad \vec{\omega} = \omega \vec{u}$$

$$\vec{d}(\vec{P}) = \tilde{q}(\vec{P}_1' + \vec{P}_2 - 2\vec{P}_o) + a\vec{u} \qquad\qquad \vec{V}(\vec{P}) = \tilde{\omega}(\vec{P} - \vec{P}_o) + \vec{V}_\omega$$

The problem is to determinate if we now the finite displacement of three points $A_i(P_i)$ i=1,3 (fig.2), the $\vec{u}$ axis of the screw of finite rotation and slip $\vec{u}$ on the axis.

The problem is similar with the associate problems

- The field of rigid velocity, definite with the instantaneous velocity of three points $\vec{V}_i(P_i)$ (fig.3)generate by the kinematic screw $\{V_\omega, \vec{\omega}\}$.

- The field of the moments in the rigid frame definite by the moments of three points $\vec{M}_i(P_i)$, decline on the central axis to the resultant screw $\{\vec{R}, \vec{M}_R\}$(fig.4).

Noted with $\vec{\delta}_i = \vec{P}_i + \vec{P}_i'$, and the three displacements

$$\vec{d}_i = \tilde{q}(\vec{S}_i - 2\vec{r}_i) + a\vec{u}(i, j, k)$$

In relative displacements

$$\vec{d}_{ki} = \vec{d}_i - \vec{d}_k$$

$$\vec{d}_{kj} = \vec{d}_j - \vec{d}_k$$

Is deleted the slide and is underline the arbitrary global representation obtain two vector equations

$$\vec{\delta}_{ki} = \tilde{q}(\vec{S}_i - \vec{S}_k) = \tilde{q}\vec{S}_{ki}$$

$$\vec{d}_{kj} = \tilde{q}(\vec{S}_j - \vec{S}_k) = \tilde{q}\vec{S}_{kj}$$

$$\vec{S}_{ki} = \vec{S}_i - \vec{S}_k \tag{12}$$

represent the two normal displacements of the nodes i,j, in the normal plane of the rotation axis (fig.5).

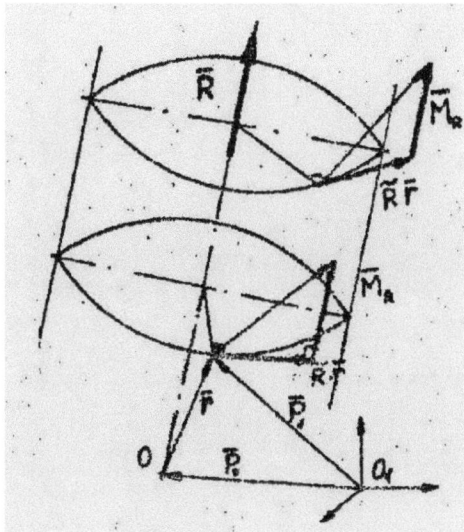

Figure 4

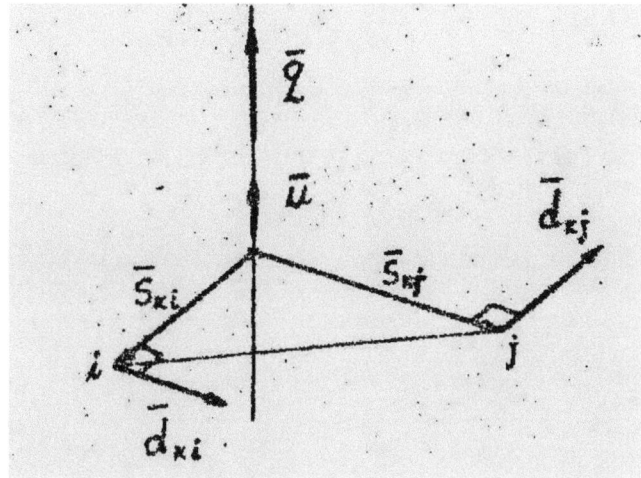

Figure.5

$$\vec{R} = R\vec{u}$$

$$\vec{M}(P) = \tilde{R}(\vec{P} - \vec{P}_o) + \vec{M}_R$$

In the equations (12) the single unknown is the vector $\vec{q}$. We obtain this vector by vector produce of the two equations:

$$\vec{d}_{ki} \times \vec{d}_{kj} = \vec{d}_{ki} \times (\vec{q} \times \vec{S}_{kj}) = (\vec{d}_{ki}\vec{S}_{kj})\vec{q} - (\vec{d}_{ki}\vec{q})\vec{S}_{kj} = (\vec{d}_{kj}\vec{S}_{ki})\vec{q} + (\vec{d}_{kj}\vec{q})\vec{S}_{ki}$$

Where $\vec{d}_{ki}\vec{q} = 0$

$$\vec{q} = \frac{\vec{d}_{ki} \times \vec{d}_{kj}}{(\vec{d}_{ki}\vec{S}_{kj})} = -\frac{\vec{d}_{ki} \times \vec{d}_{kj}}{(\vec{d}_{kj}\vec{S}_{ki})} \tag{13}$$

We can easily prove the symmetry

$$(\vec{d}_{ki}\vec{S}_{kj}) = -(\vec{d}_{kj}\vec{S}_{ki}) = \frac{1}{2}(\vec{d}_{ki}\vec{S}_{kj} - \vec{d}_{kj}\vec{S}_{ki}) = \frac{1}{2}\left[(\vec{d}_k - \vec{d}_j)\vec{S}_i + (\vec{d}_i - \vec{d}_k)\vec{S}_j + (\vec{d}_j - \vec{d}_i)\vec{S}_k\right] =$$

$$= \frac{1}{2}\left[\vec{S}_i\vec{d}_{jk} + \vec{S}_j\vec{d}_{ki} + \vec{S}_k\vec{d}_{ij}\right]$$

And we have final the complete representation

$$\vec{q} = 2\frac{(\vec{d}_i \times \vec{d}_j) + (\vec{d}_j \times \vec{d}_k) + (\vec{d}_k \times \vec{d}_i)}{(\vec{S}_i\vec{d}_{jk}) + (\vec{S}_j\vec{d}_{ki}) + (\vec{S}_k\vec{d}_{ij})} \qquad (14)$$

The angular velocity $\vec{\omega}$ $(\vec{\omega} = \omega\vec{u})$ is:

$$\vec{V}_{ki} = \vec{\omega} \times \vec{P}_{ki}, \ \vec{V}_{kj} = \vec{\omega} \times \vec{P}_{kj}, \ \vec{P}_{ki} = \vec{P}_i - \vec{P}_k$$

$$\vec{V}_{ki} \times \vec{V}_{kj} = \vec{V}_{ki} \times (\vec{\omega} \times \vec{P}_{kj}) = (\vec{V}_{ki} \cdot \vec{P}_{kj})\vec{\omega}$$

$$\vec{\omega} = \frac{\vec{V}_{ki} \times \vec{V}_{kj}}{(\vec{V}_{ki}\vec{P}_{kj})} = 2\frac{(\vec{V}_i \times \vec{V}_j) + (\vec{V}_j \times \vec{V}_k) + (\vec{V}_k \times \vec{V}_i)}{(\vec{V}_i\vec{P}_{kj}) + (\vec{V}_j\vec{P}_{ik}) + (\vec{V}_k\vec{P}_{ij})} \qquad (15)$$

and the sliding result on (AC)

$$\vec{R} = 2\frac{(\vec{M}_i \times \vec{M}_j) + (\vec{M}_j \times \vec{M}_k) + (\vec{M}_k \times \vec{M}_i)}{(\vec{M}_i\vec{P}_{kj}) + (\vec{M}_j\vec{P}_{ik}) + (\vec{M}_k\vec{P}_{ji})} \qquad (16)$$

The sliding components of the generator screw displacement $a\vec{u}$ result by any of the three projections on the direction of axis $\vec{u}$

$$a = (\vec{d}_i\vec{u}) = \frac{\vec{d}_i\vec{q}}{tg\frac{\varphi}{2}} \qquad (17)$$

Similar the sliding velocity in the momentary movement of kinematic generator screw

$$V_\omega = (\vec{V}_i\vec{u}) = \frac{(\vec{V}_i\vec{u})}{\omega} \qquad (18)$$

And the minimal torque of the reduce screw

$$C = M_R = (\vec{M}_i\vec{\omega}) = \frac{(\vec{M}_i\vec{R})}{\omega} \qquad (19)$$

One point of the generator displacement screw axis is obtain by the vector product of any three displacements (by the global referent (4) with the vector of finite rotation.

$$\vec{q} \times \vec{d}_i = \vec{q} \times \left[\vec{q} \times (\vec{S}_i - 2\vec{P}_o)\right]$$

and

$$\vec{P}_0 = \frac{1}{2}\left[\frac{\vec{q} \times \vec{d}_i}{q^2} + (1 + 2\frac{(\vec{q}\vec{P}_0)}{q^2})\vec{S}_i + \lambda\vec{q}\right]$$

and on determine the intersection with the normal plane on $\vec{q}$

$$(\vec{q}\vec{P}_0) = \alpha$$

$$\vec{P}_{0\alpha} = \frac{1}{2}\left[\frac{\vec{q} \times \vec{d}_i}{q^2} + (1 + 2\frac{\alpha}{q^2})\vec{S}_i\right] \qquad (20)$$

Similar the instantaneous axis of the generator screw

$$\vec{P}_{0\alpha} = \frac{\vec{\omega} \times \vec{V}_i}{\omega^2} + \vec{P}_i + \frac{\alpha}{\omega^2}\vec{\omega} \qquad (21)$$

respective the point where the central axis (AC) intersect the normal plane $(RP_0) = \alpha$

$$\vec{P}_{0\alpha} = \frac{\vec{R} \times \vec{M}_i}{R^2} + \vec{P}_i + \frac{\alpha}{R^2}\vec{R}$$

REFERENCES

[1] M.C.Tofan, I.Enescu, S.Dimitriu: Program Rotation Finite in D-3 Sesiune,UniversitateaConstanţa, 1993
[2] R. M. Murray, Zeniang Li: A Mathemathical Introduction to Robotics, Berkeley, 1993

**6th International Conference
"Computational Mechanics and Virtual Engineering "
COMEC 2015
15-16 October 2015, Braşov, Romania**

THE BEHAVIOUR ANALYSIS OF COMPOSITE MATERIAL MAT-450 TO THE BENDING STRAIN

Costina- Mihaela Gheorghe[1], Simona Lache[2]

[1] Universitatea *Transilvania* din Braşov, ROMÂNIA, e-mail: gheorghe.costina@unitbv.ro
[2] Universitatea *Transilvania* din Braşov, ROMÂNIA, e-mail: slache@unitbv.ro

Abstract: The outstanding properties of composite materials make them which are increasingly used in industry, is gaining more and more ground against traditional materials. Many products in the aerospace, electronics, automotive, agriculture are made of composite materials. Determining the specific characteristics of new types of composite materials and understanding how they behave are vital, because by the most of the quality of these materials can achieve high quality products. The paper analyses the behaviour of the composite material MAT-450 to the bending stress. The mechanical characteristics were determined experimentally on the stand; the specimens were tested at using the three points bending method. Tests were conducted on a material testing machine at the Faculty of Mechanical Engineering of the Transilvania University of Braşov, type-TEXTURE ANALYSER, which is produced by Lloyd Instruments. The results allow the optimization of composite structures used in construction vehicles. The paper ends with conclusions drawn from the research.
Keywords: composite materials, MAT-450, bending test, polyester resin

1. INTRODUCTION

The composite materials with polymer matrix are used in the aerospace industry, shipbuilding industry, automobile industry, because of their good mechanical properties: low density, low coefficient of thermal expansion, and high corrosion resistance.
The composite materials made of polyester resin and reinforced with glass fibre, such as fabric materials, are being used more and more. Structures with complex shapes can be made much easier of composite materials [2].
Physical and mechanical properties of the composite materials depend on the geometry and arrangement of the components [9]. Improved of mechanical properties of composite materials has led to the use of fabric as reinforcement fibre [8], [3]. The composite material is more rigid as the concentration of reinforcement is higher [6].
Determination and understanding the behaviour of this material can lead to judicious use of it and getting quality products.
A composite material increasingly used is made of polyester resin and glass fibre MAT-450.
The paper analyses the behaviour of the composite material MAT-450 subjected to static bending, to determine breaking stress and stiffness of the material.

2. THEORETICAL BACKGROUND

The calculation of strength of layered materials undergoing static tests is made using the failure criteria [7]. Theoretically, a layer failure occurs when fibres of this layer are breaking. There are programs [1], [5], which allow the calculation of strength of laminates subjected to different solicitations. During bending, normal stresses of composites materials appear in external layers, which can lead to failure of the matrix, fibres or general failure of the plate.

The bending load σ_f for a F force is calculated in [MPa], using the formula:

$$\sigma_f = \frac{3F \cdot L}{2b \cdot h^2} \tag{1}$$

where: σ_f - bending strength, [MPa];

F - the force exerted on the specimen, [N];
L - the distance between the fixed supports of specimen, [mm].
b - the section width of the specimen, [mm];
h - the thickness of the specimen section, [mm].

The modulus of elasticity is given by the following formula:

$$E_b = \frac{L^3}{4bh^3} \frac{\Delta F}{\Delta f} \tag{2}$$

where: Eb - the modulus of elasticity, [MPa];

L - the distance between fixed bearings, [mm];
b - the width of the specimen , [mm];
h - the thickness of the specimen, [mm];
ΔF - the variation of the force in the straight part of diagram Force-Sag, [N];
Δf - the variation of the sag corresponding to the force variation ΔF, [mm].

3. EXPERIMENTAL ANALYSIS

The specimens used for testing the breaking bending are rectangular.

The bending test in the three-point of specimens shall be carried like the sketch show in Figure 1. The specimen are resting on two fixed cylindrical supports to the distance of 80 mm from each other being loaded at midspan on the upper side through a cylindrical pusher with the force F. This force will cause a deformation of the specimen with the sag f. The amount of force will steadily increase until the specimen will break. The value of the force F will be recorded in daN and the value f of the sag deflection in mm.

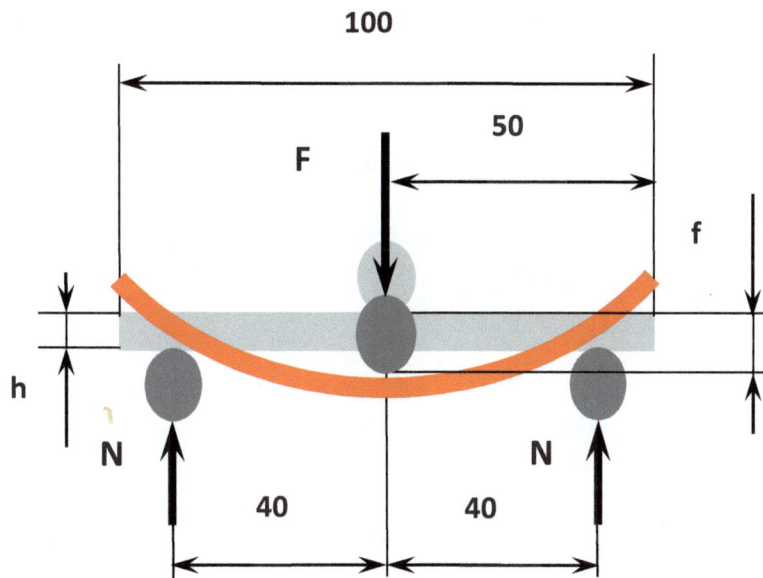

Figure 1: The scheme of the bending test of specimen

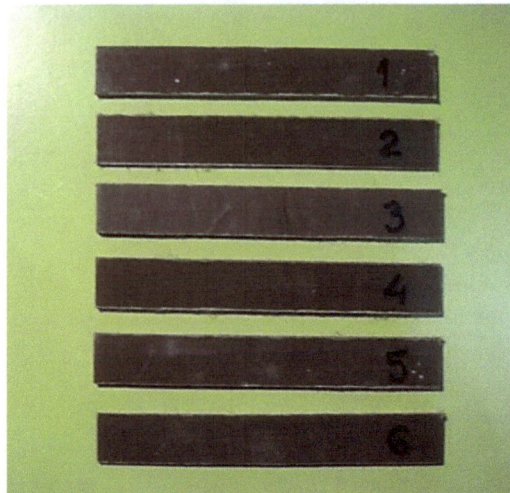

Figure 2: The specimens used for testing

There has been made a composite plate having a thickness of about 2 mm, using polyester resin and four layers of woven glass fibre MAT-450.

For testing are used six specimens obtained by cutting the plate material. The samples were numbered for identification, with the numbers 1 to 6 (Figure 2).

The dimensions of the specimens, as shown in Figure 3, were determined by measuring them with a micrometer and recorded in Table 1.

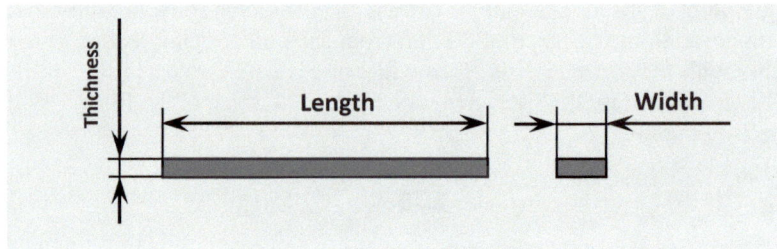

Figure 3: Specimen dimensions

Table 1: Specimen dimensions

Specimen number	Length	Width	Thickness
1	100,32	14,77	2,00
2	100,50	15,00	1,90
3	100,48	14,85	1,75
4	100,50	14,89	1,57
5	100,50	14,85	1,78
6	100,50	14,95	2,00

The tests were conducted at the Faculty of Mechanical Engineering of the *Transilvania* University of Braşov.

For the three-point bending test was used a machine manufactured by Lloyd Instruments, UK (Figure 4), TEXTURE ANALYSER LR5K Plus type, which provides a maximum force Fmax = 5 kN. The machine also complies the following conditions:

-   allow relative movement of the pressing head relative to the supports, at a speed approximately constant and adjustable;
-   the load indicated on the scale is identical to that applied to the specimen, dimensions of the scale enables reading all tasks with an error of less than 1%;
-   it has an automatic device measuring sag with a precision not exceeding 2%;
-   the supports and the pressing head are parallel to each other and larger than the specimen.

Figure 4: Specimen dimensions

The specimens were placed on the fixed supports of the testing machine and the bending force was applied to the specimen by the pusher (Figure 5). The speed pressing on the specimen was 1 mm / min. For each specimen tested bending fracture was performed a force-displacement graph.

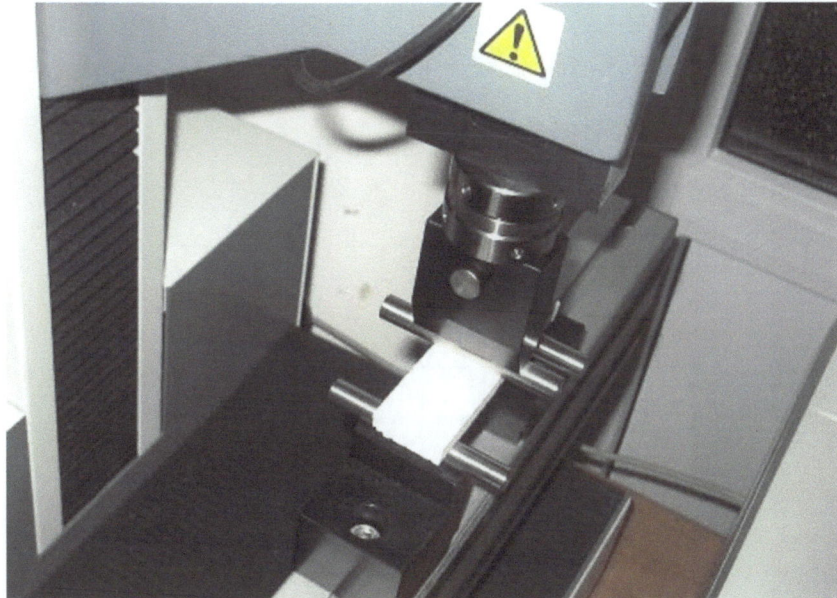

Figure 5: Settlement of the specimen

The Figure 6 presents specimens broken after attempting to bend.

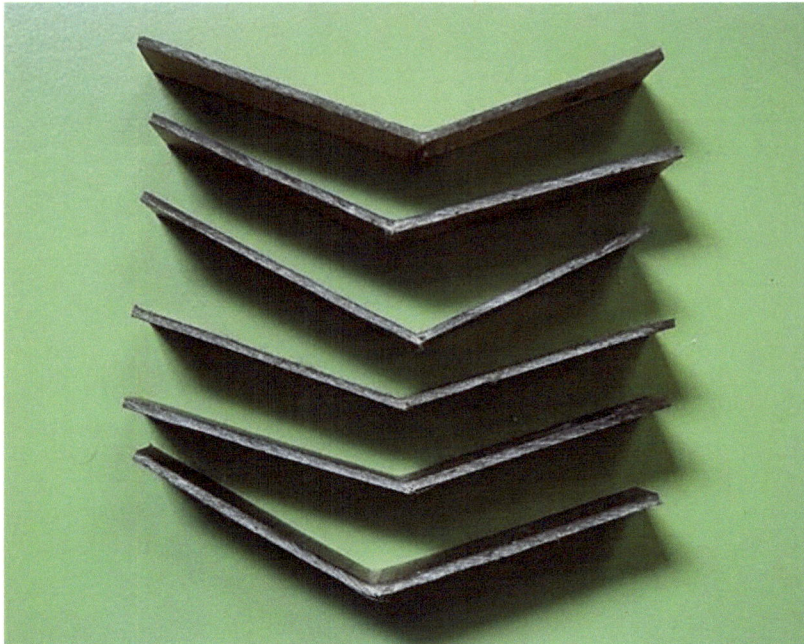

Figure 6: Specimens broken

4. EXPERIMENTAL RESULTS

Figure 7 presents the diagram recorded for specimen number 1, to the breaking bending test [4].

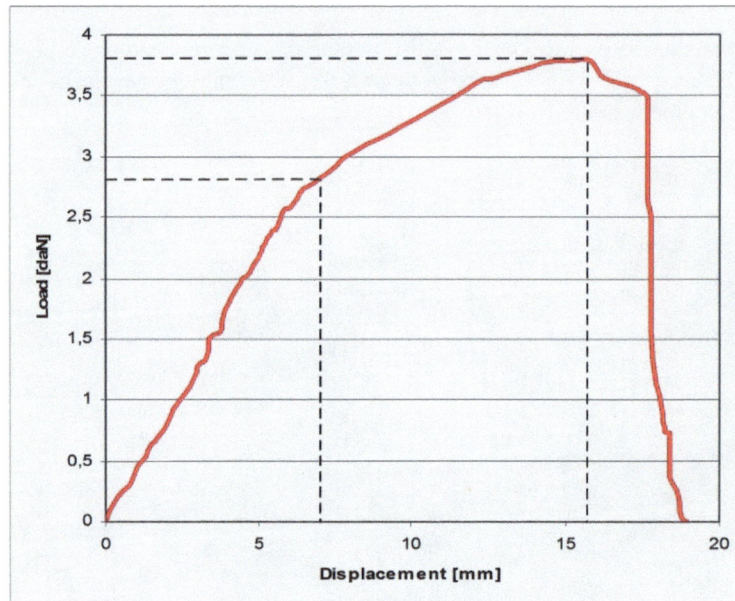

Figure 7: The Force-sag diagram to the specimen no. 1

On the horizontal axis is registered sag deformation of the specimen, in millimetres, and the vertical force exerted is recorded in daN.
The graph shows that the elastic deformation of the specimen is between 0 and 2.8 daN. The maximum force recorded is 3.8 daN.
In Figure 8 are graphs obtained for all six samples tested.

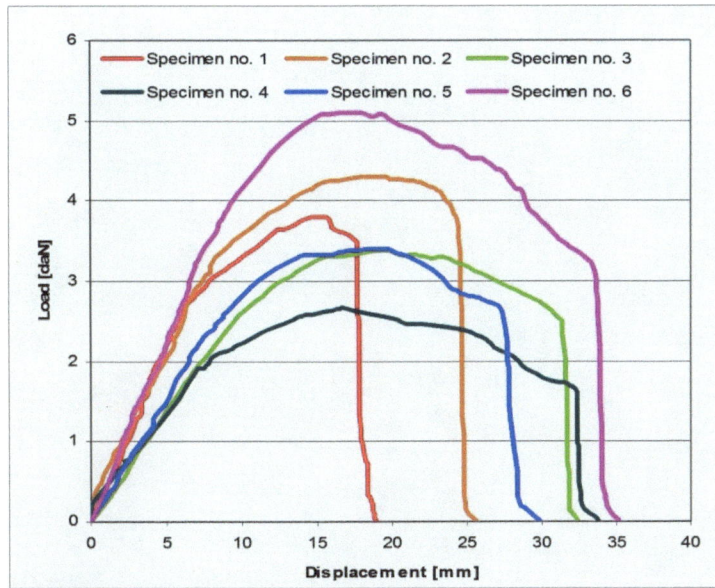

Figure 8: The Force-sag diagram to the specimens no. 1-6

In Table 2, has recorded for each specimen, *b* width and *h* thickness, the allowable stress σ_a, the breaking stress σ_r and modulus of elasticity E_b.

Table 2: Specimen dimensions

Number of specimen	*b* [mm]	*h* [mm]	σ_a [MPa]	σ_r [MPa]	E_b [MPa]
1	14,77	2,00	56,951	77,208	4332,404
2	15,00	1,90	77,437	95,316	4554,166
3	14,85	1,75	71,762	89,288	4056,248
4	14,89	1,57	61,970	83,821	5889,409
5	14,85	1,78	74,636	86,499	4193,276
6	14,95	2,00	87,055	102,23137	4275,807

In Figures 9 and 10 are shown the breaking stress and the modulus of elasticity determined for each specimen individually.

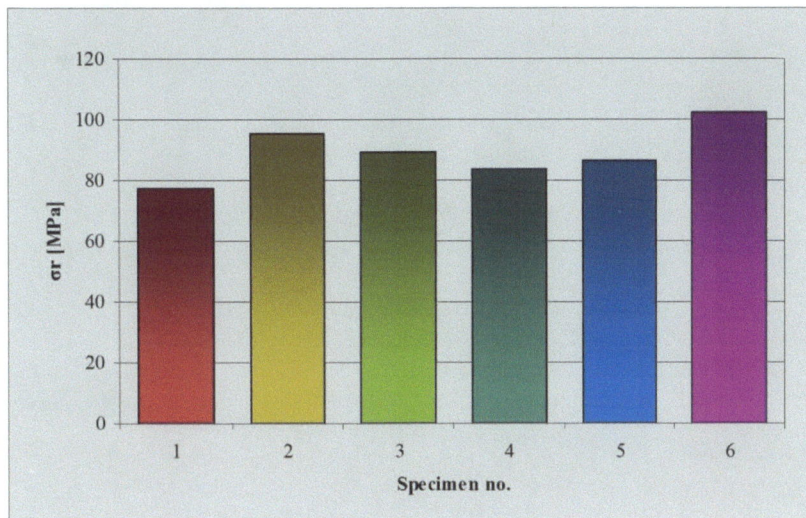

Figure 9: Breaking stress

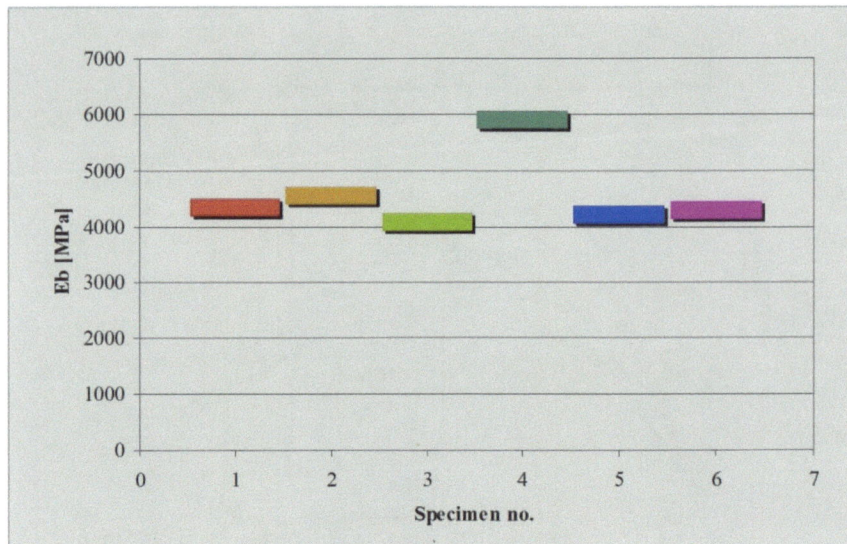

Figure 10: Modulus of elasticity

3. CONCLUSION

The research work aimed at analysing the behaviour of the composite material MAT-450 to bending strain. In this respect, experimental tests were conducted and information related to breaking stress and the modulus of elasticity of the material was collected and analysed.

From Force-sag diagrams made for each specimen tested, it is noted differences between the values determined. These differences arise because heterogeneity and structural composite material made of geometric variations specimens obtained.

The breaking force of the specimens ranged between 2.56 and 5.09 daN and sag deformation ranged between 13.88 şi19,72 mm.

The experiment was conducted to obtain information and to conduct a database on the behaviour of composite materials reinforced with glass fibres. This database will help to validate analytical models on the mechanical behaviour of composite materials.

REFERENCES

[1] N. Constantin, G. Jiga, A. Hadăr, Numerical modelling of a fibre reinforced composite, Proc. of EUROMAT'95, Padova-Veneţia, 1995, p. 521-524
[2] N.D. Cristescu, E.M. Craciun, E. Soos, Mechanics of elastic composites, Chapman & Hall/CRC, (2003).
[3] J., J., M., Decker, P., Ishikawa, T., Northolt, M., G., Picken, S., J., Schlatmann, R., Baltussen, Polymeric and Inorganic Fibres, Advances in Polymer Science, 2005.
[4] A. Hadăr, Structuri din compozite stratificate, Editura Academiei şi Editura AGIR, 2002.
[5] A. Hadăr, G. Jiga, N. Constantin, C. Mareş, Program de calcul al unui material compozit stratificat armat cu fibre, Construcţia de maşini, nr. 8-9, 1995, p. 39-43.
[6] D., Hoa, S. V., Tsai, S. W., Gay, Composite Materials – Design and Applications, CRC Press, 2003.
[7] R. M . Jones, Mechanics of Composite Materials, Mc Graw-Hill, Koga Kusha, Ltd., 1975.
[8] J., N., Reddy, Mechanics of Laminated Composite Plates and Shells.Theory and Analysis, CRC Press LLC, 2003.
[9] J.R., Sierakowski, R.L., Vinson, The Behavior of Structures Composed of Composite Materials, Kluwer Academic Publishers, 2002.

The 6th International Conference
"Computational Mechanics and Virtual Engineering"
COMEC 2015
15-16 OCTOBER 2011, Brasov, Romania

MECHANICAL INTERACTIONS BETWEEN FATIGUE CRACKS AND MICROSTRUCTURAL CONSTITUENTS

Cornel Biţ

TRANSILVANIA University, Braşov, ROMANIA, e-mail: cbit@unitbv.ro

Abstract: This paper is concentrated on some microstructural issues concerning the mechanical interactions between fatigue cracks and microstructural constituents of an aluminium alloy subjected to fatigue cycles. A diffusion process involving different manganese compounds developed within the immediate area of the fatigue cracks has been also revealed.
Keywords: crack, fatigue, microstructural constituent

1. INTRODUCTION

The present-day fatigue research is concentrated on materials cracks behavior, considering that such cracks are present to some degree in all mechanical structures. They may exist as basic defects in the constituent materials – assimilated to material deficiencies in the form of pre-existing flaws – or they may be induced in a certain engineering structure during the service life. The whole life time of a certain structure subjected to fatigue cycles (or to static loads as well) depends upon the way in which material cracks do propagate until the final failure. In Fig. 1 and Fig. 2 two fatigue crack surfaces have been represented. It is to be noted the direct influence of the fatigue crack propagation on the material grains structures.   This is why the fatigue phenomenon gives a great importance to the interaction between the fatigue cracks and the microstructural constituents of the investigated alloy. A direct and an important consequence of such a position is that Miner's linear criterion of damage mechanics loses its validity.

This paper has been focused on the mechanical interaction between fatigue cracks and microstructural constituents of an aluminium alloy (6061 T651) subjected to fatigue cycles.

2. PHYSICAL ASPECTS OF THE MECHANICAL INTERACTION BETWEEN FATIGUE CRACKS AND MECHANICAL STRUCTURE

The analysis of mechanical interaction between the microstructural components of the aluminium alloy Al 6061 T 651 and the fatigue cracks has been done within an original fatigue testing program, using specialized specimens. The specimens material used for the experimental investigations – aluminium alloy 6061 T651 was in form of rolled plates with initial crack (Fig. 3).

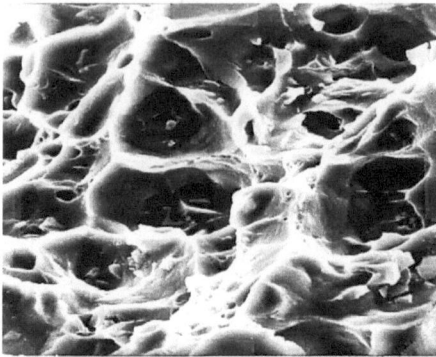

Figure 1: Fatigue crack surface (x3000)

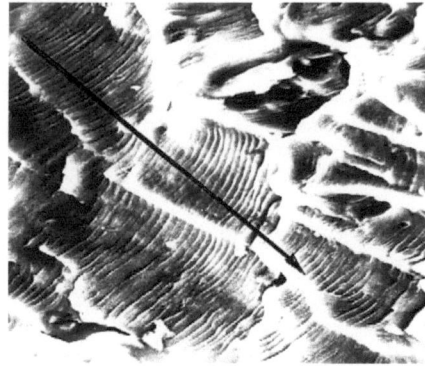

Figure 2: Fatigue crack surface (x2000)

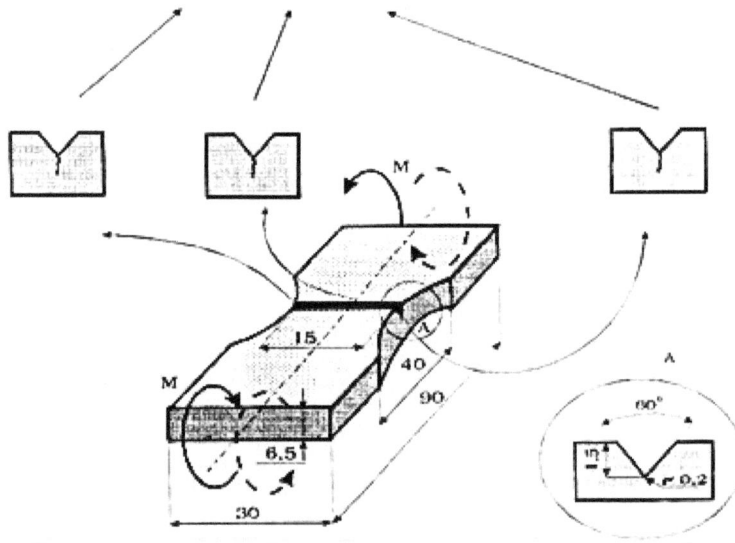

Figure 3: **Specimens used within investigations**

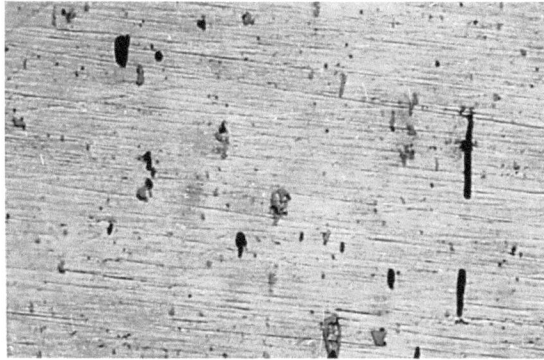

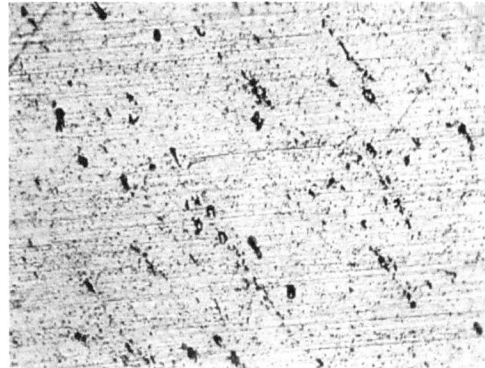

Figure 4: **Microstructure of Al 6061 T651 (x300)**   Figure 5: **Microstructure of Al 6061 T651 (x300)**

In Fig. 4 and Fig. 5 the structures of the considered aluminium alloy, before subjected to fatigue cycles have been represented. In Fig. 6 different types of fatigue cracks interacting the microstructural constituents of the investigated aluminium alloy have been presented. For a small number of fatigue cycles the cracks produced may be very fine and short (~0.04-0.07mm) - propagating along a single direction (Fig.6 a,b) or may be short and rough, with a bifurcation at end (Fig.6c). One may also notice that in Fig. 6a the crack stopped in its interaction with an above presented chemical compound. There are also cracks which are fine but long (0.1-0.3 mm) that usually propagate at the level of inter-granular area, following the grains borders and passing through the fine precipitated compounds or rounding the components.

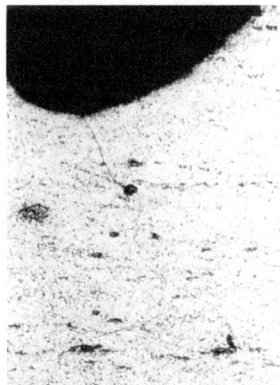

a.

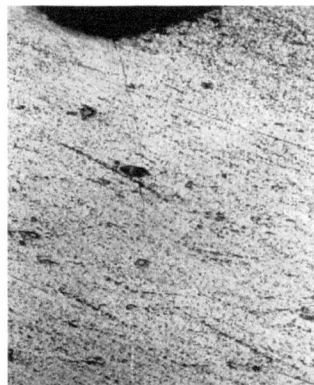

b.

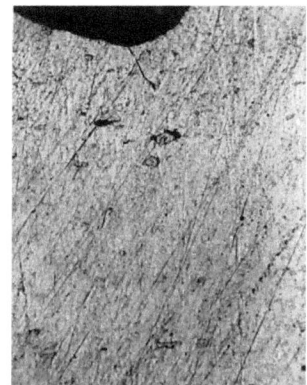

c.

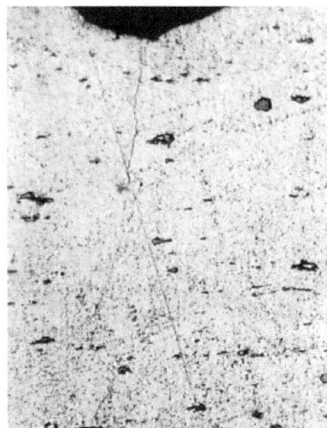

d.

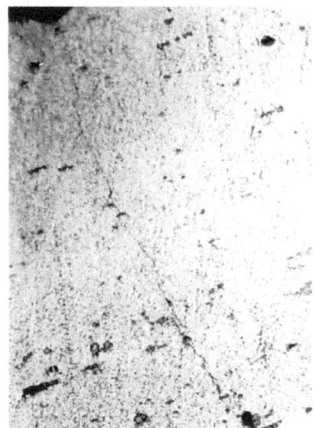

e.

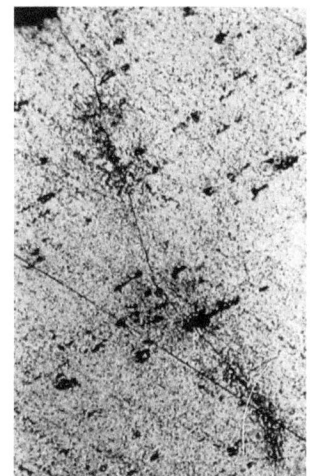

f.

g.   h.   i.

j.   k.   l.

Figure 6: Mechanical interacti‹ ‿ crostructural constituents (continued)

In Fig. 6d,e,f it is to be noticed that the crack propagated along the direction with an increased chemical unhomogeneity. For a large number of fatigue cycles the cracks are long, thick and linearly propagated (Fig.6 g,h,l) or present short bifurcations when meeting a structural constituent. In Fig. 6h,i,j,k one could observe that the crack does round the chemical compound in form of Chinese letter – identified as being $Fe_3Si_2Al_{12}$ and shown in Fig. 4i. At the same time, different manganese compounds have been observed at the level of the investigated fatigue surfaces that could be the result of a diffusion process involving different one-dimensional faults (interstitial atoms, foreign atoms, second-phase particles etc.) with very important consequences for the fatigue crack propagation and short crack growth (Fig. 6f).

3. CONCLUSIONS

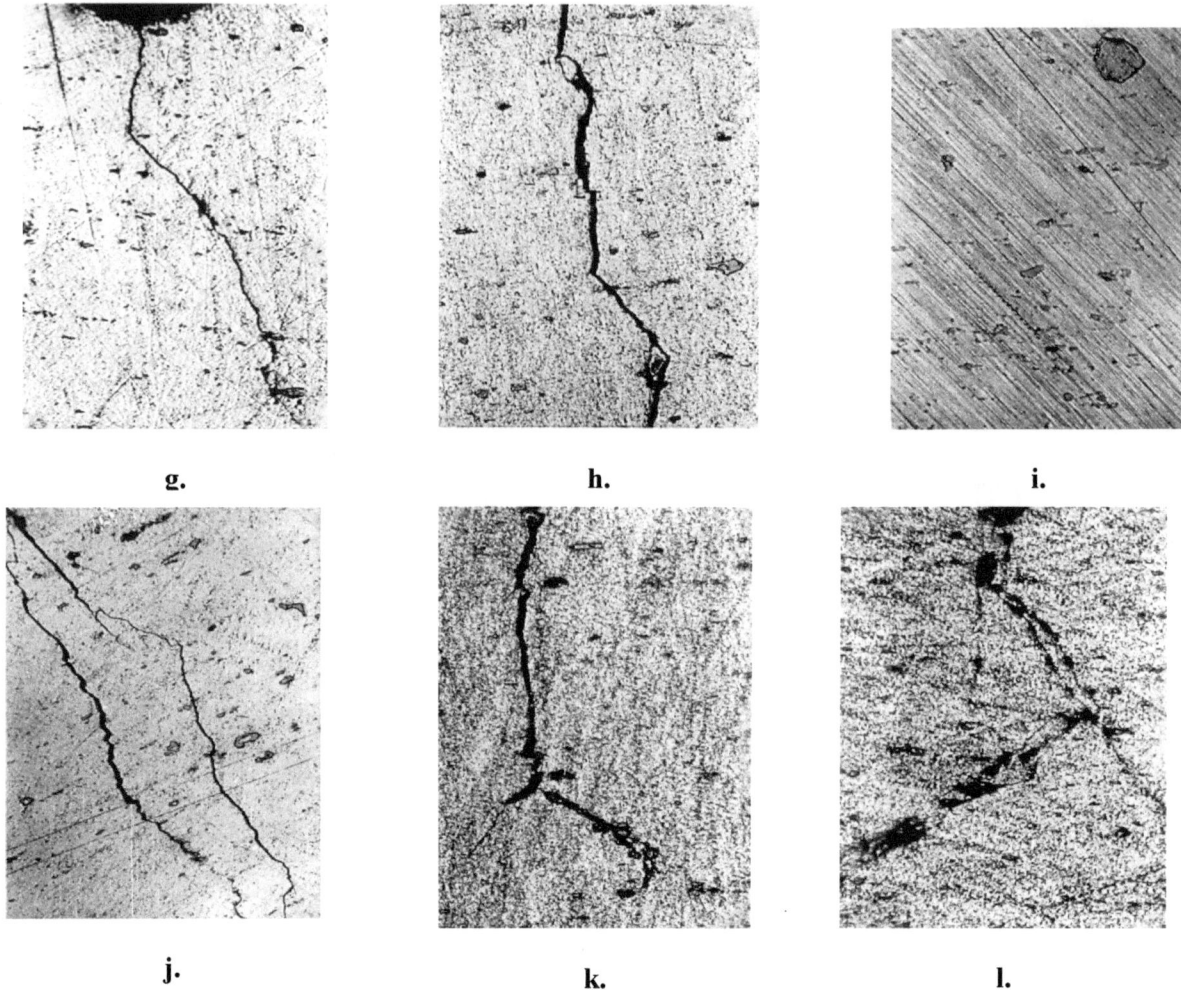

All the above presented aspects concerning the interaction between fatigue cracks and microstructural constituents of the investigated alloy may represent the physical base in creating a mathematical model for fatigue cracks propagation analysis.

REFERENCES

[1] Broek D.: Elementary engineering fracture mechanics, Martinus Nijhoff Publishers, 1982, London.
[2] Cioclov D.: Rezistenta si fiabilitate la solicitari variabile, Editura Facla, 1995, Timisoara.
[3] Bit C.: Elementary strength of materials, Risoprint Publisher, 2005, Cluj-Napoca, Romania.
[4] Bit C.: Puncte de vedere asupra oboselii mecanice, Editura Universităţii Transilvania, 2001, Brasov, Romania.

6th International Conference
"Computational Mechanics and Virtual Engineering "
COMEC 2015
15-16 October 2015, Braşov, Romania

SPECIAL PROBLEMS IN THE CASE OF HIGH SPEED BEARINGS

Traian Eugen Bolfa
Strength of Materials and Vibrations Department, Mechanics Faculty, Transilvania University of Brasov, Romania, e-mail t.bolfa@unitbv.ro

Abstract: *The paper presents the researches viewing the settle of an exact computer programme in the calculation of the moment of friction in the case of the bearing working at high speeds.*
Key words: bearings, moment of friction in bearings, high speed bearings.

1. INTRODUCTION

The calculation relations for the moment of friction in bearings, indicated in different catalogues of bearings, are based on a series of coefficients which are different from catalogue to catalogue in quite large limits. Because these catalogues don't give definitions viewing the limits if the speeds on which the relations can be used, it was imposed the neccesity of performing some studies in this way. The relations for the calculation of the moment of friction given in catalogues don't consider a series of peculiarities of the internal geometry specific for the high speed bearings and also the unneglected effect of the centirfugal forces at high speeds.

2. THEORETHICAL MODEL FOR THE CALCULATION OF THE MOMENT OF FRICTION

Starting from this faults it has been elaborated a model of calculation for the moment of friction in the case of radial roller-bearings which work at high speeds. The model is based on the Houpert's relations (Houpert 1984), for the evaluation of different local moments of friction, completed by the setting of the centrifugal forces and the variation of the lubricant viscosity with the temperature.
So, for a radial roller-bearing, the total moment of friction can be written:

$$M_{tot} = \sum_{i=1}^{z} dT_R + \sum_{i=1}^{z} dT_C + \sum_{i=1}^{z} dT_{ER} \tag{1}$$

where dT_R is the resistant moment at rolling on a ball given by the relation:

$$dT_R = (T_{R_i} + T_{Re})(1 + \frac{D_w}{d_{c_i} + D_w}) \cdot \frac{d_{c_i}}{2} \cdot 10^{-3} \tag{2}$$

dT_C is the moment caused by the glidings on the contact surfaces between the ball and the bearing races and it is calculated:

$$dT_C = M_{C_i}(1 + \frac{d_{c_i}}{D_w}) + M_{C_e} \cdot \frac{d_{c_i}}{D_w} \tag{3}$$

and dT_{ER} is the moment caused by the elastic hysteresis calculated with the relation:

$$dT_{ER} = M_{R_i}(1 + \frac{d_{c_i}}{D_w}) + M_{R_e} \cdot \frac{d_{c_i}}{D_w} \tag{4}$$

In the relation (1) it is neglected the effect given by the ball cage. The relations (2) - (4) are deduced on the base of the ball stability stressed by the stresses and moments shown in figure 1.

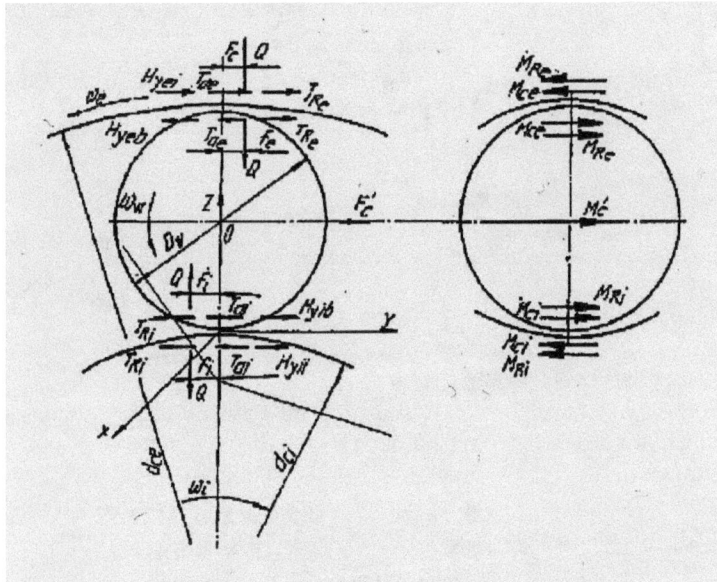

Figure 1: Schematic of the base of ball stability stressed by the stresses and moments.

$T_{Ri(e)}$ are the resistance forces at rolling caused by the lubricant presence and they are calculated with Hamrock's relation:

$$T_R = 2.86 \, E \, R_y^2 \, K^{0.348} \, G^{0.022} \, U^{0.66} \, W^{0.47} \tag{5}$$

where R_y is the equivalent radius of curvature for a ball/ball race contact in a running direction and it is calculated with the relation:

$$R_{y_{i(e)}} = \frac{D_w(1 \mp \gamma)}{2} \tag{6}$$

and K is the ellipticity factor.
M_C is the resistance moment caused by the curvature of the contact surface and it is calculated with the relation:

$$M_C = 6.58 \cdot 10^{-2} \, \overline{\mu} \, \frac{Q \cdot a^2}{\overline{R}} \tag{7}$$

where $\overline{\mu}$ is the medium friction coefficient on the contact surface, a is the major semi-axis of the contact ellipse, and $\overline{R}$ is the radius of the curvature of the deformed surface.

$$\overline{R} = \frac{2 f D_w}{(2f + 1)} \tag{8}$$

M_R is the moment given by the elestic resistance to the pure rolling (hysteresis) and it is calculated with the relation:

$$M_R = 7.48 \cdot 10^{-7} \left(\frac{D_w}{2}\right)^{0.33} Q^{1.33} \cdot [1 - 3.519 \cdot 10^{-3} (K - 1)^{0.806}] \tag{9}$$

H_y is the hydrodinamics force developed because of the lubricant's presence and can be expressed in function of T_R (Houpert 1984);

T_a is the traction force in the lubricant film and it is expressed in function of T_R, M_C and M_R based on the stability of the ball (Houpert 1984);

F'_C and M'_C is the force and moment caused by the ball – cage interaction which can be considered with negligible weights.

Because of the high speed the normal loads Q at the ball-ball race contact have another distribution than in the case of low/depressed speeds.

The centrifugal forces weight becomes essential at high speeds and low radial loads. Plus it is obtained a load distributed on the whole circumference of the exterior bearing race.

The estimation of all the parameters which interfere with relation (1) was made with the helps of a calculation program. In this program the viscosity which interferes with the speed parameter v is corelated with the normal seed through the intermediate parameter temperature.

For Tb A 57E oil, to which the experimental tests have been made with this corelation is presented in figure 2 (Bolfa T. 1991).

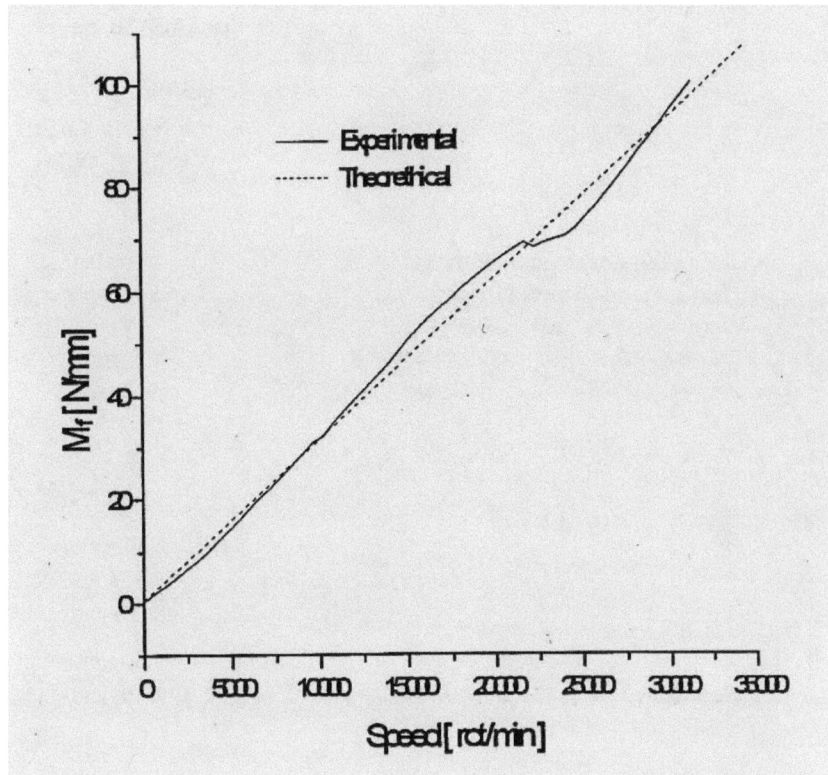

Figure 2: Correlation of the viscosity with the normal speed of rotation.

Also, the exponent of viscosity variation with the pressure α_p was correlated with the temperature (Bolfa T.& Neamtu T. 1985).

3. CONCLUSION

The model was applied to the 6306 MAUP bearing radial loaded with a 26 N force, on the range of revolutions/speed from 0 to 31.500 [rot/min].
The results obtained by simulation on the computer are shown in figure 3.
It is observed an explicit increwasing of the moment of friction, explained by the occurrence of the speed parameter v in the general calculation relation of the moment of friction, with some disturbances in the zone of 18.000÷23.000 [rot/min] speeds, in facts explained by the loss of the cage stability.
It is observed a very good concordance between the two curves, the experimental one and the theoretical one.

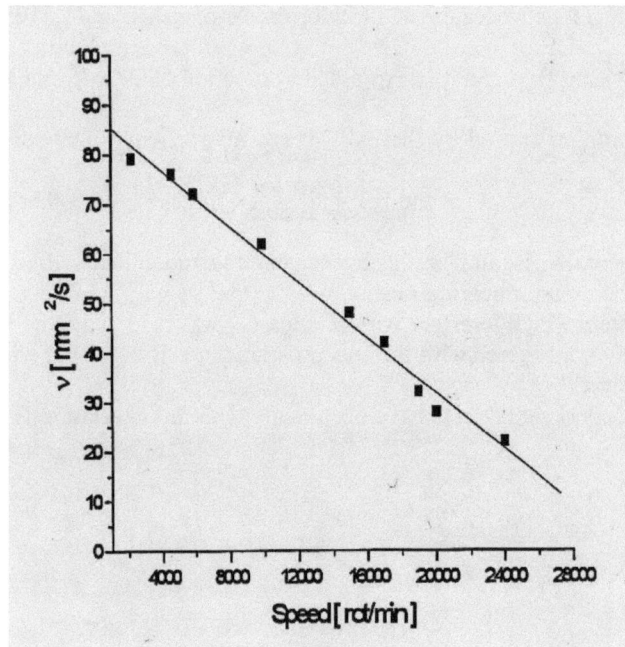

Figure 3: Moment of friction obtained by computer simulation (full line) and experimentally versus speed of rotation.

Knowing the moment of friction in the bearing it is possible to calculate the lost power through friction using the relation:

$$P = M_f \, \omega \qquad (10)$$

The criterion which uses the demanded power in bearing can be considered as certain criterion viewing the comparison of different constructive solutions adopted for high speed bearings.

REFERENCES

[1] Bolfa T. (1991) PhD Thesis, Iasi, Romania.
[2] Bolfa T. & Neamtu T. (1985). Scientific Research Contract no. 89, CIROA, Brasov, Romania.
[3] Houpert R. (1984) The Film Thickness in Piezoviscos-Rigid Regime Film, Journal of Tribology, July.

6th International Conference
"Computational Mechanics and Virtual Engineering "
COMEC 2015
15-16 October 2015, Braşov, Romania

MECHANICAL AND ACTUATION SOLUTIONS FOR A ROBOTIC GLOVE

Dorin Popescu, Sorin Manoiu-Olaru, Livia-Carmen Popescu, Cosmin Berceanu, Anca Petrisor
University of Craiova, Craiova, ROMANIA, dorinp@robotics.ucv.ro

Abstract: *This paper presents the research activities and results for robotic glove development with application in post-stroke rehabilitation. In order to implement a good movement of robotic glove fingers a human hand motion was analysed by image processing, a tele-operated robotic hand and some robotic gloves were developed. The mechanical structures and actuation systems were tested and compared.*
Keywords: *robotic glove, mechanical structure, actuation system, rehabilitation*

1. INTRODUCTION

According to the data from the European Union, cardiovascular disease (CVD) remains the leading cause of death in Europe (more than 40% of men and 50% of women). In Romania, according to Romanian Society of Cardiology and the statistics provided by the Ministry of Health for 2011, CVD is responsible for 62% of deaths, ranking the third in the world. The first place is coronary heart disease, followed by cerebrovascular accident (stroke).
A cerebrovascular accident (CVA) represents the loss of brain function due to disturbance in the blood supply to the brain. In case of a disturbance for more than a few seconds, brain can neither receive blood nor oxygen causing permanent damages.
Available hand rehabilitation devices vary greatly in structure and mechanical properties but all have the general purpose of assisting with finger extension. The majority of devices currently on the market are active systems powered by electric or pneumatic motors. A survey on rehabilitation exoskeleton hands and on robotic devices for upper limb rehabilitation, including hand rehabilitation is achieved by Maciejasz et. al. [1]. That paper brings together some theoretical and practical points of view on the implementation and control of such a system while highlighting the growing need.
In the last decade, numerous concepts and techniques to enable assessment of physiological properties of the hand, its structure, especially the anatomy and functional features were presented. Some researches [2, 3] allowed the development of kinematic structures to replicate as much as the kinematics of the human hand as possible. These configurations have approached both empirical simulation methods and techniques enabling sophisticated, more accurate, more precise animation of the basic functions.
A study on the state of research on the mechanisms and developments that models or substitutes the human hand was done. Thus we could identify two categories of systems that models or substitutes the human hand:
- gripping anthropomorphic mechanical systems for prosthetics (prosthesis: i-Limb, Luke, DECHEV, RTR II) [4, 5, 6];

- gripping anthropomorphic mechanical systems for robots (robotic hand: HRP-P, UTAH-MIT, SHADOW, DLR II, UBH III, NASA-DARPA) [7, 8, 9, 10, 11].

Stroke rehabilitation for hands includes passive movements or exercises that are movements done with the help of a therapist and more active exercises you do with little to no assistance. We received a challenge from the doctors to develop a device for helping the patients with CVA to recover their hand movement. The goal of our work was the design and develop an Intelligent Haptic Robot-Glove (IHRG) for the patients' rehabilitation, that have had a cerebrovascular accident. In order to implement good movement for the robotic glove fingers, a human hand motion was analysed by image processing, a tele-operated robotic hand and some structures of robotic glove were developed. The video-based motion capture and motion analysis are used in areas like sport [12], medicine, biomechanics [13], but may find application in any field treating motility, movement.

2. JOINTS AND FINGERS BIOMECHANICS ANALYSIS

Design of the IHRG consists of three parts: mechanical exoskeleton, actuation with motors, and local control using minimum number of sensors and observers. The human hand rehabilitation function can be counted as one of the most difficult systems to emulate with a mechatronic device. This difficulty is mainly due to two reasons. On one hand is the unavailability of large space for components placement while on the other hand is the high number of DOF. These requirements can be summarized as: low mass/inertia, unconstrained range of motion, minimum complexity, comfort, compliance.

The first phase of research was dedicated to the study and design of the exoskeleton model for an assistance hand (a robotic glove) and associated actuation system. This system must be attached to the human hand and allow the hand and fingers to move. Based on these requirements (the movement in different planes adapted to the patient's hand, the possibility to touch and grasp, the opening / closing of the hand) specific biomechanical design of the components was achieved, taking into account the new criteria and requirements:

a) a mechatronics approach is necessary, the mechanical structure cannot be designed independently of the sensory system, the actuation or control system;

b) mechanical compliance should be introduced to increase dexterity of the movement, manipulation of objects.

The proposed system has to provide three features for grasping force:

1) The system has to allow a grasping force proportional to the human grasping force;

2) The system shall not disturb human finger movement;

3) Robotic glove has to allow a variable compliance as the human finger so that the dexterity and stability of the grasping is preserved.

In the context of developing an exoskeleton structure to assist the development of techniques for the rehabilitation of the main functions of the hand, in the context of different patients with various problems of malfunction, the developed architecture should cover the diversity of issues and anatomical structures. An example of this variety is shown in Figure 1 (left side) in which two different hands can be analyzed from the point of view of dimensionality of the phalanges and also anatomically.

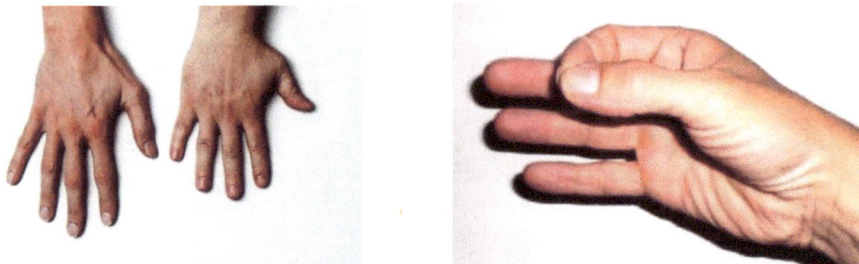

Figure 1: Hand configuration (left side). Hand in a grasping operation (right side) [2]

Therefore, developing an exoskeleton that wishes to achieve desired rehabilitation operations will require completion of several milestones:

• a clear understanding of the basic functions of the hand and the implementation of these functions with a corresponding kinematic structure;

• accurate simulation of the proposed prototype and the analysis of the kinematics obtained by sets of appropriate tests;

• adaptive verification of the performance and adjustment of model parameters.

For a fair assessment of hand functions, we need to analyze in detail the three main components: the palm, the fingers (four) and the thumb [14]. The first component (palm) plays a minor role in functional rehabilitation operations. It serves only as a support structure enabling actuation of the fingers.

The fingers (four) perform, in principle, the function of the grasp, obtained by flexion of the phalanges of each finger (Figure 1, right side). The fingers consist of three phalanges, each phalanx is separated from the other through interphalangeal joints; MP (metacarpal joint), PIP (Proximal interphalangeal joint), DIP (Distal interphalangeal joint). The thumb has a complex function, achieving, in addition to the operation of flexion, and the operation of nearness by the hand axis.

An exoskeleton structure covering the functions required for the rehabilitation operations is shown in Figure 2. The proposed solution asks to develop a mechanical architecture consisting of a cascade of jointed elements whose design has to cover as much as possible the fingers while ensuring the support of the actuation system used.

The elements of joint and fingers biomechanics were studied and it was analysed:
• main anatomical joints of the upper limb.
• main human hand movements.
• axis of rotation of the joints and range of movement.

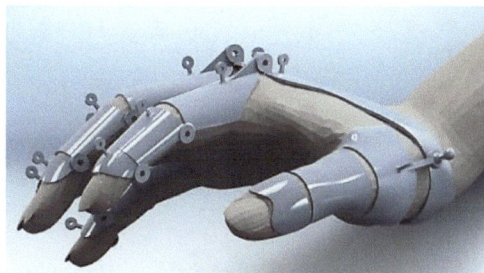

Figure 2: A virtual exoskeleton structure

The conclusions of this analysis will ground the system specifications for the human hand's rehabilitation.

After studying the anatomical model of the human upper limb and hand we can infer the following conclusions and findings:

- the anatomical model of the human hand includes at the level of phalanges 19 independent degrees of freedom;
- each finger of the human hand has a structure with four degrees of freedom, except the thumb of which structure is characterized by three degrees of freedom;
- actuation is done individually for each phalanx (anatomic joint) through tendons located both on the dorsal and palmar face of the hand;
- main actuators (muscles) of fingers are in the forearm;
- radio-carpal joint (wrist) has two degrees of freedom and allows extension-flexion and abduction-adduction of the hand relative to the forearm;
- human hand and upper limb have a huge number of sensors (tactile, temperature, vibration, etc.).

3. STUDY AND DESIGN OF THE EXOSKELETON FOR A ROBOTIC GLOVE

Some research steps have to be followed in order to design a robotic glove for post-stroke hand rehabilitation. These mean studies, analysis, interpretation, design, development, testing, and improvement. A study on the state of research on the mechanisms and achievements that models or substitutes the human hand was done.

3.1. Video-based motion capture and motion analysis

Next research step, after previous study, has implied the determining of the laws of variation of kinematic parameters of human hand fingers movement. The study and analysis of fingers movement used a video acquisition system based on hardware and software processing (SimiMotion - video motion analysis).

The data acquisition and image processing system consists of a fast video camera and a computer. The analysis procedure is based on attaching a series of markers on the hand to be analyzed. Positioning these markers represents identification points for SimiMotion software (Figure 3, right side). By attaching markers, the software automatically generates the equivalent model of the studied system and follows their movement during system operation on each frame captured by the camera, recording and analyzing the positions of markers simultaneously, which serve to obtain laws of motion.

In this paper the human index movement kinematics during flexion and extension was analyzed. A video recording was made about 5 seconds. It was analyzed (Figure 3, left side):

• metacarpal joint, denoted by MP;
• proximal interphalangeal joint (joint formed by the medial and proximal phalanx), denoted by IIP;
• distal interphalangeal joint (joint formed by the medial and distal phalanx), denoted by DIP;
• fingertip index, denoted by TF.

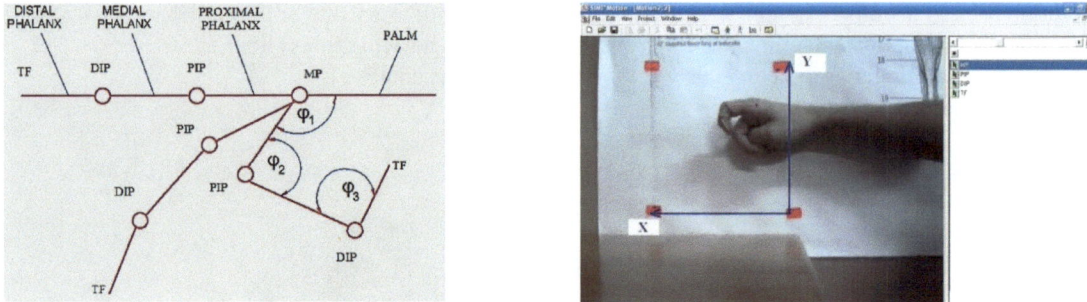

Figure 3: Index finger joints. SimiMotion software

3.2. Exoskeleton structure

To address the rehabilitation issues, we intended to construct a low-profile and lightweight exoskeleton that allows a basic motion using a natural sequence of muscle activation. First we developed virtual exoskeleton systems on the base of four-bar linkage mechanism (Figure 4). The system architecture is achieved by a cascade connection of a four-bar linkage mechanism, each mechanism associated to a joint of the finger (Figure 4). [15] Virtual model construction was performed using SolidWorks software that allows real-time simulation of the model, and then optimizing it by modifying the shape and dimensional parameters of each component separately.

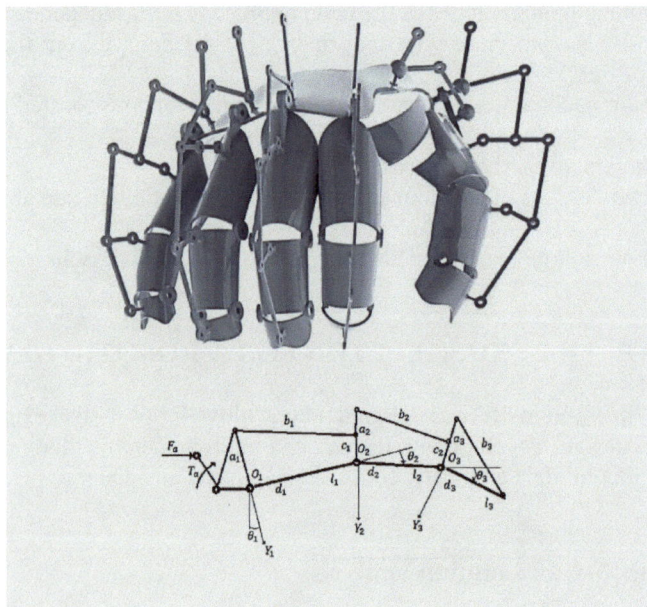

Figure 4: Virtual exoskeleton and four-bar linkage mechanism

The IHRG system has a kinematic structure similar to the natural hand with five articulated fingers. Each finger is composed of three phalanges in order to have an anthropomorphic contact with patient's hand. We tested different mechanical solutions, different motion transmission solutions and actuation systems (electrical, pneumatic and shape memory alloy). We tested two exoskeleton systems with two types of electric (rotary or linear) actuators. Since underactuated fingers have many degrees of freedom and fewer actuators, a glove will be used to constrain the finger and ensure the shape-adaptation of the finger.

In order to get an effective rehabilitation effect, the mechanical structure must allow the finger to reach the positions of a healthy finger.

Three structures were designed and developed (two as exoskeleton structure – with robotic fingers or phalanges):
- the structure with robotic fingers
- the structure with phalanges
- the glove type structure (soft robotic glove)

We designed, developed and tested two solutions in order to drive the movement from actuation system to robotic glove:
- through tendons (for the structure with robotic fingers and for the glove type structure);
- through four-bar linkage mechanism for the structure with phalanges.

First, some virtual exoskeleton structures were designed (Figure 5) with different actuation systems (linear or rotary electric servomotors). Then these virtual solutions were implemented practically (Figure 6, left side; Figure 7).

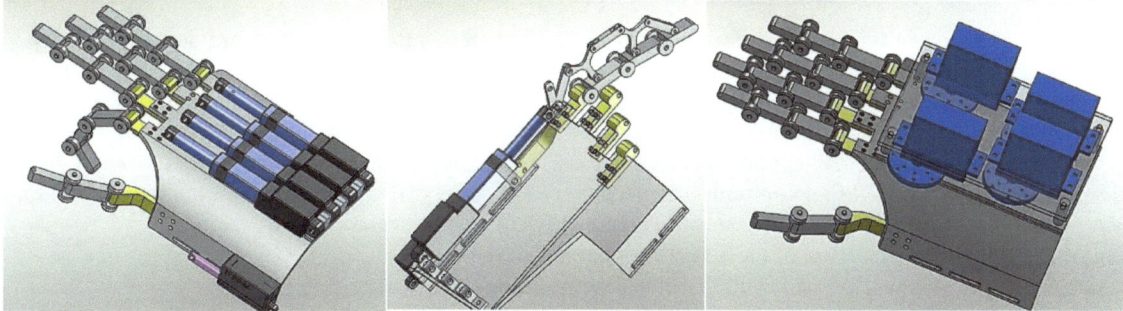

Figure 5: Virtual exoskeleton structures with different actuation systems

Most of the structural components of one of the first robotic glove have been made of hard polyacetal.

We developed a robotic hand (Figure 6, right side) for some teleoperation tests. We tried to develop a non-expansive robotic hand that follows the basic required characteristics. The basic components of our robotic hand are the silicone hand, the 5 actuators, the Arduino UNO R3 control board, the glove and the 5 bend/flex sensors. The operation of the fingers is similar to the human hand, respectively with tendons, each finger having its own tendon, namely a plastic string found in a plastic tube in order to increase the movement. At the other end, the tendon is actuated by using an actuator (servomotor). Tele-operation of the robotic hand is facilitated through the glove with bend/flex sensors.

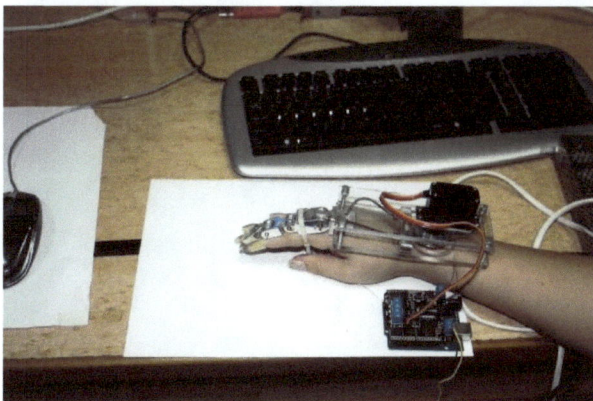

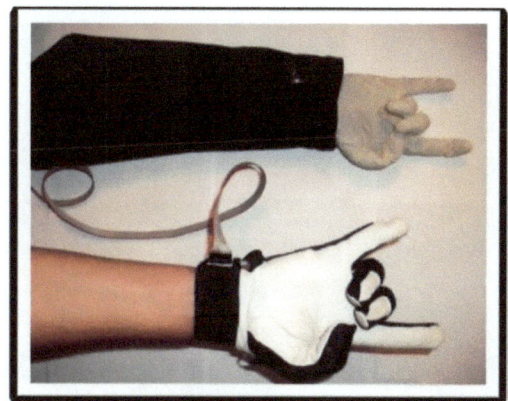

Figure 6: An exoskeleton structure. Tele-operated robotic hand

3.3. Actuation systems

Some different actuation systems (pneumatic cylinders, pneumatic muscles, shape memory alloy, electric actuators) were tested and compared. Considering the difficulties to achieve the exoskeleton with robotic fingers and to use it for a patient with CVA, after the tests with one robotic finger we decide to renounce to implementation of this solution for all five fingers.

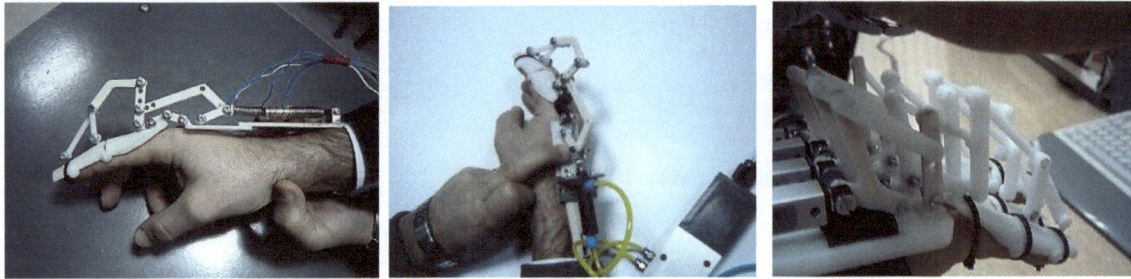

Figure 7: Exoskeleton structures and soft robotic glove

Beginning with two different types of shape memory alloy (SMA) elements we designed and developed a SMA actuator that it was tested with exoskeleton with phalanges (Figure 7, left side). Disadvantages of this type of actuation were small stroke and the complexity of the position control. Also, for exoskeleton type glove we implemented and tested two types of actuation systems: with pneumatic actuators (Figure 7, middle side) and with linear electric actuators. Glove type structure with linear electrical actuators and tendons drive was the easiest to develop and control although there are some difficulties to develop it. The exoskeleton with robotic fingers has some difficulties to develop and to use it for a patient with CVA. The structure with phalanges showed some difficulties in design and developing and it has a reduced movement of the finger. Comparison of the actuation systems shows:
- Pneumatics (cylinders or muscles): difficult to control; voluminous; expensive.
- SMA: difficult to develop; small force; small stroke and complexity of position control.
- Electric actuators:
A) rotary actuators: problems to convert rotary movement in linear movement (need of additional elements).
B) linear actuators: easy to drive the linear movement; no difficult to control.

4. STUDY AND DESIGN OF THE SOFT ROBOTIC GLOVE

Our goal is to create a portable wearable robot that assist the patient during movement. We proposed a device that uses textiles to interface with the hand in parallel with the muscles, using the bone structure, to support fingers' motion (Figure 8).

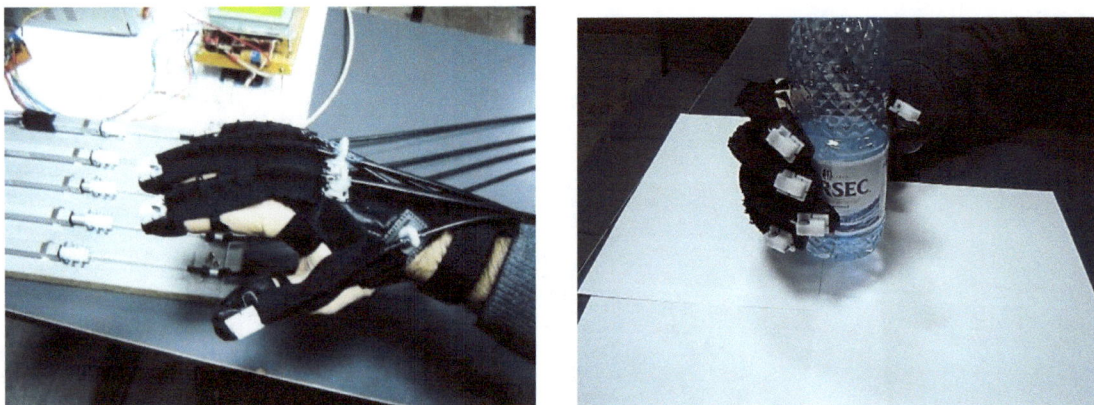

Figure 8: Soft robotic glove

The development of the robotic glove has been conducted considering the following set of guidelines: reduced weight; easy wearability; compliance; stability; high power-to-weight density and reduced energy consumptio; an embedded controller with easy programmability; reduced system costs; do not restrict the natural human kinematics or range motion;

sufficient adaptability to individual differences in patients' anthropometric dimensions (without mechanical regulation or tunings); natural/intuitive use; simplified maintenance.

In comparison with rigid exoskeleton, soft robotic glove has a number of advantages: can be very light; has extremely low inertias, which reduces the metabolic cost of wearing them; it intrinsically transmits moments through the biological joints; since it are composed of textiles, it is easy to put on and take off; it can adapt easily to anatomical variations.

5. CONCLUSION

This paper presents the hardware and software developments that have been achieved and implemented for the robotic glove for hand rehabilitation. The practical results are shown that prove the functionalities of the robotic gloves in common operating conditions.

Wearable soft robotic gloves show to be a better solution for human hand's rehabilitation as lightweight, portable, and compliant wearable systems. We envision that such systems can be further refined.

Our IHRG is a lightweight, wearable device, a system to support a rehabilitation training of hand. It is a portable device that can be worn as a glove.

ACKNOWLEDGMENT

This work is supported by PNCDI-II-PCCA 150/2012 grant of the Executive Agency for Higher Education, Research, Development and Innovation Funding (UEFISCDI).

REFERENCES

[1] Maciejasz P., Eschweiler J., Gerlach-Hahn K., Jansen-Tro A., Leonhardt S., A survey on robotic devices for upper limb rehabilitation, Journal of NeuroEngineering and Rehabilitation, 11:3, 2014.

[2] Grebenstein M., A Method for Hand Kinematic Designers, ICABB, Venice, Italy, 2010.

[3] Carrozza M. C., Vecchi F., Sebastiani F., Cappiello G., Roccella S., Zecca M., Lazzarini R., Dario P., Experimental Analysis of an Innovative Prosthetic Hand with Proprioceptive Sensors, Proc. IEEE Int. Conf. on Robotics and Automation, Taipei, Taiwan, pp. 2230-2235, 2003.

[4] Schectman L., Artificial Robotic Hand, US Patent no. 5080682, Jan. 14, 1992.

[5] Dechev N., Cleghorn W.L., Thumb Design of an Experimental Prosthetic Hand, Proceedings of the 2nd Intern. Symposium on Robotics and Automation, ISRA'2000 Monterrey, N.L. Mexico, November 2000.

[6] Massa B., Roccella S., Carrozza M. C., Dario P., Design and Development of an Underactuated Prosthetic Hand, Proc. of the 2002 IEEE Intern. Conference on Robotics and Automation, Washington, May 2002.

[7] Kaneko K., Harada K., Kanehiro F., Development of Multi-fingered Hand for Life-size Humanoid Robots, Proc. of the IEEE International Conference on Robotics and Automation, Roma, Italy, 10-14 April 2007.

[8] Jacobsen S. C., Iversen E. K., Knutti D.F., Johnson R.T., Biggers K.B., Design of the Utah-MIT Dexterous Hand, Proc. IEEE International Conference on Robotics and Automation, April 7-10, 1986.

[9] Shadow Robot Company, Design of a dexterous hand for advanced CLAWAR applications, 2003.

[10] Butterfass J., Fischer M., Grebenstein M., Haidacher S., Hirzinger G., Design and Experiences with DLR Hand II, World Automation Congress, Tenth Intern. Symposium on Robotics with Applications, Seville, 2004.

[11] Biagiotti L., Lotti F., Melchiorri C., Palli G., Tiezzi P., Vassura G., Development of UB Hand 3: early results, Proc. IEEE Int. Conf. on Robotics and Automation, pp. 4488–4493, Barcelona, 2005.

[12] Ciacci S., Merni F., Franceschetti F., Penitente G., Di Michele R., 3D Analysis with Simi Motion During a 110m Hurdles Race, 12th Annual Congress of the ECSS, Jyväskylä, Finland, .

[13] Copiluşi C., Dumitru N., Ciocan P., Motion laws determination, on experimental way, of joints from the human lower limb structure for certain activities imposed to the locomotor apparatus, Annals of the Oradea University. Fascicle of Management and Technological Engineering, Volume VII (XVII), 2008.

[14] Birglen L., Gosselin C., Kinetostatic Analysis of Underactuated Fingers, IEEE Trans. on Robotics and Automation, 20(2), pp. 211-221, 2004.

[15] Biagiotti L., Melchiorri C., Vassura G., Control of a Robotic Gripper for Grasping Objects inNo-Gravity Conditions, Proc. IEEE Intern. Conf. on Robotics and Automation, Seoul, Korea, pp. 1427-1432, 2001.

6th International Conference
"Computational Mechanics and Virtual Engineering "
COMEC 2015
15-16 October 2015, Braşov, Romania

COMPUTER AIDED DESIGN IN CASE OF THE LAMINATED COMPOSITE MATERIALS

Camelia Cerbu
Transilvania University of Brasov, ROMANIA, cerbu@unitbv.ro

Abstract: *The work firstly shows the analytical calculus model used to compute the distribution of the strains and stresses over the thickness in case of the laminated composite plates subjected to bending. Then, it is briefly described the main steps (calculus procedures) of the Matlab program that is used in design of the laminated composite plate in order to compute the stresses and to draw the distributions of the both strains and stresses. Therefore, rectangular plates subjected to the uniformly distributed force, with different boundary conditions (all sides simply supported, two sides simple supported while the others sides are free end) may be designed by using this program. The program also reports: the stiffness matrix corresponding to the plate element of the laminated composite plate that is computed; the safety coefficients are also computed by using one of the failure theories (maximum stress criterion; Tsai-Hill's criterion, Tsai-Wu's criterion).*
Keywords: *composite, Matlab program, bending, stresses, strains, safety coefficients*

1. INTRODUCTION

The Matlab program used within this work is based on both the classical theory of the laminated composite plate [1-3] and the constitutive equation of the element of the laminated composite plate with linear variation of the temperature over the thickness [4].
Firstly, the theory is briefly described and then, the main steps of the calculus methodology, used in Matlab program are presented by using the logical scheme. Finally some particular cases are considered and the results obtained by using the Matlab program are comparatively analyzed.

2. WORK METHODS

2.1. Briefly theoretical aspects

It is considered the general case of a laminated composite plate element having n layers. The coordinates corresponding to the layer k are: z_{k-1} and z_k are the distances between the median plate surface and the upper or the bottom level of the layer k; $\bar{z}_k$ is the distance between the median plate surface and the median surface of the layer k.
It is also considered that the temperature linearly varies over the thickness of the laminated composite plate:

$$\Delta T = a + bz \,, \tag{1}$$

where the coefficients a and b are computed by using the following relations:

$$a = \frac{T_1 + T_2 - 2T_0}{2} ; \qquad b = \frac{T_2 - T_1}{h} , \qquad (2)$$

where T_1 is the final temperature to the uppermost surface of the plate corresponding to the coordinate $-z_0$; T_2 - final temperature at the bottom surface of the plate corresponding to the coordinate z_n ; T_0 - initial temperature of the plate. From the scientific literature in the field of the mechanics of the composite materials, it is already known the constitutive equation corresponding to the laminated composite plate element with linearly variation of the temperature over the thickness [2, 3, 5]:

$$\begin{Bmatrix} N_x \\ N_y \\ N_{xy} \\ M_x \\ M_y \\ M_{xy} \end{Bmatrix} = \begin{bmatrix} [A] & [B] \\ [B] & [D] \end{bmatrix} \begin{Bmatrix} \varepsilon_x^0 \\ \varepsilon_y^0 \\ \gamma_{xy}^0 \\ \kappa_x^0 \\ \kappa_y^0 \\ \kappa_{xy}^0 \end{Bmatrix} - \begin{bmatrix} [A^T] & [B^T] \\ [B^T] & [D^T] \end{bmatrix} \begin{Bmatrix} a \\ b \end{Bmatrix} , \qquad (3)$$

where:

$$\left. \begin{aligned} A_{ij} &= \sum_{k=1}^{N} \left[\overline{Q}_{ij} \right]_k (z_k - z_{k-1}); & A_{ij}^T &= \sum_{k=1}^{N} \left[\overline{Q}_{ij} \right]_k \cdot \{\alpha\}_k (z_k - z_{k-1}); \\ B_{ij} &= \frac{1}{2} \sum_{k=1}^{N} \left[\overline{Q}_{ij} \right]_k \left(z_k^2 - z_{k-1}^2 \right) & B_{ij}^T &= \frac{1}{2} \sum_{k=1}^{N} \left[\overline{Q}_{ij} \right]_k \{\alpha\}_k \left(z_k^2 - z_{k-1}^2 \right) \\ D_{ij} &= \frac{1}{3} \sum_{k=1}^{N} \left[\overline{Q}_{ij} \right]_k \cdot \left(z_k^3 - z_{k-1}^3 \right) & D_{ij}^T &= \frac{1}{3} \sum_{k=1}^{N} \left[\overline{Q}_{ij} \right]_k \{\alpha\}_k \left(z_k^3 - z_{k-1}^3 \right) \end{aligned} \right\} \qquad (4)$$

In the relation (3), $[A]$ is called rigidity matrix in plane because it links the vector of strains $\{\varepsilon_x^0 \quad \varepsilon_y^0 \quad \gamma_{xy}^0\}^T$ developed within the median surface of the plate and the vector of forces $\{N_x \quad N_y \quad N_{xy}\}^T$ developed at the level of the same surface; $[D]$ - bending rigidity matrix because it links the vector of the curvatures $\{\kappa_x^0 \quad \kappa_y^0 \quad \kappa_{xy}^0\}^T$ and the vector of moments $\{M_x \quad M_y \quad M_{xy}\}^T$; $[D]$ - bending-tensile coupling matrix.

The term $\left[\overline{Q}_{ij} \right]_k$ from relations (4) is the component of the rigidity matrix $\left[\overline{Q} \right]_k$ corresponding to the layer k with respect to the global coordinate system xOy. The matrix $\left[\overline{Q} \right]_k$ depends on the both the orientation angle θ of the reinforcement fibers and the rigidity matrix $[Q]$ of that layer k with respect to the local coordinate system 12 (material coordinate system) whose 1 axis is parallel to the fiber direction while the 2 axis is perpendicular on the fiber direction.

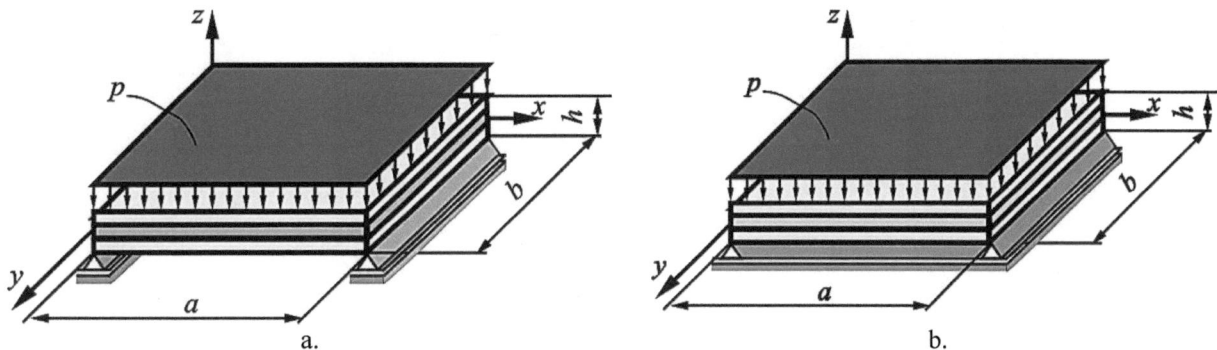

Figure 1: The cases of loading analyzed: a. The I^{st} case of loading; b. The IInd case of loading

In case of the rectangular plate made of laminated composite material that is simply supported along the edges having the dimension b while the others two edges are free (fig. 1, a), the strain component ε_x is non-zero [2]:

$$\varepsilon_x = \frac{\partial u_0}{\partial x} - \frac{\partial^2 w_0}{\partial x^2} \cdot z = \frac{A_{11}z - B_{11}}{A_{11}D_{11} - B_{11}^2},$$
(5)

and the normal stress σ_x developed at the level of layer k is computed with the following relation [2, 3, 5]:

$$(\sigma_x)_k = (Q_{11})_k \cdot \varepsilon_x.$$
(6)

The relationship between the thermal coefficients corresponding to the two coordinate systems (global and local) is the following [2]:

$$\begin{Bmatrix} \alpha_x \\ \alpha_y \\ \frac{1}{2}\alpha_{xy} \end{Bmatrix} = \begin{bmatrix} m^2 & n^2 & -2mn \\ n^2 & m^2 & 2mn \\ mn & -mn & m^2 - n^2 \end{bmatrix} \begin{Bmatrix} \alpha_1 \\ \alpha_2 \\ 0 \end{Bmatrix} = [T]^{-1} \begin{Bmatrix} \alpha_1 \\ \alpha_2 \\ 0 \end{Bmatrix},$$
(7)

where $m = \cos\theta$; $n = \sin\theta$; θ is the orientation angle of the reinforcement fibers.

Assuming that the temperature change ΔT varies linearly on the thickness of the plate, the application programme considers two cases of boundary conditions:

- case 1 of loading, rectangular plate is simply supported along the edges having the dimension b while the others two edges are free (fig. 1, a).
- case 2 of loading, plate is simply supported along all edges (fig. 1, b).

2.2. Application program

It was created an application program by using Matlab soft, based on the analytical calculus model. This application programe may be easily used to compute a thin laminated plate made of composite materials subjected to an uniformly distributed pressure p while the temperature ΔT linearly varies on the thickness of the plate.

To easily understanding of the methodology used by the application program, the logical scheme of the Matlab program is shown in the figure 2. The calculus relations are from the classical theory of the thin laminated plates made of composite material [1-3] with linearly temperature change on the thickness [2, 5].

Finally, the application programe graphically draws the distribution of the stresses over the thickness of the laminated at the level of the critical cross-section. It also computes maximum deflection $w_{0\max}$ at the midpoint of the plate and the safety coefficient c of the laminated composite plate.

Moreover, the program give us the information about the number of the layer that firstly failures. The program computes the safety coefficient c by using the three failure theories: maximum normal stress theory, *Tsai-Wu*'s failure theory or *Tsai-Hill*'s theory. The last two failure theories are dedicated for composite strength calculus.

START

1. **Data input** (dimensions of the plate; elastic characteristics and strength characteristics of all layers with respect to the the coordinate system aligned with fibre direction; external distributed pressure applied to the upper surface of the plate; temperature: T_0= initial temperature; T_1=final temperature at the lower surface of the plate; T_2= final temperature at the upper surface of the plate).

2. Calculus of the stiffness matrix $[Q]_k$ with respect to the coordinate system of the lamina

3. Calculus of the stiffness matrix $[\overline{Q}]_k$ of the lamina k, with respect to the global coordinate system for each layer of the plate

4. Calculus of the stiffness matrix of the thin laminated composite plate

Boundary Conditions

1 – Simply supported on the two edges while the others two are free

2 – Simply supporte on its edges

5. Calculus model for cilindrical bending

5. Calculus model for the case of the plate simply supported along its edges

Maximum vertical displacement w_{max} at of the middle of the plate; strain vector $\{\varepsilon_0\}$ and the curvature vector $\{\kappa\}$ at sections where w_{max} occurs

6. Calculus of the stress vector with respect to the global coordinate system for each layer of the composite plate

7. Calculus of the stress vector with respect to the local coordinate system for each layer of the composite plate

8. Calculus of the safety factors of all layers by using the failure theories: maximum stress criterion, *Tsai-Wu* failure theory and *Tsai-Hill* failure theory

9. **Output of the results** (maximum vertical displacement w_{max} at of the middle of the plate; safety factors computed with three failure theories; σ_x, σ_y, τ_{xy} developed at critical cross-section)

10. **Graphically plotting of the distributions of the stresses** σ_x, σ_y,

END

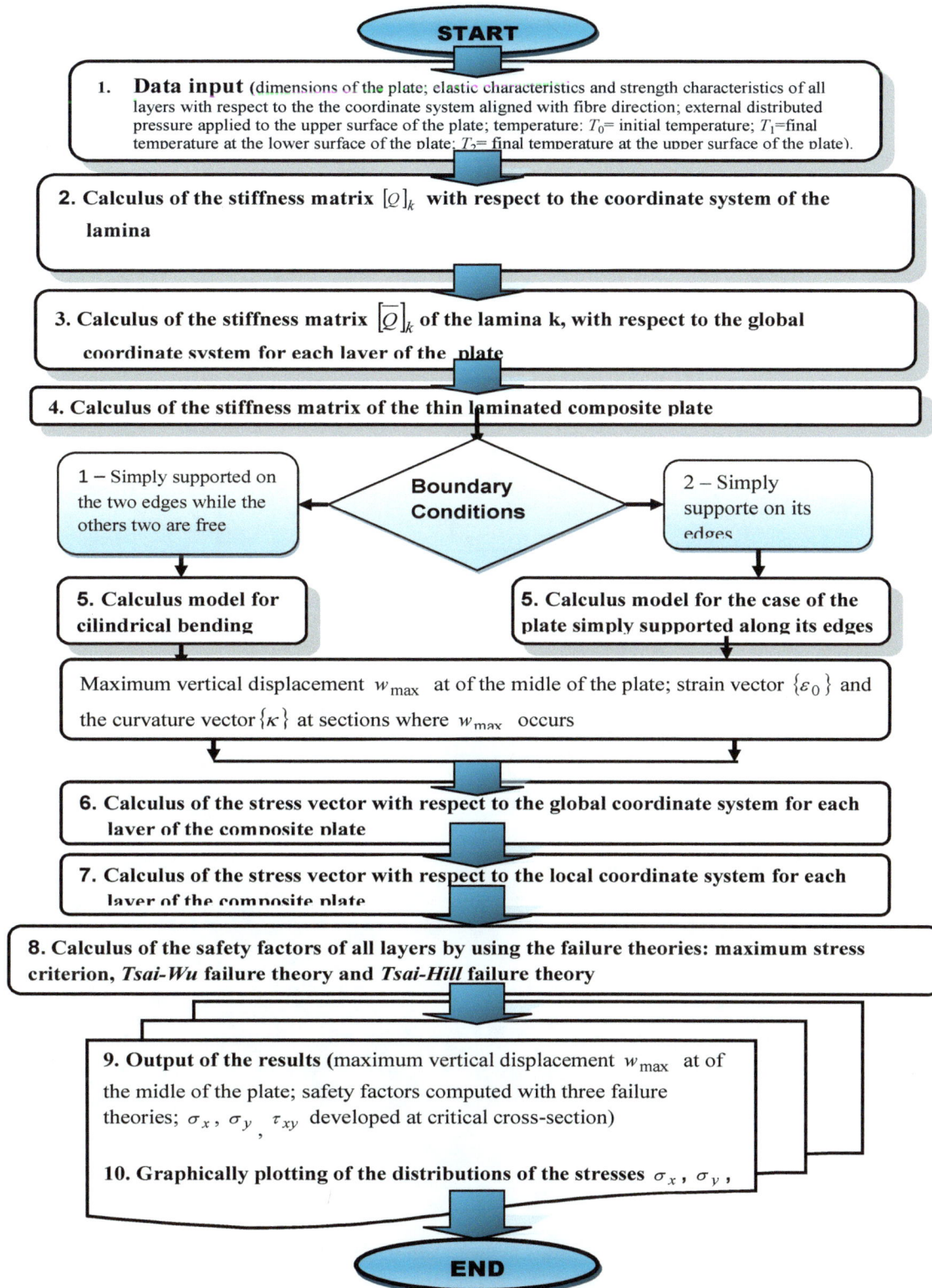

Figure 2: Logical scheme of the application programme used to compute the thin laminated plate made of composite material

2.3. Particular cases analyzed

Herein, it is studied the mechanically behavior of a laminated composite plate [0/90/0] having the dimensions $a = 800$ *mm*, $b = 400$ *mm* (fig. 3) while the total thickness of the plate is $h = 6$ *mm*. All layers have the same thickness. Each lamina is made of epoxy resin unidirectional reinforced with continuous E-glass fibres. The characteristics of the lamina are: modulus of elasticity $E_1 = 1.4 \cdot 10^5 \, MPa$ in the fiber direction; modulus of elasticity $E_2 = 5 \cdot 10^3 \, MPa$ in the direction perpendicular on fibers; transversal modulus of elasticity $G_{12} = 5 \cdot 10^3 \, MPa$ in reinforcing plane 12; tensile strength $\sigma_{1t} = 120 \, MPa$ in fiber direction; compressive strength $\sigma_{1c} = 100 \, MPa$ in fiber direction; tensile and compressive limit stresses $\sigma_{2t} = 50 \, MPa$ and $\sigma_{2c} = 120 \, MPa$ in the direction of fibers and in the direction perpendicular on fibers; thermal expansion coefficients $\alpha_1 = 0.63 \cdot 10^{-5} \, grd^{-1}$ and $\alpha_2 = 2.052 \cdot 10^{-5} \, grd^{-1}$ in the direction of fibers and in the direction perpendicular on fibers, respectively.

The composite laminated plate is subjected to an uniformly distributed pressure $p = 3 \cdot 10^{-3} \, N/mm^2$ that is applied perpendicular to the plate. The initial temperature of the composite laminated plate is $T_0 = 20 \, °C$. The final temperatures at the upper and lower surfaces of the late are $T_1 = 60 \, °C$ and $T_2 = 100 \, °C$ due to a heat flow.

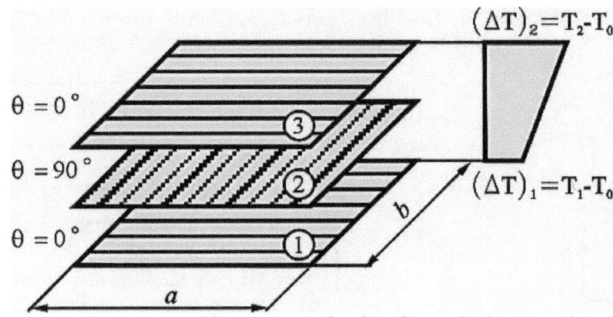

Figure 3: The composite laminated plate analyzed

3. RESULTS

By running persistently of the application program it may obtain the distributions of the both σ_x, σ_y normal stresses shown in the figures 4-8 and the results presented in the Table 1 concerning: maximum deflection $w_{0\,max}$; safety coefficient c ; number of the ply that firstly failures

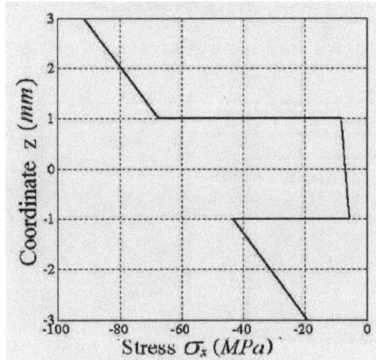

a.

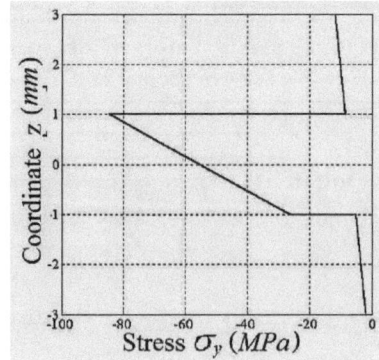

b.

Figure 4: Results obtained by using application program in *the IInd case* of loading. (T0 = 20 °C ; T1 = 60 °C; T2 = 100 °C) at the level of the critical cross-section: a. Distribution of σ_x normal stresses; b. Distribution of σ_y normal stresses

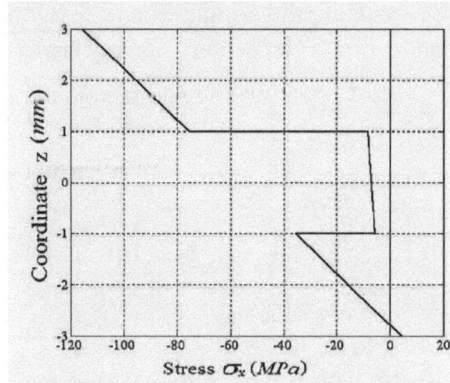

Figure 5: Distribution of the σ_x normal stresses over the critical cross-section, obtained by using application program in the I^{st} *case* of loading ($T_0 = 20$ °C; $T_1 = 60$ °C; $T_2 = 100$ °C)

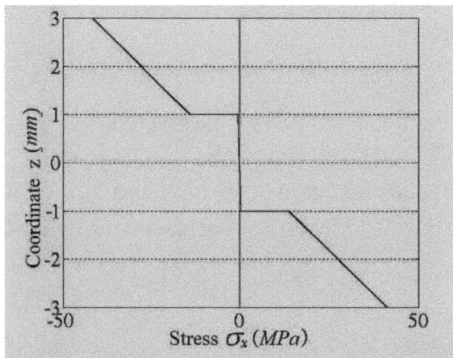

Figure 6: Distribution of the σ_x normal stresses over the critical cross-section, obtained by using application program in the I^{st} *case* of loading with $T0 = T1 = T2 = 20$ °C

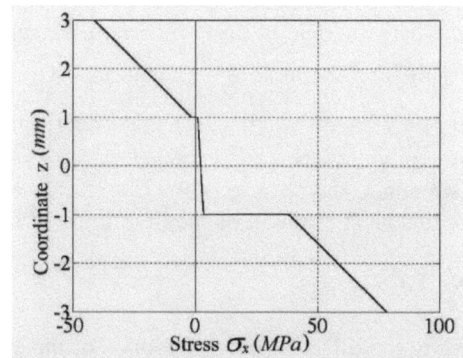

Figure 7: Distribution of the σ_x normal stresses over the critical cross-section, obtained by using application program in the I^{st} *case* of loading with $T0 = 20$ °C ; $T1 = -20$ °C; $T2 = 20$ °C)

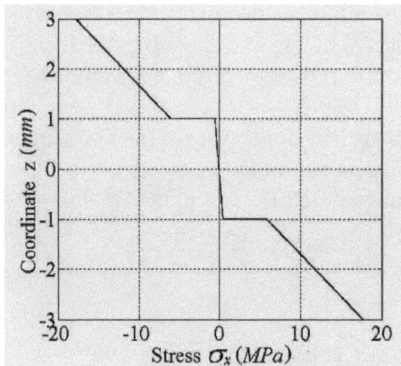

a.

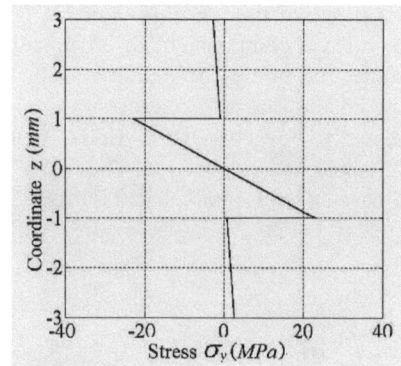

b.

Figure 8: Results obtained by using application program in *the IInd case* of loading ($T0 = T1 = T2 = 20$ °C) at the level of the critical cross-section: a. Distribution of σ_x normal stresses; b. Distribution of σ_y normal stresses

Table 1: Results obtained by using the application programme in some cases of loading

No.	Case of loading	Temperature			Deflection	Safety coefficient c of the plate			The first ply ruptured	Fig.
		T_0 (°C)	T_1 (°C)	T_2 (°C)	$w_{0\,max}$ (mm)	Max. normal stress theory	Tsai-Hill's failure theory	Tsai-Wu's failure theory		
1	The IInd case	20	60	100	2.8519	10.178	7.681	8,272	Ply 3	4
2	The Ist case	20	60	100	6.556	8.6778	-	-	Ply 3	5
3	The Ist case	20	20	20	6.556	24.1071	-	-	Ply 3	6
4	The IInd case	20	20	20	2.8519	18.7929	18.2034	18.4747	Ply 1	8
5	The Ist case	20	-20	20	6.556	15.3141	-	-	Ply 1	7

It may be observed that, in case of the using of the *Tsai-Hill*'s failure theory or Tsai-Wu's failure theory, the value of safety coefficient c is smaller than in case of the failure theory of the maximum normal stress (fig. 4,a and 5). This means that it may recommend as being better to use one of the two specific failure theories in case of the laminated plates made of composite materials.

Analysing of the values computed for the maximum deflection w_{0max}, it may observe that w_{0max} is greater in case of the cylindrical bending than in case of the simply supported along all edges.

It may easily analyze another particular cases of loading: $T_0 = T_1 = T_2 = const. = 20\ °C$ (fig. 6 and 8) and $T_0 = 20\ °C$;

$T_1 = -20\ °C$; $T_2 = 20\ °C$ (fig. 7). It may observe that if the composite laminated plate is subjected to the external pressure, by cooling only the lower layer subjected to the tensile normal stresses, this layer will be the first layer failure (fig. 7). In the same manner, the heating of the layer subjected to the compressive normal stresses (Fig. 4 and 5) only leads to the failure of this layer that will be the first ruptured layer.

4. CONCLUSIONS

The using of the different failure theories of the layer, accentuated the accuracy of the specific failure criteria described in the literature for the composite structures.

Application program developed facilitates the understanding of the aspects concerning the effects of the temperature changes on the state of stresses in case of the thin laminated plates made of composite materials.

The paper shows that the numerical applications based on the theoretical calculus model leads to interesting remarks concerning the comparatively mechanical behavior of the composite laminated plate considered when it is subjected to the external pressure and to different temperature changes at the both upper and lower surfaces of the plate, respectively. The application program used in this paper is friendly-user and useful to analyze the mechanical behavior of a thin laminated plate made of composite material, mechanically loaded with linearly change of the temperature on its thickness. It rapidly computes the safety coefficient c of the laminated plate made of composite material, the values of the stresses developed at the level of the critical cross-section.

Finally, it graphically shows the distribution of one component of the stress vectors depending of the choosing of the user, at the level of the critical cross-section. The major advantage of the application program consists in the fact that it may be easily used in case of any symmetrical structure of a composite limited plate (different number of layers, different orientation angles of the layers, layers made of different materials etc.)

REFERENCES

[1] Alămoreanu E.; Chiriță R., Bare si placi din materiale compozite. Editura Tehnica, Bucuresti, 1997.
[2] Barbero E. J., Introduction to composite materials design, CRC Publisher, USA, 1998, pp. 129-201, ISBN 978-1560327011, 1998.
[3] Berthelot J.-M. – Mechanical behaviour of composite materials and structures. Course, Institute for Advanced Materials and Mechanics, Le Mans, France, 2007.
[4] Cerbu C., Materialele compozite şi mediul agresiv. Aplicaţii special. Editura Universităţii Transilvania Braşov, pp. 112-115, ISBN 973-635-861-5, ISBN 978-973-635-861-6; 2006.
[5] Cerbu C., Curtu I., Mecanica materialelor compozite. Editura Universităţii Transilvania din Braşov, Braşov, 2007, pp.71-129, ISBN 978–973–635–951–4, 2007.
[6] Tenek L.K., Argyris J., 1998. Finite element analysis for composite structure. Kluwer Academic Publishers, Dordrecht / Boston / London, ISBN 0-7923-4899-0, 1998, pp. 135-261.

6th International Conference
"Computational Mechanics and Virtual Engineering"
COMEC 2015
15-16 October 2015, Brașov, Romania

STRUCTURAL FATIGUE ASSESSMENT OF F(P)SO UNITS CONVERTED FROM AN OIL TANKERS

Sorin Brazdis [1], Mircea Modiga [2], Lucian V. Anghel [1], Alexandru Cobzaru [1]

[1] ICEPRONAV Engineering, Galatzi, Roumania, email addresses: sorin.brazdis@icepronav.ro, lucian.anghel@icepronav.ro; alexandru.cobzaru@icepronav.ro
[2] Romanian Technical Academy, e-mail address: mircea.modiga@ugal.ro

Abstract: The paper presents essential practical approaches concerning structural fatigue of floating (production) storage and offloading (F(P)SO) units converted from oil tankers. For this units it is very important to use an accurate service history to the correct evaluation of the accumulated fatigue damage, because they are much smaller than the values based on unlimited navigation conditions. All fatigue assessments are based on the Palmgren-Miner assumption and experimentally derived S-N curves. Simplified Method based on the widely accepted assumption that the long-term distribution of fatigue stress ranges fits a two-parameter Weibull cumulative distribution is used. Are discussed fatigue assessments in Initial Scantlings Evaluations (ISE) stage and further are performed fatigue FE Analysis for exclusive hull ship areas critical details and also for F(P)SO facilities and equipment (interface areas).
Keywords: FSO, FPSO, structural fatigue, FEA

1. INTRODUCTION

ABBREVIATIONS
F(P)SO - Floating (Production) Storage and Offloading
DNV-GL - Det Norske Veritas-Germanischer Lloyd
ABS - American Bureau of Shipping
FDF - Fatigue Design Factor
ESF - Environmental Severity Factors (acc. ABS)
FEA - Finite Element Analysis
ISE - Initial Scantlings Evaluations (acc. ABS)
FEM - Finite Element Method/Model
ICE – International Contract Engineering Ltd.
ISE - Initial Scantlings Evaluations (ABS)
LC - Loading Case/Condition
RMS - Root Mean Squared
SIF - Stress influence factors
TSA - Total Strength Assessment (ABS)
WSD - Working Stress Design
VLCC - Very Large Crude Carrier

Fatigue damage of F(P)SO units is induced by cyclic loads due to environmental pressure, hull girder bending, ship motions, functional low-cycle loads (loading-offloading) and equipment operational loads. The chosen tanker has normally been subjected for short periods to fatigue loads resulting from sea states more severe than expected at the F(P)SO location, but overall the actual long term history of stresses may prove less damaging than provided for in the original tanker design. The usually milder environment at the F(P)SO site can nevertheless induce more damaging cyclic loads than those during trading life due to extended permanent exposure.

It is very important to use an accurate service history in predicting of the accumulated fatigue damage calculations for the trading period because the accumulated fatigue damage value based on the service history is, generally, much smaller than the value based on unlimited navigation conditions. For instance, the accumulated fatigue damage during the trading period, calculated for some converted F(P)SOs ([8], [9], [10], [11]) based on the service history were calculated in a range of 0.18 to 0.28. If the unlimited navigation conditions had been used, the accumulated fatigue damage for the same period would have been evaluated in a range 0.7 ... 0.9.

All fatigue assessments for F(P)SOs are based on the assumption of linear accumulation of fatigue damage due to various stress range cycles and the experimentally derived S-N curves that provide the number of cycles to failure of a structural detail for given dynamic stress ranges. For the

$$S^{m_i} N = a_i, \quad i = 1, 2,$$ (1)

formulation of the dual slope S-N curves that change slope at stress range S_0/N_0 the damage due to n_j dynamic cycles of stress range S_j is:

$$D_j = \frac{n_j}{N(S_j)} = \frac{n_j a_i}{S_j^{m_i}}, \quad i = 1 \text{ for } n_j \leq N_0 \text{ and } i = 2 \text{ for } n_j > N_0.$$ (2)

The accumulation rule of Palmgren-Miner leads to the total fatigue damage experienced by a structural detail over J classes of dynamic stresses (S_j, n_j):

$$D = \sum_1^J D_j$$ (3)

The same linear accumulation principle allows the breakdown of the total damage into unit life stages and, within each, into service periods in different loading and environmental conditions. For a FSPO typically:

$$D_{total} = D_{tanker} + D_{transit} + D_{site}$$ (4)

where the indices stand for the total damage, the damage during tanker life, the damage during the transit of the unit to the working site and the damage due to operations on site. The above damages are usually further broken down into:

$$D_{tanker} = D_{ballast} + D_{full};$$ (5)

$$D_{site} = \sum_{i=1}^{n_{LC}} \sum_{j=1}^{n_P} \sum_{k=1}^{n_H} D_{i,j,k},$$ (6)

where the site summations are for LC (loading conditions), load patterns considered in each loading conditions (n_P) and combinations of headings/loads (n_H).

In an exact approach the dynamic stress history in each stage is to be rainflow counted into stress range intervals, which is rarely feasible. For F(P)SOs and with respect to fatigue dominating wave effects even the second best approximation provided by the spectral fatigue method proves highly demanding in complexity and numerical resources, making it an option not often taken by the owners. The experience gathered in fatigue assessments for F(P)SOs relies on the so-called Simplified Method based on the widely accepted assumption that the long-term distribution of fatigue stress ranges fits a two-parameter Weibull cumulative distribution. This allow the following closed-form equation for the damage in a given unit condition formulation [5] is presented, equivalent to the one in [3]:

$$D = n \left[\frac{q^{m_1}}{a_1} \Gamma \left(1 + \frac{m_1}{h}, \left(\frac{S_0}{q} \right)^h \right) + + \frac{q^{m_2}}{a_2} \gamma \left(1 + \frac{m_2}{h}, \left(\frac{S_0}{q} \right)^h \right) \right]$$ (7)

where n is the total number of wave cycles, γ and Γ are the incomplete and complementary incomplete Gamma functions and q and h are the scale and shape parameters of the Weibull distribution.

Formulations in number of cycles are generally presented in this paper for simplicity. However these are rarely counted but rather estimated based on an average wave period/frequency T/v and the time spent by the unit in the given condition. The average period is to be taken as the average of the zero up-crossing periods of all sea states in the long-term period, weighted by the probabilities of occurrence of the sea states; empirical formulae are also available in rules, e.g. [3]. The time spent by the unit in a given condition is usually expressed by its fraction to the total life in a given unit stage, tanker or F(P)SO.

The Weibull parameters are in principle to be tuned for a best fit to the Weibull distribution through an iterative procedure. In practice good approximations for the shape parameter for F(P)SO are recommended in rules through empirical formulae, e.g. in [3] for both F(P)SO and tanker and F(P)SO life, and the necessary scale parameter results for a reference number of cycles N_R and the corresponding stress range S_R of probability of exceedance $1/N_R$ as:

$$q = \frac{S_R}{\left(\ln N_R\right)^{1/h}}$$

(8)

For F(P)SOs the reference stress level for fatigue is taken at 10^{-4} probability of exceedance (daily return period). The current S-N curves provided by classification societies are built to a 97.7% probability of survival. Therefore typical designs will take further allowance through a FDF; the fatigue criteria in damage formulation is thus:

$$D \leq 1/FDF \ , \ FDF \geq 1.0$$

(9)

Fatigue assessments are typically conducted considering the appropriate corrosion wastage for the structural details during both past tanker and future FPSO service. This is generally assumed to be linear in time and the midlife net thickness was considered for fatigue assessments in several works. For the past tanker life the calculation thickness was taken as the average of the gauged and tanker as-built thicknesses; for the F(P)SO life it was taken as the gauged/renewed thickness less half the nominal corrosion margin specified by the Class. The approach has been applied in association with FDF = 2. However, with the current provisions of [3], a net thickness approach is considered (gauged thickness for trading tanker life, gauged/new thickness less full nominal corrosion for F(P)SO service). A reduction factor of 0.95 to the reference dynamic stress range is allowed instead of considering the effects of thicker structural members in early service stages. Overall the current net thickness approach is considered to provide sufficient margin to safely consider FDF = 1.0 for F(P)SOs.

The list of the structural details to be checked for fatigue is based on the requirements of the classification societies and the designer's past experience. This list includes but is not limited to:

- end connections of longitudinal stiffeners to transverse web frames and to transverse bulkheads;
- ends or bracket ends of primary supporting members;
- flanges of transverse web frames in way of tripping brackets;
- topside and crane pedestal connections with the main deck;
- mooring integration structure with hull;
- flare tower, riser balcony or turret connection with hull.

2. FATIGUE ASSESSMENTS IN ISE STAGE

The dynamic stress ranges that occur in the hull longitudinal stiffeners are mainly due to longitudinal hull girder bending. The details of the longitudinal stiffeners' end brackets have a simple geometry and well-defined load direction which means the nominal stress approach is applicable. Therefore, a quick evaluation of the remaining fatigue life of these details may be performed during the ISE design phase. The specific software released by the classification societies for the scantling evaluation also include modules for the remaining fatigue life calculation of the longitudinal stiffeners' end brackets considering fatigue damage ratios both for the prior service to conversion and for post conversion phases. The assessment rely on the calculated α-type ESFs for the trading tanker routes history and the operational site, respectively. As a possible solution for details found not to meet the fatigue criteria new and/or modified end brackets are proposed in order to increase the fatigue life with minimum steel renewals. Figure 1 shows such improvements after the results of the ISE fatigue assessment ([9]) of longitudinal stiffeners'. The weakest details presented are the end connections of the longitudinal stiffeners placed on the splash area of the side shell.

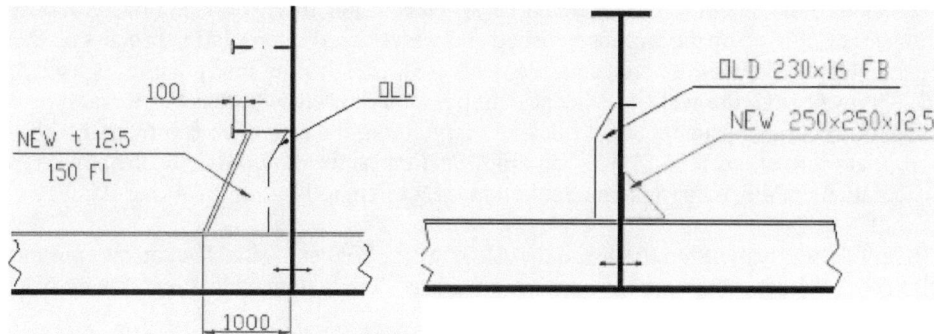

Figure 1: Typical solutions in ISE stage for end connections not meeting fatigue criteria – replacement (top) and additional (bottom) new brackets

It is to be noted that the approach is mainly applicable to stiffeners in exclusive ship areas. For hull stiffeners in way of static F(P)SO facilities and equipment (i.e. that exert only inertial loads on the hull structure, e.g. topside stools, pipe racks) which are considered in interface areas the approach can also be employed as long as the higher FDFs are considered. However more accurate damage assessment of the longitudinal stiffeners' end brackets based on FEA are usually preferred for the interface areas; for locations placed in way of F(P)SO equipment introducing dynamic operational loads (mooring and riser systems, cranes etc.) the FEA based approach is mandatory.

3. FATIGUE ASSESSMENTS FOR SHIP AREAS CRITICAL DETAILS

Further fatigue assessments are performed for the critical hull locations based on FEAs with local fatigue FEM. For the local FEA of ship critical areas the software automatically:
- builds loads (hull girder bending and shear, external and internal pressures and body accelerations) for the parent 3-tank length FEM and solves it for displacements;
- applies displacement boundary conditions on the local FEMs and rebuilds the loads on it;
- allows performing Local FEAs.

FE Analyses are performed for both trading tanker life and F(P)SO service. Additionally the software automatically calculates the fatigue damage using the long term simplified method in the following approaches.

For F(P)SO service, analyses are performed according to ABS (2015) for four loading conditions spread evenly between minimum ballast and full load. A single load pattern is considered for each loading condition and eight load combinations specially tailored for each loading condition are applied grouped in four sagging/hogging pairs for the heading angles of 0 (head sea), 30, 60 and 90°. Four stress ranges are extracted as the difference between stresses resulting from the two LCs in each pair. Damages calculated for heading conditions are accumulated based on their probabilities of occurrence, as provided from a heading analysis if available or following recommendations of ABS (2015). Damages calculated for the loading conditions are accumulated assuming a 15% probability of occurrence for the ballast/full load conditions and 35% for the intermediate ones.

For the trading tanker life also according to ABS analyses ([3]) are performed for the same base eight combinations of loading conditions and patterns and the eight LCs used for TSA. Four stress ranges are extracted from pairs of the LCs and damage is calculated for the most unfavorable one (actually only two pairs may be assessed for certain locations on ships, e.g. deck, bottom, mid-depth and transverse bulkhead).

However, in practice, in order to obtain a more accurate estimate of the damage during past voyages the full long term history is split into several shorter periods by grouping together routes of similar β factors. Damages are calculated for the shorter long-term periods of trading with conservative β factors and are accumulated by direct summation into the total tanker damage. Overall, assessment of the damage during past tanker life requires a comparable workload to the assessment of expected damage during F(P)SO service.

The damage from the transit of the unit to the operations site is in principle calculated similarly to the one for on-site operations. There is a single ballast condition to be accounted for and part of the LCs could be discarded for the absence of internal pressures. In the ABS approach, implemented in the ABS Eagle FPSO software, the loads applied on the FE are always taken at 10^{-8} probability of exceedance (20 years return period) with corresponding β factors for past trading routes and F(P)SO service on site. The 10^{-4} long-term probability of exceedance reference stress ranges are then derived by applying a 0.5 factor to the stress ranges induced by the above.

The mesh size for the fatigue FEM is refined down to the local plate thickness t in the areas of structural details of interest. Special stress recovery techniques are used for free edges - the stress ranges are fetched from rod elements of insignificant stiffness purposely added to the model (cross section area of 1 mm^2). For fillet welds at bracket toes, hotspot stresses at the weld toe location are calculated by the well-established and widely used interpolation process on the principal stresses closest to the perpendicular to the weld line on four elements ([3], [5]) - cubic interpolation is used to derive the stress ranges at $0.5t$ and $1.5t$ away from the weld toe; linear extrapolation is used with these to extract the stress range at the weld toe). In all cases corrections for members of thickness larger than the reference t_R provided with the S-N curves are applied in the appropriate format; see e.g. [4],[5] - an amplification factor of $(t/t_R)^{0.25}$ for the stress ranges is introduced typically. All considerations in this paragraph are applied in direct Femap/Nastran FEA and damage calculation outside Class Software as well.

Figue 2 present two hull details analyzed in ([9]) with the ABS Eagle FPSO software. The top part present the free edge of a hole in the centre line bottom girder connection to a transverse bulkhead. The bottom part presents the toe of a bracket connecting a side shell stiffener to a transverse bulkhead.

The charts on the right present ratios of stress ranges on site to permissible fatigue stresses used in screening the locations for damage calculations (permissible fatigue stresses can be derived using a unit damage; calculated values for the standard design life of 20 years are provided in Class rules).

The actual damage calculations ran with the *S-N* curves ([4]). For the free edge these were based on the *C*-curve and employed nominal stresses on the fictitious rods (not visible in the picture).

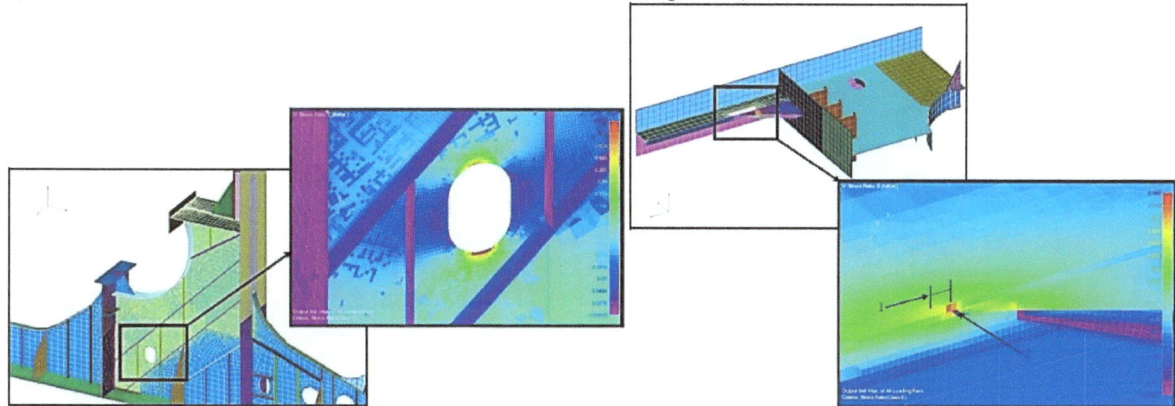

Figure 2: Hull (ship area) fatigue detail analyzed with ABS Eagle FPSO Software for a single hull FPSO

For the bracket toe the transverse weld toe on the stiffener flange was checked on the *E-curve* with the hotspot stresses interpolated from the nominal stresses on the four plate elements indicated by arrow 1; the longitudinal weld toe on the bracket was also checked on the *E* curve with the nominal stress on the element indicated by arrow 2. The bracket end was previously updated out of strength requirements and only the transit and FPSO damages are calculated; due to the update the estimated total damages for both weld toes were found below 0.5.

4. FATIGUE ASSESSMENTS FOR F(P)SO FACILITIES AND EQUIPMENT

The main effort in damage assessments for F(P)SO equipment and facilities is for their foundations and supporting structures. All of these, including the hull structures below the main deck or inside the hull shell, are classified in the latest versions of [3] as hull interface areas for which the adoption of unit *FDF* is no longer accepted. For these FDF is to be taken between 2 for non-critical, inspectable and repairable structural details and 10 for critical and non-inspectable/repairable ones.

The structures outside the hull shell will typically be new to the unit and hence the damage from tanker life is zero. The structures within the hull shell are generally inherited from the converted tanker and damage accumulated during tanker life can be fully assessed by means of the Class software.

For the assessment of expected damage during F(P)SO service of supporting structures of newly installed equipment and facilities advantage is also usually taken of the services provided by Class Software, to various levels of extent. The software provides direct and full support for the topside stools, allowing either the modeling of the topside modules including their weight or direct input of vendor supplied reaction forces. The approach can be employed also to all equipment in the midship area where 3-tank length models are supported by Class Software to find the damage from wave-induced cyclic loads on site. For static equipment, e.g. topside stools and pipe-racks, this is also the total damage on-site.

The most challenging may prove the fatigue assessment for equipment and facilities that introduce dynamic operational loads aside from the inherent inertial loads due to the vessel's motions in waves. Usually the dynamic cyclic loads from such equipment exert at frequencies different from the waves frequency, or at least with inconsistent phases. The effects of the simultaneous dynamic processes of waves and operational loads are combined into a total damage based on the separately calculated damages as:

$$D = D_w \left(1 - \frac{T_w}{T_o}\right) + \frac{1}{T_o}\left[\left(D_w T_w\right)^{1/m} + \left(D_o T_o\right)^{1/m}\right]^m . \tag{10}$$

The combination law was derived based on a single slope *S-N* curve ([5]). Here the indices "*w*" and "*o*" refer to waves and operational processes respectively, *T* are the average zero up-crossing periods of each process and *m* is the slope of

S-N curve. The law can be applied to dual-slope S-N curves and is proved to be conservative when using the large cycle range branch exponent, $m = m_2$. All damages are relative to the site operations of the unit.

The operational damage is usually assessed based on the same fatigue refined models used for waves induced damage exported from the Class software to an all-purpose linear FE solver package. The choice for previous works was the Femap/NX Nastran package and the export usually transfer also the boundary conditions.

A simpler case of equipment with distinct operational loads of relevance to fatigue are the F(P)SO on-board cranes. Usually a comprehensive analysis of the distribution of operational loads over time will not be available and the designer will have to rely on the following approach, albeit overly conservative. Stresses are to be obtained in the structural detail of interest for LCs built for a significant number of boom angles and the maximum rated crane load. A maximum stress range is to be established as the maximum difference between any pair of these (typically this will result for two boom positions at a relative angle of 180°). A number of cycles n_o is taken according to the crane certification class. The operational damage is directly calculated on the applying S-N curve with the maximum stress range and n_o above. For combining with the damage due to waves the operational loads period can be taken as the F(P)SO service life divided by n_o.

The most demanding fatigue assessments for F(P)SO facilities are usually the ones for the supporting structures of permanent mooring systems (either spread mooring or single point mooring, with either internal or external turrets), and dynamic riser systems. Operational loads from these are usually provided by the mooring/riser system vendor on a short-term basis following the systems' hydrodynamic analyses. For fatigue assessment purposes the vendor must provide also the RMS (root mean squared, usually computed as the square root of the zero-order spectral moment of the power spectrum density in a short-term sea state) and the zero up-crossing period of the loads. The data is to be provided for a number of sea states representative for the operational site long-term environment and, in an optimum scenario, for each sea state for a relevant number of combinations of headings and wave relative directions to current.

By taking a simplification assumption based on the FEA linearity the RMS of stresses in the fatigue structural details σ are extracted following a FEA with the fatigue fine FEM loaded with the RMS values of the operational loads. Considering that the short term stress variation follows a Rayleigh probability distribution, a closed-form equation for the damage in a given sea state k and heading h due to operational loads is obtained for a dual-slope S-N curve as:

$$D_o^{k,h} = n^{k,h} \left[\frac{\left(2\sqrt{2}\sigma^{k,h}\right)^{m_1}}{a_1} \Gamma\left(1 + \frac{m_1}{2}, \left(\frac{S_0}{2\sqrt{2}\sigma^{k,h}}\right)^2\right) + \frac{\left(2\sqrt{2}\sigma^{k,h}\right)^{m_2}}{a_2} \gamma\left(1 + \frac{m_2}{2}, \left(\frac{S_0}{2\sqrt{2}\sigma^{k,h}}\right)^2\right) \right]$$

(11)

$$D_o = \sum_k \sum_h D_o^{k,h}$$

(12)

The equation is usually employed in spectral based fatigue analyses ([4]). $n^{k,h}$ is the number of operational load cycles expected in the sea state and heading condition estimated based on the zero up-crossing period provided by the vendor, $T_z^{k,h}$. The total operational damage on site D_o is accumulated linearly and, when it is to be combined with the damage from waves, T_o is to be taken as the duration weighted average of $T_z^{k,h}$.

The equation above was provided for clarity considering the loads from mooring and riser systems are narrow band dynamic processes. This is generally not true, cyclic loads being exerted at both wave (high) frequency and at a lower frequency due to second-order wave effects. The vendor supplies data for both regimes (σ_H, T_{zH}) and (σ_L, T_{zL}). The combined spectrum method can be employed ([3], [6]), in which the damage calculation is to be performed with the combined parameters:

$$\sigma = \sqrt{\sigma_L^2 + \sigma_H^2}$$

(13)

$$T_z = \sigma\left(\left(\sigma_L/T_L\right)^2 + \left(\sigma_H/T_H\right)^2\right)^{-1/2}$$

(14)

The combination is actually performed in terms of operational loads instead of stresses and the resulting RMS applied as load in FEA. The approach is known to be highly conservative and one can adopt the dual narrow-banded approach which introduces an additional sub-unit factor for the RMS. The total number of sea states can be 50 or larger and within each eight or more heading and other environmental variations can be considered. Also, especially for the *riser systems*, the loads apply on a significant number of slots with strong inter-influence for the local structural details of each other and in certain cases various scenarios of combinations of used/unused slots may be required to be analyzed. In order to reduce the number of performed FEA usually a SIF approach is employed in which FEAs are performed for unit values of each load component on each slot (for mooring there will generally be a single axial load, the mooring line tension; for risers there are usually also bending and shear loads at their connection points). The stress RMS for a given finite element in the sea state k and heading condition h is then recovered as:

$$\sigma^{k,h} = \sum_s \sum_c SIF_{s,c} L_{s,c}^{k,h}$$

(15)

where $SIF_{S,C}$ is the elemental stress influence factor due to component C on slot S and $L_{S,C}^{k,h}$ is the RMS of the load component C acting on slot S in the sea state k and heading condition h.

Fig.Error! Reference source not found. presents the FEM used for the fatigue assessment of a mooring skid bracket oe ([9]) grouping the connection points of mooring lines 4 to 6. In this case the loads RMS and periods were only provided for the most unfavorable heading in each sea state; instead six combinations of lines angles were required to be scanned to find the most damaging configuration. By dropping the indexes h and C in the equation above SIF where calculated for both slots S =4, 6 and line angles A = 1, 6 and the calculation RMS for sea state k was taken as

$$\sigma^k = \max_A \sum_{S=4}^{6} SIF_{A,S} L_S^k$$

(16)

In order to avoid a numerical loss of accuracy the SIF where computed by FEA for unit loads of 1000 kN. For the hotspot detail assessed on the $S\text{-}NE$ curve the actual maximum in the equation was taken after finding the maxima between interpolated values from elements in groups marked a and b in the figure for each A. Subsequent to FEA derivation of SIF all calculations were performed based on spreadsheet for 50 sea states.

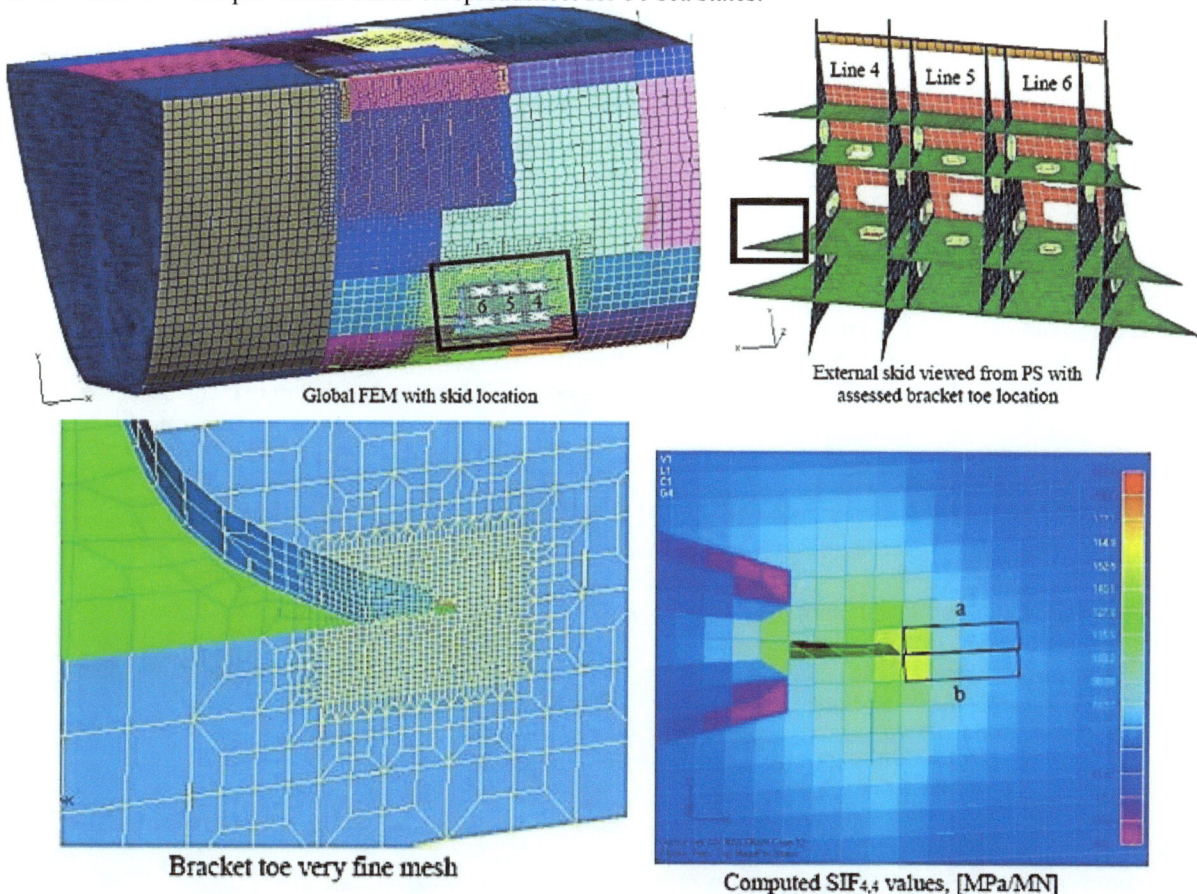

Global FEM with skid location

External skid viewed from PS with assessed bracket toe location

Bracket toe very fine mesh

Computed SIF₄,₄ values, [MPa/MN]

Figure3: Fatigue assessment for an interface area bracket toe of an aft spread mooring skid ([9])

Certain F(P)SO equipment and facilities are located at the vessel's ends, e.g. off-loading stations, flare towers and helidecks; typically no operational loads of relevance to fatigue are associated with the equipment. In this cases no support can be found in the Class software and all fatigue assessments are based on FEAs performed with all-purpose linear FE solvers. The models are built to the Class recommendations similar to chapter 5.2 and for the waves induced damage pairs of LCs of essentially opposite directions of loads are applied also per Class provisions. The loads are typically taken at 10^{-8} probability of exceedance level and include vertical bending moment, body accelerations and internal and external pressures (including slamming and green water pressures as applicable). The equipment inertial loads are applied through either concentrated masses or vendor supplied reaction forces of consistent directions to each LC. Stresses in fatigue details are established following FEAs and damage is calculated through in-house developed spreadsheets that reproduce the approach presented in p.2.

As discussed in chapter 4 above most of the structural analysis for the offshore parts of the F(P)SO equipment and facilities are performed by vendors. Usually only the helideck remains in the marine designer scope of work. When requested, fatigue checks for the tubular joints are performed for wave-induced inertial loads using in-house spreadsheets according to the requirements of [7].

5. CONCLUSIONS

In the case of the conversion from an existing tanker the work required in fatigue assessment is further enlarged, typically doubled, due to the iterative process employed in establishing the minimum reassessed scantlings that meet the hull girder strength and the need to assess the fatigue damage for both the trading tanker life and the F(P)SO service.

A trading tanker is often originally designed for harsher conditions than what it will experience as an F(P)SO hull during its remaining life; by carefully evaluating these facts as well as the trading history of the tanker the designer can in many cases reduce or eliminate the need for steel replacement even taking into account corrosion wastage and without increasing the risk for structural failure.

There are well-established methodologies to achieve the required analyses. The major classification societies have available software packages that can be used to great advantage. Particularly for integration of a wide variety of equipment and facilities needed for the service as an F(P)SO the standard packages will need to be supplemented by ad hoc analyses using standard tools.

Throughout the paper, examples of structural issues and successfully applied improvements to them are presented, and certain techniques beyond the Class requirements that allow tackling complex and extensive tasks are described.

6. ACKNOWLEDGEMENT

The authors gratefully acknowledge the grant support received to carry out the work presented in this paper as an integral part of INTERNATIONAL CONTRACT ENGINEERING LTD (ICE) funded projects. The content of the publication is the sole responsibility of the authors, and ICE is not liable for any use that may be made of the information.

REFERENCES

[1] Brazdis S., Modiga M., Dimitriu C., Cobzaru A., *Hull Condition Analysis - first stage of an oil tanker conversion into F(P)SO*, Proc. of 10th edition The days of the Academy of Technical Sciences of Romania, Galați, October 9-10, 2015

[2] Brazdis S., Modiga M., Anghel L. V., Coreschi C., *The structural evaluation of a F(P)SO unit converted from a oil tanker*, Proc. of 10th edition The days of the Academy of Technical Sciences of Romania, Galați, October 9-10, 2015

[3] ABS (2015), *Rules for Building and Classing Floating Production Installations*

[4] ABS (2014), *Guide for the Fatigue Assessment of Offshore Structures*

[5] DNV-GL (2014), DNVGL-RP-0005, *Fatigue Design of Offshore Steel Structures*

[6] DNV (2013), OS-E301 *Position Mooring*

[7] API RP 2AWSD (2007), *Recommended Practice for Planning, Designing and Constructing Fixed Offshore Platforms - Working Stress Design*

[8] ICE (2005), *VLCC to Cidade de Vitoria FPSO Conversion, Detail Engineering*

[9] ICE (2007), *VLCC to Gimboa FPSO Conversion, Detail Engineering*

[10] ICE(2010), *VLCC to FPSO Conversion Study for JUBILEE and TUPI sites*

[11] ICE (2012), *VLCC to FPSO Conversion Total Strength Assessment.*

6th International Conference
"Computational Mechanics and Virtual Engineering "
COMEC 2015
15-16 October 2015, Braşov, Romania

COMPOZITES ADHESIVES BASED ON UREA-FORMALDEHYDE RESINS AND LIGNOCELLULOSIC WASTE

IleanaManciulea[1] ,Lucia Dumitrescu[1]

[1]Transilvania University of Brasov, Research Centre Renewable Energy Systems and Recycling, Brasov, Romania,i.manciulea@unitbv.ro, lucia.d@unitbv.ro

Abstract: In order to develop a sustainable economy and due to increasing economic and environmental issues concerning the use of petrochemicals, lignocellulosic materials, as natural, abundant and renewable resources, will be extensively used in the future, as raw materials, for the production of chemicals, fuels and composite materials.
Our research has been focused on synthesis and characterization of some new, ecological composite adhesives based on woody wastes fillers ammonium, calcium and iron lignosulfonates and Salix wood aqueous extract, as partial substituents for toxic, synthetic monomers urea and formaldehyde in the synthesis of urea-formaldehyde adhesives (matrix). The improvements achieved, by using the lignosulfonates and Salix wood extract, in the properties of the new composite adhesives urea-formaldehyde-lignosulfonate-Salix extract consist on decreasing of the gel time and the content on free formaldehyde, increasing the water resistance, the pot life, the resistance to biodegradation and low costs.

1. INTRODUCTION

The growing environmental pressure caused by the synthesis of petroleum based plastics have determined the development of new, biodegradable, environmentally friendly engineering materials [1, 2]. Biodegradable polymeric materials derived from natural, renewable sources such as proteins, poly carbohydrates (cellulose, hemicelluloses, starch), lipids etc., have been regarded as alternative ecological materials, and considered as very promising materials because their availability and cost effectiveness. [3,4]. The renewable, by photosynthesis, biomass has a great chemical potential for production of green chemicals for a sustainable economy. From the chemical point of view the lignocellulosic biomass consists of natural polymers: cellulose (50-55%), hemicelluloses (15-25%) and 20-30%lignin (20-30%), along with minor content of other organic and mineral secondary compounds. Cellulose, hemicelluloses and lignin are complex polymer structures rich in chemical functional groups, sensitive to reactions, such as alcoholic and phenolic hydroxyl groups, carboxyl, carbonyl, ether, ester groups, etc. [5, 6]. These functional groups can participate to chemical reactions of etherification, esterification, alkylation, salts formation, oxidation, polymerization, polycondensation, to produce new, ecological engineering materials, such as polymers, paints, green fuels, wood preservatives and adhesives, etc.,[7, 8, 9, 10, 11]. The research in the field of recycling lignocellulosic waste to obtain new ecological, environmental friendly materials are in agreement with the sustainable development of society and environmental protection. Meantime, the production of new composites materials based on natural and synthetic polymers becomes an important way for reusing/recycling biomass/wood waste [12, 13, 14]. The major wood component lignin obtained as waste product in the paper industry is an attractive raw material for resin based adhesives. Native lignin is neither hygroscopic nor soluble in water, but, during the technical sulphite pulping of wood, lignin becomes soluble in water, due to the partial degradation and introduction of sulfonic groups [15]. The natural polymer lignin is composed of phenylpropane units linked together by carbon-to-carbon and carbon-to-oxygen ether bonds. When lignosulfonate is treated with strong mineral acid at elevated temperatures, condensation reactions leading to diphenylmethanes and sulfones take place. The reaction are of the same type as the formation of phenolic resins from phenol and formaldehyde. Lignin also reacts with formaldehyde and can be cross-linked by it, in the same manner of synthetic polyphenolic resins. Urea-formaldehyde resins are the most used class of amino resins adhesives in wood industry. These copolymers present many advantages such as low water absorption, high tensile strength, hardness, non-inflammability and adaptability to a variety of curing conditions. They are widely used for the production of wood-based composite panels, such as particleboards, fiberboards and plywood [16]. The greatest disadvantage of the amino resins is their bond deterioration, caused by water and moisture, due to the hydrolysis of the amino methylene bonds. Amino resins produced from urea are synthesized by polycondensation reaction, when urea is reacted with formaldehyde, to form the addition products, such as methylol compounds. By further reactions of water elimination, the formation of low-molecular-weight condensates (that are still soluble) is promoted. Higher-molecular-weight products (which are insoluble and infusible) are obtained by further condensing the low-molecular-weight condensates. The cured resin consists in both linear and branched polymers, as well as tridimensional networks [17, 18].

2. EXPERIMENTAL
2.1. Synthesis and characterization of urea-formaldehyde composite resins with ammonium, calcium, iron lignosulfonates and Salix wood aqueous extract

a. **Analysis of ammonium, calcium and iron lignosulfonates and Salix wood extract**
Our research has been focused on the obtaining of new composite adhesives, based on metal complexed lignosulfonates and Salix wood aqueous extract, as partial substituents for formaldehyde in the urea-formaldehyde adhesives. The calcium lignosulfonate (LSCa), from sulphite pulping), and iron (III) lignosulfonate (LSFe), obtained from the ammonium lignosulfonate and Fe $NO_3)_3$ were analysed conforming to specific methodology for lignin [5, 6]. The chemical characteristics of the ammonium, calcium and iron lignosulfonates are psented in the Table 1.

Table 1. Chemical characteristics of the metal complexed lignosulfonates.

Characteristic	Ammonium lignosulfonate	Calcium lignosulfonate	Iron lignosulfonate
Appearance	brown liquid	brown liquid	Brown liquid
pH- value	4.25	5.70	2.24
Solids, %	38.64	36.00	35.47
Density at 20^{0}C, g/cm^3	1.14	1.15	1.15
Viscosity at 20^0 C, cP	66	70	68
Ash, %	1.09	2.75	2.57
Cation, %	7.10	5.50	4.71
Functional groups:			
- OH phenolic, %	14.16	11.22	16.06
- OH alcoholic, %	13.74	14.20	13.35
- carbonyl, %	1.71	1.35	9.35
- carboxyl, %	0.60	0.56	0.74

The Salix wood was provided by Salix Cluster Green Energy (Sf. Gheorghe, Romania). As lignocellulosic material, Salix aqueous extract has the same type of chemical functional groups as ammonium, calcium and iron lignosulfonates, present into the polymeric structures with the following characteristics: 49.50% cellulose, 18.40% hemicelluloses, 27.10% lignin.

b. Synthesis and characterization of urea-formaldehyde resins with ammonium, calcium and iron lignosulfonates and Salix aqueous extract
The reaction between urea and formaldehyde are two stages. The first is an alkaline condensation to form mono-, di-, and trimethylolureas compounds. The second stage is the acid condensation of the methylolureas, first to soluble, and then to insoluble cross-linked resins.
At the alkaline pH, the reaction of urea and formaldehyde, at room temperature, leads to the formation of methylolureas compounds. At acid pH, the products precipitated from the aqueous solutions of urea and formaldehyde, or from methylolureas, are low-molecular-weight methyleneureas. Due to the methylol groups, the reaction is possible to continue, to harden the resin. Generally, an initial urea/formaldehyde molar ratio of 1:2.0 was used. Methylolation was carried out by maintaining the mixture under reflux, at temperatures of 90-95 ^{0}C. The temperature raised after 10 to 30 minutes, when methylolureas compounds formed. The reaction was completed by decreasing the pH at 5.0 to initiate the polymer-building stage. As soon as the right viscosity was reached, the pH was increased to stop polymer building and the resin solution is cooled to about 25^{0}C. More second urea was added to consume the excess of the formaldehyde, until the molar

ratio of urea to formaldehyde was in the range of 1:1.1 to 1:1.7. The resin was left to react at 25^{0}C for 24 hours. The excess of water was eliminated by vacuum distillation until a solid concentration of 65% is reached, and the pH was adjusted to achieve suitable storage life. The main strategy to reduce the formaldehyde emission of UF resins has been the change its formulation by decreasing the molar ratio of formaldehyde to urea. However, the reduction weakens the mechanical properties of particleboard and moreover it increases the time required for hardening under the action of current hardeners (latent acids) [14]. Hence it is necessary to optimize the synthesis of UF resins by changing the production process. In order to substitute the toxic monomer formaldehyde and to decrease both, formaldehyde emission and the gel time of the urea-formaldehyde adhesives, the quantities of 5%, 10% and 15% of ammonium, calcium and iron (III) lignosulfonates together with 5% Salix aqueous extract were used in the synthesis of the composite resins (see characteristics of the new composite adhesives in Table 2).

Table 2. Chemical characteristics of the urea-formaldehyde resin modified with ammonium, calcium, iron lignosulfonates and Salix wood extract

Chemical characteristic	UF resin Standard	UF resin with LSNH$_4$			UF resin with LSCa			UF resin with LSFe		
		5%	10%	15%	5%	10%	15%	5%	10%	15%
Aspect	White liquid	Brown liquid			Brown liquid			Brown liquid		
Density, g/cm^3	1.28± 0.05	1.30	1.31	1.32	1.28	1.30	1.31	1.29	1.31	1.32
Solids,%	65 ± 0,2	64.8	65.0	65.4	65.0	65.3	65.5	64.6	65.2	65.3
PH-value	7,5 ± 0,5	7.6	7.6	7.6	7.6	7.7	7.8	7.6	7.5	7.5
Viscosity, Ford cup 4, s	16 – 60	15	17	18	16	17	19	17	18	20
Free CH$_2$O % ,max	0.90	0.60	0.56	0.50	0.70	0.68	0.62	0.55	0.50	0.47
Gel time, 100^{0}C, s	60	48	44	40	51	46	43	36	33	28

Where:
LSNH$_4$ = ammonium lignosulfonate
LSCa   = calcium lignosulfonate
LSFe = iron (III) lignosulfonate

The chemical structure of the composites and the interface bonds between lignosulfonate, Salix wood extract and urea formaldehyde resin were outlined by recording their FTIR spectra with a spectrophotometer (Vertex V70, Bruker) used to record the reflectance spectra, in the 600 to 4500 cm^{-1} range, after 16 scans, with 4 cm^{-1} resolutions. In the Figure 1. are presented the FTIR spectra for the standard UF resin and for the UF resin modified with iron lignosulfonate and aqueous extract of Salix wood (UF+LSFe).

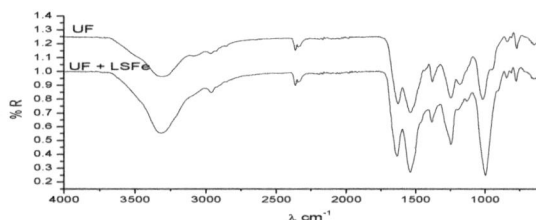

Figure: The FTIR spectra of the UF and UF+LSFe resins

In both spectra of UF and UF+LSFe resins, the broad absorption bands at 3300 cm^{-1} can be attributed to the O-H functional groups of methylol urea, the absorption band at 2900 cm^{-1} is assigned to the N-H functional group, the band at 1446 cm^{-1} represent the functional group –CH2 from methylene bridge and the band at 1086 cm^{-1} was attributed to the functional group C-O-C of ether linkage [19, 20]. It can be seen that the FTIR spectra of standard UF resin and UF+ LSFe composite resins present both two strong bands at 1500 and 1650cm^{-1}, which can be assigned to the formation of amide II and amide I compounds. Also, two more medium intensity bands at the 1150-1250cm^{-1} attributed to the lignosulfonate chemical structure are present in the FTIR spectrum of UF+LSFe composite. The band at 1350cm^{-1}, correlated with the formation of methyol groups becomes more distinct and decreases as methyol groups decrease due to the condensation reaction of UF resin with LSFe and Salix aqueous extract. All the described basic structural peaks of UF also appear in the spectrum of the new UF + LSFe composite resin, confirming the same basic chemical structure of these two amino resins. Meanwhile, the higher intensities of the peaks of the UF + LSFe resin compared with those of the standard UE resin certified the formation of chemical interactions between UF and iron lignosulfonate to form UF + LSFe composite resin. It can be also seen that better properties (especially the gel time, and the content on free formaldehyde (CH_2O) for the new adhesives based on urea, formaldehyde, metal complexed lignosulfonates and Salix wood extract are obtained for the lignin derivative LSFe with iron (III) cation as reaction partner. The gel time of urea-formaldehyde resins is essential to the establishment of effective processing parameters for applying these polymers in wood based-composites engineering. The main advantage of the new urea-formaldehyde-lignosulfonate-Salix extract resins composites is the lowering of the gel time, especially for UE resin with iron lignosulfonate and Salix extract.

Due to the content in lignin derivatives lignosulfonates and Salix wood extract, the new wood adhesives present resistance to water/weather and a lower emission of formaldehyde (the two principal disadvantages of the standard urea-formaldehyde resin).

3. CONCLUSIONS

Our research focused on the synthesis of some new composite adhesives based on urea-formaldehyde resins with lignin derivatives- ammonium, calcium and iron lignosulfonates and Salix wood aqueous extract, as ecological alternative to the existing production of commercial- urea-formaldehyde resins. The improved properties of the new composite adhesives, such as better water resistance, lower gel time and higher pot life were achieved by the contribution of the lignosulfonates and Salix extract in the polycondensation process. Due to their content on carbonyl groups, (especially for LSFe (9.35%) and to their acidic pH (especially for LSFe=2.24) the lignosulfonates and Salix aqueous extract actually participate as comonomers in the synthesis of the new urea-formaldehyde-lignosulfonate-Salix composite resins.

The reason for their application has to be seen also in the lowering of toxicity and costs, resulting from the difference in cost between monomer formaldehyde and the biomas waste lignosulfonates and Salix extract. Having in view the biocide activity of lignin it can be also expected an improvement of their resistance against the attack of the microorganisms.

The possibility of obtaining ecological, new type of wood adhesives based on lignin derivatives lignosulfonates and Salix extract presents a great interest in the future, taking into account the need for using lignocellulosic materials as non-polluting reactants for organic synthesis.

ACKNOWLEDGEMENTS

This paper is supported by the Sectoral Operational Programme Human Resources Development (SOP HRD), financed from the European Social Fund and by the Romanian Government under the project number POSDRU/159/1.5/S/134378.

REFERENCES

[1] Haveren, J.V., Scott, E.L., Sanders, J.P.M., "Review: Bulk chemicals from biomass. Biofuels Bioproduction, Biorefinery", vol. 2, no.1, p. 41-57, 2008.
[2] Zhang, Y.H.P., „Reviving the carbohydrate economy via multi-product lignocellulose biorefineries", Journal of Industrial Microbiology and Biotechnology, vol. 35, 2008, p. 367–375.
[3] Hon, D.N.S, „Chemical Modification of Lignocellulosic Materials", Mark Dekker, Publishers, New York, NY, 1996.
[4] Hon, D.N.S., Shiraishi, N., "Wood and Cellulosic Chemistry". 2nd Edition. Marcel Dekker Publishers, New York, NY., 2001.
[5] Zakis, G.F., Functional analysis of lignin and their derivatives, TAPPI Press, Atlanta, 1994.
[6] Dumitrescu, L., Research regarding the synthesis of some new wood bioprotection agents, PhD Diss., Gheorghe Asachi Technical University of Iasi, Romania, 1999.
[7] Dumitrescu, L., Perniu, D., Manciulea, I., Nanocomposites based on acrylic copolymers, iron lignosulfonate and ZnO nanoparticles used as wood preservatives, Solid State Phenomena, vol. 151, 2009, p.139-144.
[8] Bjorsvik, H.R., Liguori, L., „Organic processes to pharmaceutical chemicals based on fine chemicals from lignosulfonates". Organic Process Research & Development, vol. 6, no.3, p. 279–290, 2002.
[9] Doherty, W, Mousavioun, O.S., Fellows, C. M., Value-adding to cellulosic ethanol: Lignin polymers. Industrial Crops and Products, vol. 33, no.2, 2011, p. 259–276.

[10] Mansouri, N.E., Salvadó, J., Structural characterization of technical lignins for the production of adhesives: "Application to lignosulfonate, kraft, soda-anthraquinone, organosolv and ethanol process lignins", Industrial Crops and Products, vol. 24, no.1, p. 8–16, 2006.

[11] Dunky MA,, Pizzi A ,"Wood adhesives", in Adhesion Science and Engineering Surfaces, Chemistry and Applications, Amsterdam, p. 1039, 2002.

12] Gargulak, J.G., Lebo, S.E., "Commercial use of lignin-based materials". In: Lignin: Historical, Biological, and Materials Perspectives. ACS Symposium Series. American Chemical Society, pp. 304–320, 2000.

13] Forss, K.G., Fuhrmann, A., "Finnish plywood, particleboard, and fiberboard made with a lignin-base adhesive". In: Forest Prod. J., 29, p. 39-43, 1979.

[14] Ferra J. M.M., Ohlmeyer M. Mendes A. M. Costa M, R, N, Carvalho L, H, Magalhaes, F, D. "Evaluation of urea-formaldehyde adhesives performance by recently developed mechanical tests". International Journal of Adhesion & Adhesives 31, p.127–134, 2011.

[15] Muzaffar, A. K; Sayed, M. A and Ved, P. M., "Development and Characterization of Wood Adhesive using Bagasse Lignin". Internat. Journ. of Adhesion & Adhesives, 24, 6, p. 485-493, 2004.

[16] Schmitt,L, G, Hollis,J, W, Schoneld, Jr."Non-toxic, stable lignosulfonate-urea-formalde-hyde composition and method of preparation thereof". United States Patent Nr. 5,075,402, 1991.

[17] Xing, C., Zhang, S., Y., Deng J., Wang S., Urea–Formaldehyde-Resin Gel Time As Affected by the pH Value, Solid Content, and Catalyst. Journal of Applied Polymer Science, Vol. 103, p. 1566–1569, 2007.

[18] Gavrilović-Grmuš, Y., Nešković, O., Diporović-Momčilović, M., Popović M."Molar-mass distribution of urea–formaldehyde resins of different degrees of polymerization". Journal of Serbian Chemical Society, 75 (5) p. 689–701 2010.

[19] Osemeahon, S. A., Barminas, J. T., "Study of some physical properties of urea formaldehyde and urea proparaldehyde copolymer composite for emulsion paint formulation". International Journal of Physical Sciences Vol. 2 (7), p. 169-177, July, 2007.

[20] Gan, S., Tan, B., "FTIR studies of the curing reactions of palm oil alkyd-melamine enamels". Journal of Applied Polymer Science, 80, p. 2309-2315, 2011.

5th International Conference
"COMPUTATION MECHANICS and VIRTUAL ENGINEERING"
COMEC 2015
15-16 October 2015, Braşov, Romania

ASPECTS REGARDING SHEARING CUTTING BLADES FEATURES OF ENERGETIC PLANTS STRAINS FOR LABORATORY TESTING

Georgiana Moiceanu[1], Gheorghe Voicu[1], Mirela Dincă[1], Gigel Paraschiv[1], Mihai Chitoiu[1]
[1]University Politehnica from Bucharest, Bucharest, Romania, moiceanugeorgiana@gmail.com

Abstract: Testing shear cutting parameters of energetic plant strains it is realized in laboratories with modern equipments with two pillars and an adequate cell force.Direct shear testing of stalks it is realized with metallic blades with a V opening, the center angle being 60, 75, 90 or 120 degrees (or others). For shear cutting testing of miscanthusgiganteus strains with a testing equipment type Hounsfield H1 KS, cutting blades of the same dimensions with shearing blades were made (same length, thickness, angles openings) but with the two sharp edged at different angles (10, 20, 30, 40, 50 degrees). Experimental testing objective was to identify the most adequate sharpening angle and blades opening for a maximum efficiency in plant cutting (minimum energy consumption).. In this paper testing sheets characteristics and the results obtained by cutting plants are presented.
Keywords: shear force, cutting blades, cutting force, energy consumption.

1. INTRODUCTION

An alternative to fossils fuels is known to be biomass obtained from renewable energy sources. Biomass, as it has been shown throughout many experimental researches, has the potential to supply fuel and electricity, [1]. Research was done in order to estimate what amount of energy consumption is necessary to process biomass. Considering the constant increase of oil prices and consumption, researchers developed the legal framework conditions surrounding energy crops usage, [2].
These energy crops pass through mechanical processing operations, even from the moment of harvesting, until it becomes pellets, combustible briquettes, biogas etc. In general energy crops are subjected to a process of size reduction that consists in operations like cutting, grinding, briquetting etc, different types of forces acting on the material (shear stress, bending stress, crushing forces, cutting with or without sliding), [3].
Scientists carried out many studies in order to decrease shearing strength for different energy crops [4,5, 6]. An experimental research team tested wheat straw and obtained a shear strength between the range of 5.4-8.5 MPa. [7].
In paper (8) authors mentioned the necessity of knowing the physico-mechanical properties of energy crops (bending, shearing stress, energy requirements) in order to design properly the blades used for cutting.
Considering the design of the blades in this paper we tried to establish different aspects regarding shearing cutting blades featuresof energetic plants strains for laboratory testing.

2. MATERIALS AND METHODS

In order to conduct the experiments we analyzed the shear blades that the mechanical equipment HOUNSFIELD H1KS had from the factory. The peripheral equipment's of this apparatus consist in a cell force of 1 kN, parallel metallic plates for uniaxial compression, shear blades type V with different opening angles of 30°, 50°, 60°, 75° and also a Qmat software in order to gather experimental data.

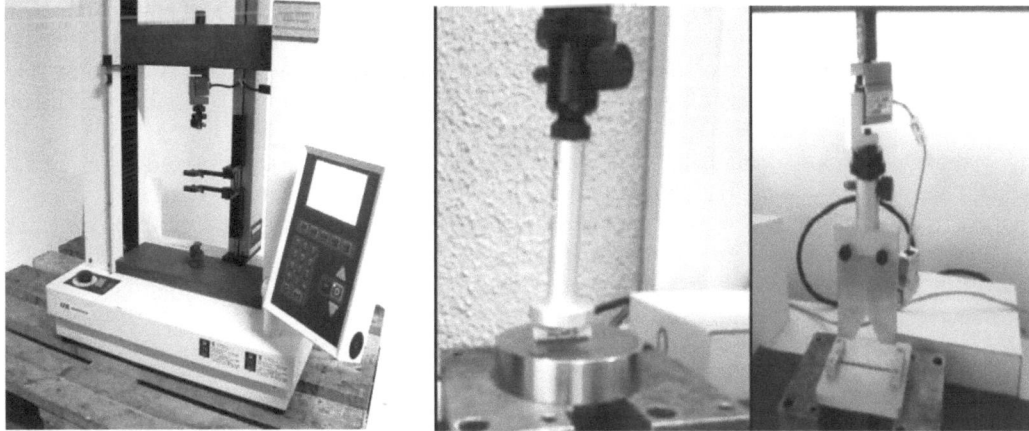

Figure 1: Hounsfield H1KS

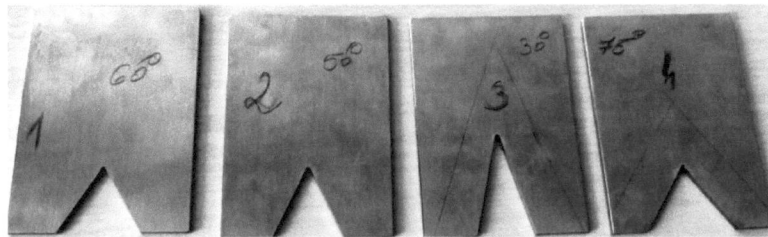

Figure 2: Shear blades

To conduct shear cutting experimental test we used type V cutting blades and straight blades.
To make these knives we went to Faculty of Engineering and Management of Technological Systems. In its laboratories out of metal plates the future knifes were cut using as references for the necessary dimensions the Hounsfield shear blades. Knifes were done with different opening angles (V = 30,50and 70°) an different sharpening angles of cut (i = 10, 20, 30, 40 and 50°), each time taking into consideration the dimensions of shearing blades of the equipment.

Figure 3: Shear cutting blades many in laboratory

After making the knives and sharpening the knives, they were taken to thematerials study laboratories in Faculty of Transports in order to treat them thermally and to find out the exact content of carbon. Steel samples were analyzed with a spectrometer with a spark optical emission SPECTROMAX (Germany), determining the chemical composition, average per sample and standard deviation. The knives hardness was determined and had a value of approximately 20 HRC. The laboratory where these test were made it is accredited RENAR.

After the spectrophotometer analyses the samples were oil-quenching in an oven type Nabetherm (Germany) with electrical resistance. The initial hardness of blades was measured 20 HRC (Rockwell) and the final hardness after the hardening process.

After heating it in the oven at approximately 840°C and then cooling the knives in water tha blades hardness was measured - 50 HRC. It could be seen that cooling the plates in oil had a lower hardness, around 44 HRC. The samples of material, considering the chemical composition, were not uniform, existing a big difference between the chemical compositions for each knife. These determinations were done in 7 points, the exteme values being eliminated, and the remaining point giving the final results. The mechanical resistance measured indicated a value above 500 N/mm^2, and the elongation tests shown that the material has a plasticity level above 10%.

In order to determine during experimental research shear resistance, we used the following equation:

$$\tau = F_s / 2A \tag{1}$$

τ - Shearing resistance;
F_s - Shearing force during disposal (N);
A - Area of the stem surface subjected to shearing (mm^2).

The material subjected to testing was miscanthus x giganteus, harvested in two different years, each time harvested during spring. It is necessary to mention that the traveled distance of the cutting blade in order to cut miscanthus stems is the plants diameter where it will be cut. Also the plant has an elliptical shape all around the stalk.

3. RESULTS AND DISCUSSIONS

Studying these blades and then using it in experimental research we obtained different variation curves for each of the blade used. The travel speed set each time was 500 mm/min (8,3*10^{-3} m/s). The Hounsfield equipment has a precision of ± 10^{-4} mm.Stalk cutting with the apparatus was done with 5 cutting blades (shearing)with an opening angle of 30, 50 degrees but with edge angles (I = 10, 20, 30, 40 50°) and also with straight knives with different bevel angles. Force – deformation curves obtained at plant cutting we could observe three distinct areas for the curve:

- Compression area until the plant is crushed, respectively the bark breakslongitudinally, and the core flattens, the curve shows a higher variationwith reaching the maximum point after which the value drops for a higherknife movement;
- Compression and cutting area presents a higher variation far exceeding thecrushing limit, the value of the force reaching a maximum that representsthe maximum cutting force;
- Third area in which the cutting takes place on the length of the plant, thecurve presenting either a slight decreasing variation, or the opposite if theplant is trapped between the margins of the orifices where a knife/plantfriction appears. [8]

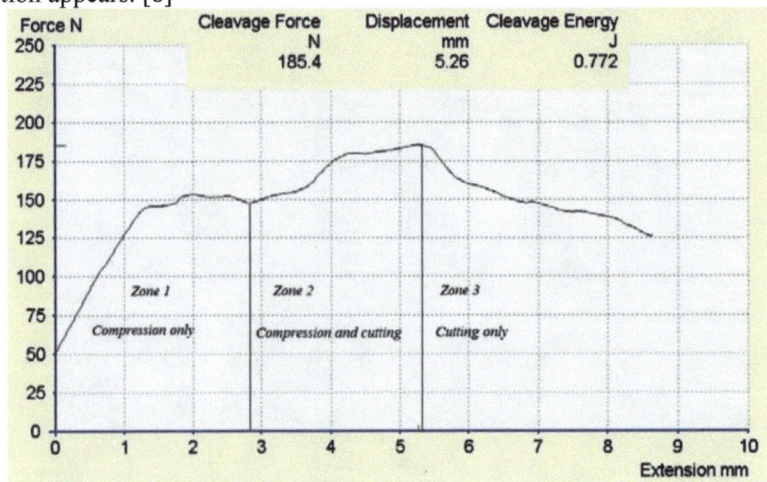

Figure 4: Knife force-displacement curve for random sample

$$y = 0.27\,x^2 - 17.30\,x + 606.3$$
$$R^2 = 0.984$$

a

$$y = 0.003\,x^2 - 0.185\,x + 4.180$$
$$R^2 = 0.936$$

b

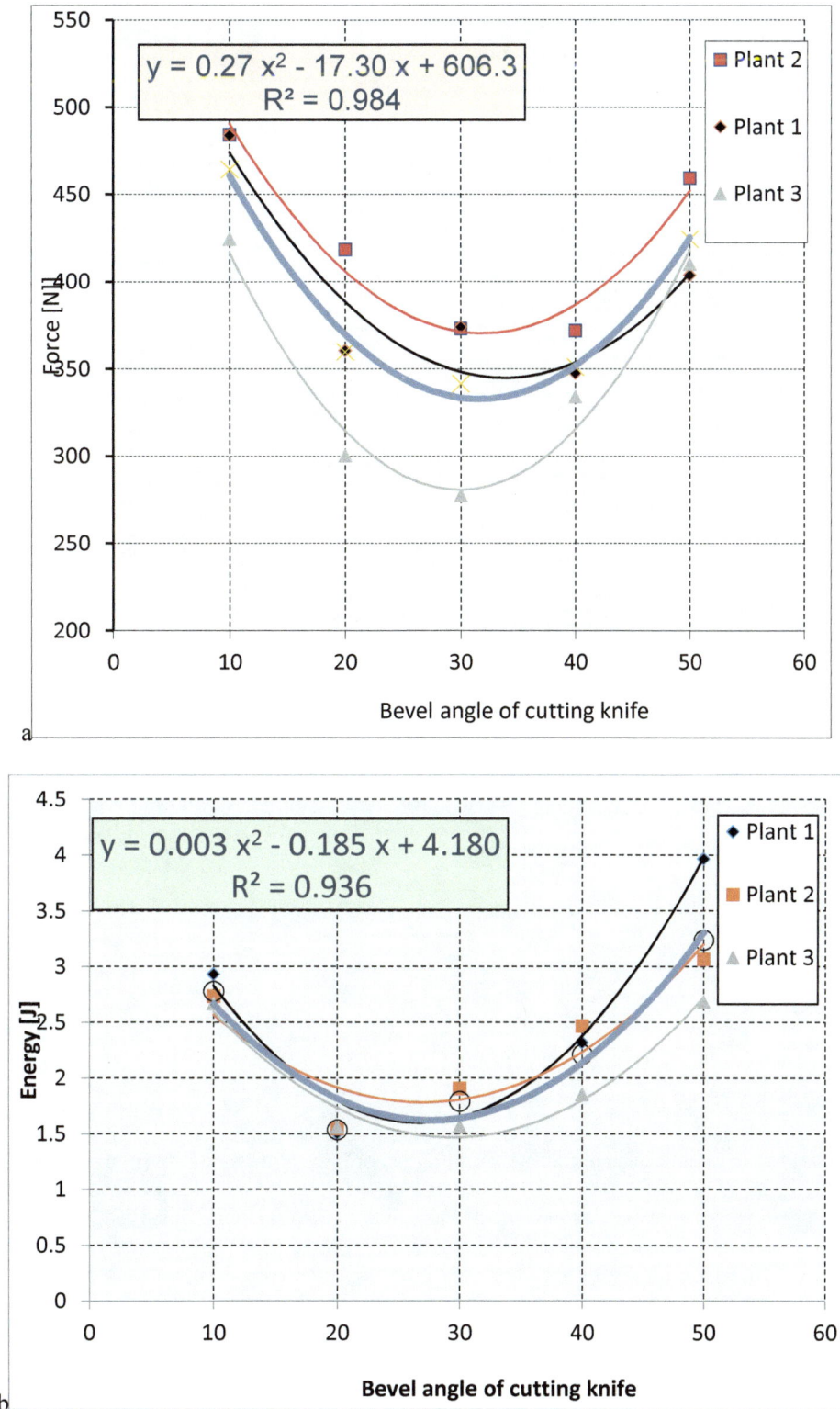

Figure 5: Variation of the force and miscanthus stalk cutting according to the blade sharpening angle (opening of 30°)
[8]

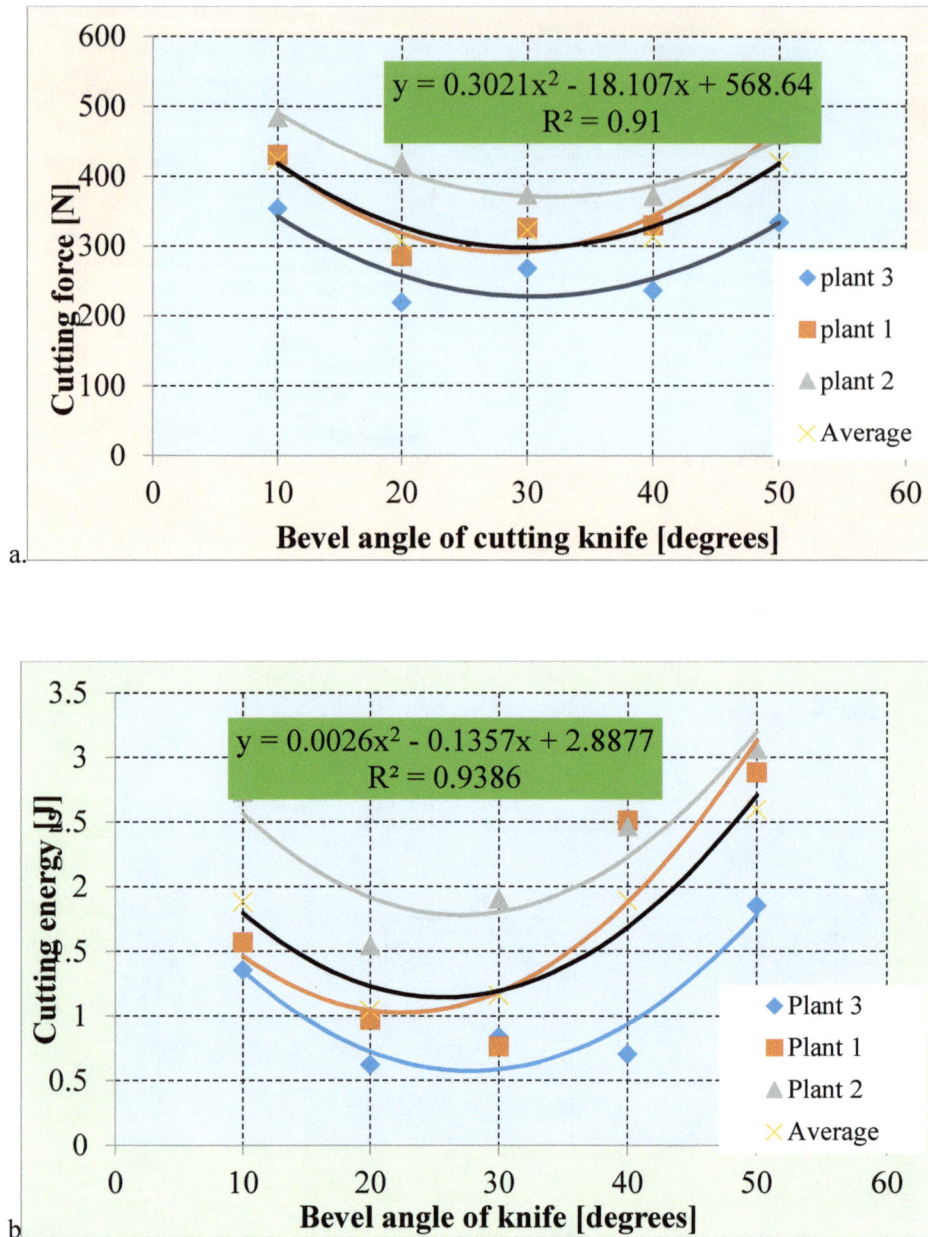

Figure6: Variation of the force and miscanthus stalk cutting according to the blade sharpening angle (opening of 50°) [9]

As it can be seen in these figures a drop in its values until an angle of 30-40 can be observed, after which the force rises forsharpening values of over 40°. The experimental data, regarding the variation of the cutting force with thesharpening angle of the blade, with the second degree polynomial variation law, isrepresented by the value of the correlation coefficient $R^2=0.984$ for figure 5a and $R^2=0.936$ for figure 5b. Shear cutting force at he speed of the knife of 500 mm/min (with support), had medium values of about 423 N for a sharpening angle of 10°and about 421 N for sharpening angles of 50°. It presented a minim between the angles of 20° – 30°the values being 308 N. It could be concluded that for a very sharped V type knives, with an opening of 50° and for sharpening angles above 35 – 40° lead to medium values of the cutting force, for a straight cut and small speeds of cutting. Thus, it is recommended that for lower cutting speeds and straight cuts with a counter knife to use knifes with sharpening angles around the values of 30°±2°. Experimental data correlation regarding the cutting force variation with the cutting angle is represented by the correlation coefficient $R^2=0.91$. The same conclusion could be drown for the energy consumption during cutting.

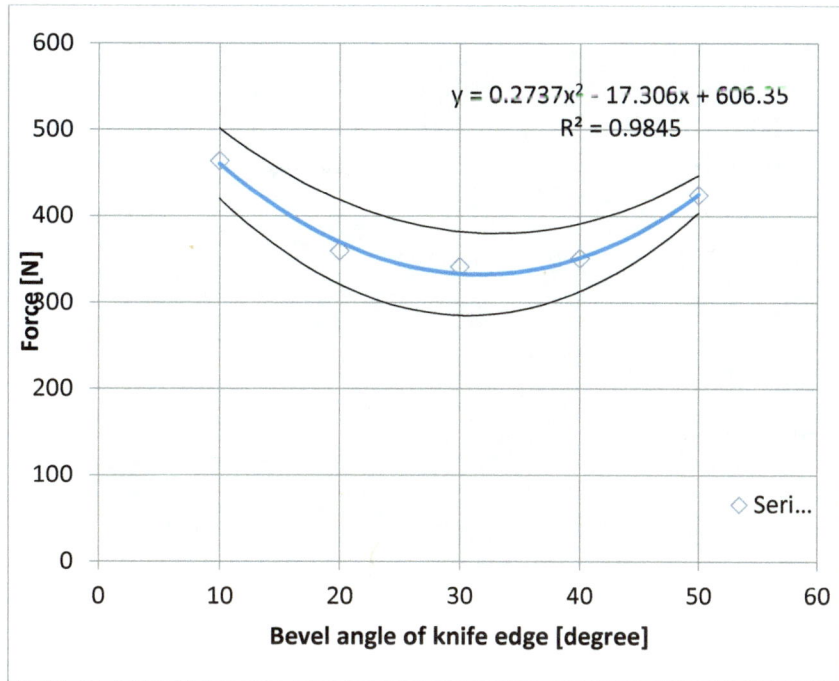

$$y = 0.2737x^2 - 17.306x + 606.35$$
$$R^2 = 0.9845$$

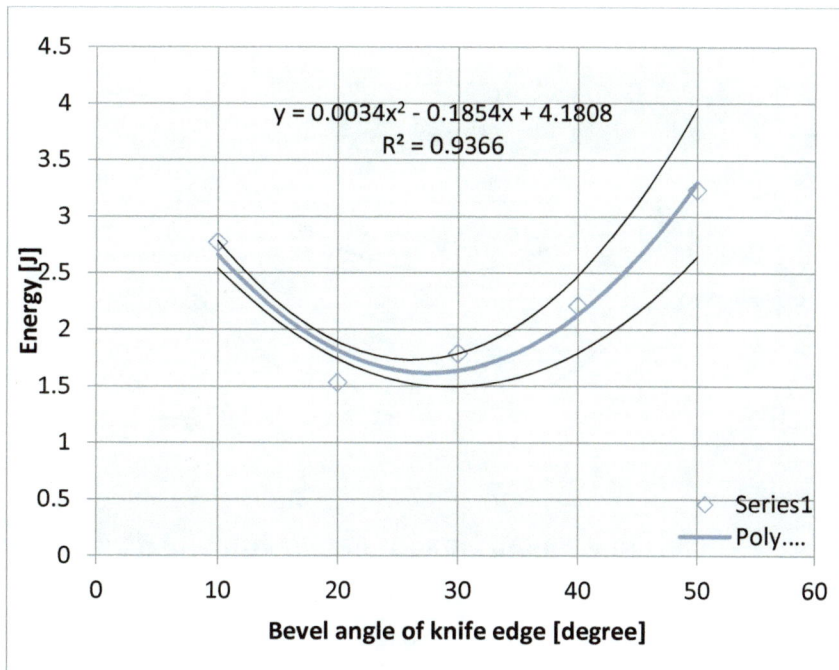

$$y = 0.0034x^2 - 0.1854x + 4.1808$$
$$R^2 = 0.9366$$

Figure 7: Variation of the average values of force and cutting energy and the mean squaredeviation (F±σF; E±σE) according to the sharpening angle of the blade (opening of 30°) [8]

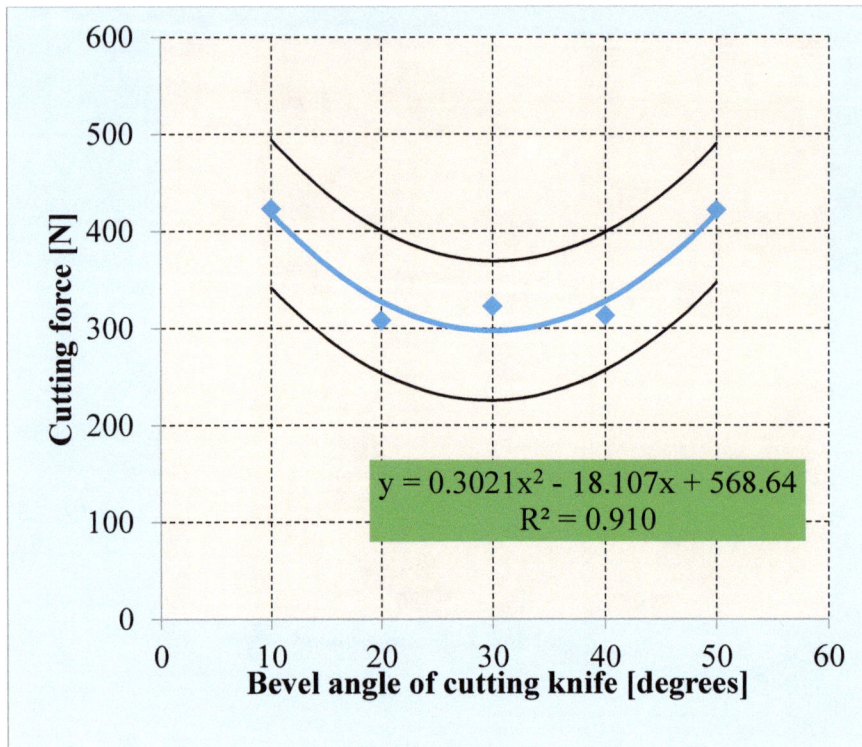

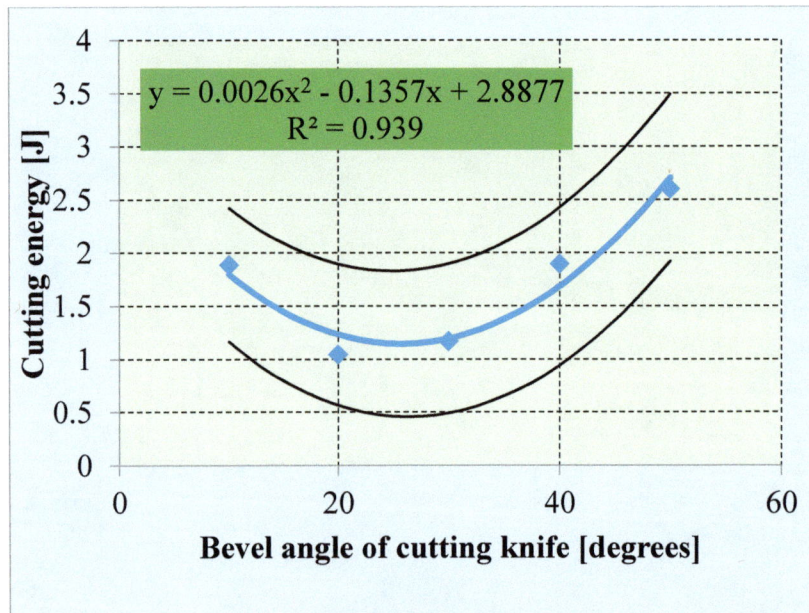

Figure 8: Variation of the average values of force and cutting energy and the mean squaredeviation (F±σF; E±σE) according to the sharpening angle of the blade (opening of 50°) [9]

Analyzing the case of cutting energy, we could see different variations of the mean square deviation fordifferent sharpening angles. For example, as it can be seen in figure 7 for sharpening angles of 10° the mean squaredeviation is σ_E=0.13J, while at angles of 20°, the mean square deviation is 0.02J. Thus we presented increasing values until the sharpening angle of 50°, where σ_E=0.65J.

Also, using the blades shown earlier we researched the cutting force and energy consumption variation obtained during cutting miscanthus straws with a straight knife but with different blades edge angles.

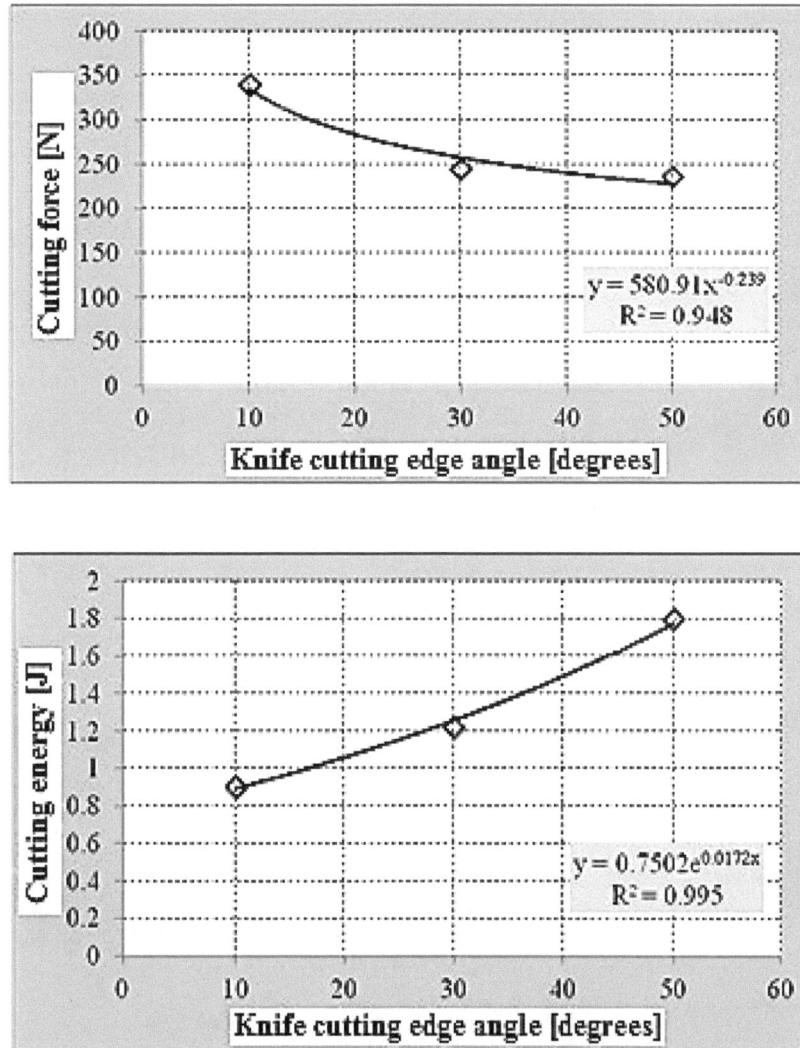

Figure 9: Cutting force and energy consumption variation of miscanthus straws taking into account the sharpening blade edge angles using a straight knife[10]

As it can be seenthe cutting force variation is contrary to the cutting energy consumption. It shows an ascending slope from a sharpening edge angle of 10° to higher values for a 50° angle unlike cutting plants energy consumption which has values 0.9 J for the 10° angle and about 1.79 J for a 50° angle, which recommends the usage of straight cutting knives with medium sharpening cutting edges, [10].

3. CONCLUSION

In conclusion as we could observe from testing we can sustain the following:
- The three stages of cutting process explained by other scientists were shown also during the experiments;
- Cutting force had values of 350 N for a sharpening angle of 10°, 235 N for angles of 50°. Also from these data it could be concluded that the optimum angle of cut is 30°;
- Each of these experiments indicated the optimum angle of cutting process each time, presenting a minim or a maximum value of the cutting force and the cutting energy;

ACKNOWLEDGEMENTS

The work has been funded by the Sectoral Operational Programme Human Resources Development 2007-2013 of the Ministry of European Funds through the Financial Agreement POSDRU/159/1.5/S/132395 and with the support of the University "Politehnica" from Bucharest.

REFERENCES

[1] P.C. Johnson, C.L. Clementson, S.K. Mathanker, T. E. Grift, A.C. Hansen – Cutting energy characteristics of Miscanthus x giganteus stems with varying oblique angle and cutting speed, Biosystems Engineering II 2, p. 42-48, 2012;

[2] Braun R., Weiland P., Wellinger, Biogas from Energy Crop Digestion, IEA Bioenergy, http://biogasmax.co.uk/media/iea_1_biogas_ A. energy_crop__007962900_1434_30032010.pdf, 2011;

[3] A.Lungescu, P.Panayotov; Ecological briquettes obtained from wood biomass, REGENT, Vol. 9,nr. 1(2), p. 51 – 54, 2008;

[4] Annoussamy, M., G. Richard, S. Recous and J. Guerif, - Change in mechanical properties of wheat strawdue to decomposition and moisture. Appl. Eng.Agric., 16(6): 657-664, 2000;

[5] Hirai, Y., E. Inoue, K. Mori and K. Hashiguchi -Analysis of reaction forces and posture of a bunch of

[6] Crop stalks during reel operations of a combineharvester. Agri. Eng. Int., the CIGR Ejournal, Manu.FP 02-002, Vol. IV, 2002;

[7] 6 Shaw, M.D. and L.G. Tabil - Compression, relaxation and adhesion properties of selected biomass grinds. Agri. Eng. Int., the CIGR Ejournal, Manu. FP 07-006, Vol. IX, 2007;

[8] 7 O'Dogherty, M.J., J.A. Hubert, J. Dyson and C.J. Marshall- A study of the physical and mechanical properties of wheat straw. J. Agric. Eng. Res., 62: 133-142, 1995;

[9] 8. Ince, A., S. Ugurluay, E. Guzel and M.T. Ozcan - Bending and shearing characteristics of sunflower stalk residue. Biosyst. Eng., 92(2): 175-181, 2005;

[10] 9. Gh. Voicu, G. Moiceanu, G. Paraschiv – Miscanthus stalk behavior at shear cutting with V cutting blades, at different sharpening angles, U.P.B Sci. Bull., Series D, Vol. 75, Iss. 3, ISSN 1454 – 2358, 2013;

[11] 10. G. Moiceanu – Research regarding energetic plant behavior during mechanical operations of cutting and grinding 2012:

[12] 11. G. Moiceanu, P. Voicu, G. Paraschiv, G. Voicu- MiscanthusBehaviour regarding shear cutting with straight knives with diffent edge angles, Enviromental Engineering and Management Journal, 2014;

6th International Conference
"Computational Mechanics and Virtual Engineering "
COMEC 2015
15-16 October 2015, Braşov, Romania

COMPARATIVE ANALYSIS OF WALKING TYPOLOGIES IN RELATION TO LOWER LIMB JOINTS LOCK

Mihaela Baritz
Transilvania University, Brasov, ROMANIA, e-mail: mbaritz@unitbv.ro

Abstract: *In this paper we present some theoretical considerations and experimental biomechanical analysis of gait typologies in subjects with foot joints blocked. The theoretical aspects of the gait cycle, parameters and limits defining the gait cycle but also an analysis of gait typologies developed by a healthy subject walking are presented in the first part of the paper. The principles and methodology for investigating structure of gait typologies vs. locking joints legs and experimental aspects highlighted during the developed analysis are also presented in the second part. In the third part of the paper are presented the results of this comparative analysis and correlation strategies and also are defined some data by determining the dynamic correlation coefficient. In the final part of the paper, conclusions of the investigations are set by records analyses. Also are presented the development of biomechanical investigations for comparative evaluations and future directions of the studies.*
Keywords: gait, biomechanics, joint, correlation

1. INTRODUCTION

Gait cycle is characterized by a series of static and dynamic parameters that can be measured during performance of. Human subject moving into the space and permanent interaction with the environment is due musculoskeletal and its components: bones, muscles and joints.

The totality of bones forms the human skeleton, helping to the shape and body posture. Bones are connected by joints making it possible human mobility and movements.

The passive part of the locomotor system is made up by bones and joints; and active part in all muscles, due to which movement is possible.

Musculoskeletal components represent a large part of the total mass of the human body respectively 52% of the total weight of an adult. From this percentage, 38% represents musculature and 14% represents skeleton. These values are variable depending on the individual's age and physical training of him. [1]

Activity of locomotion (bones, joints and muscles) is not independent but is closely related to other components of the human body (Fig.1).

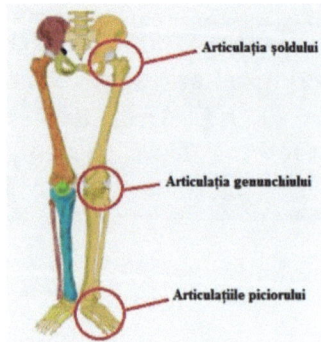

Figure 1: The human leg joints

Particularly important is the connection between the musculoskeletal system and nervous system. In the absence of this nervous system, the locomotor system would "be an inert mass or group that would operate heterogeneous and anarchic". [1]

About half of the human skeleton jointings are joints, skeletal joints very mobile.

If the lower limb joints belt they fall into the lower limb joints (pelvic belt) and lower limb joints free. Among the lower limb joints can freely include hip joint (hip), knee joints and leg joints tibio-femural.

Of all the joints of the lower limb kinematics shows great complexity hip joint, knee and foot joints. Hip joint is formed by joining, using ligaments, between the femoral head and acetabulum of the hip.

The movements they perform hip joint, from Fig. 2 are: flexion / extension (near that removing thigh abdomen), adduction / abduction (carrying thigh towards the body that sideways), external rotation (twisting thigh outwards) and circumduction (movement that combined movements listed). [2]

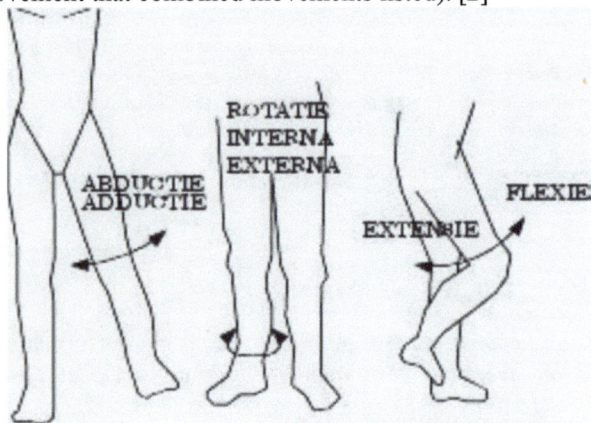

Figure 2: The movements of the hip joint [4]

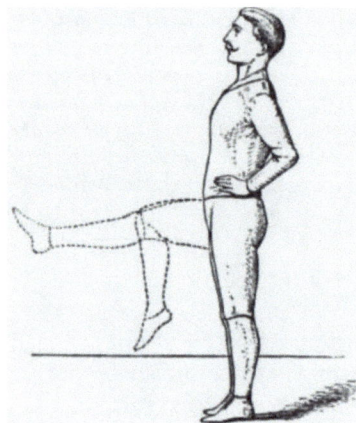

Figure 3: Knee joint movements [5]

Knee joint, considered the most voluminous and powerful body joint, unites a series of thigh leg ligaments. If a human subject, in standing position when support is distributed evenly over both legs, whole body weight is transmitted through the end of the femoral knee and hence the girls planting the direction of the femoral head, knee and ankle. [3]

The knee joint has one degree of freedom and allows two main movements: flexion and extension leg on the thigh (Figure 3). In flexion the posterior calf approaches the back of the thigh; and if extension occurs remoteness rear face of the leg of the back of the thigh.

In flexion / extension leg acts as an open kinematic chain and knee levers acting on the principle of third degree. Limiting movement of flexion is achieved by meeting face with the back hind leg and thigh extension movement is limited by ligaments. [3]

The link between the foot and lower leg ankle joint is made of. It consists of pillar tibial malleolus lateral talus and connected by ligaments. Leg joints are talocrurala articulation (neck leg) joints intertarsiene joints tarso-metatarsal joints, joints and finger joints inter-metatarsiene.

They perform leg movements are flexion / extension (talocrurale joint movements) and your truly / pronation (inter-tarsienes joints movements). [2] Movement is the movement of your truly the plant foot "concerns" are oriented toward the midline and the movement of pronation is when the plant "looks" are oriented to the side. [6]

Human walking is considered to be a cyclical movement performed by positioning a successful leg before the other. The main mechanism that relies walking is considered "alternative and constant movement of the two legs, which in turn assumes the function of support and function of propellant" [7].

This mechanism has been defined in the specialized work as an "alternative bipedalism" (Steindler) or as' a fall continues to lift its own "(Holmes). [7]

Human walking is cyclical drive unit (double step), namely the distance between the point of contact with the ground (heel) of a leg and immediately following point of contact of the same foot. [8] Loco-motor movements are learned in childhood and are settled in the central nervous system so that the driving elements are the same for all people.

The setting of the walking reflex, an important role is sensitive receptors in the muscles, tendons, ligaments and joints.

Walking is influenced by age, sex and individual body constitution, so this is personal to each individual. [6] Human walking is composed of a step in the cycle leg becomes successively as rod and pendulum while driving element (Figure 4).

Each gait cycle consists of a phase support bilateral and unilateral support phase and can be described as [6] [7]:

1. Bending the trunk forward (center of gravity will be screened before the supporting body), while the leg is thrown forward;

2. Extension of one of the legs simultaneously pendulum and the other member becomes detached from the ground up to be designed before;

3. A member pendulum will be screened again on the ground before the member support;

4. This cycle is repeated, reversing limbs and functions;

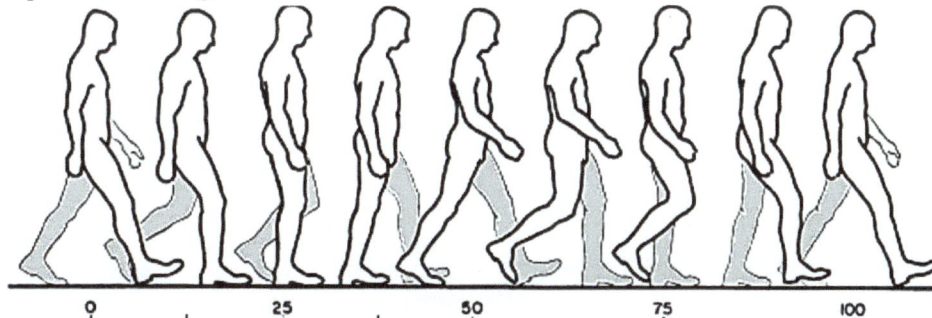

Figure 4: Gait cycle

Gait cycle is defined as a series of steps and comprises a double step that corresponds to the time between two taps on the ground by the same calculation. [6] Step defined in the literature by various authors as "the interval between two supports" or "double-step, corresponding to the series of movements which succeed one another between two positions identical to one foot". [7]

2. EXPERIMENTAL SETUP

This study aims to analyze the patient's behavior where it has ankle or knee mobilized (compared to walking in normal conditions) in order to help the patient after a pathology shows the problem or dysfunction of the gait cycle or stability Stations.

Pilot plant used in this experiment consists of RSScan pressure plate type, data acquisition and processing software " Footscan 7 Gait 2nd Generation" anthropometry and equipment (goniometer, rulers, anthropometry). (fig.5.)

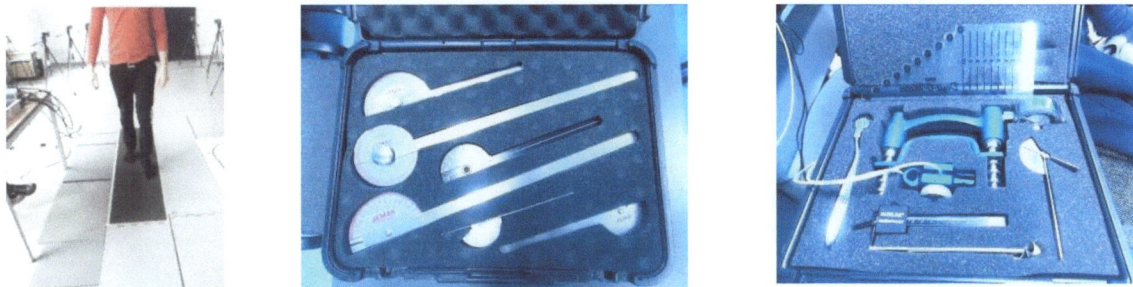

Figure 5: Equipments for experimental setup (RSScan plate, goniometer, anthropometric kit)

Measurements were recorded on RSScan board, the length of 2 m, and associated software " Footscan 7 Gait 2nd Generation. The experiment was conducted dynamically moving patient on the pressure plate in a given time. There are three measurements made on each gait cycle in order to obtain reliable and comparable data sets.

Types walk examined: normal walking, cycling step added, running knee to the chest, walking march, reversing and going sideways. They were chosen because these types of walking are considered to be most common in daily activity. For each of these types of gait measurements were performed three consecutive walk the cycle began with the left foot and right foot 3 measurements. Subjects in this experiment participants had, in turn, right ankle and then locked in these conditions were recordings of drive types listed above.

Figure 6: Gait cycles type – march, with knee up, back gait, lateral, added step

The pressure plate type RSScan recorded via the sensors 16384/2 m pressure (force) created by the soles area in contact with the ground. These values are then compared between subjects (male, female) and different forms of the gait cycle

3. RESULTS AND CONCLUSIONS

Analyses conducted on subjects aimed primarily determine normal variant walk without constraints in the joints of the foot and then established a procedure for locking joints and comparing parameters originally listed all variants walk.

The blockage was assured of all joints, knee and ankle then right foot (left foot free from constraints).

In the first case, knees locked, so if gait cycle began with the left foot and the cells start with the right foot is noted that the force with which the subject gets on his left leg in normal walking is greater than the force of the subject when walking with right knee locked.

The difference in loading of the legs when walking normally healthy and the knee locked is represented in Figure 7.

This difference in power is because the subject tends to change the center of gravity to the left when the right knee is blocked. Force left foot female subject has a value of 366 N while driving with his right foot stuck. When male subject has blocked the leg strength of 491 N value, which value for a healthy walk should be in the range 570-590 N (according to its body weight). [9]

Calculation of dynamic load on the plantar surface of each leg can be calculated with the following relationship:

$$ST(\%) = \frac{\frac{PS_1 + PS_2}{2} \cdot 100}{\frac{PS + PS_2}{2} + \frac{PD_1 + PD_2}{2}} \tag{1}$$

$$DR(\%) = \frac{\frac{PD_1 + PD_2}{2} \cdot 100}{\frac{PS + PS_2}{2} + \frac{PD_1 + PD_2}{2}} \tag{2}$$

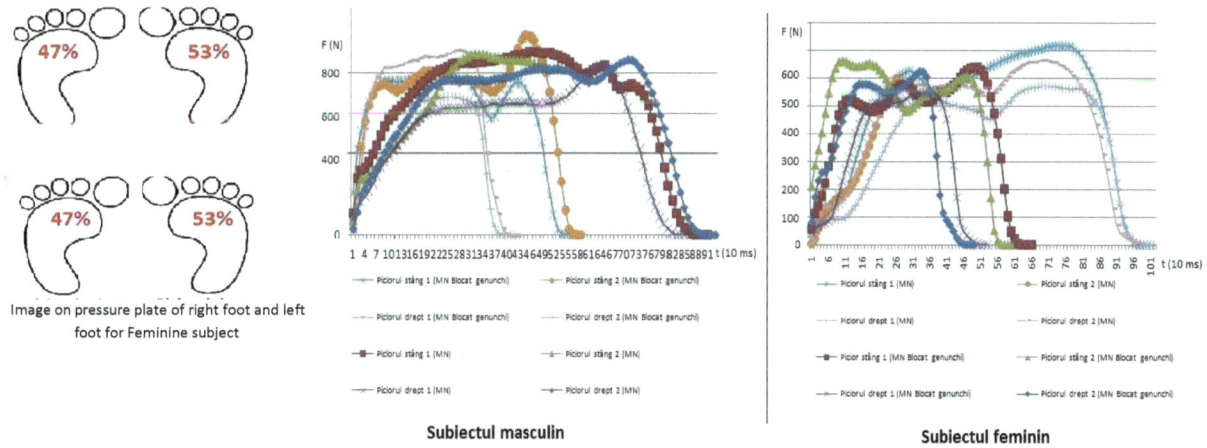

Figure 7: The plantar force distribution in normal gait with right knee locked

Walking type is the type of running march least affected by the fact that one knee is blocked because movement is executed with perfect knees stretched.

After planting measuring forces using RSScan pressure plate can say that plantar force developed by the two types of subjects in the two types of walking is different: strength in bottlenecks gait is less developed than in healthy walking.

As can be seen in Figure 8 plantar force difference occurs between left and right foot during walking with the knee locked. In marching gait with knee locked leg shows a smaller force than in healthy marching gait. The right foot is much heavier than the left leg which indicates an inclination of the body to the right and discomfort during walking. [10]

Also moving subject with joints executed without constraints is faster than the bottlenecks knee.

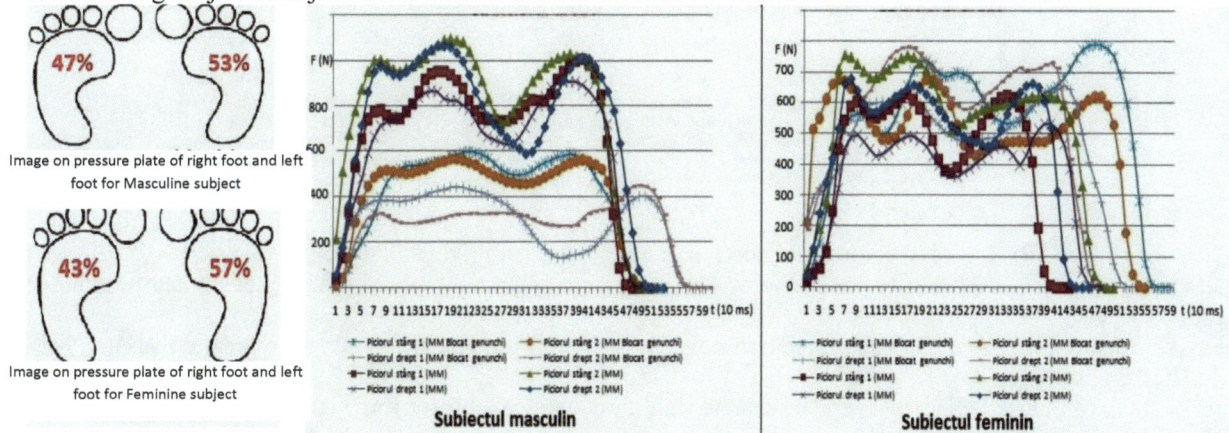

Figure 8: The plantar force distribution in march gait with right knee locked

Walking with ankle locked in the same procedure was performed (the same number of records on the same sequence of typologies walk).

Comparing walking to riding without blockages subject matter but with the man right ankle and jammed it can be seen as both a strength difference and a time difference distribution of motion. Plantar strength of the legs during walking with ankle block is greater than the force in driving without blockages. In Figure 9 can notice the difference in strength between the two plantar surfaces while walking with ankle lock: foot suffering a blockage develops a greater force than healthy leg.

As for making movement can be observed that subjects with producing member locked a move to shorter duration. When driving without blockages movement it is broader and the length is greater steps.

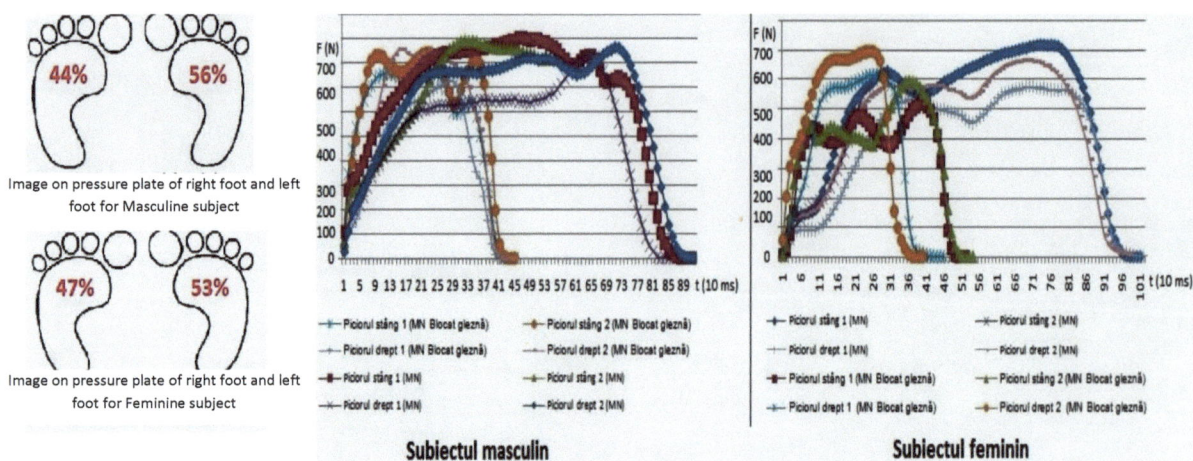

Figure 9: The plantar force distribution in normal gait with right ankle locked

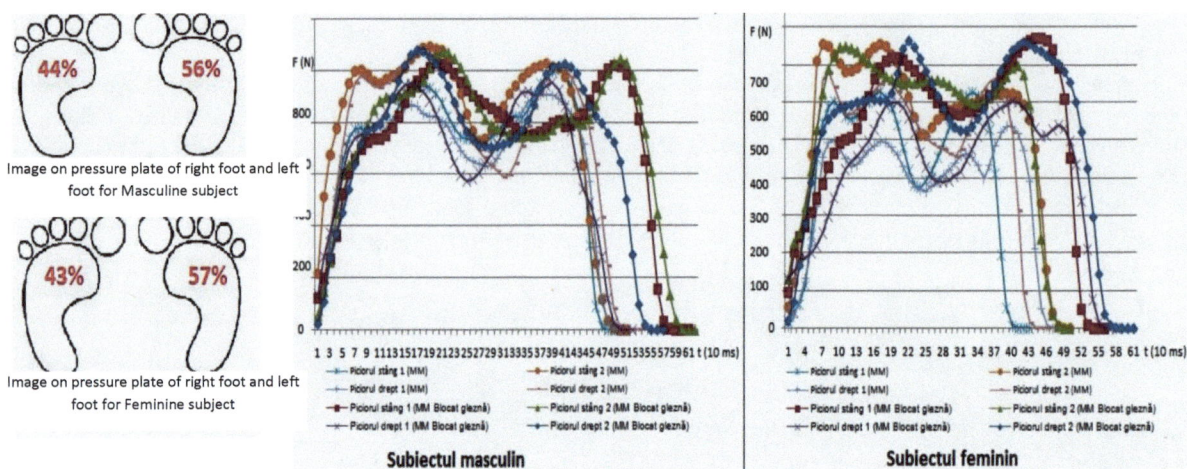

Figure 10: The plantar force distribution in march gait with right ankle locked

march gait case the developed plantar forces of the two subjects it is different as value between healthy and walking at the ankle joint locked.

It can be seen in Figure 10 that the left leg force and right leg with blocked ankle in walking is greater than the force produced in healthy walking.

Analyzing the gait with locked joints that can be seen the right foot develop a bigger force than left leg both cases: male and female subjects.

Also Figure 10 represents the distribution of the force during walking marching ankle locked. The difference between left and right legs is relatively large the right leg is longer charge than the left.

This difference in loading produce an unbalanced gait where the right leg is used as a support while advancing left foot in length.

By comparing healthy walking (without blockages) and walking with blocked knee and ankle of right leg were obtained the following conclusions:

• The right foot is more loading during gait with locked knee than the left leg.
• Normal walking resembles with blocked knee gait by step added in stride length and strength in distribution;
• The momentum is greater on leg unblocked because it behaves more dynamically during action;
• If the knee is locked, up and down stairs is made cumbersome;
• Riding with locked knee is imbalanced;
• Force on the left foot and right foot during walking increases during the gait with knee locked than in gait without blockages;
• Normal walking gait with blocked ankle resembles with the gait with added step in the distribution strength and stride length;
• The right foot is more loaded during the gait with ankle locked than the left leg;

• The momentum (in the same sort of gait-ankle blocked) is greater on leg unblocked because it behaves more dynamically during operation;
• If the ankle is locked down stairs is heavy;
• Walking with ankle block is unbalanced;
• Forces of the left foot and right foot, during walking with ankle locked increases while gait without blockages;
In future researches will consider other dynamic parameters simultaneously typologies gait constraints on both joints of the right foot to track how plantar surface load and change the temperature.

ACKNOWLEDGEMENT

These researches are part of the current researches in Applied Optometric Laboratory and *Advanced Mechatronic Researches Center* from University Transilvania Brasov.

REFERENCES

[1] Papilian, Victor: *Anatomia omului, Volumul 1. Aparatul locomotor*. Editura BIC ALL, ISBN 973-571-468-X, 2003;

[2] Cioroiu, Silviu G.: *Esential în anatomie si biomecanica*. Editura Universitatii Transilvania din Brasov, 2006;

[3] Olariu, Virgil; Rosca, Ileana; Baritz, Mihaela; Radu, Gheorghe N.; Barbu, Daniela: *Biomecanica Vol 1: Bazele biomecanicii*. Editura Macarie, Târgoviste, colectia "UNIVERSITARIA"; 1998;

[4] Tiberiu Laurian: *Contributii privind studiul proceselor tribologice din protezele de sold*. http://www.omtr.pub.ro/tlaurian/teza/teza_rez.html [accessed: 25.09.2015];

[5] *Flexion and Extensioni* http://chestofbooks.com/health/body/massage/Massage-Original-SwedishMovements /2-Flexion-And-Extension.html#.UxmH1vl_vw8 [accessed: 25.09.2015];

[6] Mircea Constantin Sora, Dan V. P.: *Mersul uman*. Editura Mirtron, ISBN 973- 661- 281- 3, Timisoara, 2004

[7] Baciu, C.: *Aparatul locomotor (Anatomie functionala, biomecanica, semiology clinica, diagnostic diferential)*. Editura Medicala, Bucuresti, 1981;

[8] Serban, Ionel: *Studii si cercetari privind influenta mediului inconjurator asupra stabilitatii si locomotiei umane*. Universitatea Transilvania din Brasov;

[9] Şerban, I.; Rosca, I; Braun, B.; Druga, C. - *Analysis parameters base of support and center of mass of the human body*, The 4th International Conference "Computational Mechanics and Virtual Engineering", COMEC 2011, Brasov, Romania, pp 429-434,

[10]   Şerban, I.; Rosca, I.; Braun, B.; Druga, C., *Environmental effects on the center's offset of the Kistler force plate*, International Conference on Medicine, and Health Care through Technology, MediTECH 2011, Cluj – Napoca, 2011;

6th International Conference
"Computational Mechanics and Virtual Engineering "
COMEC 2015
15-16 October 2015, Braşov, Romania

RESEARCHES REGARDING AUTOMOTIVE ENGINE FUNCTIONING

Eng. Ramona-Monica Stoica [1], Eng. Marius Simionescu [2], Eng. Virgiliu-Justinian Rădulescu[3], Eng. Cătălin Trocan [4], Eng. Viorel Mardari [5], Univ. prof. eng. Ion Copae, PhD[6]

[1]Military Technical Academy, Bucharest, Romania, monyk_dep@yahoo.com
[2]Claims Support, Bucharest, Romania, office@claimssupport.ro
[3]Metrorex, Bucharest, Romania, radulescu1961@yahoo.com
[4]S.A-R City Insurance, Bucharest, Romania, catalin.trocan@gmail.com
[5]Akka Romserv, Bucharest, Romania, viorel.mardari@gmail.com
[6]Military Technical Academy, Bucharest, Romania, copaeion@yahoo.com

Abstract. *The paper presents experimental results obtained from tests conducted with an automotive fitted with electronic control Diesel engine. There are highlighted some engine functional characteristics. It is aimed the matter of engine energy efficiency growth. It is approached the matter of parameters influence on engine functioning by using sensitivity analysis, information theory and variance analysis.*
Keywords: *vehicle engine, energy efficiency, sensitivity analysis, variance analysis*

1. INTRODUCTION

Fitting of automotives with on-board computer, electronic control systems, embedded sensors and actuators, the increasingly development of analysis equipments and the theoretical developments of another disciplines have represented the main influence factors regarding the theoretical and experimental study theories of engines functioning [3]. The endowment of engine automotive with lot of actuating and control systems was also possible due to remarkable progress in electronic field and software development.

2. EXPERIMENTAL RESEARCH

The experimental research were conducted with a Ford Focus automotive fitted with supercharged Diesel engine and the "common rail" Diesel fuel injection plant. According to technical specification, the automotives has a fuel consumption of 5.9 liters/100 km for an urban cycle, 4.0 liters/100 km for an extra-urban cycle and a 4.7 liters/100 km fuel consumption for combined cycle [8].
The automotive's Diesel engine is DOHC (Double Over Head Camshaft), with four inline cylinders, 4 valves per cylinder, 1.56 liters capacity displacement, 18 compression ratio, and "common rail" fuel injection. The engine develops a maximum power of 66 kW (90 CP) at 4000 rev/min and a maximum torque of 215 Nm at 1750 rev/min. The engine is equipped with embedded sensors and actuators and the on-board computer assures the functioning, the electronic control and diagnostic [9].

256

For the acquisition of automotive and engine functional parameters based on sensors have been used the interface and Ford's FoCOM software [9]. The FoCOM software (from the acronym **Ford COM**pany) represents a specialised software for the diagnostic of automotives like Ford, Mazda, Lincoln, Mercury and Jaguar and assures during experiments the following: reading functional parameters from the automotive on-board computer memory by connecting the interface with the diagnostic socket (figure 1); viewing data in a graphical or tabular form; data transfer on a computer with FoCOM application (figure 1).

From obtained data have been kept the more significant 100 experimental research, which marks a normal driving during exploitation. The experimental research have taken place on the same rolling track (dry asphalt in good keep) and during the same weather conditions, in the absence of wind.

During experiments, various parameters have beeen measured, among which are mentioned: automotive's speed V [km/h]; cyclic flow of fuel c_c [mg/cycle]; engine speed n [rev/min]; throttle shutters' position, percentage from their maximum opening p [%]; engine's torque M_e [Nm]; fuel pressure from common rail p_c [kPa]; intake air pressure p_a [kPa]; engine's hourly intake air consumption C_a [kg/h]; relative amount of recirculated gas γ [%]; relative tilt of turbocharger's vanes α [%].

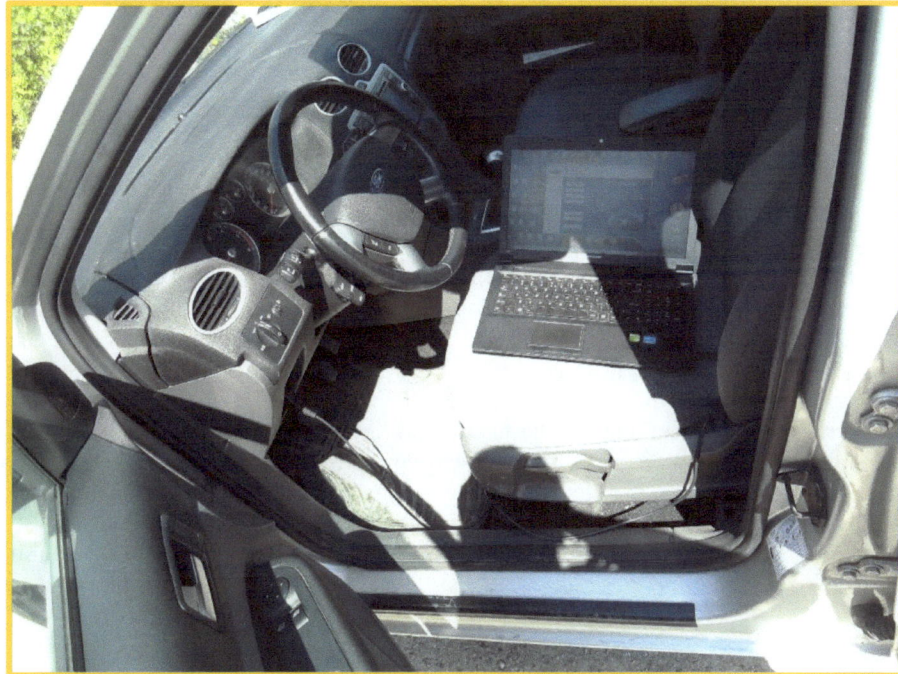

Figure 1: Experimental research

At these measured parameters we add others obtained by calculus: engine power P_e [kW], fuel consumption at 100 km C_{100} [liters/100 km], engine hourly fuel consumption C_h [liters/h], specific fuel consumption c_e [g/(kWh)], engine efficiency η_e [-], travel speed variation dv/dt [m/s^2], so acceleration/deceleration, traveled space S [m] etc.

In figure 2 are shown the values of fuel cyclic flow and engine torque, in continuous representation (usually used, but unreal) and in discrete representation (rarely used, but real, the experimental data having a discrete character). Discrete representation, the real one, allows a more easy viewing of ranges with the most values for the targeted parameter. For example, from figure 2a results that more than a half of cyclic flow of fuel values (54.6%) are in the range 10-20 mg/cycle. Likewise, from figure 2b we can see that few values (4.3%) are in the range where engine torque is high, in graph are shown both maximum experimental value and maximum value from engine technical specification [8].

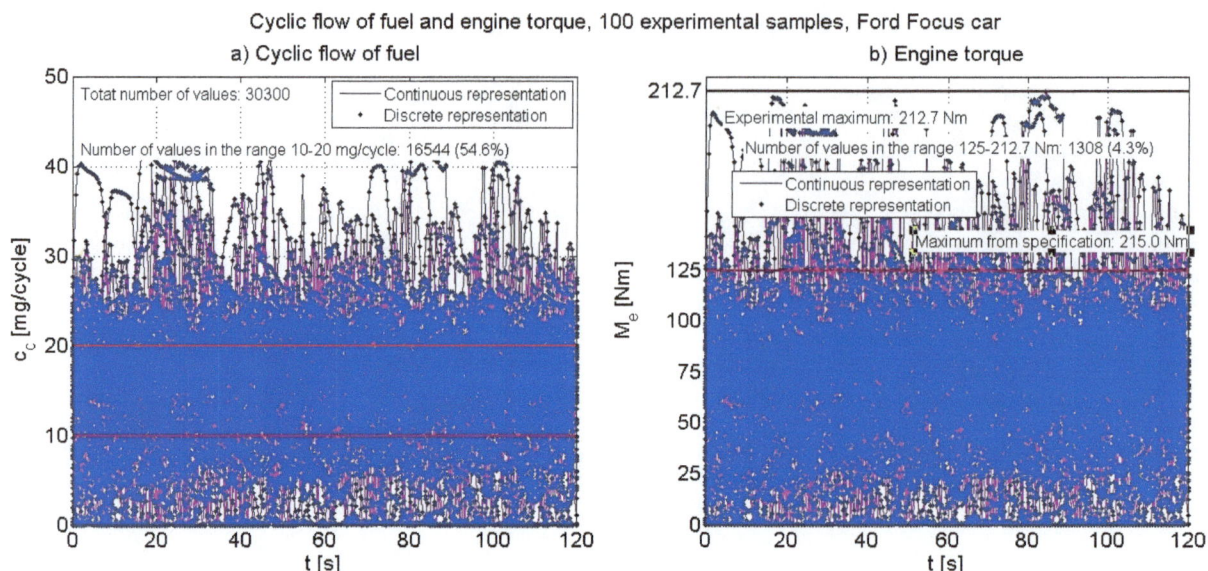

Figure 2: Cyclic flow of fuel and engine torque

3. THE STUDY OF FACTORS INFLUENCE ON ENGINE FUNCTIONING

The more and more drastic requests regarding dynamics and fuel saving performances that are currently assessed to engine automotives, imply a thourough study regarding the influence of various factors on engine functioning. In specialty literature can be found quantitative and qualitative estimations towards the influence of functional parameters, adjustment, design and automotive operation performances. It must be mentioned that in specialty literature the study of different parameter's influence is made based on a restrictive methodology such as: for the study of one parameter influence it is considered that the others are constant, which is obviously unreal.

The study performed during this paper eliminates the mentioned limitation regarding the hypothesis that other factors, beside the one targeted, remain constant, especially that in case of automotives equipped with on-board computer there are pronounced functional interfaces. For quantitative highlighting of parameters influence on engines functioning, there are used the correlation analysis, dispersional analysis, information theory and sensitivity analysis [1; 2; 4; 5; 6; 7].

In *correlation analysis*, for highlighting if functional parameters are independent or not and for establishing the character (linear or non-linear), it is frequently used the *correlation coefficient* ρ, defined by ratio [3]:

$$\rho_{xy} = \frac{R_{xy}(0)}{\sqrt{R_{xx}(0)R_{yy}(0)}} \qquad (1)$$

with the values $\rho \in [-1;1]$, a perfect linear dependency is for $\rho^2=1$. If $\rho=1$, then there is a perfectly direct linear dependency, and if $\rho = -1$ then there is a perfectly indirect linear dependency; if $0 < \rho \leq 1$ there is a directly non-linear dependecy, and if $-1 \leq \rho < 0$ there is an directly non-linear dependency. To conclude, if $\rho =0$, then the two parameters x and y are independent; obviously, if $\rho \neq 0$ parameters are dependent. In addition, in formula (1), at numerator is the intercorrelation function with the origin of discrete time, meaning for $\tau=0$, and under the radical are intercorrelation functions still for $\tau=0$.

For example, in figure 3 are presented the values of correlation coefficients for the 100 experiments of Ford Focus automotive; the factorial sizes (the influence factors) are fuel pressure p_c and recirculated gas γ, and the resulted parameter is engine torque M_e.

Samples values for correlation coefficient between engine torque, fuel pressure
and quantity of residual gas, 100 experimental samples, Ford Focus car
a) Engine torque - fuel pressure

b) Engine torque - quantity of residual gas

Number of the sample

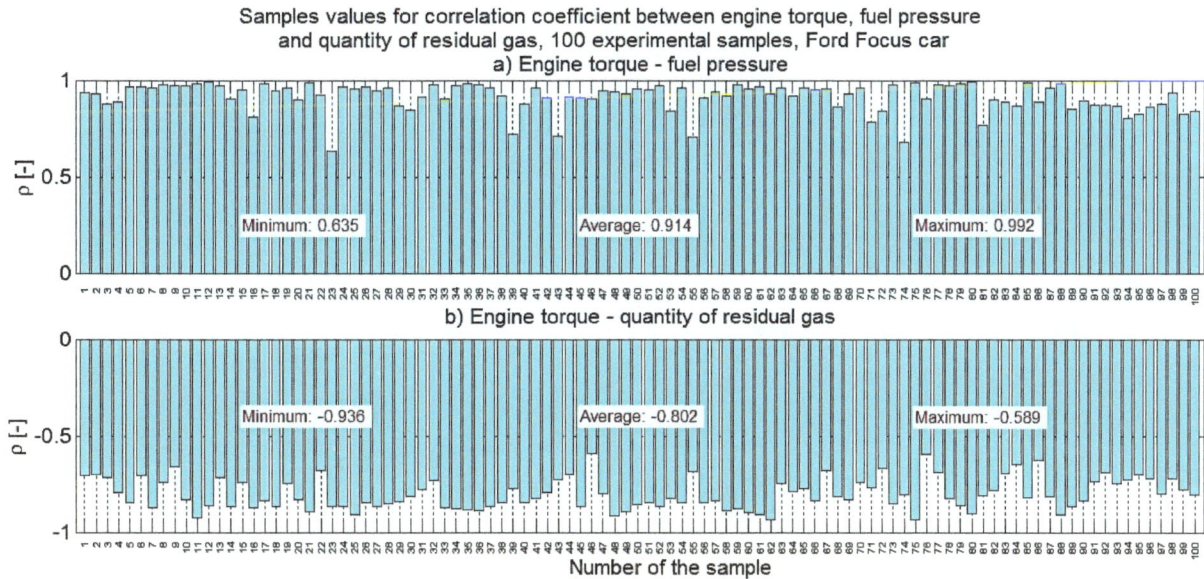

Figure 3: Correlation analysis

As it can be seen from figure 3, for every experimental sample the correlation coefficient is not null so the targeted parameters are dependent, which would be expected from a functional point of view. Also, because the correlation coefficients are subunitary for every test, it results that between these parameters are more or less non-linear dependencies; from the two, the most pronounced non-linear dependency is between engine torque and the quantity of residual gas, the correlation coefficients having smaller values, including the average for all the tests $\rho = -0.802$ as against to $\rho = 0.914$ in the first case. The existence of non-linear dependencies leads to the conclusion that engine functioning must be described by using non-linear mathematical models for a higher accuracy.

In addition, the graph from figure 3a shows that between engie torque and fuel pressure there is a direct dependency, the correlation coefficients have positive values; so, when the fuel pressure increases, engine torque increases and backwards. But the graph from figure 3b shows that between the engine torque and the quantity of residual gas there is an indirect dependency, the correlation coefficients having negative values; so, when the quantity of residual gas increases, the engine torque decreases and backwards.

Informational analysis of engine functioning, based on experimental data, allows establishing relevant parameters, so those factors with the highest influence on the targeted parameters. Informational analysis of engine functioning is based on two main concepts from information theory: entropy and information [1; 5; 7]. So, *the entropy* represents the product of probability p and information I for the entire events $x_i \in X$:

$$H(X) = \sum_{i=1}^{n} p(x_i)I(x_i) \quad \Rightarrow \quad H(X) = -\sum_{i=1}^{n} p(x_i) \log_2 p(x_i) \tag{2}$$

and *mutual information* is established with a formula that also contains conditional entropy:

$$I(X;Y) = H(X) - H(X|Y) \tag{3}$$

Mutual information represents a basic concept for the study of systems dynamics (engine functioning) and *represents a measure of interdependency* between variables. Because of this, the variables with the highest influence are defined by the highest mutual information; these are called relevant variables, attached to the concept of relevance. From mentioned reasons, it is considered that information theory represents an extension of classical correlation, and the mutual information represents a measure of relevance.

For example, in figure 4 is presented the graph with the results of informational analysis when the engine torque is the resulting parameter (placed in the upper part) and by taking into account the 6 factorial sizes (influence factors): throttle shutters' position, engine speed, fuel pressure, intake air pressure, quantity of recirculated gas and relative angle tilt of turbocharger's vanes.

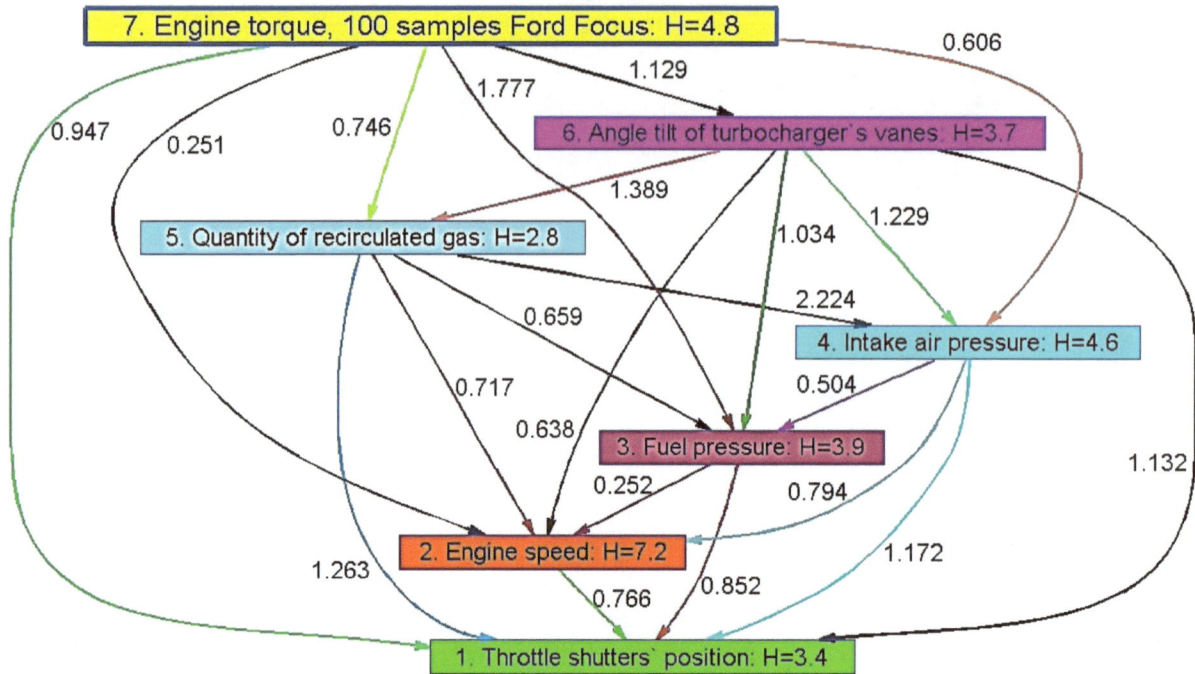

Figure 4: Informational analysis

In graph's assembly from figure 4 are shown the values of H entropy. On the graph's bend are written down the values of mutual information between two I parameters. It can be seen that the first two relevant variables (with the highest inflence on engine torque) are fuel pressure (mutual information with the engine torque of 1.777 bits) and tilt angle of vanes (I=1.129 bits).

In figure 4 are also shown the mutual informations between the 6 factorial sizes; it is determined that the biggest mutual information is between intake air pressure and quantity of recirculated gas, of 2.224 bits, which exceeds the ones mentioned above; this aspect confirms the necessity that the study of engine functioning should also target the mutual influence between different factors, not only between them and the targeted parameter.

Sensitivity analysis allows highlighting the attribute of a resulting parameter to modify it's value because of a factorial sizes. Similar to correlation analysis, if there is a single factorial sizes, then is targeted the simple sensitivity, otherwise it is a multiple sensitivity; in the first case is defined local sensitivity (the classical one, which uses the sensitivity function), in the second case is defined the global sensitivity, which is measured by Sobol [4]. In this second case, Sobol coefficient (noted S) represents the proportion between parameter's dispersion and total dispersion of resulting parameter; in conclusion, there is the ratio:

$$\sum_i S_i + \sum_i \sum_{j>i} S_{ij} + \sum_i \sum_{j>i} \sum_{k>j} S_{ijk} + \dots = 1 \tag{4}$$

where for the influence factor i there is the Sobol first order coefficient S_i (or main Sobol coefficient), for the interaction between factors i and j there is the Sobol second order coefficient S_{ij} etc. As it can be seen, global sensitivity also considers interactions between targeted factors, similar to informational analysis.

For example, in figure 5 is presented the Sobol first order coefficient for engine torque M_e, engine power P_e, cyclic flow of fuel c_c and hourly fuel consumption C_h as resulting parameters; the influence factors are throttle shutters' position p, engine speed n, fuel pressure p_c, intake air pressure p_a, quantity of recirculated gas γ and relative tilt of turbocharger's vanes α.

As it can be seen, for example, for figure 5a and figure 5c, engine torque and cyclic flow of fuel are the most sensitive to fuel pressure variation (S=0.837 and S=0.864).

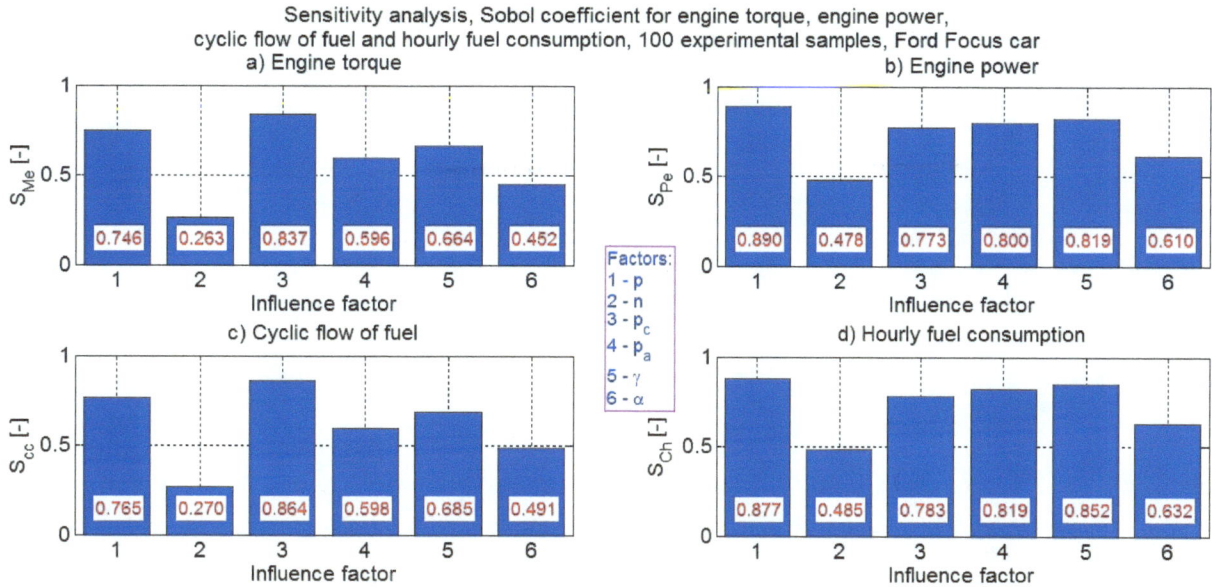

Figure 5: Sensitivity analysis

The study of different factors influence also uses *variance analysis* (ANOVA – **ANalyse Of Variance**, MANOVA – **Multivariate ANalyse Of VAriance**); dispersion has a significant relevance in the analysis of different factors influence on the progress of a dynamic process, here is engine functioning [2; 6].

The english statistician and mathematician Ronald Fisher, founder of variance analysis, proved that by estimating dispersion of a characteristic, influenced by a factor, then by eliminating the influence and comparing the two dispersions, there are obtained quantitative informations about this influence. In consequence, variance analysis means comparing two types of dispersion, factorial and residual. If factorial dispersion is bigger than residual dispersion, then the current factor has a noticeably influence on the targeted process. Backwards, if factorial dispersion (individual or interaction with another factor) is smaller than the residual one, then the factor has a sloppy influence on the targeted process. Practically, this comparison can be made by establishing percentage contribution for every factor and the residual for total dispersion.

For example, in figure 6 are presented the results after applying generalized MANOVA algorithm (are being considered the targeted factors and interactions between them), through the study of influence on cyclic flow of fuel and hourly fuel consumption for 4 factors, including the interactions between them.

From figure 6 it can be determined that in both cases, the residual dispersion values (with 1.7% and 0.74%) are smaller than dispersions values associated with fuel pressure (66.8% and 50.9%), tilt angle of the vanes (18.5% and 19.5%), quantities of residual gas (9.49% and 27.4%) and interactions between throttle shutter's position and quantity of residual gas (2.18% and 0.96%).

So, in both cases, fuel pressure has the biggest influence on cyclic fuel variation and hourly fuel consumption.

Generalized MANOVA algorithm - contribution of different factorial sizes to total dispersion, cyclic flow of fuel and hourly fuel consumption, 100 experimental samples, Ford Focus car

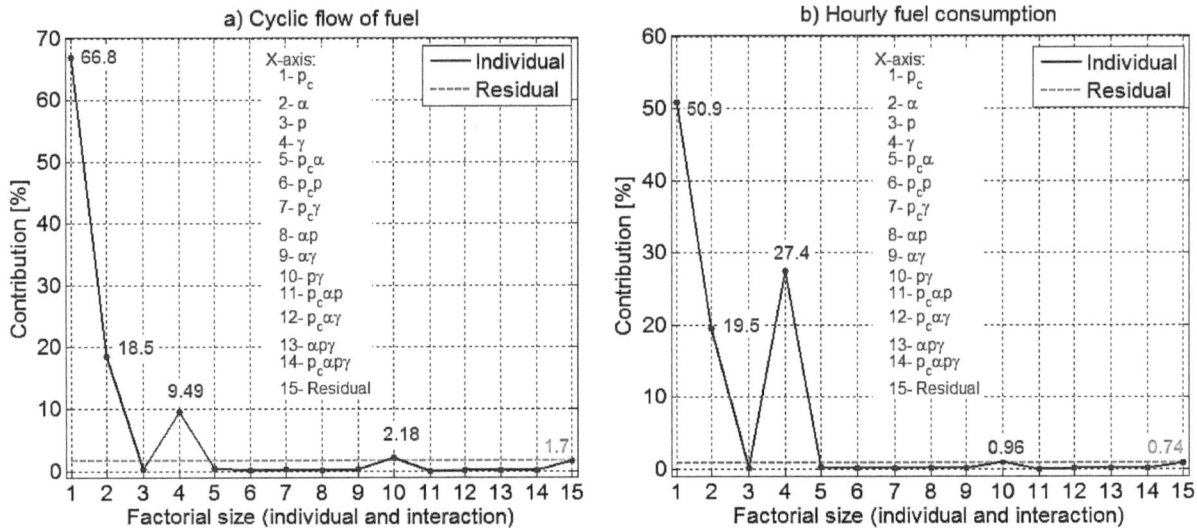

Figure 6: Variance analysis

4. CONCLUSIONS

The study of automotive engine functioning based on experimental data allows highlighting different features if there are used methods and algoritms specific to systems dynamics.

From the above results that the study of different factors influence on engine functioning must also take into account the interactions between factors, as well as the fact that factorial sizes vary simultaneously during functioning, these two represents main differencies towards classical study from specialty literature.

REFERENCES

[1] Brissaud J., The meanings of entropy. Faculty of sciences, Rabat, Morocco, 2005.

[2] Carey G., Multivariate Analysis of Variance (MANOVA). Colorado State University, 1998.

[3] Copae I., Lespezeanu I., Cazacu C., Dinamica autovehiculelor, Editura ERICOM, Bucureşti, 2006.

[4] Glenn G., Isaacs K., Estimating Sobol sensitivity indices using correlations, Environmental Modelling&Software, 37 (2012), pp. 157-166, 2012.

[5] Gray R., Entropy and information theory, Stanford University, New York, 2007.

[6] Langsrud Ø., MANOVA for Matlab, Division of Statistics, Departament of Mathematics, University of Oslo, Norway, 2003.

[7] Schneider T., Information theory primer. University of Colorado, 2007.

[8] ***Technical note of Ford Focus car, with engine 1.6 TDCi, 66 kW.

[9] *** Diagnostics software FoCOM, http://www.obdtester.com/focom.

6th International Conference
"Computational Mechanics and Virtual Engineering"
COMEC 2015
15-16 October 2015, Braşov, Romania

POSTERIOR CERVICAL PLATES - DESIGNED AND FABRICATION

Cornel Drugă

Transilvania University, Brasov, ROMANIA, druga@unitbv.ro

Abstract: *This paper presents two implants spine (cervical area) made in 2015 in the Faculty of Product Design and Environment (in the Medical Engineering Laboratory). In the first part presents a series of issues related to the anatomy of the spine. The following are the most important diseases of the spine. The last part presents a few examples of posterior cervical plates from literature but the two models made in the Laboratory of Medical Engineering, from their design and production deal with them.*
Keywords: *spine, cervical vertebrae, plates, posterior.*

1. INTRODUCTION

The spine consists of discrete bony elements (vertebrae) joined by passive ligamentous restraints, kept separated by intervertebral discs and articulating joints, and dynamically controlled by muscular activation. The spine is broadly divided into 5 regions: the cervical spine, the thoracic spine, the lumbar spine, the sacrum, and the coccyx (Figure 1). Each has it's own unique set of kinematic functions, pathologies, and treatments. In fact, the cervical, thoracic, and lumbar regions are further divided based on kinematic and clinical considerations [1].

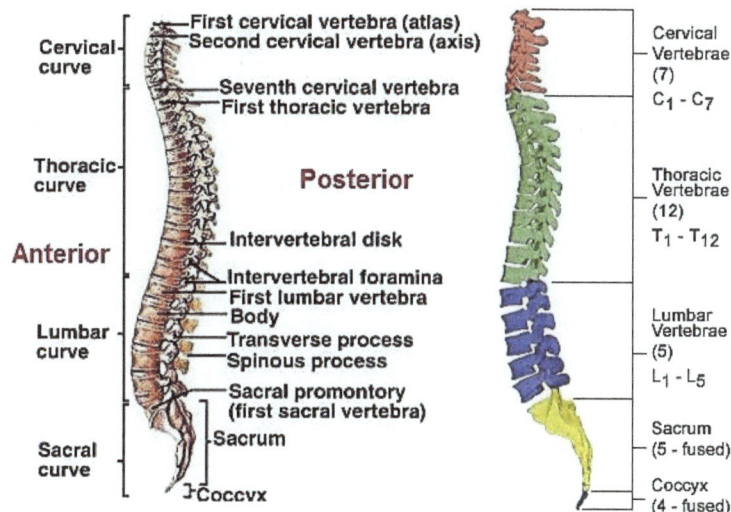

Figure 1: Anatomy of the spine. Spinal Column with vertebrae. [2]

263

The cervical spine (C-spine) consists of 7 vertebrae (C1-C7) in all mammals and the base of the skull, the occiput (C0) and is divided into upper (C0-C2), middle (C2-C5), and lower (C5-T1) regions. A natural, slight lordotic curve exists in the C-spine, meaning that the middle lies farther anterior than the ends. The thoracic spine (T-spine) consists of 12 vertebrae (T1-T12) and is divided into upper (T1-T4), middle (T4T8), and lower (T8-L1) regions [1]. The ribs attach to the thoracic vertebrae. A natural, slight kyphotic curve exists in the T-spine, meaning that the middle lies farther posterior than the ends. The lumbar spine (L-spine) consists of 5 vertebrae (L1-L5) in humans (some mammals, e.g., goats have 6 lumbar vertebrae). The sacrum and coccyx consists of 5 fused vertebrae each (for the coccyx, 4 or 5) - Figure 1.

2. CERVICAL SPINE DISORDERS

2.1. Degenerative Cervical Spine Disorders

Intervertebral disc herniation is also known as herniated nucleus pulposus (HNP). The intervertebral discs make up approximately one-fourth of the cervical spine's height. Over time the water content within the nucleus pulposus of the disc decreases from approximately 90% at birth to 70% by age 70 [3]. The diminished water content, along with changes due to the effects of proteoglycan, collagen, keratin sulfate, and chondroitin sulfate, results in degeneration [3]. As the degenerative process continues, the nucleus pulposus cannot generate the intradiscal force required to keep the annulus fibrosus expanded. In turn, the annulus is subjected to excessive compressive and shear forces, causing weakening and tears in its layers. The weakness puts the annulus at risk of nucleus pulposus bulging, protrusion, or herniation [3].

Spondylosis is generally defined as age- and use-related degenerative changes of the spine. This diagnosis includes degenerative disc disease and the progressive changes that occur as a result of disc degeneration, such as osteophyte formation, ligamentous hypertrophy, and facet hypertrophy (Figure 18). As the degenerative cascade continues, changes in normal spinal curvature occur [3].

Cervical spondylotic myelopathy is defined as "*spinal cord dysfunction accompanying typical age-related degeneration of the cervical spine*". Spondylosis is the most common etiology, and spondylotic myelopathy is the most common cause of spinal cord dysfunction in persons older than 55 years [3]. However, cervical spondylosis is commonplace in the aging spine, and most patients will not develop myelopathy. Radiographically, cervical spondylotic myelopathy is considered when the central canal is less than or equal to 13 mm (normal = 17 mm) or when patients have greater than or equal to 30% narrowing of the cross-sectional area of the canal with associated symptoms [3].

Cervical stenosis, classified as either congenital or acquired, is a result of either being born with a narrow spinal canal or developing a narrow spinal canal as a result of degenerative changes. Ossification of the posterior longitudinal ligament (OPLL) is a specific condition that causes cervical spinal stenosis [3]. OPLL is characterized by calcification and thickening of the PLL. This results in narrowing of the spinal canal and potential spinal cord compression as well as increased spine rigidity. With any cause of stenosis, the degree of spinal canal narrowing determines the significance of the clinical implications [3]. If spinal cord compression is evident, the patient will be counseled on operative management options, alternatives to surgery, and the risks involved with both operative and nonoperative management. Stenosis may exist throughout the cervical spine or may be limited to a few adjacent segments. In severe spinal cord compression, even with no neurologic deficit, there is a potential for catastrophic neurologic impairment [3].

Rheumatoid arthritis is a "chronic systemic autoimmune disease characterized by erosive synovitis that infiltrates and destroys multiple joints in the body" [3]. Synovitis, an acute inflammatory response,
is a result of antibody-antigen complex formation. Eventually, this can lead to complete destruction of the joint. The acute process is followed by a chronic granulomatous process of pannus formation. This produces collagenase and other enzymes that destroy surrounding cartilage and bone. The cervical spine is at risk
because the atlantooccipital (occiput and C1) and atlantoaxial (C1 and C2) articulations are purely synovial.

Ankylosing spondylitis, a seronegative spondyloarthropathy associated with the human leukocyte antigen B27, causes inflammation in the synovial joints, beginning in the sacroiliac joints [3]. As the disease progresses, ossification and ankyloses occurs in an ascending manner. The patient eventually develops a rigid, brittle, and immobile spine. This leaves the individual very susceptible to deformity (loss of normal spinal curvature) and fractures.

2.2. Neoplastic Cervical Spine Disease

More than 95% of the clinically significant spinal column tumors are metastases, and 60% of those are from cancers of the breast, lung, and prostate; myelomas; or lymphomas [3]. Approximately 8%–20% of spine metastases are in the cervical spine. In addition, 11%–17% of breast cancer patients will suffer metastases to the cervical spine; the percentage increases to 40% in patients with advanced disease [3]. Cancers of the lung, of the prostate, renal, and thyroid glands, as well as gastrointestinal and gynecologic cancers, and melanoma, in descending order of frequency, commonly metastasize to the cervical spine. Spinal involvement of metastatic cancer can lead to vertebral collapse and instability, causing pain and potential neurologic compromise. Nerve root or spinal cord compression also can be caused by the infiltration of the tumor mass, resulting in neural element compression. Although surgical intervention may not cure these patients, it may be indicated to treat tumor-induced neurologic compromise or fracture. Surgical intervention is aimed at stabilizing the spine and optimizing neurologic function.

2.3. Deformity of the Cervical Spine

Deformities develop from either anterior or posterior vertebral element disruption. This can be caused by a number of conditions, such as congenital anomalies, surgery, osteoporosis, tumor, or inflammatory or degenerative processes [3]. The underlying pathology and biomechanical imbalances it creates will determine the extent and significance of the deformity. The most common cervical spine deformity is *kyphosis*. As the deformity forms, the head is shifted forward, which increases compression on the anterior vertebral bodies. The
posterior neck muscles become less effective at holding up the head. As the cycle continues, kyphosis, unfortunately, worsens over time. Common signs and symptoms are neck pain, muscle fatigue, radiculopathy, myelopathy, potentially poor posture because of looking down, and poor nutritional status because of the patient's inability to look up.
Osteoporosis, the most common metabolic bone disease, is characterized by low bone mass and structural deterioration of bone tissue. These events occur when bone resorption happens too quickly or replacement occurs too slowly. Structural deterioration leads to increased susceptibility to fractures, which are
related to increased bone fragility most often seen in the hip, spine, or wrist. A current definition of osteoporosis is based on gradations of low bone mass; that is, bone mineral density (BMD) is more than 2.5 standard deviations below the peak BMD of gender- and ethnicity-matched healthy, 30-year-old Caucasian women [3]. Certain risk factors are linked to or contribute to the likelihood of an individual developing osteoporosis. Some
of these factors are genetically determined and others are related to lifestyle. Increasing age plays a significant factor as the resorption of bone surpasses its formation, putting both sexes at increased risk [3]. Persons may not be aware that they have developed osteoporosis because bone loss occurs without symptoms. The first sign may be pain, spinal deformity, loss of height, or fracture.

3. POSTERIOR CERVICAL INSTRUMENTATION

3.1 Some examples of posterior cervical plates

Instrumentation used in interventions on the cervical spine has evolved alike for both types of approaches, the posterior and anterior of the basic method to strengthen interspinous processes with wires. Posterior stabilization techniques of the atlanto-axial complex historically have made use of various wiring techniques.
In 1939, Gallie described a relatively simple technique for controlling situations in which anterior instability of the atlanto-axial complex exists. The only requirement for this technique is that integrity of the C2 arch is maintained [4].
Using wires in posterior cervical column was first described by Rogers in 1942 to treat fractures and dislocations of the cervical spine. This operation involved connecting multiple law suits with wires and incorporation of bone grafts for intensifying merger. In 1959 Forsyth spoke posterior internal fixation technique cervical spine using 20 stainless steel wires to splice spinous processes [5].
In 1980 Dr. Roy-Camille was pioneer posterior cervical stabilization using lateral mass screws to the vertebrae. His technique of stabilizing the fusion induced spontaneous deck facets and therefore required no additional grafting [5].

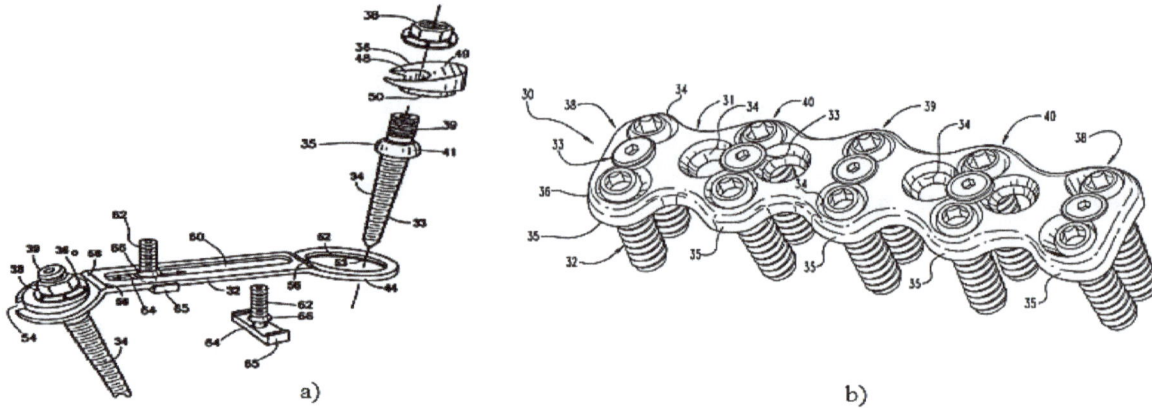

Figure 2: a) The system developed by dr. Roy-Camille; b) AME Haid Universal Bone Plate System. [5]

In 1989 Dr. Haid has developed a posterior cervical plate (AME Haid Universal Bone Plate System) made of titanium alloy to reduce artifacts in CT scans. It had a slightly concave cross-section to accommodate anatomical shapes of the articular processes (Figure 2). One of the advantages lateral mass plate fixation is independent of the integrity of the posterior elements and the fact that this positioning by normal cervical lordosis store. Fusion success rate was 98% [5].

Using anterior cervical plates provided by fusion stabilization capacity increase selective cervical segments, eliminating the need for external immobilization. A large number of anterior cervical plates have been developed and used clinically for over the years.

In 2002, Haid proposed a nomenclature for describing and labeling of anterior cervical plates (ACPs) based on biomechanical properties and load capacity of these systems graft. In order to reduce the normal stress (*stress shielding*) against graft, but also preventing unscrewing, semi-constrained plates were created. The objective was to develop nameplates takes only a percentage of uniformly distributed load and graft difference loading [5]. A system that was built for this purpose was DOC Rod (DePuy Spine, Rayham, MA). It allows translations vertical sliding bolt controlled by the over bars, a phenomenon that leads to compaction graft (Figure 3.a). This controlled slide by rigid rods implanted maintain the alignment of the anatomical structure [5]. Another example of a dynamic system is developed by Aesculap ABC plate, and presents slots and variable angle screws to reduce positioning of stress shielding. This allows the movement of translation and rotation of the screws in the screw-plate interface.

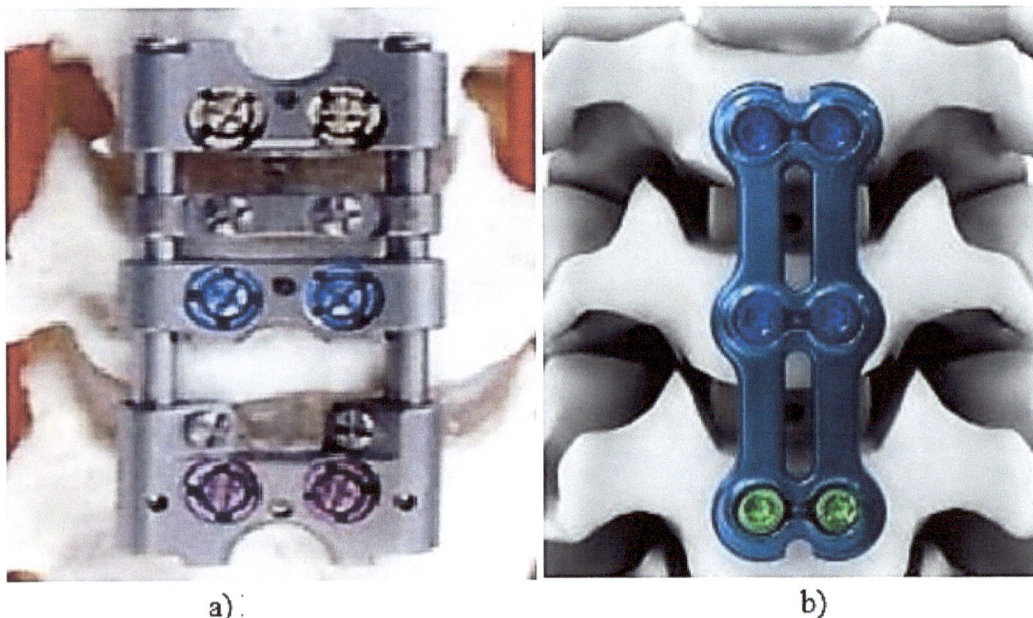

Figure 3: a) DOC Rod System; b) Solstice NEO-SL Plate. [5]

EO-SL System of anterior cervical plate is a slim and easy to use. It has an integrated locking mechanism that allows bone screws to be securely fixed without the need for additional locking components. Its key features include plates of different lengths for procedures from level 1 to level 5. Bone screws are available that are designed for quick twist configurations in both variable and fixed angles. Moreover, the system has a locking mechanism integrated in a single step.

3.2 New models designed and fabrication in the Medical Engineering Laboratory

In the Laboratory of Medical Engineering (from Faculty of Product Design and Environment) from 2010 until now, they have been designed and produced a series of implants for hip joints, knee, shoulder and spine. In this paper are present two types of posterior cervical plates. As with Rod DOC system it allows translations vertical sliding bolt controlled by the over bars, a phenomenon that leads to compaction graft (Figure 4). This controlled slide by rigid rods implanted maintain the alignment of the cervical vertebrae [5]. The systems (posterior cervical plates) were designed using CATIA V5R19 and SOLID WORKS 2015 software. Data on the overall dimensions and aspects of strength of materials found in reference [5]. The second system (PCP-02-F) is a bit like the Solstice NEO-SL and is shown in Figure 5. The systems were produced from 316L steel in first step. Technological process to achieve systems are detailed in reference [5].

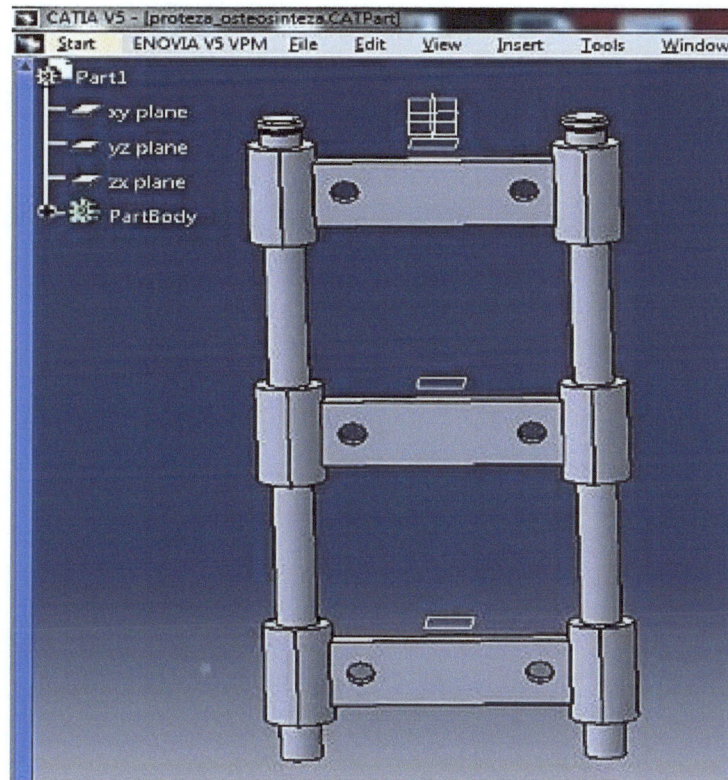

Figure 4: The first system (PCP-01-M) [5].

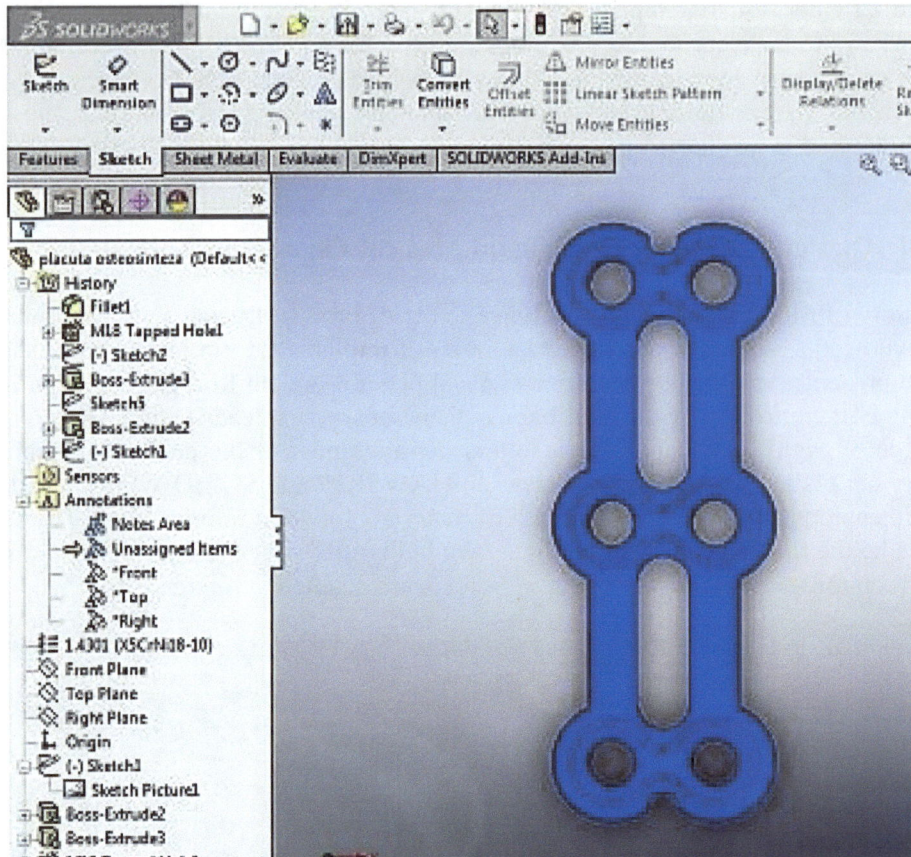

Figure 5: The second system (PCP-01-M) [5].

If the development of the first system were used for classical technologies (such as cutting, drilling, turning and welding) for the second preferred system to use nonconventional technologies (cutting water jet machine) - Figure 6. Cutting Water Jet Machines is at Research and Development Institute of *Transilvania* University (ICDT) in L3 Laboratory.

Figure 6: Cutting Water Jet Machine (*ICDT*, Laboratory –L3).

Figure 7: PCP-02-F and PCP-01-M Systems.

4. CONCLUSION

PCP-02-F device was carried out on a CNC milling of the same material (316L steel) but the time for achievement was triple. The devices were subjected to a Finite Element Analysis (FEA) using CATIA V5R19 software (only in static) to observe the behavior of the material at peak loads that occur in the cervical area; the results were good.

REFERENCES

[1] Mer/Bio Soft Tissue Mechanics., Spine Biomechanics, pp.SB1-5, available at: https://www.google.ro/?gws_rd=cr,ssl&ei=0O0LVu6YDoayswHo16vABQ#q=36-Reading+Rapoff.pdf
[2] Anatomy of the Spine, available at: https://www.ivanchengmd.com%2Fanatomy-of-the-spine.php
[3] Cervical Spine Surgery: A Guide to Preoperative and Postoperative Patient Care, AANN Reference Series for Clinical Practice, pp.12-16, available at: www.aann.org.
[4] Kramer D.L.,Vaccaro A.R.,Albert T.J., Posterior Cervical Instrumentation, Seminars in Spine Surgery, Vo19, No 3 (September), pp 240-249, 1997.
[5] Grigore S., Spinal vertebral prosthesis, paper license, Transilvania University of Brasov, pp.65-67, 2015.

6th International Conference
"Computational Mechanics and Virtual Engineering "
COMEC 2015
15 – 16 October 2015, Brasov, Romania

COMPARISON STUDY USING IMAGISTICAL TOOLS BEFORE AND AFTER SURGICAL INTERVENTION

Ionel Şerban

Transylvania University, Brasov, Romania, serban_ionel1984@yahoo.com

Abstract: The purpose of this paper is to demonstrate that using a tomographic image processing program can emphasize its health improvements after appropriate treatment (medical or surgical). This study regards the realization of an abstract model of a knee joint disease, synovitis. For this there was used a set of tomography images from an anonymous patient. After the realization of the model it was done some measurements to highlight the results obtained after the surgery compared to ones before surgery. The software that was used is MIMICS and the set of tomographic images are from a sitting knee before and after surgery.
Keywords: MIMICS, synovitis, CT

1. INTRODUCTION

Since the dawn of modern medicine, doctors were concerned about studying in detail the human body in terms of composition and functionality, as well as the links between systems and subsystems that compose it, to intervene specifically for each condition in part.

The first method of investigation was the dissection of corpses, method condemned by the church.

The crucial moment in the field of non-invasive medical investigations is the discovery of X-rays this time has developed a wide range of investigations based on the same principle.

Currently it is very important that the images acquired using the apparatus of medical imaging can be processed in such a way that the result is easily understood primarily by the doctor (to establish diagnosis and treatment) and from the point of view of the patient, the relationship between doctor and patient is an important aspect of modern medicine.

The role of medical imaging processing software is to: help the doctors in better understanding of the human anatomy and the pathology [1, 2, 3]; help in developing new technics and optimization of surgery; help in prosthetic prototyping; help in case of a high risk surgery it allows to the surgeons to exercise for better understanding of the organ anatomy before the surgery starts

MIMICS is a software specialized to generate anatomical models [4] and analysis of CT (Computer Tomography) images. It allows to import different types of image formats like: DICOM, TIFF, JPEG, and BMP.

Synovitis is an autoimmune disease that causes an inflammation of the synovial membrane. Synovitis has several causes such as arthritis or rheumatic resulting from trauma.

Diagnostics condition of synovitis, in addition to clinical presentation (hot joint, reddened and swollen), consist of the analysis of synovial fluid.

Synovitis is treated with anti-inflammatory drugs. Another treatment is surgical, as in this case of patients suffering from left knee.

2. EXPERIMENTAL METHOD

Mimics (Medical Image Segmentation for Engineering on Anatomy) is a software developed by Materialise for processing medical images. This program is used for 3D medical image segmentation from CT, MRI (Magnetic Resonance Imaging), micro -CT, etc. The software has its applications in different medical areas like: orthopedics, cranium - maxillofacial, cardiovascular, etc.

After the 3D modeling, the virtual model can be printed using a 3D printer offering the possibility to obtain the physical model.

In what follows we used tomographic images for viewing results before and after an operation on left knee, suffering from synovitis.

To carry out the practical test we followed certain steps, knee scans both before the operation and after the knee surgery to finally compare them.

The steps are:

1. Importing images. Importing tomographic imaging is performed in software "Mimics "by going to "File" menu and selecting "Import Images".

2. Select tomography directions and area of interest. These directions are selected based on anatomy. After their selection we will see the three dimensional planes (sagittal, transverse and front (figure1, figure 2)) of the knee.

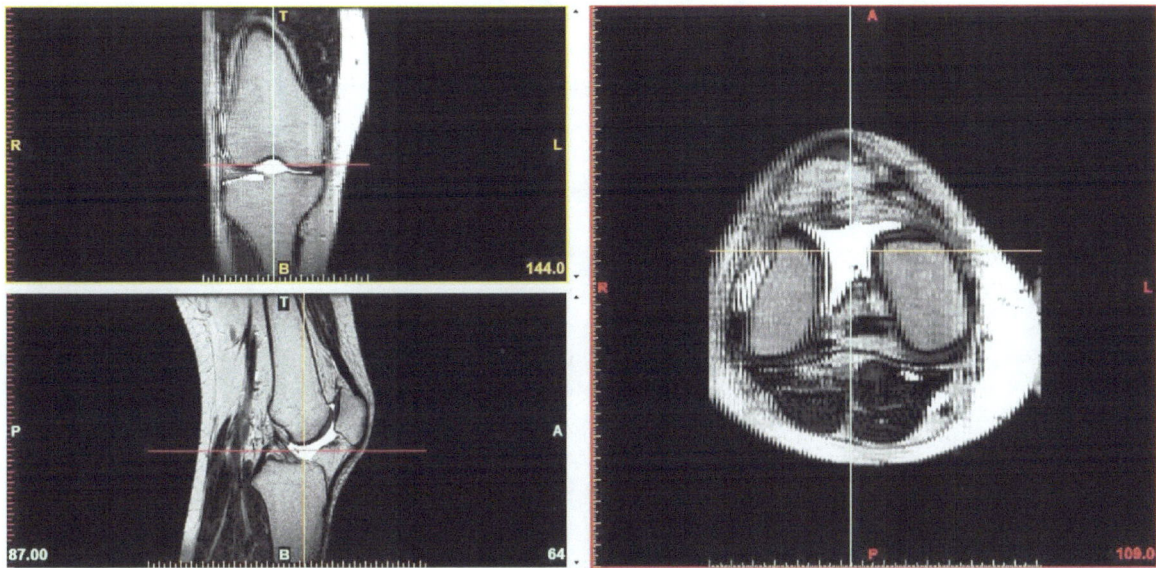

Figure 1: Before surgery. The area of interest left knee.

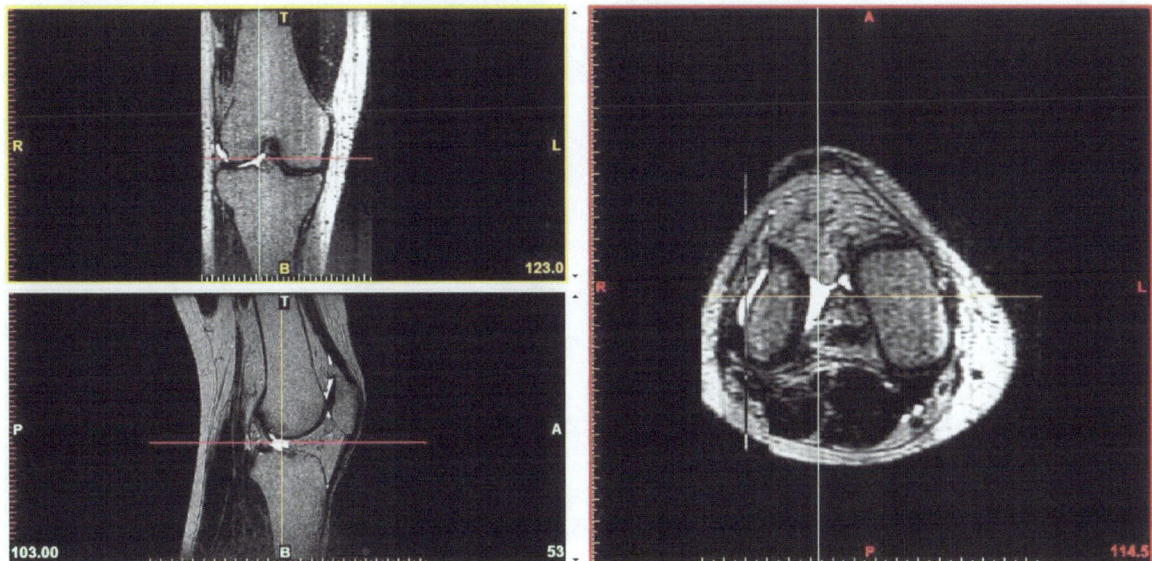

Figure 2: After surgery. The area of interest left knee.

271

3. "Thresholding". Following the selection tool select the area of interest "thresholding" from the "Segmentation". This tool is used for classification of pixels which is a target value measured in Hounsfield. By selecting these pixels it results a "mask" (figure 3, figure 4) containing all the pixels with the same Hounsfield value selected (2207-3056 HU).

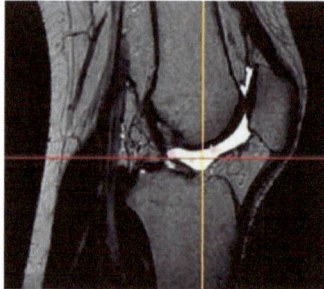

Figure 3: Before surgery. Sagittal plane with green "mask" (selecting the area of interest).

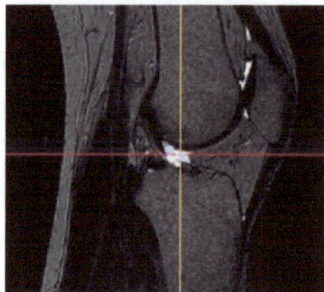

Figure 4: After surgery. Sagittal plane with green "mask" (selecting the area of interest).

4. Create 3D model. With "Region Growing" (figure 5, figure 6) and "Calculate 3D" (figure 7, figure 8) successively tool the software generates the necessary model of the previously selected area of interest from the CT images of the knee.

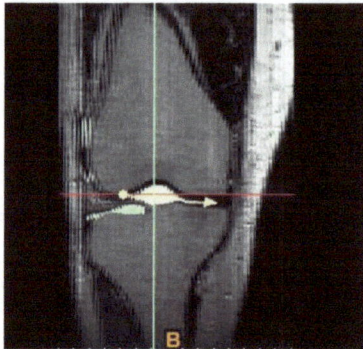

Figure 5: Before surgery. The result (yellow mask) after using the command "Region Growing" in the frontal plane.

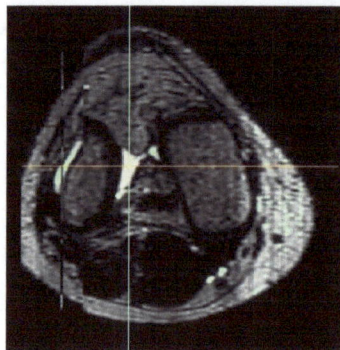

Figure 6: After surgery. The result (yellow mask) after using the command "Region Growing" in the transversal plane.

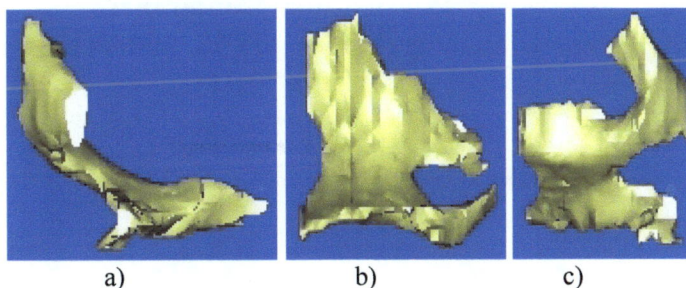

a)   b)   c)

Figure 7: Before surgery. Different perspectives of the 3D model: (a) – sagittal plane left view; (b) – transversal plane above view; (c) - Front plane front view.

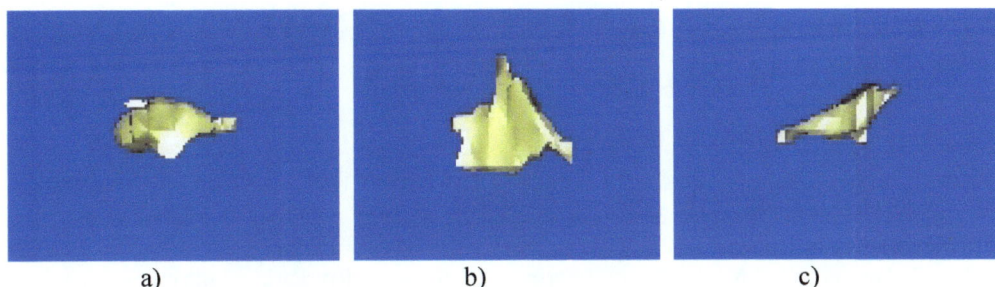

a)   b)   c)

Figure 8: After surgery. Different perspectives of the 3D model: (a) – sagittal plane left view; (b) – transversal plane above view; (c) - Front plane front view.

5. Measure the 3D model. Measuring 3D model is necessary because in the end we will compare results of scans before and after the operation. The tool used is: "3D Measure Distance".
6. Comparing the morphology of the patient's knee.

3. CONCLUSION

According to the measurements performed on the model it was found a reduction of 65% of the synovial liquid. Besides this there is a visual difference between the two situations suggesting that the surgery had a great impact on the outcome of the initial diagnosis.

Tomographic imaging program has fulfilled the main function of which was to highlight the surgical removal of inflamed synovial fluid.

CT images in imaging software are increasingly used by the doctors. Creating databases structured on organs and diseases would allow, in the future, doctors to make diagnosis more quickly based on comparing patient stored images. This comparison can be done using computer models created and based on the images in the database.

In conclusion tomographic imaging software have an important role in every stage: diagnosis, treatment and control. The diagnostic accuracy of the image allows the doctor to take the best decisions for each case.

4. ACKNOWLEDGMENT

We like to thank student Meşco Lucia for all the help during the course of this research.

REFERENCES

[1] Willi A. Kalender (2005) Computed Tomography. Ed. Corporate Publishing;
[2] Cornelius T. Leondes (2005) Medical Imaging Systems Technology; Methods in General Anatomy. Ed. World Scientific;
[3] Otto H. Wegener (1992) Whole Body Computed Tomography. Ed. Blackwell Scientific Publications;
[4] Anderson, N. (2007) *Empirical direction in design and analysis*. Mahwah, NJ: Erlbaum.

6th International Conference
"Computational Mechanics and Virtual Engineering"
COMEC 2015
15 – 16 October 2015, Brasov, Romania

A STATISTICALLY RELEVANT EXPERIMENT CONCERNING THE COLLOIDAL SILVER INFLUENCE ON HUMAN BODY

Ioan Curta[1], Ileana-Constanta Roşca [2], Z. Marosy [3], A.C. Micu[4], I. Mohirta[5]

[1-2]Transylvania University, Brasov, Romania, icurta@yahoo.com; roscaileana@yahoo.com
[3]Ecologic University of Bucharest, Romania, marosy.zoltan@gmail.com
[4] Carol Davila University of Medicine and Pharmacy, Bucharest, Romania, cmidrmicualexandru@yahoo.com
[5]Romanian Association of Transpersonal Psychology, Bucharest, Romania, mohirta@tanatopsihologie.ro

Abstract: *The experiments presented in this article are meant to determine certain changes in the characteristic parameters of the bio-field that can be demonstrated through measurements with the GDV Camera device (gas discharge visualisation). The study made on a statistically significant lot of 32 persons presupposes the ingestion of a small amount of electro-colloidal silver solution (ECS), followed by repeated measurements made with the GDV camera and the statistical processing of the results. The method and the measuring intervals were previously established through two pilot studies made on groups of 3 persons. The statistical method used was the test-retest method. Statistical results obtained from this experiment have shown that the ECS solution is a lot more than a mere solution with antiseptic effect and they are opening new research paths towards finding the manner in which the body reacts to the ingestion of different substances.*
Keywords: *GDV camera, electro-colloidal silver (ECS), bio-energy field, Chakras*

1. INTRODUCTION

For the past 20 years, alongside the increased interest for the study of the properties of nanomaterials, numerous studies have appeared of silver, nanoAg (AgNPs), and fine silver dispersed in colloidal solutions in which the size of the silver particles ranges between 0.5 and 100 nm. So that, over 22,000 references can be found on the internet for the medical application of Ag and over 160,000 references about its bactericide effect. In this respect, we believe it is useful to remind of studies conducted for the last two years to demonstrate that colloidal silver amplifies the action of different antibiotics [6]. Other several studies demonstrate the proportionality of ECS effect to its concentration and inverse proportionality to the particle size. Their authors, so far, focused the study of ECS exclusively on its bactericide, antifungal and virucidal properties [7, 8, 9], neglecting other properties such as the anti-depressive/calming effect upon the nervous system [10], the pain killing effect [5], and the accelerated wound curing effect (initiated by Dr. Becker [3]). This study is an attempt of remedy of this gap.

2. EXPERIMENTAL METHOD AND DEVICE

The main scope of the research can be synthesized like follows:
observation and interpretation of the changes occurring in the human energy informational matrix under the influence of ingested colloidal silver solution, by measuring the changing of the parameters that may be highlighted by the GDV camera, followed by statistical interpretation and analysis;
rigorous determining if ECS has any (physical and subtle) influence upon the human body, even in the case of extremely low ingested amount, close to a homeopathic dilution (10 ml of 25 ppm concentration was used);

determining the manner of action of ECS by measuring, at 4 different moments of time, the evolution of different human parameters using the GDV camera. The three data series obtained from the measurements were compared with control measurement and the results are then processed and interpreted statistically.

The compact GDV camera and the afferent software manufactured by Kirlionics Technologies International Ltd is a practical device developed by Dr. Korotkov based on the well-known Kirlian effect (Electro - Photonic Capture - EPC or Gas Discharge Visualization - GDV). GDV camera is homologated as a medical device by the Russian federation and the EU, and it is in course of homologation by the US relevant authorities.

The investigation/measurement with the aid of the GDV camera is simple and it consists of the photographing of the photon emissions that the 10 fingers produced through stimulation in high tension electromagnetic field, followed by the computed processing of the photos by some dedicated software that use non-linear fractal analysis. All beings are surrounded by a bio-energy informational field – a type of bio-radiation that has direct contact to the physical body [11]. The energy informational field reacts more quickly to stimuli than the physical body, which is the ideal condition for GDV Camera research [12]. After this processing, an image is created of the distribution of the human energy informational field based on the connexion between different parts of the fingers and different organs and systems of the body through the energy meridians described by Oriental medicine[13, 14, 15]. This idea was first advanced by Dr. Voll and then developed by Dr. Mandel in Germany. Subsequently, the idea was clinically tested, verified and corrected by a Russian team coordinated by Dr. Korotkov. Introducing a polymer foil filter we can make the difference between the psycho-physiological field and the physical field. The higher the level of stress experienced by a person, the bigger the difference between the two images. The filter disables the influence of all the processes that are directly connected to the surface of the skin – the first being the sweat. *"We can assume that the filter distinguishes between the activity of the sympathetic and parasympathetic nervous system. To analyze the psycho-physiological field it is necessary to take the GDV-grams without the filter"* [12, pg. 34]. The GDV-associated software offers information about the general level of homeostasis of the body, of the subsystems and organs, and the stress level, and it may reveal the subjects' reaction to different procedures and treatments.

The parameters that may be analyzed by the GDV camera ([12], p. 273-275):
average area - scale 0-16210 pixels on the PC screen [Joules/cm^2];
frontal FC - (scale 0 to 63.2);
stress Index (T) - (scale from 0.3 to 7.37);
average Js - (normal values -0.6/ +1.0) (the person's health index);
average RMS (root mean square) - standard deviation of intensity, is a statistical measure with positive values;
average entropy - scale (0-4);
value of the Chakras - the average of the whole GDV area of right and left hand;
asymmetry of the chakras - deviation of the positions of the chakras from the vertical axis.

Among *the advantages* of using the GDV is that it is a simple, quick and painless procedure that does not require the sampling of bodily substances, that is clean, septic and that consists of the mere photographing of the 10 fingers.

3. EXPERIMENT

The test/retest was done to verify the hypotheses for pair samples with repeated measurements as the same variable, measured at four different moments on the same lot of subjects will be compared. An Experimental Plan (with a single independent variable) was used on dependent correlated groups. The independent variable is the challenge of certain modifications occurring in the entirety of the energy informational field of the human body after the ingestion of a controlled amount of ECS (10ml/25 ppm), quantified through the modification of the parameters measured with the GDV camera. The data are processed and interpreted through SPSS 16.0.

The electro colloidal silver used in the experiment (ECS) is a liquid dietary supplement with the concentration of 25 ppm. It tastes bitter sweet and it consists of distilled and structured water and high purity (99.99%) nanometric silver particles and Ag+ silver ions. The particle size ranges between 0.5 and 4.5 nm, with an average value of 2.7 nm, and it has been measured with the aid of a TEM microscope.

32 subjects aged between 22 and 74 participated in the experiment: 21 female and 11 male subjects. Their age classification was as follows: 5 subjects were aged between 20 and 30, 14 between 30 and 40, 10 between 40 and 50, 1 between 50 and 60, and 2 subjects were over 60. The subjects were explained the working protocol that consisted of 4 series of measurements for each of the 32 subjects: the first measurement was the control measurement – M_0, made before the ECS ingestion, the second measurement – M_1 – was made within 3-5 minutes after the procedure, the third measurement – M_2 was made 30 minutes after the procedure, and the last measurement – M_3, 90 minutes after the procedure. These images were processed using the GDV Aura, GDV Chakras and GDV Diagram software and 21 parameters resulted for each subject. These parameters were analysed

statistically. The research complied with the requirements specified in the protocol established by Dr. Korotkov [12]. If all the protocol conditions are met, the accuracy of the determining made with the GDV camera will range between 5 and 10%. At the base of the experiment was the hypothesis that after ECS ingestion T (the difference between the intensity of the luminosity of the physical body field and the emotional one) stress level decreases, the average FF area, the Js health index and symmetry will increase, the position of the Chakras (centers of consciousness) will change and, there will be a restructuring of the energy informational field (aura).

The analyzed average values of measured parameters are presented in the tables below.

Table 1: Final average values of the physical parameters (with a filter).

Variables	Sample volume	M_0	M_1	M_2	M_3
Average Area FF	32	12259.84	12446.12	13211.56	12744.22
Average FC	32	35.89	35.4	34.13	34.08
Stress index	32	3.24	3.06	2.71	2.89
Average Js	32	-0.49	-0.48	0.47	-0.44
Average RMS	32	0.52	0.5	0.42	0.46
Average entropy	32	1.94	1.92	1.93	1.97
Symmetry	32	81.18	80.87	81.93	81.04

Table 2: Final average values of the psycho-emotional parameters (without a filter).

Variables	Sample volume	M_0	M_1	M_2	M_3
Average Area FF	32	16167.06	16660.81	17072.68	17479.56
Average FC	32	23.92	23.31	21.85	22.75
Stress index	32	3.24	3.06	2.71	2.96
Average Js	32	-0.15	-0.12	-0.1	-0.1
Average RMS	32	0.34	0.33	-0.11	0.31
Average entropy	32	1.99	2.01	2.01	2.01
Symmetry	32	90.26	90.87	90.93	91.45

Average area *measured with a filter* increased constantly after the procedure, going from 16167.06 during the M_0 control measurement to 17479.56 during M_3, and going through intermediate values such as 16660.81 at M_1 and 17072.68 at M_2. The relative increase measured 90 minutes later (the rate of the curve indicates that the average area might be greater at a subsequent measurement) is 8.1%. The correlation coefficients of the first 2 measurements SET!, M_1 and M_2 indicates some significant correlations r_1=0.82 (p_1=0.00001, r_1^2= 0.67); r_2= 0.4 (p_2=0.020, r_2^2=0.016) and M_3 non-significant statistic r_3= 0.13 (p_3=0.490, r_3^2=0.05), the r_2 determining coefficients indicating a strongly differentiated effect at M_1, an average one at M_2, and a very low or insignificant one at M_3 of the technique over the averages of the measured values, which allows us to presume that the ECS effect was the greatest at the M_1 moment and it started to fade immediately afterwards, even though the average value of the series is always greater than the initial value!

Average area *measured without a filter* has increased constantly after the ingestion, going from 12259.84 at the M_0 control measurement to la 13211.56 at M_2, and then decreasing to the value of 12744.22 at M_3 (in this case a maximum point was reached at M_2, 60 min. after the ingestion). Next, there was a decrease to the value determined during M_3. The correlation coefficients of all the 3 measurements indicate a strong correlation as compared to the control measurement: r_1=0.71 (p_1=0.00001, r_1^2= 0.50); r_2= 0.67 (p_2=0.0001, r_2^2=0.44) and r_3= 0.47 (p_3=0.006, r_3^2=0.22), the r_2 determined coefficients indicating a strong effect of the applied procedure. The correlation analysis allows to find that the effect upon the parasympathetic system starts to diminish at the time of M_3.

Average FC *measured with a filter* dropped from 23.92 at the M_0 control measurement to 21.85 at M_2, passing through the intermediate value of 23.31 at M_1, then slightly increasing to 22.75 at M_3, but without reaching its initial value. The correlation coefficients of all the 3 measurements indicate strong correlation as compared to the control measurement: r_1=0.81 (p_1=0.0001, r_1^2=0.65); r_2=0.62 (p_2=0.0001, r_2^2=0.38) and r_3= 0.55 (p_3=0.001, r_3^2=0.30).

Average FC *measured without a filter* constantly decreased from 35.89 at the M_0 control measurement to 34.08 at M_3, passing through the intermediate values of 35.4 at M_1 and 34.13 at M_2. The correlation coefficients of all the 3 measurements indicate strong correlation as compared to the control measurement: r_1=0.76 (p_1=0.0001, r_1^2= 0.57); r_2= 0.69 (p_2=0.0001, r_2^2=0.47) and r_3= 0.57 (p_3=0.001, r_3^2=0.32).

In both types of measurements, with and without a filter, the r_2 determined coefficients indicate a strong effect of the procedure upon the averages of the measured values, which allows us to claim that the measured differences of the values of this parameter may certainly be accounted for by the effect of ECS.

The **Stress index** had a most special evolution, as there was a significant and constant drop from an average of 3.23 at the M_0 control measurement, to 2.92 at M_3, passing through intermediate values of 2.87 at M_1 and 2.71 at M_2. All the values are in the normality interval (Korotkov 2002, page 271-275 [11]). The correlation coefficients resulted from the statistical analysis of 3 sets of data compared to the initial measurements (Test-ReTest$_1$, Test-ReTest$_2$, Test-ReTest$_3$) show average (r) correlations and statistically significant averages, with average r_2 determining coefficients: $r_1 = 0.45$ ($p_1 = 0.010$, $r_1^2 = 0.20$), $r_2 = 0.40$ ($p_2 = 0.040$, $r_2^2 = 0.16$), $r_3 = 0.37$ ($p_3 = 0.034$, $r_3^2 = 0.13$), which indicates the fact that the modifications of the values of this parameter may be certainly accounted for by the independent variable, represented by the ingestion of ECS.

We consider this *as a special result*. This parameter relies on the hypothesis that the difference between the physical field and the mental one represents the measure of anxiety. Bearing in mind that the stress index is obtained as the ratio between the parameters, representing the measure of the activity of the sympathetic (physical) system and the parasympathetic (emotional) system, we may conclude that this constant drop in the stress index average may most probably occur through an activity of increase/ decrease of the two systems. This may mean that the ECS has a slightly calming effect upon the subjects and, by extension; it may have applicability in psychology for example. In the article/study [16], the authors developed this aspect as compared to other ways of reducing stress presented in different PhD theses (hypnosis, music therapy, dance therapy), all validated through statistical analyses.

The **average JS** is considered as the health index. *Measured with a filter* index increases by 15% (from -0.15, to -0.12) as compared to the control measurement. This rate of the increase is maintained until the next measurement, made 30 min after the ingestion (from -0.12 to -0.10) and then it remains stable until the measurement made 90 min after the ingestion. As compared to the control measurement, the correlation coefficients of the 3 measurements indicate oscillating values and correlations: $r_1 = 0.78$ ($p_1 = 0.0001$, $r_1^2 = 0.60$) statistically significant; $r_2 = -0.15$ ($p_2 = 0.396$, $r_2^2 = 0.20$) statistically insignificant, and $r_3 = 0.58$ ($p_3 = 0.001$, $r_3^2 = 0.34$), statistically significant, which can mean that, between the measuring intervals, there is a reorganization, restructuring and resetting of the bio-energy informational field.

Measured without a filter, it has increased slightly during M_1, JS=-0.48 as compared to M_0, Js=-0.49, the increase being very significant during M_2, Js=+0.47, and then going back close to the control value during M_3, Js=-0.44. The (r) correlation coefficients of the 3 measurements as compared to the control measurement indicate *great, statistically significant correlations* $r_1 = 0.75$ ($p_1 = 0.0001$, $r_1^2 = 0.56$); $r_2 = 0.69$ ($p_2 = 0.0001$, $r_2^2 = 0.47$) and $r_3 = 0.64$ ($p_3 = 0.001$, $r_3^2 = 0.13$). We may conclude that the action of ECS upon JS starts rather slowly, but this consistent increase occurring during the first hour after the ingestion, measured both with and without a filter, indicates the time limited effect of ECS. A second conclusion, deriving from the first, might show that the average duration of the maximum effect of ECS is *one hour*. Then, the effect starts diminishing, a new dose being useful.

Average RMS *measured with a filter* has dropped constantly after the procedure, from 0.34 during M_0 to 0.33 during M_1. Then it was followed by a drastic drop to the value of -0.11 during M_2, and then by an increase during M_3 to the value of 0.31, i.e. a value close to the initial one. As compared to the control measurement, the correlation coefficients of the 3 measurements indicate average, statistically significant correlations, such as: $r_1 = 0.50$ ($p_1 = 0.003$, $r_1^2 = 0.25$); $r_2 = 0.50$ ($p_2 = 0.003$, $r_2^2 = 0.25$) and $r_3 = 0.29$ ($p_3 = 0.104$, $r_3^2 = 0.008$) without any statistical significance.

Average RMS *measured without a filter* dropped after the procedure, from the average of 0.52 during M_0 to 0.42 during M_2, passing through the intermediate value of 0.50 during M_1 and then slightly increasing to the value of 0.46 during M_3, but without reaching the initial value. The correlation coefficients of all the 3 measurements indicate a strong correlation as compared to the control measurement: $r_1 = 0.68$ ($p_1 = 0.0001$, $r_1^2 = 0.46$); $r_2 = 0.50$ ($p_2 = 0.003$, $r_2^2 = 0.25$) with an average statistical significance and $r_3 = 0.11$ ($p_3 = 0.546$, $r_3^2 = 0.012$) statistically insignificant. The two series of results measured with and without a filter, present some major similarities in terms of behavior of this parameter in time. The result of both types of measurement was the fact that, at the time of M_3, the changes of the values can no longer be accounted for by the independent variable, or, to put it differently, the effect of ECS is certain and validated through statistical methods for only *one hour* after the ingestion.

Average entropy (physical parameter), *measured without a filter* increased slightly from 1.99 at the time of M_0 to 2.01 at M_1, and then it remained constant at this value during the following measurements M_2 and M_3. The correlation coefficients of the 3 measurements indicated very low, statistically insignificant correlation as compared to the control measurement: $r_1 = -0.08$ ($p_1 = 0.646$, $r_1^2 = 0.0064$); $r_2 = 0.27$ ($p_2 = 0.127$, $r_2^2 = 0.007$) and $r_3 = 0.13$ ($p_3 = 0.490$, $r_3^2 = 0.01$), the determine coefficients r^2 indicating a very low or *insignificant effect* of the technique upon the measured average values. **Average entropy** *measured with a filter* (psycho-emotional parameters) value was 1.94 at the time of M_0, it had intermediate average values of 1.92 and 1.93 and then it reached the average value of 1.97 at the final M_3. As compared to the control measurement, the correlation coefficients of the distributions of the measured averages, i.e. $r_1 = -0.03$

(p_1=0.885, r_2^2=0.0009), r_2=0.007 (p_2=0.707, r_2^2=0.0049); r_3=0.007 (p_3=0.710, r_3^2=0.49) indicate very low and statistically insignificant correlations, thus indicating a reduced effect of the technique.

Note: In the researches made within the framework of their doctoral studies, the authors Manolea [17] and Mohirta [18] too obtained increases of the entropy by applying other stimuli – modified states of consciousness, and music therapy. Bearing in mind that, in a (biological) system Dulcan [19], entropy increases at the same time as the intake of energy and/or information, we may presume that the procedure determined some energy intake. The authors consider that more study is required of this aspect.

Symmetry *measured without a filter* (physical parameters) increased slightly and constantly from 90.26 at the time of M_0 control measurement, to 90.87 at M_1, value remaining 90.87 at M_2 and finally reaching 91.45 at M_3. The correlation coefficients of the 3 measurements indicated very low, statistically insignificant correlation as compared to the control measurement: r_1=0.24 (p_1=0.185, r_1^2= 0.05); r_2=0.24 (p_2=0.880, r_2^2=0.05) and r_3=0.05 (p_3=0.788, r_3^2=0.025), thus indicating a weak effect of the technique upon the averages of the measured values.

Therefore, we may conclude that, despite the fact that the average values of the symmetry parameter indicate an increase/improvement of this parameter, the values are not supported by statistical results.

The values of the average **Symmetry** *measured with a filter* (psycho-emotional parameters) dropped from 81.18 at the M_0 control measurement to 80.87 at M_1, and then they increased to 81.93 at M_2. Next, there was a decrease till 81.04 at M_3, i.e. a lower value than the initial one. *The system oscillated as if it had been searching a balance value.* The modulus of the maximum variance was 0.89, i.e. approximately 1%. As compared to the control measurement, the correlation coefficients of the 3 measurements are oscillating too: r_1=0.69 (p_1=0.0001, r_1^2= 0.47); r_2=0.54 (p_2=0.001, r_2^2=0.29) *statistically significant* and r_3=0.27 (p_3=0.130, r_3^2=0.07), statistically insignificant.

Partial conclusion 1) is that the parasympathetic system evolves more rapidly and the effect is stronger, statistically significant for a period of at least *30 minutes*. Then the effect fades, as the correlations indicate.

Partial conclusion 2): Indirectly, we have obtained a threshold value that can be useful in the applications in psychology, i.e. the *effective time of action of ECS statistically and certainly determined.*

Value of the Chakras *measured with a filter* showed that all the 7 parameters that indicate the values of the virtual chakras had some statistically significant changes, high correlation coefficients without exception.

Value of the Chakras *measured without a filter*: during all the measurements made without using a filter, all these parameters that show the values of the virtual chakras had some statistically significant changes, with high correlation coefficients, except for the values of the Anja Chakra during the M_3 measurement, for which the differences from the values measured during the control measurement had no statistical significance. This fact shows that the effect of ECS had faded by the time of M_3.

Asymmetry of the Chakras *measured with a filter:* the value of the asymmetry of Muladhara Chakra decreased constantly *as compared to the control measurement* (M_o=-0.11, M_1=-0.04, M_2=-0.03, M_3=-0.02). The correlation coefficients indicate a strong correlation: Test-ReTest r_1=0.59 (p_1=0.0001, r_1^2=0.34); on limit in the case of Test-ReTest2, r_2= 0.35 (p_2=0.050, r_2^2=0.125), and strong and statistically significant again in the case of Test-ReTest3, r_3= 0.46 (p_3=0.008, r_3^2=0.21) statistically insignificant. We also find statistically relevant results in the case of the asymmetry of the Swadistana Chakra – average effect (r_2=0.38, p_2=0.032, r_2=0.15), Manipura (weak effect (r_2=0.36, p_2=0.043, r_2^2=0.13), both upon the analysis of the results of M_2 measurement as compared to M_0, and of the Anahata Chakra at M_3 (r_3=0.34, p_3=0.05, r_2^2=0.12) on limit effect.

Asymmetry of the Chakras *measured without a filter:* statistically significant results were obtained only in 4 of the 21 cases studied, i.e. the asymmetry of the Muladhara Chakra at the time of M_1 measurement, Vishudi M_2, and Ajna at moments M_1 and M_2. The value of the asymmetry of the Muladhara Chakra dropped drastically from the value of -0.24, at M_0 (control measurement), to -0.1 at M_1, the values of found averages at the following measurements being on the increase: 0.14 at M_2 and -0.21 at M_3. In the case of M_2, the r correlation index indicates a strong and significant correlation: r_1=0.54 (p_1=0.001, r_1^2=0.25). There are statistically significant results also in the case of the decreasing asymmetry of the Vishuddhi Chakra in the case of the Test-Retest 2, with the asymmetry going down from -0.34 at M_0 to -0.09 at M_2, in which case r_2=0.37 (p_2=0.040, r_2^2=0.14).

The asymmetry of the Ajna Chakra decreased drastically at M_1 (0.310) as compared to M_0 (-0.28), and then it went back up to 0.05 at M_2. In both cases we have statistical significance.

4. DISCUSSION

After analysis of the results of the measured parameters, their modification through comparison to the control parameters, the following conclusion may be drawn: *all the 32 subjects reacted, within certain limits, to the ingestion of AECI, the value of the investigated parameters being, without exception, positively modified. From another viewpoint, we may claim that the effect that ECS has upon the subjects and, by extrapolation upon the human body, has been demonstrated, thus confirming the initial hypothesis.*

At the M_1 measurement made within the first minutes after the ingestion we noticed a rapid evolution of the modification of the values of all the analyzed parameters. The statistical comparison of the series of measurements made at the time of M_1 as compared to the M_0 control measurements indicates that the modification of 12 of the 21 analyzed parameters is statistically significant and that it thus confirms the initial hypothesis, i.e. *the action of ECS is very quick, quasi-instantaneous.*

In the emotional field, the parameters were measured without using a filter. The psycho-emotional GDV-measured parameters were very strongly influenced from a statistical point of view ($p < 0.05$) in the all Test –ReTest: average Js, FC, and Area. The average Js ranked the first terms of intensity of the effect.

In the physical field, the parameters measured with using a filter. The physical GDV-measured parameters that were very strongly influenced from a statistical point of view ($p < 0.05$) in the all Test-ReTest were only the average FC.

In the case of all the 3 series of average entropy and symmetry measurement, the results are not statistically significant.

Therefore, we may conclude that ECS has a more rapid, powerful and sustained effect upon the parasympathetic system, thus proving Prof Korotkov's statement that "influences/disturbances first bear an effect upon the emotional body and only afterwards upon the physical body".

5. CONCLUSION

The statistical analysis made upon the 21 parameters measured with filter and without filter at the M_1 (3-5 minutes after the procedure), M_2 (30 minutes after the procedure) and M_3 (90 minutes after the procedure) moments, by way of comparison to the M_0 control measurement that bears statistical significance $p<0.05$, is presented in the two rows below, which allows us to come up with the final conclusions;

Physical (using a filter): Test –ReTest₁ 13/21, Test-RTest₂ 14/21 , Test-ReTest₃ 12/21
Emotional (without using a filter): Test –ReTest₁ 15/21, Test-Retest₂ 15/21 , Test-ReTest₃ 10/21

1) The effect of ECS is extremely rapid, quasi-instantaneous, the general effect fading in time.
2) ECS effect on the parasympathetic nervous system (measurements without filter) is strong and steadily in the first 30 minutes, and rapidly decreases thereafter, while ECS effect on the sympathetic nervous system (filter measurements) is less strong, but steadily throughout the period in which measurements were made.
3) Among the average area, FC, Js, RMS, entropy, and Stress Index, symmetry parameters, the most powerful effect of ECS is upon the FC and the stress index. This appeared in all measurements, with or without a filter. The weakest effect, quasi-inexistent, is on the average entropy parameter, where the statistical analysis reveals that the modifications of this parameter values of cannot be accounted for by the independent variable with certainty.
4) The most powerful and constant effect is on the system of energy vortexes/chakras: the highest is on *Vishudi* chakra, whereas the weakest effect (with filter) is on the Ajna Chakra.
5) The effect of ECS on Chakras asymmetry is rather weak, and acts especially in reducing it for the 1ˢᵗ Chakra (Manipura), for Vishudi and Anja in all the measurements.

The action of ECS determines a redistribution of the energy informational field, the intensity of which gets uniform (the shape coefficient decreases, the average range of the iso-line increases, the stress index drops). The general level of homeostasis of the body increases, a fact that has also been found in the studies [18] and [17].

The results of the study show that the substances that we ingest interact first with the energy body fields and only later with the physical body. The authors consider this as a very important finding, opening new research directions on the manner in which food, drinks, remedies and even the water we consume must be considered.

Most definitely, the future laborious studies will determine the way in which every substance acts on the body as to find feedback methods able to establish the alimentary perfect combination adapted to the needs of each of us.

REFERENCES

[1] Pies J. Immun mit colloidalem silber, Vak Verlag, (2013).
[2] Becker R.O. The Effect of Electrically Generated Silver Ions on Human Cells, Proceedings of the First international Conference on Gold and Silver in Medicine, Silver Institute, WS (1989).
[3] Becker R.O., Shelden G. The Body Electric, Morrow, NY, (1985), chapter 5.
[4] Kulsy, L.A. Cerebrianaia voda, Naukova Dimka, Kiev 1987.
[5] W. Kühni, W. von Holst, Koloidales silber als Medizin. Das gesunde Antibioticum, AT Verlag, 2011R.
[6] Morones J.R., Ramirez et al. - Silver Enhances Antibiotic Activity against Gram-Negative Bacteria, 2013/ Sci. Transl. Med. DOI: 10.1126/scitranslmed.3006276R.

[7] Schier Holz J.M., Wachol-Drewek Z., Lucas L.J., Pulverer G. Activity of silver ions in different media, *Zentralbl Bakteriol*, may 1998, 287: 411-20.

[8] Foran S.M. Therapeutic Properties of Silver–An Historical and Technical Review, Qanta, (2009).

[9] Morones J.V. et al. Interaction of silver nanoparticles with HIV-1, J. of Nanobiotechnology 2005, v. 3, nr 6.

[10] Mungiu, O.C. Metalale pretioase-o abordare farmacologica, Ed. Gr. T. Popa, 2009.

[11] Korotkov K., Jakovleva E. Electrophotonic Applications in Medicine: GDV Bioelectrography (2013).

[12] Dobson, P. Tchernychko, E. Investigations into Stress and its management using GDV Technique. International, Journal of Alternative and Complementary Medicine, 2000.

[13] Korotkov, K. G. Human Energy Fields. Study with GDV Bioelectronography, (2002).

[14] Motoyama, Theories of the chakras, New Age Books, (2003).

[15] Gerber, R. Medicina vibrationala, Ed Elit (2011).

[16] Sheldrake, H. Morfic resonance, Park Street Press, Rochester/Vermont 2009, chapter, 5, I.

[17] Curta I., et al. Summary of the methods used to lower the anxiety parameter-stress index (T)-According to the measurement made with the GDV Camera, Rad 2014.

[18] Mohirtă, I. Psihologia Sonoluminică, Ed Ars Docendi, 2011.

[19] Manolea, A. Conditionarea psihosomatica. Psihodiagnoza si interventie psihoterapeutica folosind starile modificate de constiinta, University of Bucharest, 2012.

[20] Dumitru, C.D. Inteligenta Materiei, ed III, Eicon 2009.

6th International Conference
"Computational Mechanics and Virtual Engineering "
COMEC 2015
15-16 October 2015, Braşov, Romania

A COMPARATIVE EXPERIMENTAL STUDY OF IMPACT DAMPING USING HIGHLY COMPRESSIBLE POROUS MATERIALS IMBIBED WITH FLUIDS

Mihaela Radu [1], T. Cicone [1], M.D. Pascovici [1], Petrica Turtoi [2], Marcel Istrate [3]
[1] University POLITEHNICA of Bucharest, Bucharest, ROMANIA, mihaela.radu@upb.ro
[2] ACTTM, Bucharest, ROMANIA, turtoipetrica@gmail.com
[3] STIMPEX, Bucharest, ROMANIA, marcel.istrate@stimpex.ro

Abstract: An innovative solution for damping of residual energy after the impact of a bullet on a protective vest, based on the flow of a fluid through a porous material that is placed under the protective Kevlar layers, has been recently proposed. The fluid flow through the open pores is produced by the compression of the porous material, during impact, i.e. the deformation of the Kevlar layer under the impact with the bullet. The simulation of the impact is done using a pendulum test rig that with a spherical cap. The acceleration and force under impact is analyzed on the impacted body that is protected by the porous material. Combinations of candidate porous materials and fluids were tested. The comparison is done in terms of peak-acceleration and peak-force but also in combination with the thickness and the porosity of the material. The best combination of porous material and fluid was determined. The experiments were also focused on the evaluation of various sealing (encapsulating) membranes necessary for bulletproof application.
Keywords: damping, impact, porous, Kevlar, lubrication

1. INTRODUCTION

Hostile-Aggressive behaviors and the politics of some countries opened new threats and intensified armed or bomb attacks. In this circumstance, protection is a matter of major interest in situations where different constructions, transportation vehicles or human lives (civilian or military) are subjected to high energy transfer phenomena, like explosions or high speed impacts of projectiles, bullets, shrapnel etc. The effectiveness of a protective system depends not only on the capability to stop the penetration, but also on blunt trauma protection – the blow suffered by the body from the impact of the bullet/ projectile/ explosive with the protective system. For personnel involved in high risk operations, blunt trauma can be a matter of life or death.
During the last decade the development of improved protection equipment for human personnel which contains an encapsulated, highly compressible porous layer imbibed with a liquid, has been analyzed. Placed in the backing of a bulletproof vest, it will absorb the energy left after the bullet's impact and reduce the risk to injury due to blunt trauma.
Viscous friction of a fluid that flows through the pores of a highly compressible material shows great potential in shock absorbers applications. The fluid flow is produced by the compression of the porous material, with open pores, under the action of external loads. The first mention of this mechanism was in 2000, by Prof. Mircea D. Pascovici [1], under the name of Ex-Poro-HydroDynamic (XPHD) lubrication. It is based on the interaction between the fluid and the porous structure, coupling the effect of the compression of the material with the flow of the fluid through its pores. During compression, the porous material is highly deformed, the porosity decreases, and so, the permeability. As a consequence, the resistance to flow increases and generates high pressures. Impact damping based on XPHD lubrication mechanism is difficult to study, with few experimental investigations so far.

The experimental studies found in literature are scarce, with very different approaches, and reduced significance. Regarding the porous materials used and imbibing fluids, there is an extremely wide variety of solutions, but very few were considered experimentally. Dawson studied theoretically and experimentally the force generated by squeezing in dynamic conditions reticulated foam imbibed with high viscosity glycerine [2]. The results show, that when compressing the foam up to 50%, the fluid inside has an important contribution to the lift force.

Later, Vossen [3] studied the behaviour of polyurethane foams imbibed with cells of glycerine as application for motorcycle helmets.

This paper presents the experimental work done to find a compatible combination of porous material and fluid that provides the best damping capacity relative to the volume occupied.

2. EXPERIMENTAL SET-UP

Experimental set-up of shooting bullet-proof vests can be very expensive and time consuming, which is why, for analyzing the behavior of the porous material imbibed with fluid, the Kevlar layers deformed by the bullet was associated with a sphere of a larger radius (Fig. 1).

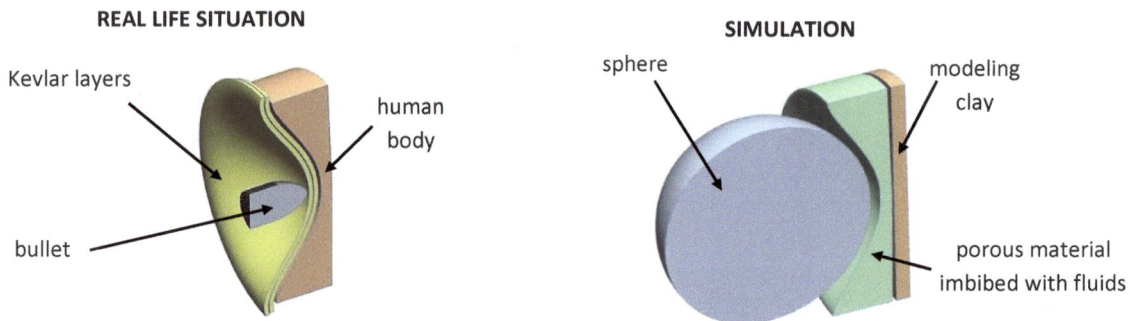

Figure 1: Analogy between real life situation and simulations of a bullet's impact

Two sets of simulating experiments have been performed:
   (a) Tests done on a instrumented pendulum test-rig.

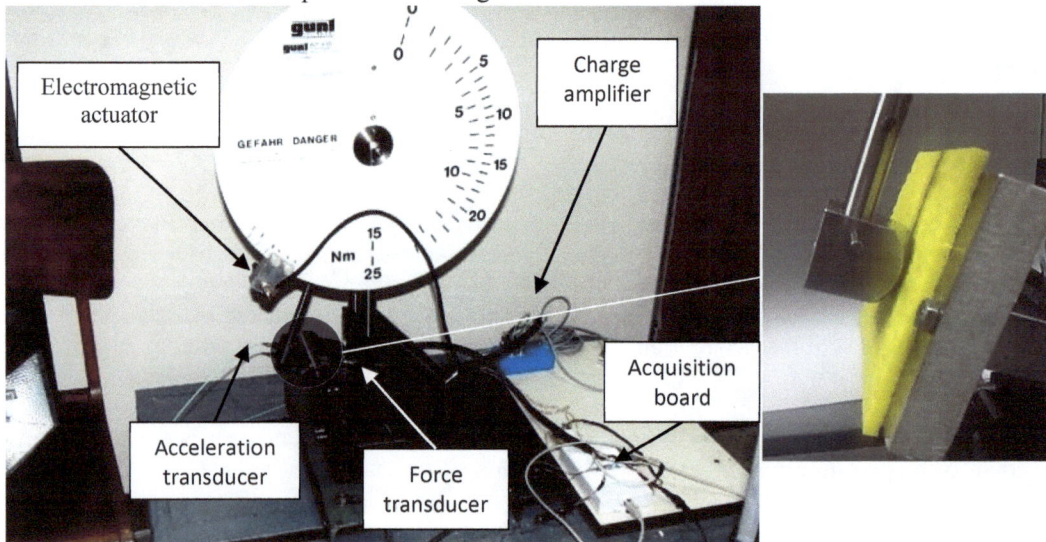

Figure 2: Pendulum test-rig

The experimental test-rig (Fig. 2) is composed of a pendulum with a spherical cap (radius 25 mm), that drops from a certain height H, on a rigid plane where the porous material is fixed [4]. An uni-axial acceleration piezoelectric sensor (KISTLER 8274A5) of maximum 20000 m^2/s screwed on the impacting body, tangent to the trajectory at the point of impact, and a force transducer (KISTLER 9301 B) of maximum 2500N, screwed in the middle of a bracket supporting the impacted body, record the two main characteristics of the impact through a data acquisition system. The bracket is free to slide against the force transducer and is inclined with an angle calculated to ensure that the line of contact is tangent to the pendulum trajectory, in the point of impact. The results presented in this paper are focused only on the sphere-on-plane configuration. A series of three tests have been done in the same experimental conditions in order to check the repeatability of the results.

(b) Falling-ball drop tests were made in a second series of experiments.
A 30mm dia sphere drops from H over a witness material that replaces the human body: water-based modeling clay. Each test is made on a different sample of clay which solidifies after 24 hours.
Between the impacting sphere and the modeling clay was placed one or two layers of the tested specimen (porous material imbibed with fluid). Some tests include also the presence of the Kevlar layer(s) in order to simulate better the real situation.

3. TESTED SPECIMENS

The experiments presented in this paper were done on different porous materials, commercially available for purposes like: cleaning, clothing, filtering etc., imbibed with various fluids All of the materials are highly compressible (deformable), homogeneous, with open pores. The structure of the solid materials can differ, having either ordered or disordered fibbers, consolidated or nonconsolidated.
The main **candidate materials** (Fig. 3) that meet the necessary characteristics were:
- Textile materials:
  o *Woven materials*(T03, T04);
  o *Unwoven materials*(NT01, NT03, NT04);
- Reticulated Polyurethane foams (F133);
- 3D spacer (S3D), a double wall woven fabric interconnected with yarns perpendicular to the walls.

Prior to experiments all these materials have been characterized in terms of structure (morphology) thickness h_0, and initial porosity, ε_0.

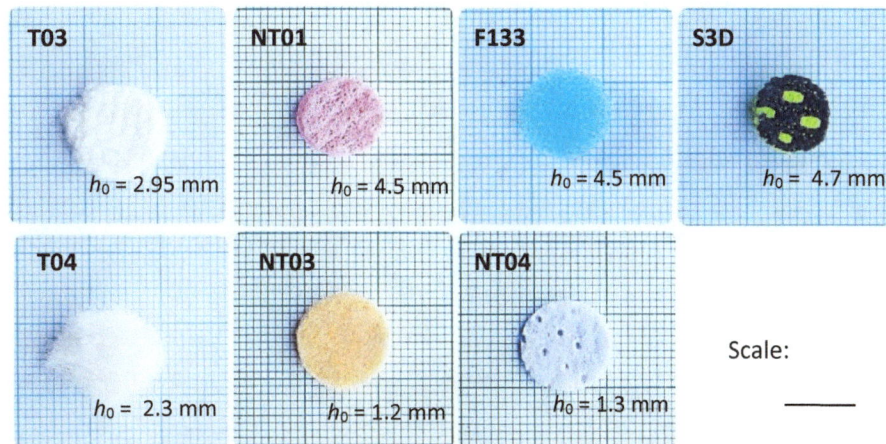

T03	NT01	F133	S3D
$h_0 = 2.95$ mm	$h_0 = 4.5$ mm	$h_0 = 4.5$ mm	$h_0 = 4.7$ mm

T04	NT03	NT04	Scale:
$h_0 = 2.3$ mm	$h_0 = 1.2$ mm	$h_0 = 1.3$ mm	

Figure 3: Candidate materials

The fluid imbibed in the porous material can be any liquid adhering to the surface of the pores and/or absorbed in the fibbers (water, glycerine, oil, pastes, gels etc). They can be more or less viscous, Newtonian or non-Newtonian. The possibilities of combinations of porous materials and fluids are endless, considering the variety of products commercially available. Of course, many of these combinations are not compatible because of the particular properties of the materials, such as chemical or physical incompatibilities (e.g. hydrophobia), or large porosities that cannot sustain low viscosity fluids.

4. EXPERIMENTAL RESULTS

Generally, the damping capacity of materials [5], can be evaluated using different methods:
- Peak acceleration of the impacting/impacted body;

- Peak force on the impacting/impacted body;
- Deformation of the impacted body (footprint);
- Vibration damping after impact;
- The evaluation of the rebound of the impacting body (restitution coefficient).

Experiments on the pendulum test-rig were done from a height of H=200mm, that gives a velocity of approximately 2m/s at impact. A comparative study (Fig. 4) for different materials imbibed with fluids was done in order to establish the best pair that has the lowest peak acceleration and force. The materials are compared in terms of peak-value of acceleration (a_{max}), and peak-value of force (F_{max}). To overcome the difference in thickness and porosity of the materials, two relative indicators have been proposed, using thickness h and fluid fraction, $h \cdot \varepsilon$, respectively.

The best damping capability of the materials is there obtained by the following criteria:

- lowest values of peak-acceleration a_{max}, relative acceleration $a_{max} \cdot h$ and $a_{max} \cdot h \cdot \varepsilon$;
- lowest values of peak-force F_{max}, relative force $F_{max} \cdot h$, $F_{max} \cdot h \cdot \varepsilon$

The results show that the foam F133 imbibed with tooth paste has the lowest values in the measured parameters, but taking into account the relative indicators, the S3D material imbibed with paste seems to be a better choice.

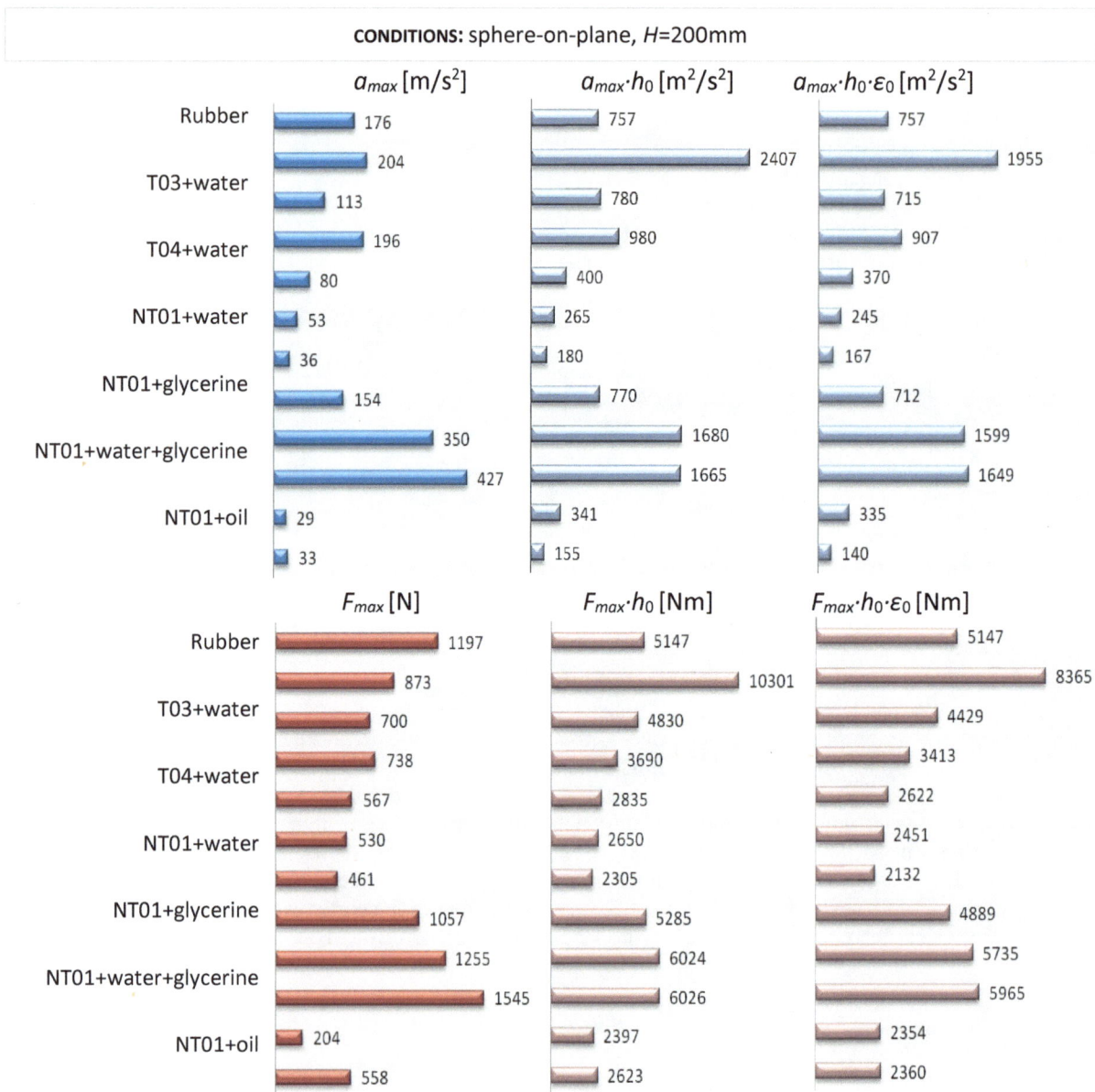

Figure 4: Comparative study of impact behaviour of different porous materials imbibed with fluids

Since from the comparative study, the F133 material has the most promising results, therefore, it was used for further experiments in an encapsulated form (with plastic wall). In Fig. 5 to Fig. 8 are presented the acceleration and force variation in time for different materials. Comparing Fig. 5 and Fig. 6, one can observe that adding a layer of Kevlar, increases the reactions on the impacted body. This can be explained by the fact that the Kevlar layer, which is not compressible, acts as an external layer of the sphere, increasing its radius. Hence, a larger zone is deformed. This doesn't imply that the Kevlar is harmful, because currently it is the strongest layer known that can decrease the velocity of the bullet.

By comparing Fig. 6 and Fig. 7, we can demonstrate that the presence of the glycerin in the porous material can decrease the acceleration by 60% and the force by 30%, a very conclusive result that demonstrates the effectiveness of the XPHD mechanism. Fig 8. demonstrates that two layers of porous material are very efficient in damping the impact.

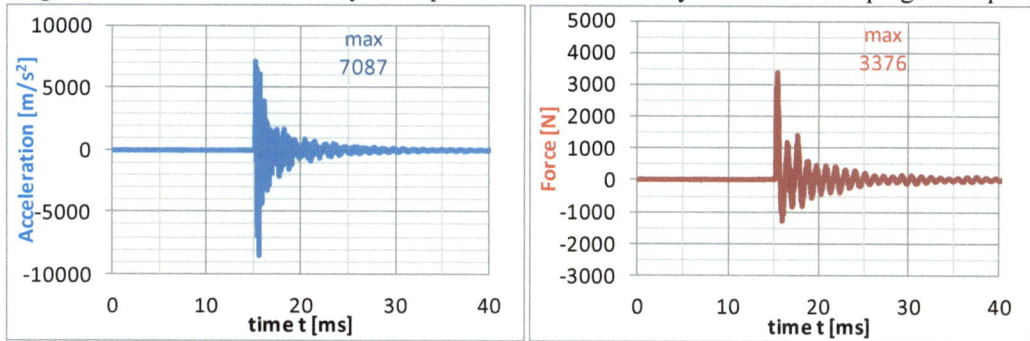

Figure 5: Acceleration and force variation in time of an impact using one layer of encapsulated dry F133

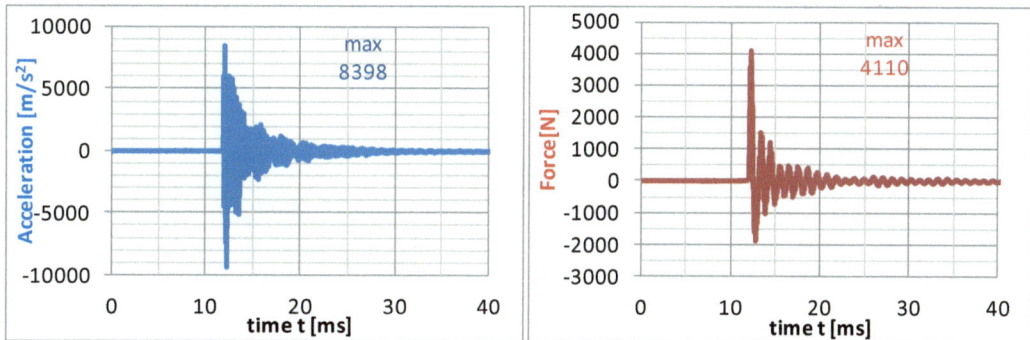

Figure 6: Acceleration and force variation in time of an impact using one layer of encapsulated dry F133 and a layer of Kevlar

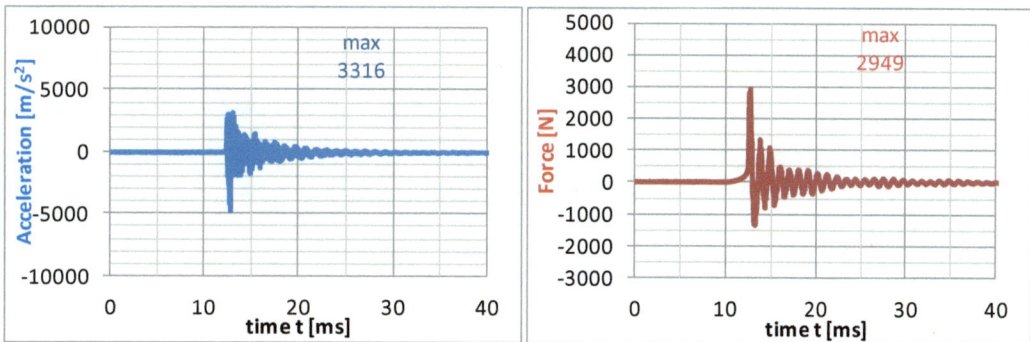

Figure 7: Acceleration and force variation in time of an impact using one layer of encapsulated F133 imbibed with glycerin and a layer of Kevlar

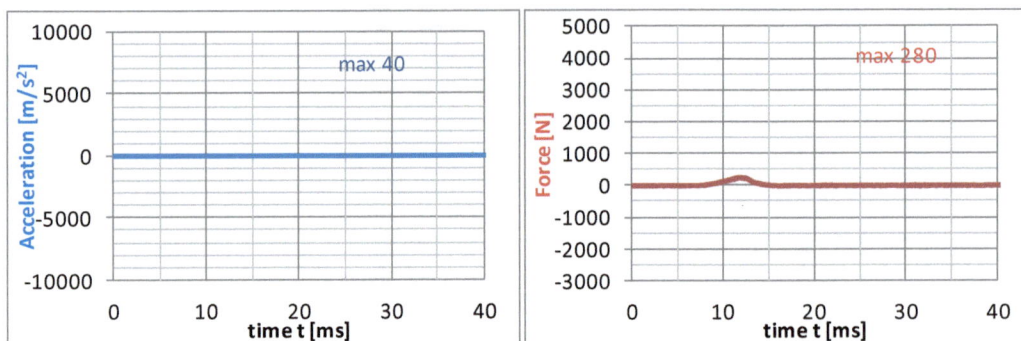

Figure 8: Acceleration and force variation in time of an impact using two layers of encapsulated F133 imbibed with glycerin and a layer of Kevlar

Three tests were done by dropping a ball from H=4.91m that gives a velocity of 9.8m/s at impact:
- A - undamped;
- B - damped with one layer of F133 imbibed with glycerin; .
- C - damped with one layer of F133 imbibed with glycerin and with four layers of Kevlar on top.

The 3D scans of the three clay samples are presented in Fig. 9.

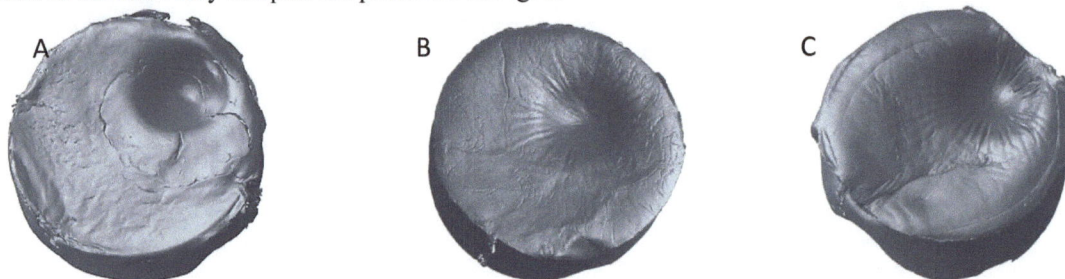

Figure 9: 3D scans of modelling clay after impact of a sphere: A, B and C

The footprint was evaluated with three methods: by measuring the mass of water or wax that can be placed inside the footprint and calculating the mass and by using a software calculation of the 3D scan.
The results, presented in Table 2, are very close.

Table 2: Porosity measurement results

Method	Volume of sample A [mm^3]	Volume of sample B [mm^3]	Volume of sample C[mm^3]
Water	2513	1759	3015
Wax	3011	2444	3322
Software	3150	2355	3358

The footprint of the sample B is smaller than the one of sample A, which is a clear proof of the damping capacity of the glycerin. Also the increased footprint of sample C with respect to sample B shows that the Kevlar layer acts like a cover of the impacting sphere.

3. CONCLUSION

This experimental study is relevant and constitutes an important basis for further studies on XPHD lubrication, as two types of impact tests validate the lift force generated and confirms the damping capacity of this mechanism. The low energy impact, with velocities under 2m/s, can be completely damped and show great effectiveness when compared with a contact without any material interposed between the impacted bodies. The tests done with and without fluid prove that the damping effect is produced by the viscous friction generated by the flow of the fluid and not the elasticity of the material, although this has its contribution. Also, the results in terms of peak acceleration and peak force are smaller in comparison with classical damping materials (like rubber).

Testing the impact behavior of HCPL material imbibed with different fluids proved to be a quick approach for comparison in terms of damping capacity, due to the simplicity of the experimental stand and time required for experiments. The best pair of HCPL + fluid tested was reticulated polyurethane foam imbibed with tooth paste, but a lot of other combinations (based on Newtonian fluids) have promising results.

ACKNOWLEDGEMENT

This work was partially supported by MEN-UEFISCDI Partnership Program PN-II Contract No 287/2014 (PROTHEIS). The 3D scans were realized by SPECTROMAS S.R.L.

REFERENCES

[1] Pascovici M.D., Lubrication by Dislocation: A New Mechanism for Load Carrying Capacity, Proceedings of 2nd World Tribology Congress, Vienna, vol. 41, 2001.

[2] Dawson M.A., McKinley G.H., Gibson L.J., The Dynamic Compressive Response of Open-Cell Foam Impregnated With a Newtonian Fluid, ASME J. of Applied Mechanics, vol. 75, 2008, 041015-1.

[3] Vossen B.G., Modeling the application of fluid filled foam in motorcycle helmets, MSc. Thesis, Eindhoven University of Technology, 2010.

[4] Radu M., Cicone T., Experimental determination of the damping capacity of highly compressible porous materials imbibed with water, BALKANTRIB'14 8th International Conference on Tribology, October, 30 - November, 1, 2014, Sinaia, Romania.

[5] Harris C.M, Crede C.R., Harris' Shock and Vibration Handbook 5th Edition McGraw-Hill, NY 2002.

6th International Conference
"Computational Mechanics and Virtual Engineering"
COMEC 2015
15-16 October 2015, Braşov, Romania

MEASUREMENT METHODS FOR THE PATH DEVIATION OF MOBILE ROBOTS

Radu Ţărulescu1, Stelian Ţărulescu [2]

[1] Transilvania University, Braşov, Romania, radu.tarulescu@unitbv.ro
[2] Transilvania University, Braşov, Romania, s.tarulescu@unitbv.ro

Abstract: Accuracy of mobile robots movement is vital to their orientation in the environment. Indoor mobile robots or automated guided mini-vehicles need accurate path tracking to avoid collisions with obstacles. In this paper are presented two methods for determining the deviation from the path, in order to eliminate errors that may occur due to imperfections of robot locomotion system. A method is based on measuring the linear deviation from path, using a route with bands of different sizes and colors. The other method determines the angular deviation using a device, with rotary motion sensor, mounted on a rail.
Keywords: *path deviation; mobile robots; locomotion system; ultrasonic sensor.*

1. INTRODUCTION

Locomotion is the process through the mobile robot moves in the environment with certain forces actuating on it [1]. Locomotion system has a crucial role in achieving the aim pursued in mobile robot function. It contributes to this both, embodiment of locomotion (wheel tracks, legs etc.) and structural aspects (degrees of mobility, orientation, maneuverability) of the variant admitted for the robot [2].
To determine the influence of the locomotion system in the workspace orientation were performed tests with mobile robots that moves using wheels, tracks and with walking robots. For measure the linear deviation from path it was use a route with bands of different sizes and colors, and for measure the angular deviation, it was use a device with rotary motion sensor, mounted on a rail.

2. MOBILE ROBOT USED FOR MEASUREMENTS

For the experimental study have been used six mobile robots presented in Fig. 1, namely: Pro Bot 128, Spy Video TRAKR, KSR4 - Escape, Maxibot, Hexbug Delta, and Remote-controlled mini-vehicle. Of those six mobile robots, depending on the locomotion system, three are wheeled, one tracked and two are walkers.
The Pro Bot 128 robot uses a wheeled differential locomotion system and has a high level of autonomy, having the possibility to sense and avoid obstacles with infrared sensors detection system, without the intervention of the human operator. For the navigation the robot was programmed to use its anti collision infrared sensorial system.
KSR4–Escape mobile robot uses a locomotion system with six wheels and for the navigation it is equipped with infrared detection sensors. The Escape Robot uses three infrared emitting diodes and one infrared receiving module to send and receive signals and detect obstacles.
The remote-controlled mini-vehicle uses for locomotion a four-wheel system with Ackermann steering type.

Spy Video TRAKR uses a tracked locomotion system, being a programmable robot, equipped with a color camera with infrared detection possibility, microphone and speaker. The robot can be controlled by remote control or can be programmed to travel along a selected route.

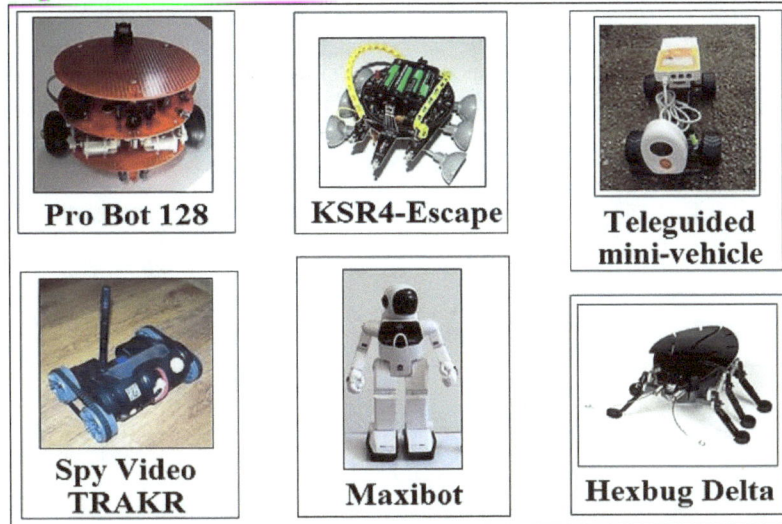

Figure 1: Mobile robots used at experimental research

Maxibot robot is a humanoid walking robot and for navigation uses an anti collision infrared sensorial system. The robot programming is performed using the program module located in the rear. This module has four keys and by pressing them it is determined how many steps are forward, backward, left or right.

Hexbug Delta is a walking robot that reacts to touch. His antennas are tactile sensors that modify the direction of travel to the touch. When the right antenna is touched, the robot turns to the right and left when the antenna is touched, the robot turns to the left. The robot is powered by the legs in the middle. The other four legs provide stability while driving. When the antennas are not reached, the direction of movement is forward.

3. DETERMINATION OF LINEAR DEVIATION FROM PATH

To determine the deviation from the path of mobile robots, it was set up a trail with bands of different sizes and colors like in Fig. 2. The narrow band, in the middle is green with width of 10 mm, left and right bands are yellow, with width of 15 mm, which are bordered by red bands with a width of 25 mm.

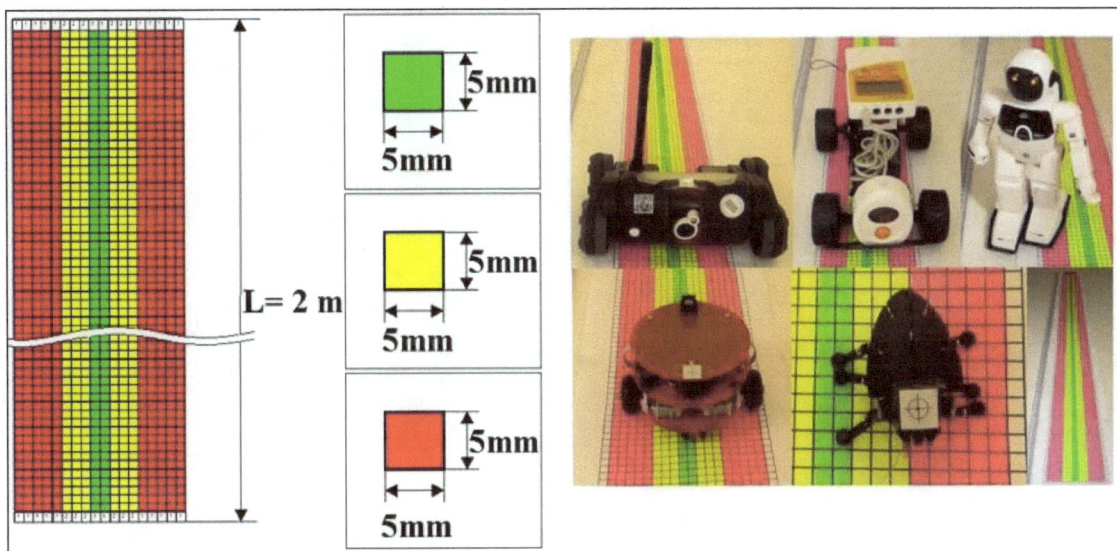

Figure 2: Trail chosen to determine the deviation from the trajectory

In order to determine the deviation from linear trajectory, each robot was separately measured and marked the center of symmetry like in Fig. 2.

The deviation from the straight path was determined by measuring the distance between the symmetry axis of the track and the center of each robot [3]. As long as the symmetry of each robot falls in the green band deviation is negligible, the measurement being made on yellow and red bands.

The robots displacement was filmed with a camera mounted on ceiling of the room where was arranged the workspace. Based on movie recording, has been determined accurately the linear deviation at every 200 mm of a total distance of 1800 mm (1.8m).

Table 1 shows the linear deviation values, determined for the six mobile robots.

Table 1: Linear deviation from the path

Distance	Wheeled			Tracked	Walkers	
	Pro Bot 128	Remote-controlled vehicle	KSR4-Escape	Spy Video TRAKR	Maxibot	Hexbug Delta
	Linear deviation					
d	d_1	d_2	d_3	d_4	d_5	d_6
[mm]	[mm]	[mm]	[mm]	[mm]	[mm]	[mm]
200	3.2	4.2	5.2	2.8	5.5	5.8
400	6.1	8.8	11.1	5.3	10.1	12.3
600	8.5	12.2	15.8	7.2	18.2	22.2
800	11.7	15.1	23.6	9.8	24.5	28.9
1000	13.8	17.3	28.8	11.6	30.2	36.3
1200	15.9	20.2	32.3	13.9	38.8	48.2
1400	18.3	23.8	38.6	15.3	48.3	63.3
1600	21.3	26.1	45.8	17.6	63.2	82.2
1800	22.2	28.6	52.3	19.1	91.3	112.1

Fig. 3 presents the variation of linear deviation from the path based on the distance traveled. It is noted that robots Pro Bot 128, Remote-controlled vehicle and Spy Video TRAKR deviation increased linearly with the distance. At walking robots it is noted that the deviation values increased exponentially to the distance traveled [4].

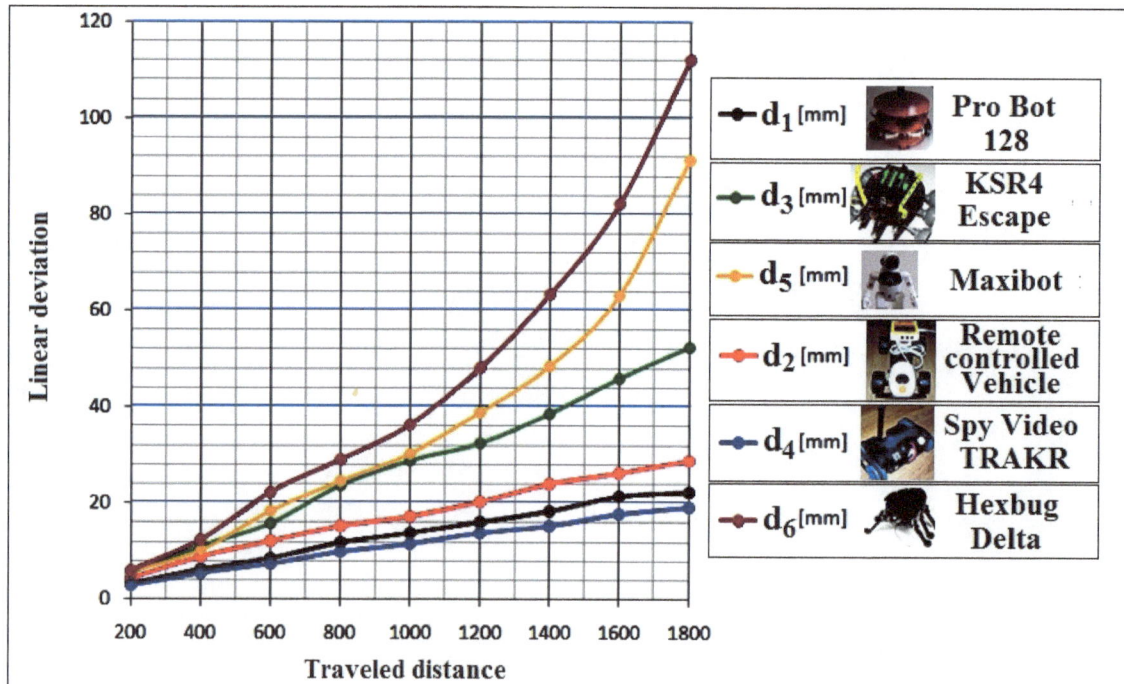

Figure 3: Linear deviation for the six mobile robots

Fig. 4 shows the distance traveled of all six mobile robots, on the three different color bands. It is noted that the robot Spy Video TRAKR traveled the greatest distance to the green and was the only one who did not passed on red band. Delta Hexbug walking robot has traveled the shortest distance on the green and the biggest distance on red.

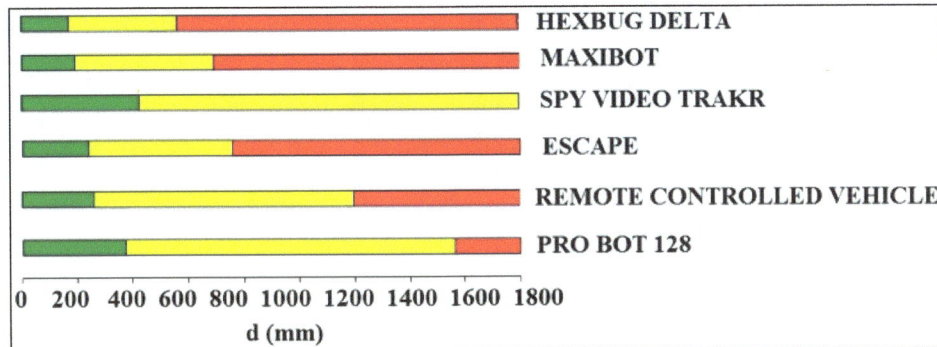

Figure 4: Distance traveled on the bands of different colors

4. DETERMINATION OF ANGULAR DEVIATION FROM PATH

In order to determine angular deviation from the path has been designed a device that runs on rails and incorporates a sensor that measures the angular deviation [5].

The sensor mounted on the device is connected to the tested robot and to the data acquisition system MultiLogPRO witch provide real-time readings. The device movement will be rectilinear and friction between the wheels and the two rails will be minimized.

The used rotary motion sensor, DT148A monitors the angular position and indicates the direction of travel, having positive or negative values. Measuring the amount of angle deviation from the path, the sign is irrelevant. Sensors accuracy is 0.125° with a sampling rate of 10 samples per second. For data acquisition it was used a MultiLogPRO system. The MultiLogPRO system is a standalone 12-bit data logger with a clear LCD graphic display and a 128K internal memory. Recorded data are displayed in the form of graphs, tables, meters or digital displays, and can be analyzed with a number of pre-programmed analysis functions. This system internal memory stores experiment notes and instructions for carrying out the experiment, which can be edited or expanded at any time. MultiLogPRO can record data from up to 8 sensors simultaneously and it's capable of recording at rates of up to 21000 samples per second, and of collecting up to 100000 samples in its internal memory [6].

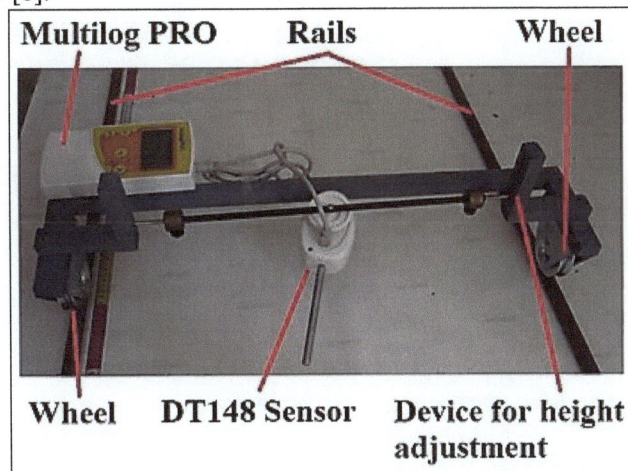

Figure 5: System for determining the angular deviation

In Fig. 5 it is presented the whole system obtained to determine the angular deviation at robots movement. With that system were tested the robots Spy Video TRAKR (tracked), Pro Bot 128 (wheels), Remote-controlled vehicle (wheels) and Maxibot (walking), like in Fig. 6. The sensor was attached to each robot in part with a fixed grip.

Figure 6: System used connected to the mobile robots

Table 2 presents the values obtained from the measurements performed with the device. Values of the measured angles were recorded from 200 to 200 mm, over a distance of 1.8 m.

Table 2: Linear deviation from the path

Traveled distance	Wheeled		Tracked	Walker
	Pro Bot 128	Remote vehicle	Spy Video TRAKR	Maxibot
	Angular deviation			
d	β_1	β_2	β_3	β_4
[mm]	degrees	degrees	degrees	degrees
200	0.92	1.20	0.80	1.58
400	1.75	2.52	1.52	2.89
600	2.43	3.49	2.06	5.20
800	3.35	4.32	2.81	6.98
1000	3.95	4.94	3.32	8.59
1200	4.54	5.77	3.98	10.98
1400	5.23	6.79	4.37	13.58
1600	6.08	7.44	5.03	17.54
1800	6.33	8.14	5.46	24.54

After the data analysis, we can say that the values obtained for Maxibot walking robot are different from values obtained for the others robots. The angular deviation from the path is greater in comparison with the deviations determined for the other robots and is showed in Fig.7.

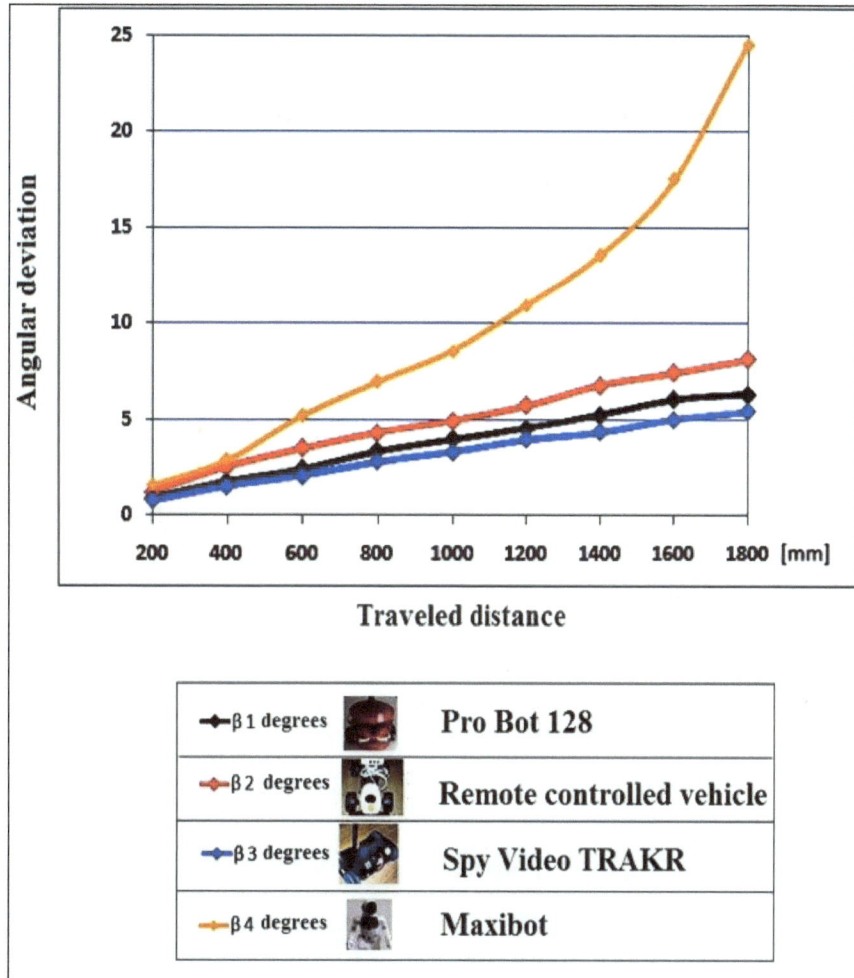

Figure 7: Angular deviation for the tested robots

5. CONCLUSION

In terms of deviation from the path, the largest differences were obtained with walking robots Maxibot and Hexbug Delta. To correct the way that travels the Maxibot walking robot, will be set like that: after four steps forward, the robot shall be made one step to the left. This will correct its trajectory, the robot initially having a tendency to move to the left.
At Hexbug Delta, if the robot, during movement tends to turn left, then the left leg position will be adjusted, and if it tends to turn right, then right leg position will be adjusted. The steps to cancel the malfunction are [7]: determine the leg that must be adjusted; supports the robot so that adjustment can be made as simple as possible; adjust the leg position from its pivoting point.

REFERENCES

[1]   Abdalla T., Hamzah M., Trajectory tracking control for mobile robot using wavelet network, International Journal of Computer Applications, Volume 74– No.3, pp. 32–37, July 2013.
[2]   Korodi A., Dragomir T., Correcting odometry errors for mobile robots using image processing, Proceedings of the International MultiConference of Engineers and Computer Scientists, Vol II, IMECS 2010, Hong Kong, March 17-19, 2010.

[3]   Sas C., O'Hare G., Reilly R., Virtual environment trajectory analysis: a basis for navigational assistance and scene adaptivity, Future Generation Computer Systems, vol. 21 (7), pp. 1157–1166, July 2005.[4]   T. M. Howard, A. Kelly, Optimal rough terrain trajectory generation for wheeled mobile robots,The International Journal of Robotics Research, vol. 26, pp. 141–166, 2007.

[5] Tarulescu R., Benche V., Horizontal micro-carrier of heat and mass, ANNALS of the ORADEA UNIVERSITY, University of Oradea Press, ISSN 1583-0691, 2013.

[6] Tarulescu R., Olteanu C., Navigation system optimization for mobile robot Pro Bot 128, 5th International Conference "Advanced Composite Materials Engineering" and The 3rd International Conference "Research & Innovation in Engineering" COMAT 2014, Romania, pp. 135-141, 16-17 October 2014.

[7] Correll N., Sempo G., Halloy J., Deneubourg J.L., Martinoli A., A tracking tool for multi-unit robotic and biological systems., IEEE/RSJ International Conference on Intelligent Robots and Systems, pp. 2185 – 2191, October 2006.

6th International Conference
"Computational Mechanics and Virtual Engineering "
COMET 2015
15-16 October 2015, Braşov, Romania

RESEARCH REGARDING DAMAGE DETECTION IN THE INERTIAL DAMPERS OF THE ELECTRIC VEHICLE

Calin ITU[1], Maria Luminita SCUTARU[1]
[1] Transilvania University of Brasov, Brasov, ROMANIA, e-mail: lscutaru@unitbv.ro

Abstract: *The paper presents some research on inertial dampers wich ensure, through the structural design, mechanical energy dissipation due to vibration or noise generating source. These dampers offer a good vibration energy dissipation in volume of plastics on the one hand and a dissipative higher power efficiency and stability on the other hand (due to the high energy absorption distributed system).*
Keywords : *inertial damper, structural design, mechanical enery, vibration*

1. INTRODUCTION

Inertial dampers ensure, through the structural design, mechanical energy dissipation due to vibration or noise generating source. Also these dampers allow, as result of vibration energy dissipation in volume of plastics on the one hand and as result of dissipative higher power efficiency and stability on the other hand (due to the high energy absorption distributed system) the use of corresponding nonlinear zone flexibility based plastics. Moreover, the spatial distribution of inertial masses ensure a more uniform thermal loading plastic base and can be achieved without breaking conditions thereby avoiding pregnancy and the development of concentrators zones in basic plastic. The same spatial distribution in conjunction with inertial masses form, their surface roughness and especially the flexibility of plastics allow obtaining anisotropic features extremely useful in terms of optimal adaptation in relation to the type of shock absorber damping designed for this task. In [11] such a solution is presented that allows vibration damping electronic control circuit due to the commutation of a stepper motor power. If evoked system consists of silicone plastic housed in a hermetic housing, and moving inertial masses is limited mechanically. If plastic studied system consists of a silicone housed in a hermetic housing and the motion of the inertial masses is limited mechanically.

In [1] are shown the theoretical equations of motion corresponding inertial mass immersed in basic plastic. In the paper [2] such dampers are included in the class of 'meta-materials', mechanical hybrid structures respectively (composite) that allow "predicting the mechanical properties of the design phase" and damping properties in relation to specific needs in system which the damper is included. Such systems are found in the living world that can fulfill both dissipation function but also amplification and therefore accelerations of vibration transmitted through the damper. These issues are extremely important in case of electric vehicles, as a result their shareholders through either the power of synchronous motors or asynchronous motors with the control based on the electric commutation circuit occurring two categories of sources of vibration or noise frequency bands with two distinct areas. First, lower, corresponding to the fundamental frequency applied to the electric motor that is dependent on variable angular speed.

The second frequency range of vibration is due to electrical switchgear and circuit characteristics and correspond to higher frequencies such as those produced by electrical switching circuit. This second field will vary very little in relation to the angular speed of the engine or the velocity of the vehicle. This second area is intended to be covered by the damping characteristics (band pass filter) that can perform inertial damper.

Alternative solutions, involving the development of an annulator of oscillations in electric commutation type Wiener filter [8], which both in terms of reliability, energy consumption, especially high computing power necessary to implement satisfy only partially the needs of applications involving a rotating as uniform for electric vehicle drive system. The Wiener filter operation is based on fundamental control signal adding an estimated signal so that the resulting signal at the output compensate the vibrations due to the commutation electric power applied to the electric motor. The use of a computer system in real time requires high computing power or specialized circuits but also requires a certain energy and so strictly administered in case of an electric vehicle. Furthermore, if for some reason practically disappears the power control system of the torsional vibration, disappears too the possibility to decrease the vibrations.

2. TECHNICAL REQUIREMENTS

In figure 1 a standard model of an inertial damper is presented. The damper is made by rubber reinforced with balls. In the paper two cases are considered: the balls are made by steel and, alternately, by aluminium (Al). The results are similar, which is why we only present the case where steel balls are. The properties for rubber and steel are those currently used in engineering

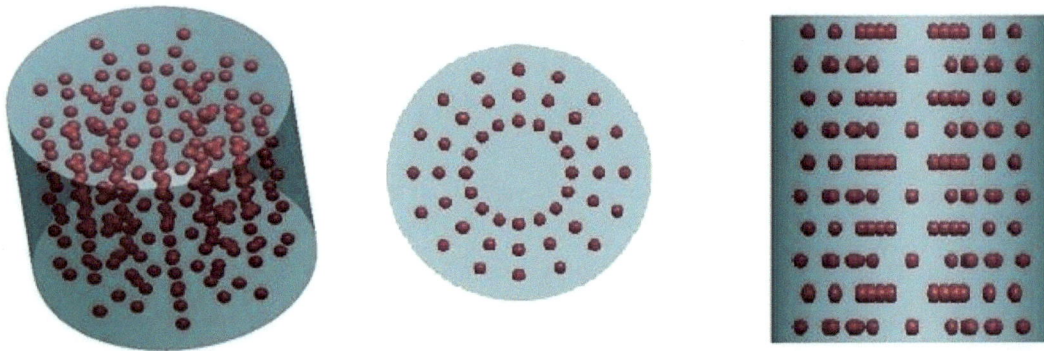

Figure 1. Physical model of the inertial damper

A finite element model for the damper is presented in Fig.2. The main goal of the paper is to determine the eigenvalues for a functional structure and for the damaged structure. Making an analysis in the eigenfrequencies changes is possible to identify the moment when the damage is produced. The method was presented in previous paper as [3],[4],[7]. Analytical methods in the field of the eigenvalues can be found in many paper (see [5],[6]).

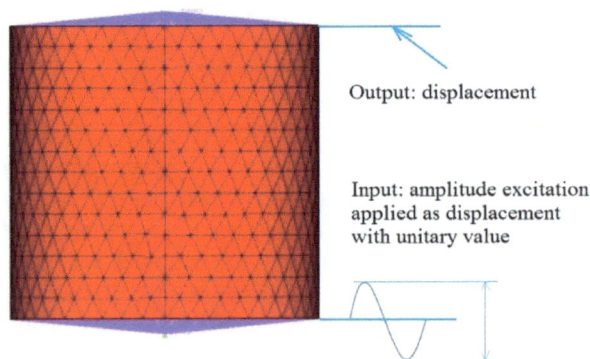

Figure 2. Finite element model of the damper

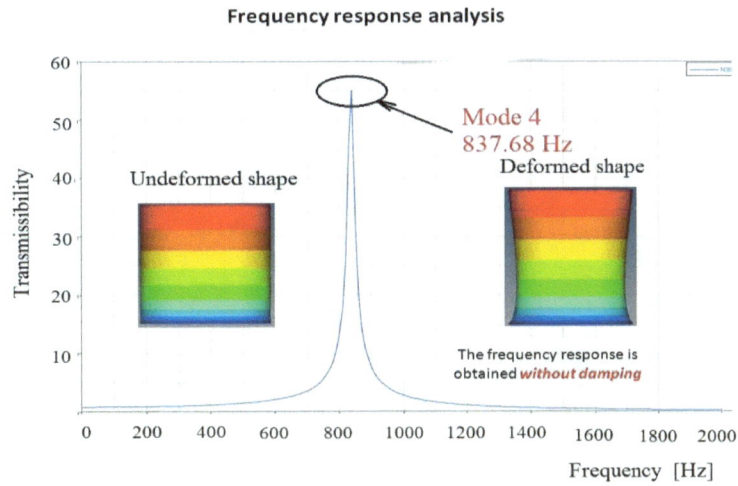

Figure 3. Frequency response for a rubber model

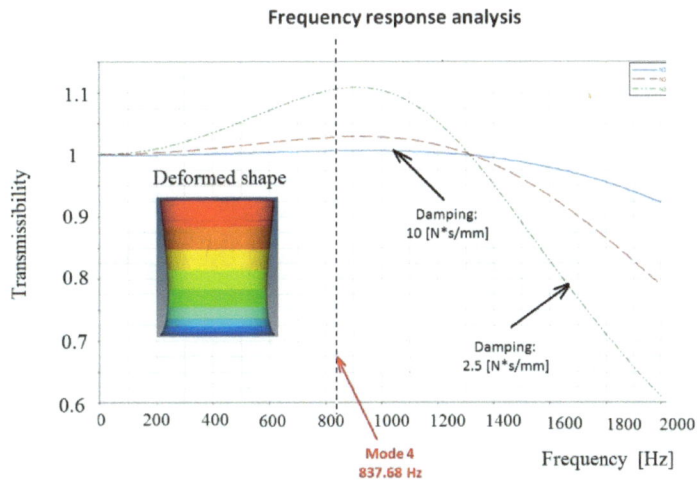

Figure 4. Frequency reponse for an inertial damper with steel balls

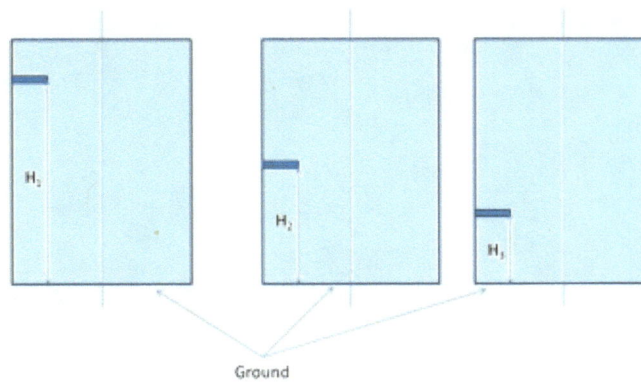

Figure 5. Damage scenarious for the damper

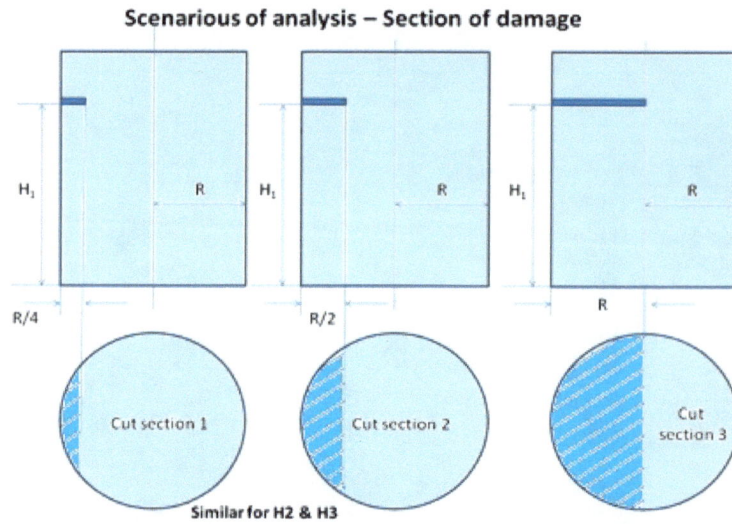

Figure 6. Damage scenarios for the damper

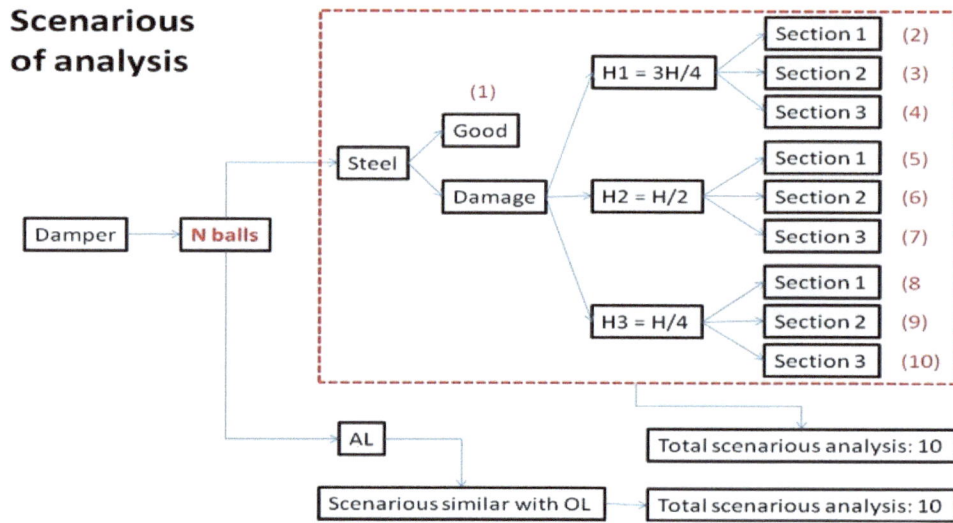

Figure 7. Scenarious of analysis

3. EIGENVALUES FOR THE DAMAGED DAMPER

The eigenvalues and eigenmodes were computed using a finite element soft. The result for the eigenvalues computed for the scenarious presented in Fig. 7 are presented in Table nr.1.

Table nr. 1. Eigenvalues for the first 10 scenarious

Mode no.	Frequency [Hz]									
	Good (1)	D1 (2)	D2 (3)	D3 (4)	D4 (5)	D5 (6)	D6 (7)	D7 (8)	D8 (9)	D9 (10)
1	320	320	320	305	318	302	212	312	280	170
2	320	321	321	311	320	310	243	318	305	223
3	459	460	458	437	456	438	383	451	427	349
4	837	835	814	554	824	711	461	822	730	500
5	879	874	859	727	860	852	711	863	835	671
6	879	876	864	843	880	858	844	872	854	803
7	1379	1356	1245	934	1366	1270	1009	1381	1383	1375
8	1559	1541	1460	1140	1547	1487	1232	1559	1555	1495
9	1560	1554	1519	1289	1562	1515	1369	1561	1560	1505
10	1600	1600	1588	1382	1602	1590	1548	1600	1592	1570

A representation of the first ten eigenvalues for the scenarious considered is made in the Fig. 8-12

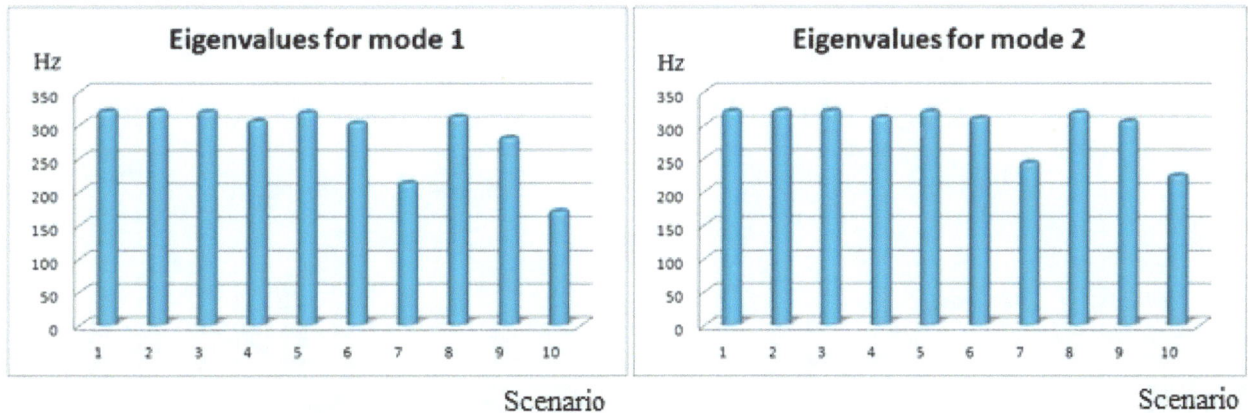

Figure 8. The first two eigenvalues for the ten scenarious

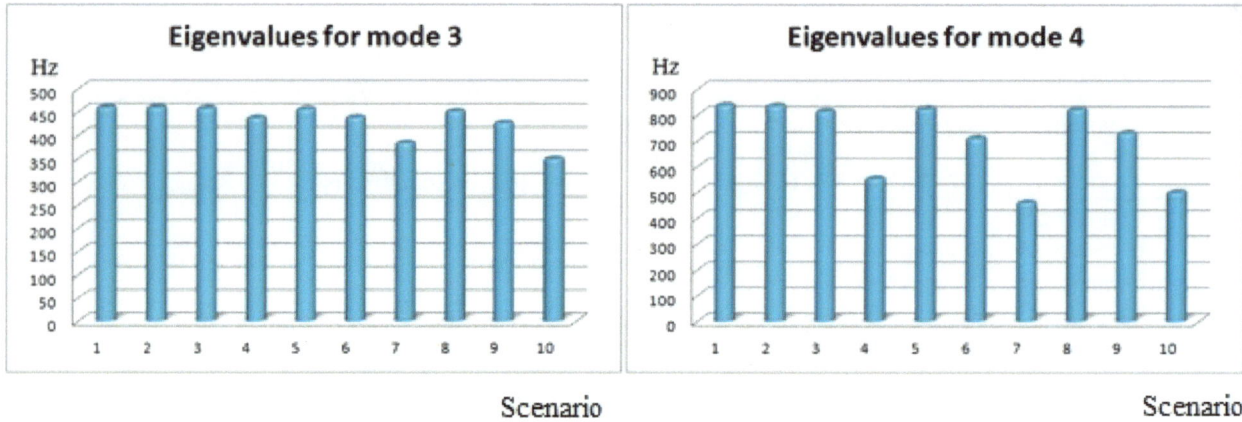

Figure 9. The 3rd and 4th eigenvalues for the ten scenarious

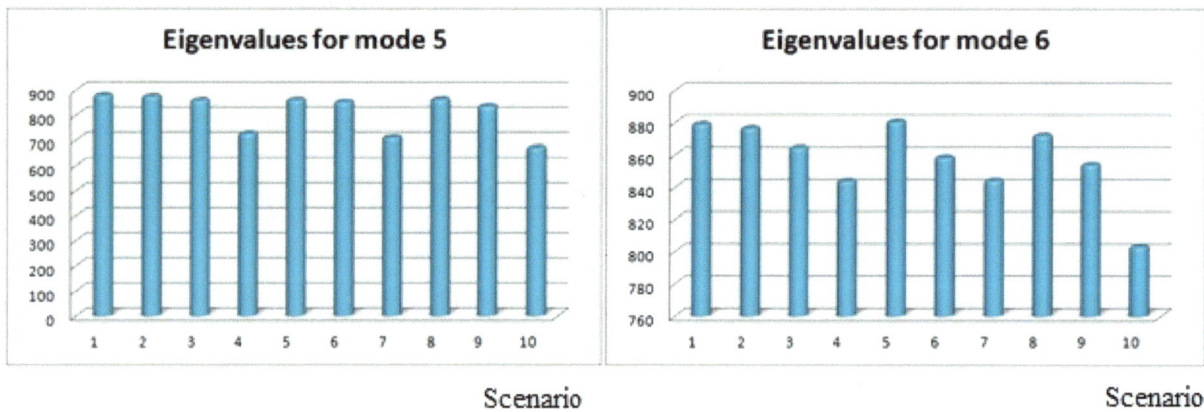

Figure 11. The 7th and 8th eigenvalues for the ten scenarious

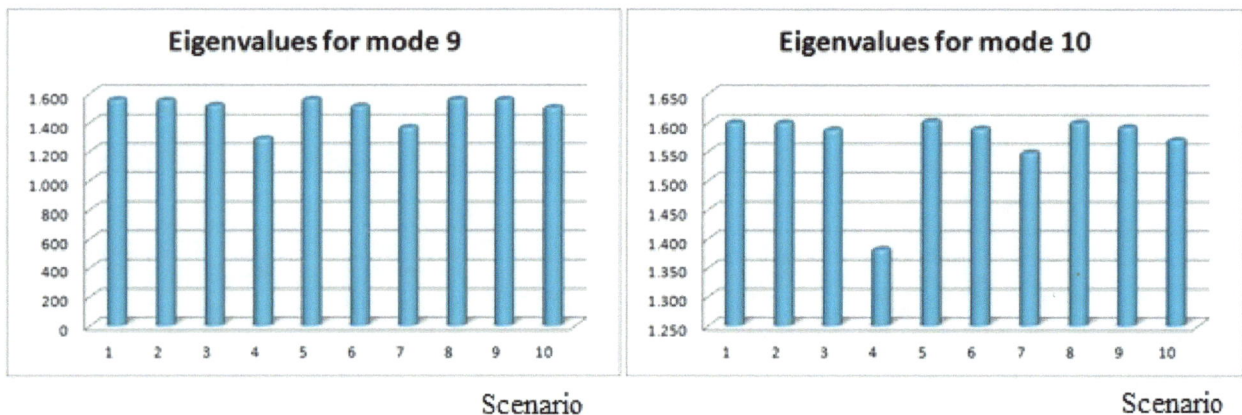

Figure 12. The 9th and 10th eigenvalues for the ten scenarious

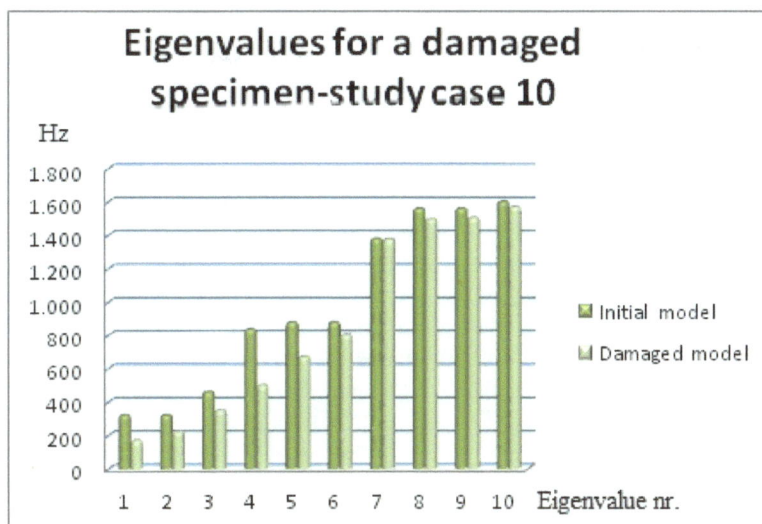

Figure 13. The first 10 eigenvalues for the scenario nr. 10 (initial and damaged model)

Symmetric modes gives equal eigenvalues [9], [10] but a crack in the material makes one of the eigenmodes to keep his value while the other will change. That is why it is necessary to consider various eigenmodes each time to make comparison. Some eigenvalues, despite the emergence of crack and damage of the damper, do not change (it can compare, for example, line 1 and 2 of Table 1).

If we analyze one of the cases, for example, the last case (version 10 - Figure 13) you can see that the eigenmode 7, for example, is not changed although the case study shows a consistent fissure. In contrast to most other eigenvalues are severely affected. It will result so that some eigenvalues are not altered by a change in material continuity while others give some consistent differences. It follows therefore that for detecting a beginning of damage is necessary to consider several vibration modes to be comparable to the initial situation.

Table 2. Eigenvalues for the next 10 scenarious (balls made of aluminium)

Mode no.	Frequency [Hz]								
	Good (11)	D1 (12)	D2 (13)	D3 (14)	D4 (15)	D5 (16)	D6 (17)	D7 (18)	D8 (19)	D9 (20)
1	327	327	326	312	325	309	217	318	286	174
2	327	328	328	317	326	316	248	325	311	228
3	467	467	465	445	464	446	390	459	434	356
4	855	853	831	565	841	725	471	839	745	511
5	895	891	876	741	877	870	725	879	851	684
6	895	893	881	861	897	875	861	888	871	819
7	1401	1379	1265	951	1389	1291	1029	1404	1406	1398
8	1592	1574	1489	1164	1579	1516	1258	1592	1588	1526
9	1592	1587	1550	1315	1594	1546	1394	1594	1593	1536
10	1634	1634	1621	1408	1637	1624	1580	1634	1626	1605

When we use balls made by aluminum the results are presented in Table 2. The conclusions analyzing the values presented in Table 2 offer similar conclusions as for the values presented in Table 2, from a qualitative point of view. The density of aluminum being smaller as the density for steel, the quantitative results differ.

4. CONCLUSION

In the operation of a vibration damper, insert mode of the balls inertial in the mass damper rubber makes it vulnerable to damage. It is, of course, be very useful to detect early failures of this damper before damage. It can thus avoid significant material expenses. A periodic visual inspection requires expenses and waste of time. For this reason it can be very useful to find a method of early detection of cracks in the rubber body. In this paper we propose the use of their eigenfrequencies damper measurement and comparison to baseline. If there are significant differences the damper may be replaced. As seen in the presented figures, some eigenfrequencies can differ more or less, depending on the place and dimension of the fissure. For this reason it should be analyzed and compared multiple frequencies simultaneously. We must not confine ourselves to the study of a single frequency of failure because there may be cases where the eigenfrequency corresponding to a particular mode of vibration, not be influenced by the mechanism of destruction, but other frequencies are affected. The analysis was carried out for a single shock absorber, attached at one end. In practice it may be more useful to consider the damper mounted in the mechanical system. This arrangement makes its eigenfrequencies of the damper to change. They can be determined from a more complex calculus but the findings of this study do not change. From the results presented it can be concluded that a change of eigenfrequency more than 5% can pull the alarm on the possibility of cracks in the structure while a difference of more than 10% should lead to a revision of the whole assembly to determine the cause of this change.

REFERENCES

[1] Carotti,A., Turci,E., A tuning criterion for the inertial tuned damper. Design using phasors in the Argand–Gauss plane. Applied Mathematical Modelling, 23(3), 199–217, 1999.
[2] Claeys,C. C., Pluymers,B., Sas,P., Desmet,W., Design of a resonant metamaterial based acoustic enclosure, Proceedings of ISMA2014, 3351–3358, 2014.
[3] Gillich,G.-R., Praisach,Z.-I., Iavornic,C.M., Reliable method to detect and assess damages in beams based on frequency changes. ASME 2012 International Design Engineering Technical Conferences and Computers and Information in Engineering Conference. pp. 129-137, 2012.
[4] Gillich,G.-R., Praisach,Z.-I., Damage-patterns-based method to locate discontinuities in beams. SPIE Smart Structures and Materials - Nondestructive Evaluation and Health Monitoring. 869532-8, 2013.
[5] Neubert, V.H., Design of Damping and Control Matrices by Modification of Eigenvalues and Eigenvectors, Shock and Vibration, Volume 1, Issue 4, Pages 317-329, 1994.
[6] Nobari, A.S., Ewins, D.J., On the Effectiveness of Using Only Eigenvalues in Structural Model Updating Problems, Shock and Vibration, Volume 1, Issue 4, Pages 339-348, 1994.
[7] Praisach,Z.-I., Gillich,G.-R., Birdeanu,D.E., Considerations on natural frequency changes in damaged cantilever beams using FEM. International Conference on Engineering Mechanics, Structures, Engineering Geology, International Conference on Geography and Geology-Proceedings. pp. 214-219, 2010.
[8] Shimkin,N., Estimation and Identification in Dynamical Systems. Israel Institute of Technology, Department of Electrical Engineering, Lecture Notes, 2009, Chapter 3, Fall 2009.
[9] Vlase,S., Danasel,C., Scutaru,M.L., Mihalcica,M., Finite Element Analysis of a Two- Dimensional Linear Elastic Systems with a Plane "Rigid Motion", Romanian Journal of Physics; 59(5-6), 2014.
[10] Vlase,S., Teodorescu,P.P., Itu,C., Scutaru,M.L., Elasto-Dynamics of a Solid with a General "Rigid" Motion Using FEM Model. Part II. Analysis of a Double Cardan Joint. Romanian Journal of Physics, 58 (7-8), 2013.
[11] ***, "Inertial Damper for Stepper Motor" Focus, I. (2012). Dmp 20 / 29 / 37, 42. Phytro Corp. see on www.phytron.com/DMP April 2015.

6th International Conference
"Computational Mechanics and Virtual Engineering "
COMEC 2015
15-16 October 2015, Braşov, Romania

CONSTRUCTIVE SOLUTION, USING FINITE ELEMENT METHOD FOR GLASS CERAMIC TOOTH

Elena Anca Stanciu

Transilvania University of Brasov, Brasov, Romania, anca.stanciu@unitbv.ro

Abstract: The paper presents a theoretical approach regarding the mechanical behavior of glass ceramic for tooth, this has inside FILTEK Z250 . As the most exposed part of the tooth is the sight hole, the research was focused upon the stress of this part, which was subjected to the action of the stored materials pressure. The results can be easily accessed and the input/output values of the required parameters may be identified at any point of the geometric domain. The replacement in the ideal physical system might identify the modeling errors.

Keywords: finite element method, glass ceramic, stress.

1. INTRODUCTION FEM

Finite element method (FEM) becomes more and more a general method used for solving different types of complex problems concerning both stationary and non-stationary phenomena from all engineering fields but also in other activity and research areas.

As far as the stress and deformation are concerned we may observe that the internal mechanical work is linked to three components of the stress in 2D coordinates, the normal plane component of the stress does not involve the canceling of other strains or stresses.

From mathematical point of view, the problem is very similar to that of plane stress and deformation analysis, this is why the situation may be regarded as two dimensional.

By symmetry, the two components of the displacements in any 2D section of the body along the symmetry axis, completely defines the deformation state and obviously the stress state.

In order to control the complexity of the problem and "filter" the irrelevant aspects we need to accomplish a suitable mathematical model. This model should consider the fact that we are dealing with an anisotropic material, consisting of several layers and also that the loads and deformations along the contours are difficult to be obtained.

The internal stress and deformation field is locally influenced by the relative difference between the constituents' properties, their size, shape and relative orientation as well as by the geometry of the repeating structures that form the glass ceramic.

2. MATHEMATICAL MODELLING WITH FEM

The main part of the process is, as shown in the diagram, the mathematical model. This is mostly an ordinary equation or a differential one, developed in space and time. A discrete model with finite elements is generated by help of the variation form of the mathematical model. This stage is called meshing. The FEM equations are solved using an equation solver that will provide a discrete solution.

For example, if the mathematical model is represented by a Poisson equation, the physical achievement can be represented by the heat conduction in a bar or a problem of electric charge distribution. This stage is not always necessary and may be eliminated. The FEM meshing can be done without any reference to the modeled process physical aspects.

The concept of error occurs when the discrete solution is replaced in the mathematical model. This is generally named checking. The solution error represents the extent to which the discrete solution does not check the discrete equations. More relevant are the meshing errors, representing the extent to which the discrete solution does not check the mathematical model.

3. FEM ANALYSIS FOR GLASS CERAMIC TOOTH

The model was achieved using MSC Patran preprocessor/postprocessor and MSC Nastran processor. In the preprocessing stage, the finite elements geometric modeling requires the finite element model, which will be finally solvable by help of the programs kit meant for this purpose.

A finite element modeling requires the material behavior modeling, selection and personalization of finite elements, finite elements structure generation, introduction of boundary conditions and loads.

The analysis and solution of the finite element model, elaborated during preprocessing requires the preliminary setting of the solving parameters and the execution of the specific program modules. During this stage, the information and error messages occurred while the program is running should be carefully monitored.

The post processing of the results obtained after solving the finite elements model assumes the visualization of the deformed and animated state of the studied structure and also the visualization of various parameters using lists, diagrams and fields.

The generation of the geometric model using elementary entities was achieved by maintaining the continuity in the passing areas between one entity and the other.

The geometric modeling previous the meshing requires the generation of closed contours consisting of lines for plane areas or surfaces. In figure 1 we presented the detailed model geometry.

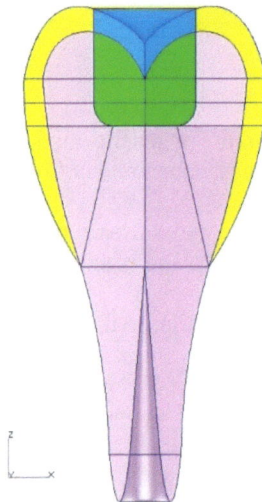

Figure 1: Geometric model of the glass ceramic tooth

Then the model will be analyzed by help of MSC Nastran processor but before running the file we need to do some previous checking in order to validate the finite elements model, as follows:

- determination of the distance between two locations or nodes;
- determination of the angle between two directions determined by three point, one of them being considered as origin;
- identification of common points;
- identification of common lines;
- identification of common nodes and joining them;
- identification of nodes belonging to a selected plane, with the possibility of moving to this plane of the nodes from the adjacent area;

- identification of the common finite elements;
- determination of a finite element distortions;
- identification of the normal in a plane finite elements group and comparing them to a given direction;
- determination of mass properties for the finite elements;
- checking the geometric boundary conditions;
- determination of the loading forces sum in a node.

Then, during post processing, the output data will be associated to both the nodes and the finite elements.
The output data corresponding to the nodes, usually include the problem unknowns, like displacements, temperatures, pressures, velocities.
The output data corresponding to the finite elements are different from one element to another, for example the internal forces, strains, deformation energy.
As material placed in the cavity (material for fillings) is Filtek Z550. These classic composite tubes are inserted into the cavities but treviscoasa is then polymerized, becoming solid.
The thickness of the ceramic crown (porcelain) is about 1 mm, we're talking about modern cearamics (Galss ceramic, zirconia) not metal cearmics. The root of remains untouched and removed from the coronal tooth.

Table 1: Materials properties used in FEA model

Material	Elastic modulus (GPa)	Poisson's ratio	Density (g/cm³)
Glass ceramic	70	0.19	2.40
Zirconia	205	0.19	2.40

Table 2: Properties of restorative materials and tooth tissue

	Young's Modulus	Poisson's ratio	Compressive strength (Mpa)	Tensile strength (Mpa)	Shear strength (Mpa)
Composite	19 GPa	0.24	277	45	122
Amalgam	50 GPa	0.29	388	50	188
Porcelain	69 GPa	0.25	172	110	34
Enamel	80 GPa	0.30	384	10.3	90
Dentin	20 GPa	0.31	297	98.7	138
Pulp	2.07 MPa	0.45	N/A	N/A	N/A

MSC Patran model is analysed in 3D 2008R2 for tooth, containing dentin, enamel, ceramic filler above, material properties and strengths (100 N applied distributed on opposite sides of the tooth means) attached to the bottom (Figure 1).
We analyzed one case, considering the materials to dentin and enamel (enamel) of Table 2, stuffing Filtek Z250 values of material from Composite in Table 2, and pottery (above the filling) with values from Glass ceramic table 1. [2]

Patran 2008r2 15-Sep-15 10:02:55
Fringe: Default, A1 Static Subcase, Stress Tensor. , von Mises. (NON-LAYERED)

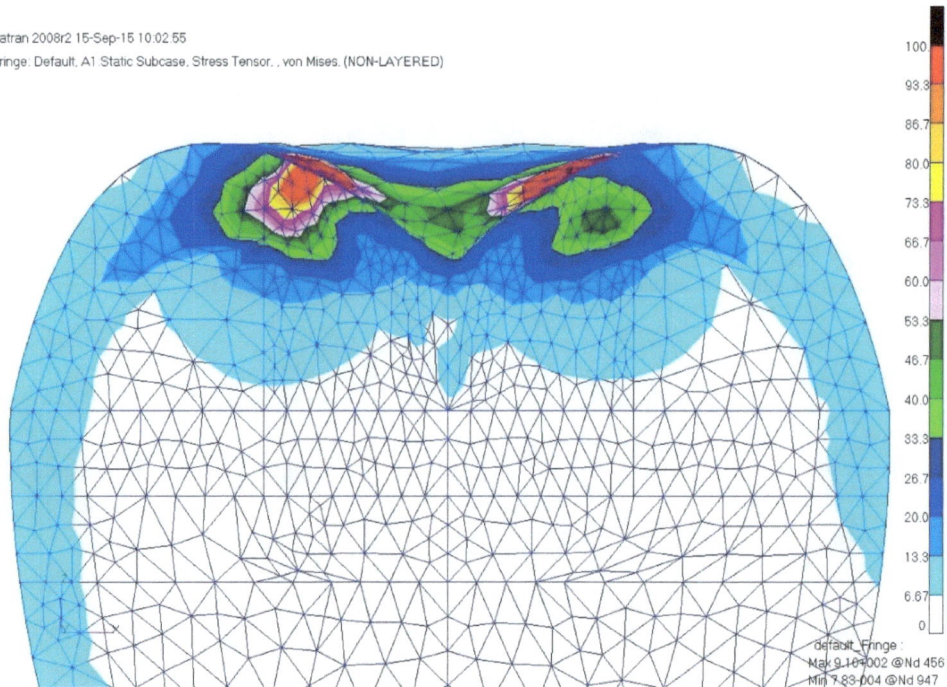

Figure 2: The maximum Von Misses stress distribution

Table 3: Results t
ooth compression stress

Properties	Results
Stiffness	1004000 N/m
Young's Modulus	390,59 MPa
Load at Maximum Load	1095,2 N
Stress at Maximum Load	38,736 MPa
Machine Extension at Maximum Load	0,761 mm
Extension at Maximum Load	0,761 mm
Strain at Maximum Load	0,06915
Percentage Strain at Maximum Load	6,915
Work to Maximum Load	461 Nmm
Load at Maximum Extension	-30,360 N
Stress at Maximum Extension	-1,0738 MPa
Machine Extension at Maximum Extension	0,766 mm
Extension at Maximum Extension	0,766 mm
Strain at Maximum Extension	0,069636
Percentage Strain at Maximum Extension	6,9636
Work to Maximum Extension	463 Nmm
Load at Break	1067,0 N
Stress at Break	37,738 MPa
Machine Extension at Break	0,761 mm
Extension at Break	0,761 mm
Strain at Break	0,069156
Percentage Strain at Break	6,9156
Work to Break	461 Nmm
Tensile Strength	38,736 MPa

3. CONCLUSION

The maximum stress was at Young's Modulus 910 MPa, for Von Misses in FEM comparation with material testing at copresion it is 390 MPa.
Following the experimental researches and also the studies based on FEM we conclude that the material thickness is oversized.

ACKNOWLEDGMENT

This paper is supported by the Sectorial Operational Programme Human Resources Development (SOP HRD), financed from the European Social Fund and by the Romanian Government under contract number POSDRU/159/1.5/S/134378

REFERENCES

[1] Guillaume Couegnata, Siu L. Fokb, Jonathan E. Cooperb, Alison J.E. Qualtroughc, , Structural optimization of dental restorations using the principle of adaptive growth, doi:10.1016/j.dental.2005.04.003
[2] Enescu,I., Stanciu, A., Finite elements, Transilvania University Publishers, 2007, ISBN 978-635-947-7.
[3] MSC/NASTRAN for WINDOWS, Version 2.0, Users manuals.
[4] Thompson, E. G., Introduction to the Finite Element Method. Theory. Programming and Applications J., New York, Wiley & Sons Publishers, 2004.

6th International Conference
"Computational Mechanics and Virtual Engineering "
COMEC 2015
15-16 October 2015, Braşov, Romania

REGULATIONS IN THE FIELD OF USING MEDICAL DEVICES. OVERVIEW

Al. Bejinaru Mihoc[1], A. M. Botez[1], L. G. Mitu[1]
[1] Transilvania University of Brasov, Braşov, ROMANIA,

Abstract: Medical devices are medical-technical products of great complexity in terms of the patients health, of functionality and constructiveness. They are built in a wide range of categories, groups of categories and typo dimensions. They are used worldwide on a wide variety of patients: children, youth, adults, elderly, both men and women. In this context, the paper offers the definitions of medical devices, presented at an international level. Also presents in a systemic manner issues regarding basic rules from the field of using medical devices, developed at an international level. The book is useful for specialists and those interested in this field.
Keywords: medical device, medical diagnostics, directive, classification

1. INTRODUCTION

Medical devices are medical technical-products particularly complex, from sophisticated computers to lingual wooden spatulas [30], important for the health and behavior of humans and living beings in general [1]. They have a very important role in any medical diagnosis (Fig.1). The action of a medical device is specific and personalized, differing from that of a drug, meaning that it is not metabolic, immunological or pharmacological.

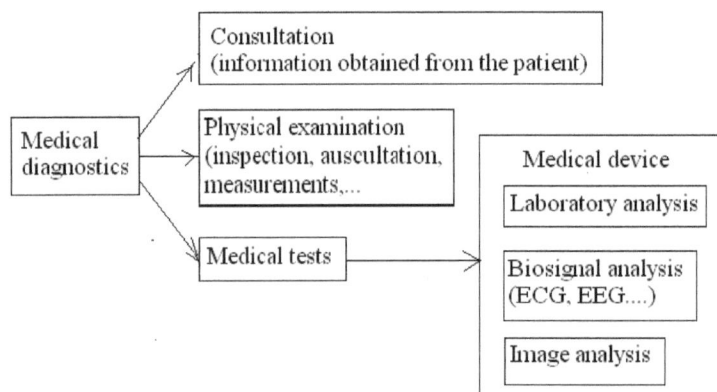

Figure 1 Medical diagnostics and medical device, after Strzelecki [12]

There is a wide range of categories and types are medical devices. For example, in 2011, the World Health Organization (WHO) estimated using 10,000 categories of medical devices in medical practice and manufactured between 90,000 and 1.5 million construction types Cynober [3].

Medical devices are designed, manufactured and used with the help of extensive theoretical and practical knowledge in multiple areas, Lewiner and Le Pape [5]: biomechanics, medicine, biology, chemistry, physics, electronics, computer science etc. At this moment we can talk about the medical device industry, Chriqui [5] as a sub-field of biomedical engineering peak and also a strategic sector in terms of economic and medical health industry.

The definition of the medical device is governed by rules elaborated at European, national, and USA level etc. For instance is presented , selectively, definitions used in the USA, Europe and Romania:

- the Global Harmonization Task Force (USA) [30]: „Medical device - means any instrument, apparatus, implement, machine, appliance, implant, *in vitro* examination (reagents, calibrators, sample collection devices, control materials, and related instruments or apparatus), software, material or other similar or related article, intended by the manufacturer to be used, alone or in combination, for human beings for one or more of the specific purposes of:

  • diagnosis, prevention, monitoring, treatment or alleviation of disease;
  • diagnosis, monitoring, treatment, alleviation of or compensation for an injury;
  • investigation, replacement, modification, or support of the anatomy or of a physiological process;
  • supporting or sustaining life;
  • control of conception;
  • disinfection of medical devices;
  • providing information for medical purposes by means of *in vitro* examination of specimens derived from the human body and which does not achieve its primary intended action in or on the human body by pharmacological, immunological or metabolic means, but which may be assisted in its function by such means'';

- in Europe the medical device has the following definition [21]: „any instrument, apparatus, appliance, material or other article, whether used alone or in combination, including the software necessary for its proper application intended by the manufacturer to be used for human beings for the purpose of:

  • diagnosis, prevention, monitoring, treatment or alleviation of disease;
  • diagnosis, monitoring, treatment, alleviation of, or compensation for an injury or handicap;
  • investigation, replacement or modification of the anatomy or of a physiological process;
  • control of conception, and which does not achieve its principal intended action in or on the human body by pharmacological, immunological or metabolic means, but which may be assisted in its function by such means" Brolin [2];

- according to the Food and Drug Administration (FDA)(USA) the medical device is defined as: „A device is an instrument, apparatus, implement, machine, contrivance, implant, in vitro reagent, or other similar or related article, including any component, part, or accesory, which is:

  • recognized in the official National Formulary, or the United States Pharmacopoeia, or any supplement to them;
  • intended for use in the diagnosis of disease or other conditions, or in the cure, mitigation, treatment, or prevention of disease, in man or other animals;
  • or intended to affect the structure or any function of the body of man or other animals, and which does not achieve its primary intended purposes through chemical action within or on the body of man or other animals and which is not dependent upon being metabolized for the achievement of its primary intended purposes'';

- ISO 13485 defines the medical device: „ Any instrument, apparatus, implement, machine, appliance, implant, in vitro reagent or calibrator, software, material or other similar or related article, intended by the manufacturer to be used, alone or in combination, for human beings for one or more of the specific purpose [31] of diagnosis, prevention, monitoring, treatment or alleviation of disease, diagnosis, monitoring, treatment, alleviation of /or compensation for an injury, investigation, replacement, modification, or support of the anatomy or of a physiological process, supporting or sustaining life, control of conception, disinfection of medical devices, providing information for medical purposes by means of in vitro examination of specimens derived from the human body and which does not achieve its primary intended action in or on the human body by pharmacological, immunological or metabolic means, but which may be assisted in its function by such means" [25],

In our country, Romania, Law no. 176/2000 defines the medical device: „ any instrument, apparatus, machinery, material or other article used alone or in combination, including the software necessary for it to be applied correctly , intended by the manufacturer to be used for human beings and which does not achieve its principal intended in / or on the human body by pharmacological , immunological or metabolic means, but which may be assisted in its function by such means in order:

  • diagnosis, prevention, monitoring, treatment or alleviation of pain;
  • diagnosis, monitoring, treatment or compensation for an injury or disability;
  • investigation, replacement or modification of the anatomy or of a physiological process;

control of conception''.

From the definitions presented can be noticed that medical devices are defined as items used for medical purposes, being characterized by a high constructive and functional diversity. The conditions of design, manufacture and use [10], [15] are determined primarily by operational requirements and the specific properties and behavior of materials (reliability, mechanical, thermal and chemical behavior, etc.) from which they are made, generically referred to as biomaterials. For these reasons, the issue of designing, manufacturing, clinical use and rehabilitation of medical devices requires collective made up of specialists from different fields: physics, mechanical engineering, metallurgy, bio-engineering, veterinary medicine, industrial design etc. Figure 2 shows, in a systemic form, the issue of manufacturing a medical device.

2. TECHNICAL REQUIREMENTS

For identification and classification of medical devices various criteria are used [4], [6], [8], [28], [29]. The most criterion used is ISO 15225 [24] and the GMDN code (The Global Medical Device Nomenclature). The GMDN code is a comprehensive system of internationally recognised coded descriptors used to generically identify medical devices products'' (Tab.1).

Table. 1 Classification of medical device, after standard ISO 15225 [24]

Category	Description
01	Active implantable devices
02	Anaesthetic and respiratory devices
03	Dental devices
04	Electro mechanical medical devices
05	Hospital hardware
06	In vitro diagnostic devices
07	Non-active implantable devices
08	Ophthalmic and optical devices
09	Reusable devices
10	Single use devices
11	Assistive products for persons with disability
12	Diagnostic and therapeutic radiation devices
13	Complementary therapy devices
14	Biological-derived devices
15	Healthcare facility products and adaptations
16	Laboratory equipment

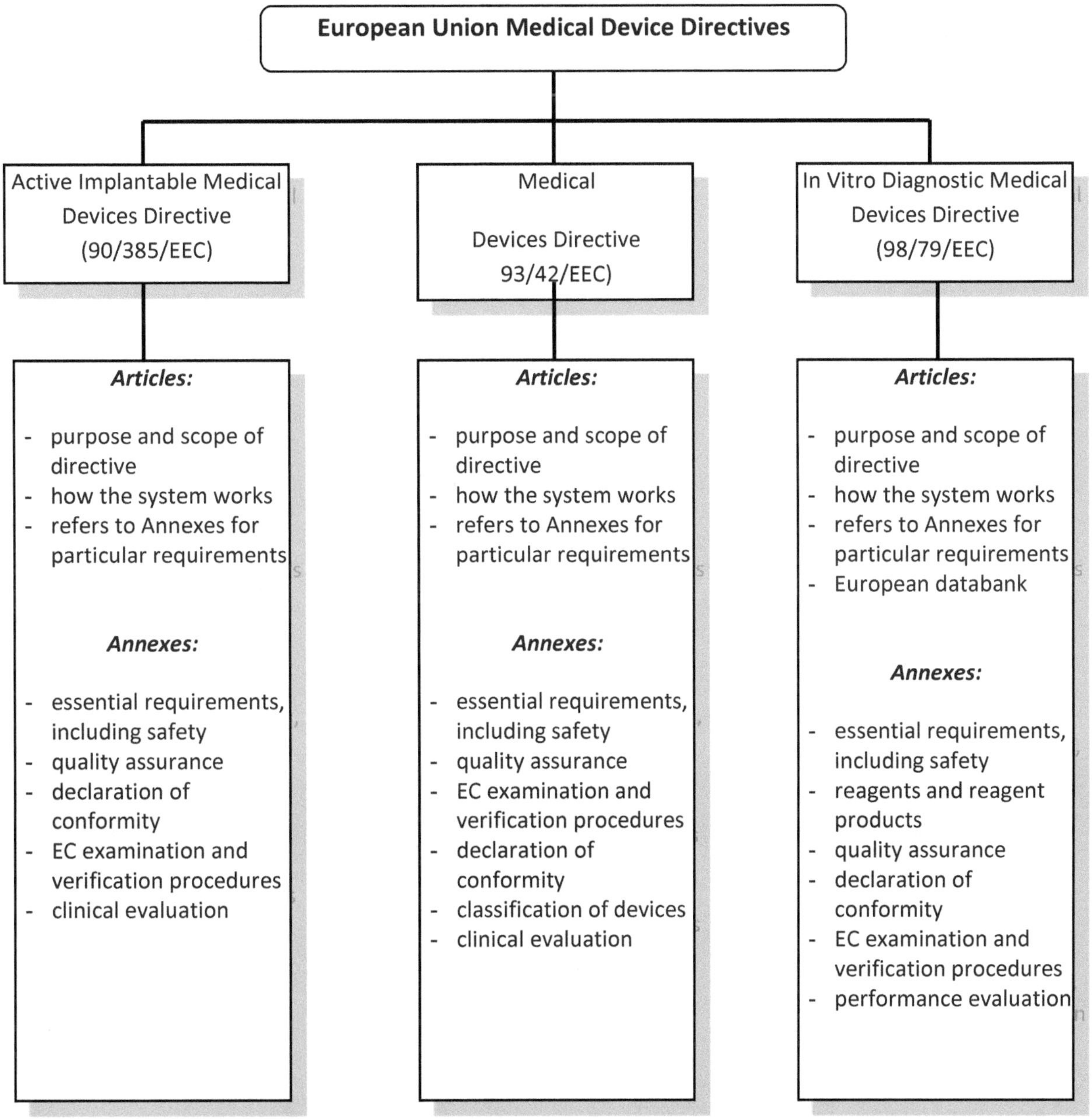

European Union Medical Device Directives

Active Implantable Medical Devices Directive (90/385/EEC)	Medical Devices Directive 93/42/EEC)	In Vitro Diagnostic Medical Devices Directive (98/79/EEC)
Articles: - purpose and scope of directive - how the system works - refers to Annexes for particular requirements *Annexes:* - essential requirements, including safety - quality assurance - declaration of conformity - EC examination and verification procedures - clinical evaluation	*Articles:* - purpose and scope of directive - how the system works - refers to Annexes for particular requirements *Annexes:* - essential requirements, including safety - quality assurance - EC examination and verification procedures - declaration of conformity - classification of devices - clinical evaluation	*Articles:* - purpose and scope of directive - how the system works - refers to Annexes for particular requirements - European databank *Annexes:* - essential requirements, including safety - reagents and reagent products - quality assurance - declaration of conformity - EC examination and verification procedures - performance evaluation

Figure 2 The systemic approach to European regulatorsin the field of medical devices, after Shefelbine et al.[11]

In Europe, the legislation on medical devices is based on three basic directives (Fig.2) [17], [18], [19], [20]:

• Directive 90/385 / EEC. It includes regulations on active implantable medical devices (Active Implantable Medical Devices Directive-AIMD) [18], [14];

• directive 93/42/EEC, Medical Devices Directive MDD [27], [14]. It is amended by Directive 2007/47/EC). Directive 93/42 / EEC is based on the systemic definition of two main terms:,, medical device "and ,,accessory " of the medical device. In the acceptation of the Directive, a product may be considered accessory to a medical device only if specified by the device manufacturer;

• directive 98/79/EC, *InVitro* Diagnostic Medical Devices IVD [19], [31]. According to this directive, in vitro diagnostic medical devices are made [20] of reagents, reaction products, apparatus, equipment, tools, etc. intended for medical examination of samples taken from the human body (living organism) such as tissue, blood, urine, etc. for the purpose of diagnosing the existence of pathological conditions, namely disease, congenital malformations, to monitor the state of health of the individual being investigated, for therapeutic treatment, etc. For example [20]: Hepatitis or HIV tests, Clinical chemical tests, Coagulation test systems , urine test strips etc. In essence [31], directive 98/79 / EC regulates areas and rules that govern the use of medical devices from the conceptul and medical point of view for in vitro diagnostic use in the human body respectively a living organism. In figure 2 are shown, synthetically, issues that are regulated in the three directives.

At international level, many standards are developed [30], [31], [29] more than 100 [26] in which are covered the issues regarding quality characteristics of a medical device Ramakrishna et al. [9], Zhong [13]

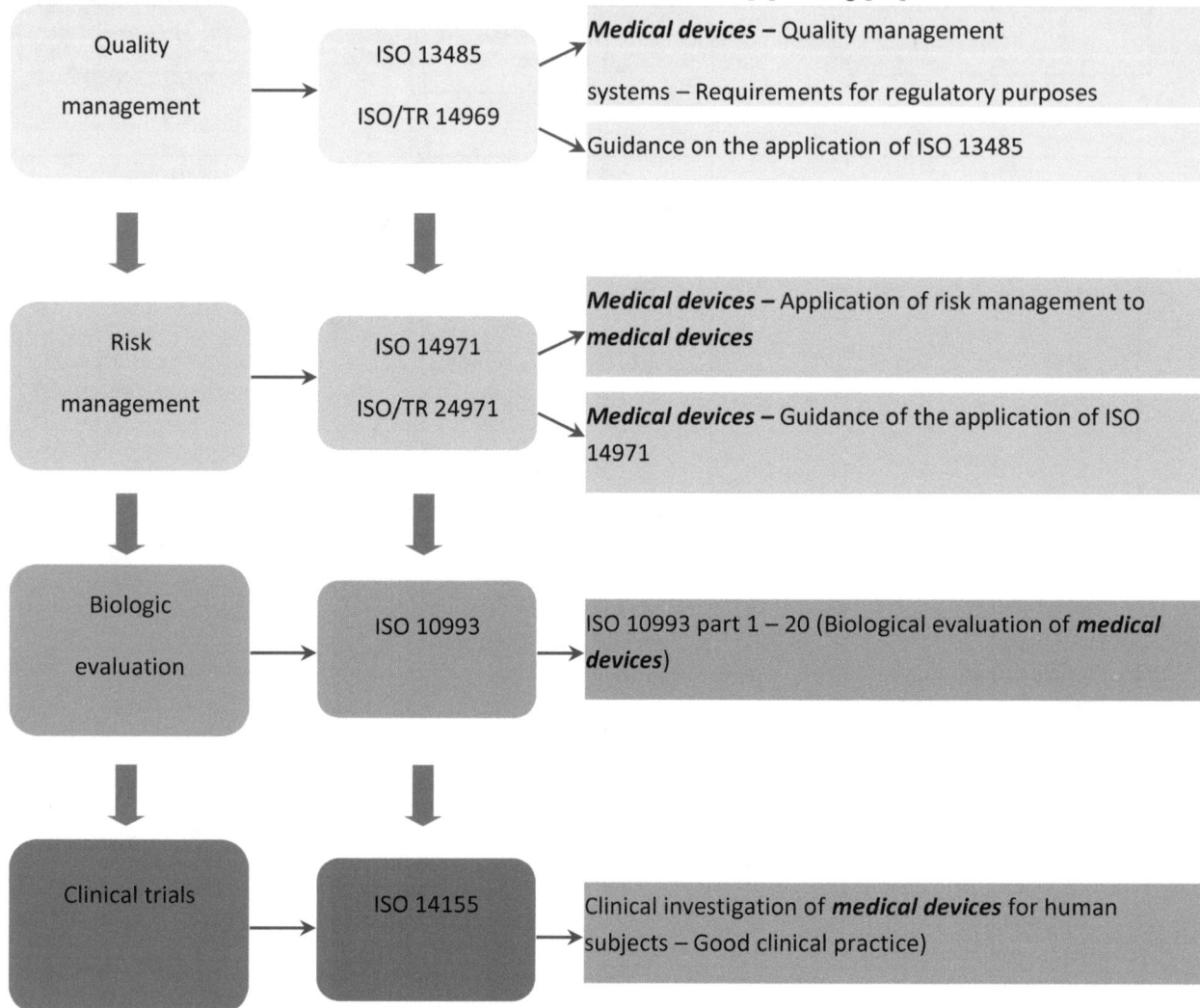

Figure 3 Basic standards ISO for medical devices, after Ramakrishna et al. [9]

- quality management;
- risk management;
- biologic evaluation;
- clinical trials.

In essence, each of the standards mentioned in Figure 3 must be applied in systemic interdependence. The ISO 10993 standard Part 1-20 has a special role regarding biological evaluation of any medical device regardless of manufacturer.

3. CONCLUSION

Medical devices are defined by the following main functional purposes:
- diagnosis, prevention, control, treatment and mitigation of a disease;
- diagnosis, control, treatment, mitigation or compensation for an injury or handicap;
- the study, relocating or modification of the anatomy, or of a physiological process, etc.

The manufacturer of a medical device and the user is obliged to observe the legislation. It builds on the European Union Medical Device Directives: Directive 90/385 / EEC; Directive 93/42 / EEC and Directive 98/79 / EEC. Each of these directives are accompanied by annexes showing the specific requirements from which stands out essential requirements, including safety.

REFERENCES

[1] Beaudet, Th., Couty, E., La place des dispositifs médicaux dans la stratégie national de santé. available from: www.opengrey.eu/item/display/10068/747939, accesed: 2015.

[2] Brolin, S., Global regulatory requirements for medical devices. pp. 1-51, Mälardalen University Department of Biology and Chemical Engineering, available from: https://www.diva-portal.org/smash/get/diva2:121327/FULLTEXT01.pdf, accesed: 2015.

[3] Cynober, M., Le marché des dispositifs médicaux. Analyse et recommandations. Agence d'intelligence économique, Feanche-Comté, 2011, available from: www.aiefc.org/.../le-march-des-dispositifs-m-dicaux-et...

[4] Lemaitre, J., *Biomatériaux*. Dispositifs médicaux et réglementation. Laboratoire de Technologie des Poudres-EPFL MX Ecublens, CH-1015 Lausanne Suisse, ltp2.epfl.ch/Cours/Biomat/BioMat-12.pdf, 2007.

[5] Lewiner, J., Le Pape., Le dispositif médicale innovant. Attractivité de la France et développement de la filière. Centre d'analyse stratégique, France, 2012, available from: biodesign.stanford.edu/.../18_dispositifs_medicaux_fi...

[6] McCormack, F., Medical Device Identification. Understanding the trends and developments. 2012, available from: www.gs1.org/.../healthcare/.../TUE_7_1_Fiona_McCormack_GS1_Healt...

[7] McKeen, W. L., 3 Plastics used in medical devices. În: Handbook of polymer applications in medicine and medical devices, (eds.) Modjarrad and Ebnesajjad, ISBN: 978-0-323-22805-3, Elsevier Inc., Amsterdam, 2014.

[8] Paleos, A. G., What is Biocompatibility?. Pittsburgh Plastics Manufacturing Inc. 2012, available from: www.pittsburghplastics.com/.../Biocompatibility%20for%20website.pdf.

[9] Ramakrishna, S., Tian, L., Wang, Ch., Liao, S., Teo, E. W., Medical devices. Regulations, standards and practice. ISBN: 978-0-08-100291-9, Elsevier B.V., 2015.

[10] Ratner, B. D., Hoffman,S. Al., Schoen, J. Fr., Lemons, E. J., Biomaterials science: A multidisciplinary endeavor, În: Ratner, B. D., Hoffman, Al. S., Schoen, Fr. J., Lemons, J. E., (eds), Biomaterials science. An Introduction to materials in medicine, 2nd Edition, ISBN 0-12-582463-7, pp. 1-9, Ed. Elsevier, Academic Press, San Diego, California, 2004.

[11] Shefelbine, S., Clarkson, J., Farmer, R., Good design practice for medical devices and equipment – A framework. ISBN 1-902546-10-5, University of Cambridge Engineering Design Centre, 2002, available from: https://www-edc.eng.cam.ac.uk/.../gooddesignpractice2/requirements%2...

[12] Strzelecki, M., Medical Imaging. Introduction to medical imaging. Biomedical engineering, IFE, 2013.

[13] Zhong, H., Primer: The medical device industry. American Action Forum, 2012, available from: americanactionforum.org/sites/default/files/OHC_MedDevIndPrimer.pdf.

[14] Yaneva-Deliverska1,M., Deliversky, J., Maya Lyapina, M., Biocompatibility of medical devices – legal regulations in the European Union. Journal of IMAB - Annual Proceeding (Scientific Papers), vol. 21, no. 1, 2015.

[15] * * * Assessing biocompatibility. A guide for medical device manufacturers. Pacific BioLabs, Inc. – The Service leader in bioscience testing. 2013, available from: www.pacificbiolabs.com/downloads/Booklet%20Biocompatibility.pdf.

[16] * * * Biomerics biocompatibility guide. Confidential Reviews, pp. 1-10, 2011, available from: biomerics.com/pdfs/Biocompatibility-Guide-2011-Quadrathane-ALC.pdf.

[17] * * * Certification and registration. Medical devices on the European market - ttopstart spotlight. 2013, available from: www.slideshare.net/.../certification-and-registration-medical-devices-on-t...

[18] * * Directive 90/385/EEC - EUR-Lex - Europa. available from: eur-lex.europa.eu/LexUriServ/LexUriServ.do?uri=CONSLEG...en:PDF

[19] * * * Directive 98/79/EC - EUR-Lex - Europa. available from: eur-lex.europa.eu/LexUriServ/LexUriServ.do?uri=CONSLEG...en:PDF

[20] * * * Directive 98/79/EC on a *vitro* diagnostic Medical Device. TÜV SÜD Product Service GmbH, Med-Info, EU, 2012, available from: .de/uploads/images/.../ivd-directive-98-79-ec.pdf

[21] * * * European Commission, Directive 93/42/EEC, published 14/06/1993 and valid to 31/12/2008, available from: http://eur-lex.europa.eu.

[22] * * * International Standard Organisation. ISO 13485:2003, available from: http:// www. iso. org/iso/home/store/catalogue_tc/catalogue_detail.htm?csnumber=36786

[23] * * * Introduction Biocompatibility - Dotolo, available from: www.dotolo.eu/pdf/understanding%20iso%2010993.pdf, accesed: 2015.

[24] * * * ISO 15225: 2010. Medical devices -- Quality management -- Medical device nomenclature data structure.

[25] * * * Lege nr. 176 din 18/10/2000 privind dispozitivele medicale. Publicat in Monitorul Oficial, Partea I nr. 544 din 02/11/2000.

[26] * * * List of international standards for medical devices. available from: w.mdb.gov.my/mdb/documents/.../international%20standard.pdf, accesed: 2015.

[27] * * * Medical Devices Directive 93/42/EEC – Requirements for..., pp.1-17, 2012.

[28] * * * Medical devices: Guidance document - Classification of medical devices. European Commission DG Health and Consumer, MEDDEV 2. 4/1 Rev.9, 2010, available from: ec.europa.eu/.../medical-devices/.../meddev/2_4_1_rev_9_classification_...

[29] * * * Medical devices : Guidance document. European Commission DG Enterprise, MEDDEV 2. 1/1, 1994, available from: ec.europa.eu/health/medical-devices/files/meddev/2_1-1___04-1994_en..

[30] * * * Medical device regulations : global overview and guiding principles. ISBN 92 4 154618 2, World Health Organization, Geneva, 2003, available from: www.who.int/medical_devices/publications/en/MD_Regulations.pdf.

[31] * * * UK Standards for Microbiology Investigations. European Directive on In Vitro Diagnostic Medical Devices (98/79/EC). UK Standards for Microbiology Investigations | Issued by the Standards Unit, Public Health England, 2013.

6th International Conference
"Computational Mechanics and Virtual Engineering"
COMET 2015
15-16 October 2015, Braşov, Romania

THE ANALYSIS OF CYCLICALLY SYMMETRIC STRUCTURES – HISTORICAL REMARKS

Andrei Vasilescu

Technical University of Civil Engineering, Bucharest, ROMANIA, e-mail: andrei.vasilescu@utcb.ro

Abstract: *This paper aims to review and explain the path of evolution of the concepts and techniques to analyze symmetric structures. First domes designed by Schwedler imply a new theory of spatial truss system in the static analysis of determinate and indeterminate problems. The system of equilibrium equations obtained has a circulant determinant. This specific form conducted to complex numbers and some savings appeared. The development of methods to exploit the symmetry properties have depended on achievements in mathematics and informatics. The use of electronic computers within the development of finite element structural analysis package had a spectacular increased at work done in this area. Two main advanced approaches in the analysis of symmetric structures have used: the discrete Fourier transformation (DFT) for rotational symmetry, and group representation theory to exploit any type of symmetry.*

Keywords: *cyclic symmetry, history, Schwedler domes, spectral decomposition, group representation theory*

1. INTRODUCTION

The author of these notes will present his point of view and his conviction about what we can be seen and learned from the history of cyclic symmetry implication in structural analysis. Obviously, it is not possible to mention explicitly all of the important results and their authors. The number of approaches and contributions in this field after 1990 is overwhelming. As a consequence of industrial progress and technologies, the addresses time is divided into periods. One covers the beginnings of the use of steel in constructions like the domes, till the era of digital computers, and one after it.

2. SYMMETRY AND CYCLIC SYMMETRY

Symmetry commonly conveys the idea of harmony and proportion. Symmetry is a special kind of transformation - a way to move an object. In plane geometry, symmetry denotes a balance of the parts of a figure to a central point, line, or plane. Axial *symmetry* is a case in which a figure can be divided by a line into two mirror-image halves. Another common type is *radial symmetry*, in which a figure can be made to coincide with itself if it is rotated about a point. Thus, a symmetric structure is a structure that is left unaltered, geometrically and mechanically, after a symmetry operation. These operations may be reflections, rotations, improper rotation (an improper rotation is a reflection in a plane, followed by a rotation about an axis perpendicular to the plane), translations or dilations/contractions. [1].

3. BEGINNINGS IN THE LATE OF THE 19TH CENTURY

3.1. Golden age of Schwedler dome

Johann Wilhelm Schwedler (1823-1894) was one of the most important Prussian civil engineers of the 19th century. Starting from 1851, Schwedler developed a new chapter of structural analysis: the truss theory. Karl Culmann (1821-1881), Squire Whipple (1804-1888) and Dimitry Ivanovich Jouravski (1821-1891) contributed to this theory, but the truss theory was confined to the late 19th century at the plane systems. Spatial structural systems of buildings such as industrial buildings, railway stations and bridges had an orthogonal structure, so that a breakdown in planar systems was sufficient. Added to this was that the spatial engineering thinking was trained since the beginning of the 19th century by the Descriptive Geometry, where the dominant method was the orthogonal projection put in the form of technical drawings. Schwedler was the first to exceed this stage with the extraordinary strength and clarity of his spatial intuition - named "*stereometric imagination*". Here only two Schwedler domes may be mentioned, which can be admired today.

Figure 1: Johann Wilhelm Schwedler (1823-1894)
Source: Zeitschrift für Bauwesen, 1895.

In 1863, Schwedler completed the dome above the gas container in Holzmarktstrasse 28, Berlin. He was the first engineer for the transition to a spatially dome that should be remembered as "*Schwedler dome*" in the literature. The second is the roof of the municipal gasworks in the Fichtestraße, in Berlin-Kreuzberg, erected in 1875 (Fig. 2). The latter dome has a diameter of 54,9 m and a rise of 12,2 m with an iron consumption of 28,7 kg /m²! This was a remarkable performance that remained unsurpassed in his lifetime [2].

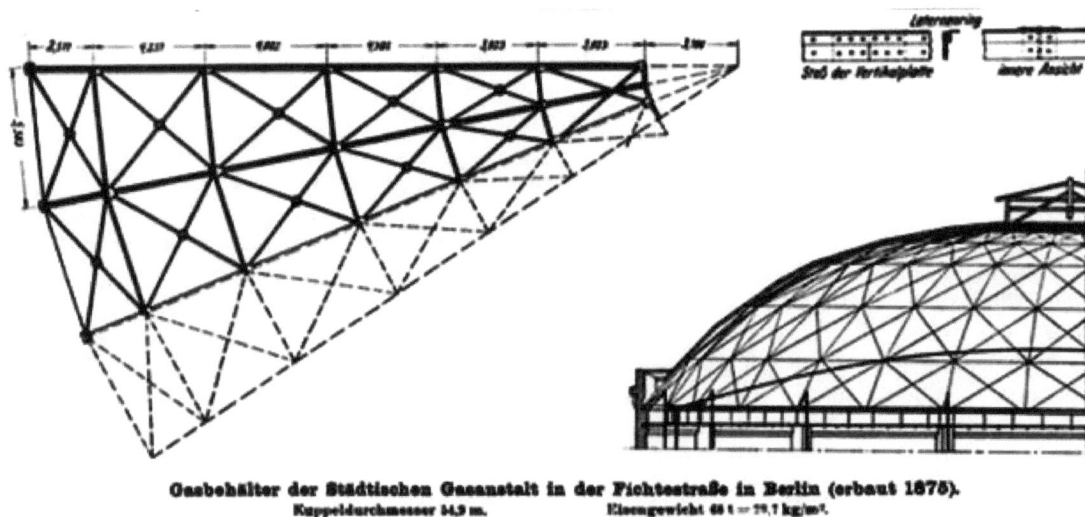

Gasbehälter der Städtischen Gasanstalt in der Fichtestraße in Berlin (erbaut 1875).
Kuppeldurchmesser 54,9 m.   Eisengewicht 68 t = 28,7 kg/m².
Figure 2: Schwedler dome; Source: [2]

Three years later, in 1866, Schwedler has published *"The Construction of the Dome Roof"* [3] in which he reported not only about his first truss domes, but also the theory of a simplified calculation method. The membrane stresses within the dome were projected on the longitudinal and latitudinal lines. The three-dimensional framework, which was statically indeterminate, has particular system form due to the rotationally symmetrical configuration. The main structural elements of Schwedler dome were made of iron. The constructive concept of the rigid structure has two main elements: radially arched rafters and horizontal rings. They are stiffened by trusses in the form of tension crosswise and between rafters and fixed rings. The advantage of spatially cross-braced construction system was that the stability was ensured even with asymmetrical loads and each concentric ring constitutes a solid system that could be prefabricated.

The gained knowledge in this time, permitted to other engineers to design important domes. An example is the existing dome of Romanian Athenaeum in Bucharest, Romania. This building was inaugurated in 1888. One year later, the calculation notes of the dome were published by the Romanian engineer Elie Radu (1853-1931) [4]. He considered only one sector under the axisymmetrical loads.

3.2. The emerging mathematical solution

The equilibrium conditions for trusses with cyclic symmetry are studied by Hans Jacob Reissner. He published in 1908 the paper *"About trusses with cyclic symmetry"* [5]. Reissner considered series of trigonometric functions to analyze the stresses of the elements of a cyclic symmetrical structure.

One qualitative step forward in the analysis of cyclically symmetric domes was performed by Ludwig Mann [6]. He used of the forces method for the static calculation. The cyclicality of unknowns' coefficients of the system of equations is put into correlation with the properties of the nth root of unity.

The two-story Schwedler dome has succeeded Kaufmann in 1921 [7], using the properties of cyclically symmetric systems to calculate the unknowns efforts from the equations system, by means of a cyclic determinant of order n, where n is the number of identical segments. Kaufmann divided the analysis problem into two independent specific tasks of analysis:

   a) Calculation of the dome in a plane with bend rigid circuit ring;
   b) Calculation of the dome in the meridian planes bending stiff continuous rafters.

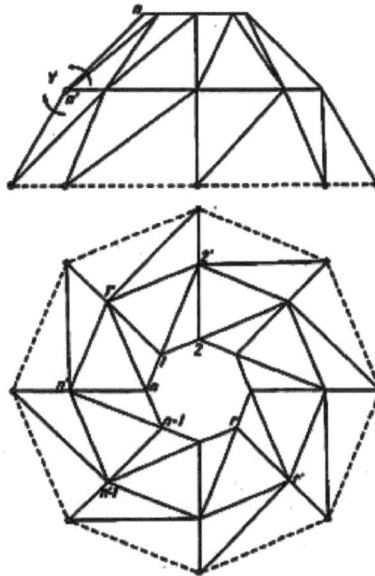

Figure 3: Schwedler dome with *n*=8 order of cyclicity.

The remarkable idea of symmetrical component three-phase systems has been published by Fortescue [8] in 1918. This was one of the first paper considering cyclic symmetry in the field of electrical engineering. Starting from the experiments, he carried out a mathematical model based on harmonic analysis. Later, this type of Fourier analysis will be one of the main instrument in the systematical approach of cyclical symmetric problems.

3.3. Richard Courant and the "birth" of finite element method (FEM)

The publication of Richard Courants's paper "*Variational Methods for Problems of Equilibrium and Vibration*" was a crucial contribution to the development of the FEM [9]. His work is unique.

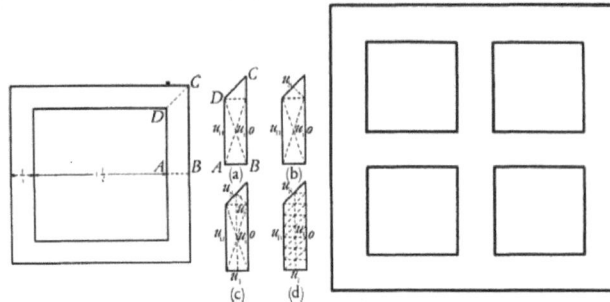

Figure 4

He gave us a body of purely mathematical work, and simultaneously offered the first example in numerical analysis that occur in applying the finite element method. In the appendix entitled "*Numerical treatment of the plane torsion problem for multiply-connected domains*" (Fig. 4), he referred also to the symmetry of his model to reduce the number of the unknowns:

"We exploit the symmetry of the domain and the resulting symmetry of the solution; thus we may confine ourselves to considering only one-eighth of the domain B^*, namely the quadrangle $ABCD$ " [10].

The Conference "*Finite Element Method-fifty years of the Courant element*" organized in Jyväskylä – Finland (1993), was dedicated to celebrate his prior recognition in FEM theory

4. DIGITAL COMPUTERS EPOCH

4.1. Matrix methods of structural analysis

The spearhead of the advanced theories in the structural analysis was carried out by the aircraft industries, in the sixties. The achievements of new technologies in modern digital computers have created the opportunity to elaborate extended and complex mathematical models of aircrafts. In this way, it was possible to obtain more precise results to design lighter and more resistant aircrafts. At that time, it was realized the potential of solving problems in continuum mechanics by using discrete elements.

By using matrix transformation methods it was clearly shown that most structural analysis methods could be categorized as either a force or a displacement method. The displacement method offers the advantage to work with a unique base system of the model and was adopted later like a standard in computing programmes for structures analysis.

The matrix methods of structural analysis developed for use on modern digital computers were systematized in parallel by Argyris in Germany and England [11-13], and Martin and Clough in America [14-15], based on the earlier theoretical papers of well-known scientists: Richard Courant (1943) [9], Walter Ritz (1908) [16], and J. W. S. Rayleigh (1870).

A different approach to the solution of continuum mechanics problems was realized by discrete element idealization. Therefore, analysis models for both continuous structures and frame structures were modeled as a system of elements interconnected at joints or nodes. Ray William **Clough** coined the terminology **finite element method** (FEM) of this new approach, in 1957.

4.2. Techniques and methods to exploit the symmetry

We consider symmetric structures subjected to a general loading. From the beginning of the structural analysis, the exploitation of bilateral or reflection symmetry is well known in structural mechanics by the use of "symmetry and anti-symmetry" techniques [17].

A different class of symmetry is the rotational or cyclic symmetry. Examples are the large radio telescope or space dome structures which can possess high degree of rotational symmetry.

This type of physical symmetric entity can be exploit by **discrete Fourier transform methods**. The first Fourier approach was developed by Fortescue [8] for the analysis of polyphase electrical circuits, as we shown. Renton [18] was the first in application discrete Fourier transformation methods to structural analysis. The "*General theory of cyclically symmetric frames*", is done by Hussey (1966-67) [19-20] in order to investigate the buckling under cyclically symmetric loading. Later, Thomas (1979) have deduced exact methods for solving eigenvalue problems and have demonstrate that any forced vibrations can be decomposed into independent rotating components [21]. The wave propagation in periodic structure was the subject in earlier papers of Mead in 1970 [22], and Orris and Petyt in 1974 [23].

In this type of analysis, the governing matrix (stiffness, mass, etc.) has a circulant or block-circulant form. A circulant matrix is a special matrix where each row vector is rotated one element to the right relative to the preceding row vector. An example of a circulant matrix is:

$$\mathbf{C} = \begin{bmatrix} 1 & 2 & 3 & 4 & 5 \\ 5 & 1 & 2 & 3 & 4 \\ 4 & 5 & 1 & 2 & 3 \\ 3 & 4 & 5 & 1 & 2 \\ 2 & 3 & 4 & 5 & 1 \end{bmatrix} \tag{1}$$

The theory of circulant matrices has been presented by Pipes "*Circulant matrices and the theory of symmetrical components*" [24] in 1966. Davis has published in 1979 his book "*Circulant matrices*" [25]. Characteristic for a mathematician fallen in love with "symmetry", he adopted the motto: "*What is circular is eternal; What is eternal is circular*" expressing in this way the magic of the results from his work.

Several problems of physics involve circulant matrices and block-circulant matrices. A pre-requisite of the Fourier method is that the coordinate system for the load and displacement vectors must have the same rotational symmetry as the structure.

By matrix transformations using the Fourier matrix, the governing matrix is reduced to a block-diagonalise form. This is the way to split the large system in sub-systems of equations. The characteristic of this type of analysis: every $(n \times n)$ circulant matrix has the same eigenvectors, and these are given by the columns of the Fourier matrix $\mathbf{F}$ [26]:

$$\mathbf{F} = \frac{1}{\sqrt{n}} \begin{bmatrix} 1 & 1 & 1 & . & 1 \\ 1 & w & w^2 & . & w^{n-1} \\ 1 & w^2 & w^4 & . & w^{2(n-1)} \\ . & . & . & . & . \\ 1 & w^{n-1} & w^{2(n-1)} & . & w^{(n-1)^2} \end{bmatrix} \qquad \text{where } w = e^{2\pi i/n} = \sqrt[n]{1} \tag{2}$$

Kangwai, Guest and Peliegrino have presented that the Fourier method is a special case of the group representation theory method [26]. The same remark we can find in the works of Bossavit [27-30]. Dinkevici has considered in 1972, the method of spectral analysis, essentially identical to the discrete Fourier method [31].

4.3. Group theory and group representation theory

The best way to describe the full symmetry of any entity is the group theory. The properties of a symmetric structures are valorized at maximum by the group representation theory.

"The mathematical language of symmetry is group theory, and group representation theory is the vehicle for exploiting symmetry in linear problems" have stated Healey and Treacy in 1991 [32]. Another early example is *"Group Theory and its Applications to Physical Problems"* by Hammermesh (1962) [33]. The applications of group theory was studied by Willard Miller in *"Symmetry Groups and Their Applications"* [34], but the publications of Zloković in 1973 and respectively in 1989, are considered to be the first application of group representation theory to structural analysis [35-36]. A good reference (1976) is the doctoral thesis of Fässler *"Application of group theory to the method of finite elements for solving boundary value problems"* [37].

Nowadays, we can observe the increasingly number of articles using symmetry group operations. This is doubtless only the consequence of the universal feature of this method applied by researchers working in advanced new technologies domain. To have an idea of other diversity of domains (molecular vibrations, biological anisotropic hyperelastic materials, spectroscopy of crystals, crystallographic group theory in crystal chemistry, application of point-group symmetries in chemistry and physics), we can quote some relative recent articles [38-42].

4.4. FEM packages and cyclic symmetry

The Fourier approach to be applied for cyclically symmetric structure has been implemented first in the general purpose finite element package NASTRAN, in 1974 [43]. An example of a study in order to reduce computational effort is the paper *"Skyline solver for the static analysis of cyclic symmetric structures"* [44]. The procedure of Skyline has an efficient node numbering result and consequently, the time of processing data is significantly reduced. New techniques to reduce the hardware resources was the substructure method. The programming by means of a basic substructure for the cyclic symmetric analysis also improved the performances in the harmonic (Fourier) analysis [46].

The analysis of cyclically symmetric structures has become a common feature of FEM packages, e.g. ANSYS [47] and ABAQUS [48].

We can not omit the particular case when we consider a structure with a certain type of symmetry subjected to a loading with the same symmetry. The analysis is resume only of one identical substructure: Zienkiewicz in 1972 [49] and Glockner in 1973 [50].

5. INSTEAD OF CONCLUSIONS

The cyclic symmetry remains on-going subject in the theoretical and practical field of scientists. Even in this small field, there is already an enormous amount of work to be done to unearth inestimable treasures. The basic idea is to obtain a reduced problem before analyzing, to construct a problem having fewer unknowns than the original system.

Symmetry and group theory are important tools in analyzing physics and mechanics problems. The exploitation of symmetry via group invariance also yields an efficient computational approach to many problems in mechanical engineering: static, dynamic, stability, wave propagation, etc. Some comments of famous scientists reveal us the fascination of symmetry.

Hermann Klaus Hugo Weyl (1885 - 1955) concludes in his book *"Symmetry"* [51]:

"Symmetry is a vast subject, significant in art and nature. Mathematics lies at its root, and it would be hard to find a better one on which to demonstrate the working of the mathematical intellect.

I hope I have not completely failed in giving you an indication of its many ramifications, and in leading you up the ladder from intuitive concepts to abstract ideas."

Ian Nicholas Stewart (born 24 September 1945) pleaded for the beauty of mathematics in his book *"Why Beauty is Truth: A History of Symmetry"* [52]:

"The implications of symmetry for physics, indeed for the whole of science, are still relatively unexplored. There is much that we do not yet understand. But we do understand that symmetry groups are our path through the wilderness-at least until a still more powerful concept (perhaps already waiting in some obscure thesis) comes along.

In physics, beauty does not automatically ensure truth, but it helps.

In mathematics, beauty must be true-because anything false is ugly."

REFERENCES

[1] Kangwai R.D., Guest S.D., Pellegrino S., "An introduction to the analysis of symmetric structures", Computers and Structures, 71, 671-699, 1999.

[2] Hertwig A., Johann Wilhelm Schwedler, "Sein Leben und sein Werk ", Berlin Wilhelm Ernst & Sohn 1930 tafel. 7.

[3] Schwedler W., "Die Construction der Kuppeldächer ", Zeitschrift für Bauwesen, Bd. 16, 1866, pp. 7-34.

[4] Radu E., "Dome of Romanian Athenaeum in Bucharest" (published in Romanian: "Cupola Ateneului român din Bucureşti"), Buletinul Societăţii Politecnice, Ed. Societatea Politecnica, 399-419, 1889.

[5] Reissner, H. "Über Fachwerke mit zyklischer Symmetrie, Archiv der Mathematic und Physik", III Reihe, XIII, 317-325, 1908.

[6] Mann L., "Über *zyklische Symmetrie* in der Statik mit Anwendungen auf das räumliche Fachwerk". Der Eisenbau 2, Verlag von Wilhelm Engelmann in Leipzig, 18-27, 1911.

[7] Kaufmann W., "Fachwerke von zyklischer Symmetrie mit biegungssteifen Ringen und Meridianen", Zeitschrift für Angewandte Mathematik und Mechanik, Volume 1, Issue 5, 345–424, 1921.

[8] Fortescue C.L., "Method of symmetrical co-ordinates applied to the solution of polyphase networks", Trans. AIEE, 1027-1140, 1918.

[9] Courant R., "Variational methods for the solution of problems of equilibrium and vibrations", Bull. Amer. Math. Soc., 49, 1-23, 1943.

[10] Vasilescu A., "The symmetry and the Courant element", in: M. Krizek, P. Neittaanmaki, R. Stenberg, Eds., Finite Element Method-fifty years of the Courant element, Series: lecture notes in pure and applied mathematics, 164, Marcel Dekker, Inc., 461- 465, 1994.

[11] Argyris J.H., Kelsey S., "Energy Theorems and Structural Analysis", Butterworth, London, 1960.

[12] Argyris J.H., Kelsey S., "Modern Fuselage Analysis and the Elastic Aircraft", Butterworth, London, 1963.

[13] Argyris J.H., "Recent Advances on Matrix Methods in Structural Analysis", Oxford Pergamon Press, 1964.

[14] Turner M. I., Clough R. W., Martin T H. C., Topp L. I., "*Stiffness and Deflection Analysis of Complex Structures*", Journal of the Aeronautical Sciences, Vol. 23, No. 9, pp. 805-823, *1956*.

[15] Clough R. W., Wilson E. L., "Early Finite Element Research at Berkeley", Fifth U.S. National Conference on Computational Mechanics, Aug. 4-6, 1999.

[16] Ritz W., "Über eine neue Methode zur Lösung gewisser Variationsprobleme der mathematischen Physik ", Journal für die Reine und Angewandte Mathematik, vol. 135, pages 1–61, 1909.

[17] Vanderbilt M. D., "Matrix structural Analysis", Quantum Publishers , New York, 1974.

[18] Renton J. D., "On the stability analysis of symmetrical frameworks", Quarterly J. of Mechanics and Applied Mathematics 17, Part 2, 175 – 197, 1964.

[19] Hussey M. J. L., "General theory of cyclically symmetric frames", J. of Structural Division ASCE, 93, No. ST2, 163 - 176, 1967.

[20] Hussey M. J. L., Humpidge H. B., "The elastostatic analysis and elastic critical loads of extensive cyclic domes", Intern. Conf. on Space Structures, University of Surrey, 424 - 438, 1966.

[21] Thomas D. L., "Dynamics of rotationally periodic structures", Int. J. of numer. Methods eng., 4, 81-102, 1979.

[22] Mead D.J., "Free wave propagation in periodically supported infinite beams", J. of Sound and Vibration, 11, 181 – 197, 1970.

[23] Orris R.M., Petyt M., "A finite element study of harmonic wave propagation in periodic structure", J. of Sound and Vibration, 33, No.2, 223 – 256, 1974.

[24] Pipes L.A., "Circulant matrices and the theory of symmetrical components", The Matrix and Tensor Quarterly, 17, Part 2, 35-50, 1966.

[25] Davis P.J., "Circulant Matrices", John Wiley & Sons, 1979.

[26] Kangwai R. D., Guest S. D., Pellegrino S., "An introduction to the analysis of symmetric structures", Computers & structures 71 (6), 671-688, 1999.

[27] Bossavit A., "Symmetry, groups and boundary value problems, a progressive introduction to noncomutative harmonic analysis of partial differential equations in domains with geometrical symmetry", Computr Methods in pplied Mechanics and Engineering, 56, 167-215, 1986.

[28] Bossavit A., "On the exploitation of geometrical symmetry in structural computations of space power stations", Space Power, 7, No.2, 199 – 209, 1988.

[29] Bossavit A., "On the computation of strains and stresses in symmetrical articulated structures", in: E.Allgower, K. Georg, R. Miranda, Eds., *Exploiting Symmetry in Applied and Numerical Analysis* - (Proc. AMS-SIAM Seminar, Fort-Collins, Aug. 1992), Lectures in Applied Mathematics, 29, 111-123, 1993.

[30] Bossavit A., "Boundary value problems with symmetry and their approximation by finite elements", SIAM J. Appl. Math. 53, No. 5, 1352 – 1380, 1993.

[31] Dinkevici S.Z., "Rascet Ticliceskih Constructii - Spectralinii metod", Moskva Stroiizdat, 1972.

[32] Healey T. J., Treacy J. A., "Exact block diagonalization of large eigenvalue problems for structures with symmetry", Internat. J. for Num. Meth. in Engineering, 31, 265-285, 1991.

[33] Hammermesh M., "Group Theory and its Applications to Physical Problems", Addison Wesley Publishing Company, Inc., Reading, Massachusetts, USA, London, 1962.

[34] Miller W. Jr., "Symmetry Groups and Their Applications", Academic Press, New York, 1972.

[35] Zloković G., "Group Theory and G-Vector Spaces in Vibrations, Stability of Structures", ICS, Beograd, 1973.

[36] Zloković G., "Group Theory and G-Vector Spaces in Structural Analysis, Vibrations, Stability and Statics", Ellis Horwood Ltd., 1989.

[37] Fässler A., "Application of group theory to the method of finite elements for solving boundary value problems", Dr. Math. thesis, Eidgenössischen Technischen Hochschule Zürich, 1976.

[38] Laane, J. & Ocola, E.J., "Applications of Symmetry and Group Theory for the Investigation of Molecular Vibrations", Acta Applicandae Mathematicae, vol. 118, no. 1, 3-24, 2012.

[39] Anh-Tuan Ta, Holweck F., A., Labed N., Thionnet A., Peyraut, F., "A constructive approach of invariants of behavior laws with respect to an infinite symmetry group - Application to a biological anisotropic hyperelastic material with one fiber family", Intern. J. of Solids and Structures, vol. 51, no. 21-22, 3579-3588, 2014.

[40] Kustov, E.F., Yarzhemsky, V.G. & Nefedov, V.I., "Application of symmetry groups of four-dimensional space in spectroscopy of crystals", Optics and Spectroscopy, vol. 102, no. 6, 857-866, 2007.

[41] Fritzsche S., "Application of point-group symmetries in chemistry and physics: A computer-algebraic approach", Intern. J. of Quantum Chemistry, Special Issue: Mathematical Methods and Symbolic Calculation in Chemistry and Chemical Biology, Volume 106, Issue 1, 98–129, 2006

[42] Müller, U., "Symmetry relationships between crystal structures: applications of crystallographic group theory in crystal chemistry". Oxford University Press, 2013.

[43] MacNeal R, Harder R, Field E, Herting D., "The NASTRAN theoretical manual", National Aeronaut. Space Administrat. NASA-SP-221, 154–160. Section 4.5., 1981.

[44] Balasubramanian P., Suhas H.K., Ramamurti V., "Skyline solver for the static analysis of cyclic symmetric structures", Computers & Structures, 38 No.3, 259 – 268, 1991.

[45] Dickens J. M., Wilson E. L., "Numerical method for dynamical substructure analysis", Report No. UCB/EERC 80/20, University of California, Berkeley, 1980.

[46] Samartin A., "Analysis of spatially periodic structures- Application to shells and spatial structures", Intern. Symp. on Innovative Applications of Shells and Spatial Forms, 205-221, 1988.

[47] ANSYS User annual, Revision 5.0, Swanson Analysis Systems, Inc., 1992.

[48] ABAQUS Verification and Example Problems Manual, Version 5.3, Hibbitt, Karlsson and Sorensen, Inc., 1993.

[49] O.C. Zienkiewicz, F.C. Scott, "On the principle of repetability and its application in analysis of turbine and pump impellers", Internat. J. of the Num. Meth. in Engrng., 9, 445 – 448, 1972.

[50] Glockner P.G., "Symmetry in structural mechanics", J. of Structural Division ASCE, 99, No. ST1, 71 - 89, 1973.

[51] Weyl H., "Symmetry", Princeton University Press, 1952.

[52] Steward I., "Why Beauty is Truth: A History of Symmetry", Basic Books, 2007.

6th International Conference
" Computational Mechanics and Virtual Engineering"
COMEC 2015
15-16 October 2015, Braşov, Romania

ESTABLISHING TO CONFIGURATION AND MATHEMATICAL MODEL OF SYSTEM BY AUTOMATICALLY ADJUSTMENT FOR HUMIDITY AND AIRFLOW IN GREENHOUSES FOR VEGETABLES AND FLOWERS

Gh. Brătucu[1],C.G. Păunescu[1], D.D Păunescu[1], C.E. Badiu[1]
[1]Transilvania University of Braşov, Braşov, ROMANIA, gh.bratucu@unitbv.ro

Abstract: The paper proposes a scheme of automated control humidity and airflow greenhouses for vegetables or flowers. For the mathematical modeling of these parameters is considered a greenhouse cylindrical cone the air contains a lot of infinitesimal elements on which acts inlet and outlet pressures at a distance x along airflow ventilation. To construct the auto were passed equations of the time in Laplace, the initial conditions null knowing the transfer function, as well as the entry and exit of air, get representation system functional automatic adjustment of MIMO system . With these functions built in Simulink model to transfer control of the proposed system, but before it was examined stability of each transfer functions in MATLAB, observing that the system is fully observable and controllable.
Keywords: vegetable and flower greenhouses, humidity and airflow, mathematical and dynamic modeling

1. INTRODUCTION

The greenhouses for vegetables and flowers are special constructions which must provide shelter, optimal climate conditions for the development of vegetables and flowers throughout the year. To this end research focuses on the most precise and reliable systems capable of automatically adjust temperature, humidity and airflow inside greenhouses in line with the needs of plants at a time regardless of outside weather conditions. Implement a system of automatic adjustment of this type involves numerous researches and expenses, which can simplify or reduce if we rely on simulation and modeling of processes related to climate. On the other hand, the results of this research will be more reliable and powerful, how modern methods of study, and the variables are as close to real situations.

2. MATERIALS AND METHOD

2.1. Establishing scheme and a mathematical model to automatically adjust the humidity and airflow in the greenhouse

Ventilation and air conditioning, active or passive, can be achieved in several ways: as radial tree or as ladder airflow. These configurations are important if we take into account the overall effectiveness of the ventilation system and energy required for continuous ventilation [1], [5], [9], [10]. Ventilation and air conditioning systems should operate continuously, regardless of changes the inner and outer dynamic environment (which may have significant variations of pressure, temperature and humidity).

It was considered a greenhouse that has two recirculation fans, and an area where the air is humidified to the optimal percentage. A filter is used and the recycled air introduced from the outside. It is also necessary control air temperature and humidity obtained by mixing outdoor air with the recirculation. Figure 1 introduced such a system is used in modern greenhouses [2], [3], [4].

Diffusivity ventilated air volume has the effect of changing indoor air quality and temperature variation also stored products. Typically, the entrained air flow is described by two different air flow patterns.

Normally these two models of the air diffusers are present within the cells of storage under steady state mixture, with the ability to adjust the proportion of each in order to provide the desired movement of the air, and to maintain its quality [6], [16].

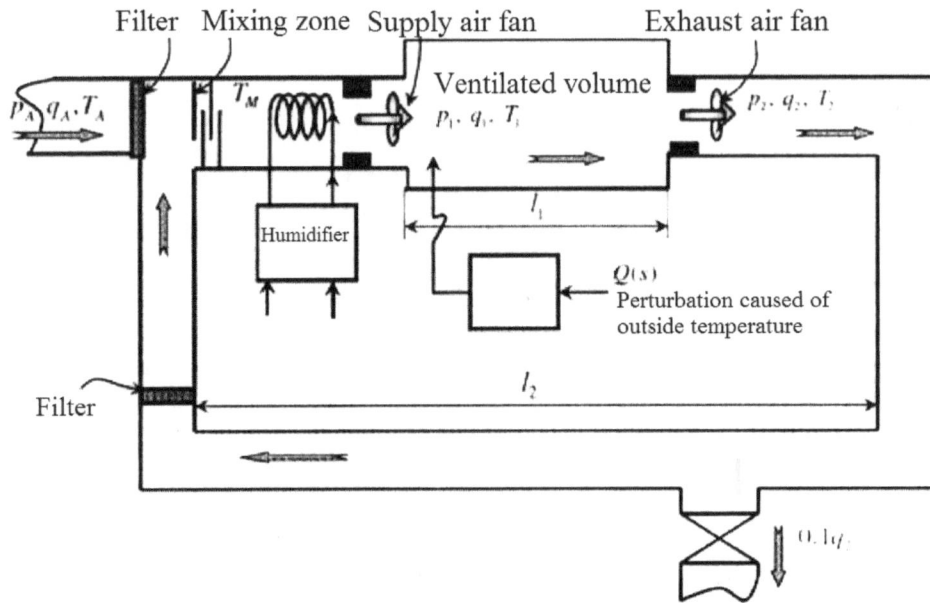

Figure 1 The scheme of automatic control humidity and airflow in greenhouses

The two fans used will be modeled as parametric focus, so the pressure changes of input and output can be controlled by varying the voltage of the fan motors, so they will be two of the actuators that will receive the order from the control system. However, the size and volume of ventilation air conditioning will be modeled by incorporating distributed energy storage parameters and the effects of inductance and capacitance of the air.

To achieve mathematical model of air flow pressure with distributed parameters is considered as length and diameter storage cell, which is considered to have a cylindrical shape. Air cone comprises an infinite series of infinitesimal elements on which it acts *dx* inlet pressure *p(t, x)* and output *p(t, x+dx)* at a distance x along the flow of the ventilation air.

The volume of these elements is properly flow *q(t, x)* for input and *q(t, x+dx)* to exit. Each element has an associated inductance L, which represents the inertia of the gas per unit of length, and a capacitance loads representing the conformity of gas flow per unit length. These are expressed mathematically by the relationship [8]:

$$L = \frac{l_1}{\pi \cdot r_1^2} ; \tag{1}$$

$$C = \frac{V}{R \cdot T_n}, \tag{2}$$

where: V is the room volume, $V = \pi \cdot r_1^2 \cdot l_1$ in m^3; l_1- length of the room in m; r- the radius of the room in m.

In order to perform the analysis is considered that the air flow consists of an infinite series of infinitesimal elements, as shown in Figure 2.

The differential equations that describe the process are:

$$p(t,x+dx) - p(t,x) = -L \cdot \frac{\partial q}{\partial t}(t,x+dx)dx ; \tag{3}$$

$$q(t,x+dx) - q(t,x) = -C \cdot \frac{\partial q}{\partial t}(t,x+dx)dx \, . \tag{4}$$

At the limit, the equations (3) and (4) are [7] [13]

$$\frac{\partial p(t,x)}{\partial x} = -L \cdot \frac{\partial q(t,x)}{\partial t} \, ; \tag{5}$$

$$\frac{\partial q(t,x)}{\partial x} = -C \cdot \frac{\partial p(t,x)}{\partial t} \, . \tag{6}$$

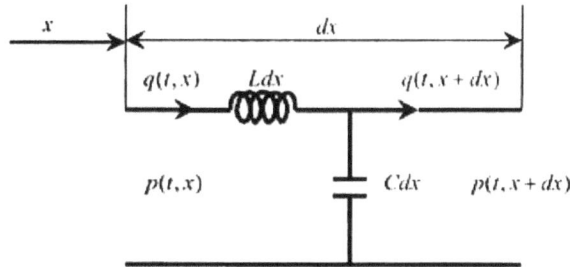

Figure 2 Schematic representations of the elements of the air flow

For building automatic control system will pass the time in the equations of Laplace, at baseline void. Then equations (5) and (6) become:

$$\frac{dp}{dx} = -L \cdot s \cdot q \, ; \tag{7}$$

$$\frac{dq}{dx} = -C \cdot s \cdot p \, , \tag{8}$$

where $p = p(s,x)$ and $q = q(s,x)$.
These equations longer derives once again depending on the variable x and obtain:

$$\frac{d^2 p}{dx^2} = -L \cdot s \frac{dq}{dx} \, ; \tag{9}$$

$$\frac{d^2 q}{dx^2} = -C \cdot s \frac{dp}{dx} \, . \tag{10}$$

Insert equations (1) and (2) in equations (3) and (4) to give:

$$\frac{d^2 p}{dx^2} = L \cdot C \cdot s^2 \cdot p \, ; \tag{11}$$

$$\frac{d^2 q}{dx^2} = L \cdot C \cdot s^2 \cdot q \, . \tag{12}$$

It is obvious that the equations (1) and (2) have the same shape and the same general solutions. If defined in Laplace domain propagation function of the form:

$$\Gamma(s) = s \cdot \sqrt{L \cdot C} \, , \tag{13}$$

then general solutions are in the form:

$$p(s,x) = A \cdot \cosh \Gamma(s)x + B \cdot \sinh \Gamma(s)x \, ; \tag{14}$$
$$q(s,x) = C \cdot \sinh \Gamma(s)x + D \cdot \cosh \Gamma(s)x \, ; \tag{15}$$

For x = 0 we obtain:

$$A = p(s,0); \ D = q(s,0). \tag{16}$$

If derives equations (14) and (15) as a function of the variable x and then calls the equations (7) and (8) we obtain:
$$-L \cdot s \cdot q(s,x) = A \cdot \Gamma(s) \sinh \Gamma(s)x + B \cdot \Gamma(s) \cosh \Gamma(s)x \, ; \tag{17}$$
$$-C \cdot s \cdot q(s,x) = C \cdot \Gamma(s) \cosh \Gamma(s)x + D \cdot \Gamma(s) \sinh \Gamma(s)x \, . \tag{18}$$

For x = 0 equation (17) becomes:

$$-L \cdot s \cdot q(s,0) = B \cdot \Gamma(s) \; ; \tag{19}$$

$$\Rightarrow B = -\frac{L \cdot s}{\Gamma(s)} \cdot q(s,0) = -\sqrt{\frac{L}{C}} \cdot q(s,0) . \tag{20}$$

If the characteristic impedance is expressed as in the following relationship:

$$\xi = \sqrt{\frac{L}{C}} , \tag{21}$$

where equations (14) and (15) become:

$$p(s,x) = \cosh \Gamma(s)x \cdot p(s,0) - \xi \cdot \sinh \Gamma(s)x \cdot q(s,0) ; \tag{22}$$

$$q(s,x) = -\xi^{-1} \cdot \sinh \Gamma(s)x \cdot p(s,0) + \cosh \Gamma(s)x \cdot q(s,0) . \tag{23}$$

From the equations (12), (14) and (15) we obtain

$$B = -\xi \cdot q(s,0) \; si \; C = -\xi^{-1} \cdot p(s,0) . \tag{24}$$

At a distance l along the air flow is obtained equations:

$$\begin{bmatrix} p(s,l) \\ q(s,l) \end{bmatrix} = \begin{bmatrix} l \cosh \Gamma(s) & -\xi l \sinh \Gamma(s) \\ -\xi^{-1} \sinh \Gamma(s) & l \cosh \Gamma(s) \end{bmatrix} \begin{bmatrix} p(s,0) \\ q(s,0) \end{bmatrix} . \tag{25}$$

This equation may also be set from the point of view of impedance:

$$\begin{bmatrix} p(s,l) \\ q(s,l) \end{bmatrix} = \begin{bmatrix} -\xi l ctnh \Gamma(s) & \xi l cschl(s) \\ -\xi l cschl(s) & \xi l ctnh \Gamma(s) \end{bmatrix} \begin{bmatrix} q(s,l) \\ q(s,0) \end{bmatrix} , \tag{26}$$

where: $ctnh \Gamma(s,)l = \dfrac{e^{2\Gamma(s)l} + 1}{e^{2\Gamma(s)l} - 1} = w(s)$, and $csc\, h\Gamma^2(s,)l = (ctnh^2 \Gamma(s)l - 1)^{1/2} = (w^2(s) - 1)^{1/2}$

With this new notation equation (7) and (6) become:

$$p(s,0) = p_1(s); \, p(s,l) = p_2(s); \tag{27}$$

$$q(s,0) = q_1(s) \; si \; q(s,l) = q_2(s), \tag{28}$$

or as the matrix is obtained:

$$\begin{bmatrix} p_1(s) \\ p_2(s) \end{bmatrix} = \begin{bmatrix} \xi w(s) & -\xi(w^2(s)-1)^{\frac{1}{2}} \\ \xi(w^2(s)-1)^{\frac{1}{2}} & -\xi w(s) \end{bmatrix} \begin{bmatrix} q_1(s) \\ q_2(s) \end{bmatrix} . \tag{29}$$

If considered as $p_2(s) = f(p_2)q_2$ constituting pressure to distance l then obtain:

$$\begin{bmatrix} p_1(s) \\ 0 \end{bmatrix} = \begin{bmatrix} \xi w(s) & -\xi(w^2(s)-1)^{\frac{1}{2}} \\ \xi(w^2(s)-1)^{\frac{1}{2}} & -\xi w(s) - f(p_2) \end{bmatrix} \begin{bmatrix} q_1(s) \\ q_2(s) \end{bmatrix} . \tag{30}$$

To be found MIMO transfer function system control and adjustment will be deemed to humidity variation inside the greenhouses is a linear function of pressure, according to Dalton's Law. Also, fan speed variation by automatically adjusting system will ensure a laminar flow of air.

To be properly described the physical phenomena that occur when ventilation air masses are taken into account the following assumptions:

• introducing air fan power required is the sum of the following components: the power required to drive air through filters and flow through the mixing humidified air to the interior of the greenhouse; power necessary to overcome resistance to the entry into interior volume; power required to drive airflow to the volume of greenhouse;

• ventilator power required evacuating air and recirculates it is the sum of the following components: power output required to overcome resistance inside the greenhouse; the power needed to drive the flow of air inside the greenhouse to air recirculation system.

For expression of the equations that describe mass of air entering the greenhouse and the air mass at the outlet of the gases are introduced coefficient of friction f_1, which occurs due to resistance to entry of air into the volume vented and the coefficient of friction f_2, which occurs due to the resistance at the outlet volume ventilated and entry into the circulation system.

Laplace equation written in describing the air mass m_1 entry into the greenhouse vegetable or flower is:

$$\Delta p_1(s) - \frac{(m_1 s \Delta q_1(s) + f_1 \Delta q_1(s))}{a_2} = \xi w_1(s) \Delta q_1(s) - \xi(w_1^2(s) - 1)^{\frac{1}{2}} \cdot \Delta q_2(s), \qquad (31)$$

and the equation describing the air mass m_2 of the volume extracted and inserted into the ventilated circulation system is:

$$-\Delta p_2(s) + \frac{(m_2 s \Delta q_2(s) + f_2 \Delta q_2(s))}{a_2} = \xi(w_1^2(s) - 1)^{\frac{1}{2}} \cdot \Delta q_1(s) - \xi w_1(s) \cdot \Delta q_2(s). \qquad (32)$$

For the calculations are the following notations:

$$\gamma_1(s) = \frac{m_1 s + f_1}{a_2}; \qquad (33)$$

$$\gamma_2(s) = \frac{m_2 s + f_2}{a_2} \qquad (34)$$

where a_2 is recirculating air duct section.

This system of equations is written in matrix form in order to more easily work with him and get:

$$\begin{bmatrix} \Delta p_1(s) \\ -\Delta p_2(s) \end{bmatrix} = \begin{bmatrix} \xi w_1(s) + \gamma_1(s) & -\xi(w_1^2(s) - 1)^{\frac{1}{2}} \\ \xi(w_1^2(s) - 1)^{\frac{1}{2}} & -\xi w_1(s) - \gamma_2(s) \end{bmatrix} \begin{bmatrix} \Delta q_1(s) \\ \Delta q_2(s) \end{bmatrix}. \qquad (35)$$

Below are the following notations:

$$w_1(s) = \frac{e^{2l_1 r_1(s)} + 1}{e^{2l_1 r_1(s)} - 1} \qquad (36)$$

and: $$\Delta(s) = \xi(\gamma_1(s) + \gamma_2(s)) w_1(s) + \gamma_1(s) \gamma_2(s) + \xi_1^2, \qquad (37)$$

where $\Delta(s)$ is the determinant of the matrix.

Equation (35) is rewritten:

$$\begin{bmatrix} \Delta q_1(s) \\ \Delta q_2(s) \end{bmatrix} = \frac{\begin{bmatrix} \xi w_1(s) + \gamma_1(s) & -\xi(w_1^2(s) - 1)^{\frac{1}{2}} \\ \xi(w_1^2(s) - 1)^{\frac{1}{2}} & -\xi w_1(s) - \gamma_2(s) \end{bmatrix}}{\Delta(s)} \begin{bmatrix} \Delta p_1(s) \\ \Delta p_2(s) \end{bmatrix}. \qquad (38)$$

When including the dynamics of the fan and the effect of pressure which is directly proportional to the change in the voltage $v_1(s)$ and $v_2(s)$, the effect of pressure of air flow on the moisture $ph_1(s)$ and $ph_2(s)$ and make the following replacements [15] :

$$p_1(s) = \frac{K_1 v_1(s)}{\tau_1 s + 1} + p_{h1}(s); \qquad (39)$$

$$p_2(s) = \frac{K_2 v_2(s)}{\tau_2 s + 1} + p_{h2}(s), \qquad (40)$$

where: τ_1 and τ_2 are time constants of introducing air fan motor or discharging air; k_1 and k_2 - amplification factors of fan motors, we get the following relationship:

$$\begin{bmatrix} \Delta q_1(s) \\ \Delta q_2(s) \end{bmatrix} = \frac{\begin{bmatrix} \xi w_1(s) + \gamma_1(s) & -\xi(w_1^2(s)-1)^{\frac{1}{2}} \\ \xi(w_1^2(s)-1)^{\frac{1}{2}} & -\xi w_1(s) - \gamma_2(s) \end{bmatrix}}{\Delta(s)} \left(\begin{bmatrix} \dfrac{K_1}{\tau_1 s + 1} & 0 \\ 0 & \dfrac{K_2}{\tau_2 s + 1} \end{bmatrix} \begin{bmatrix} \Delta v_1(s) \\ \Delta v_2(s) \end{bmatrix} + \begin{bmatrix} \Delta p_{h1}(s) \\ \Delta p_{h2}(s) \end{bmatrix} \right) \quad (41)$$

Equation (41) gives the transfer function between input and output automatic control system, that is:

$$\begin{bmatrix} \Delta q_1(s) \\ \Delta q_2(s) \end{bmatrix} = H(s) \cdot \begin{bmatrix} \Delta v_1(s) \\ \Delta v_2(s) \end{bmatrix}. \quad (42)$$

If we consider the time constants of the two fans to be equal to 10 s, $(\tau_1 = \tau_2 = 10\ s)$ the transfer function becomes:

$$H(s) = \frac{\begin{bmatrix} \xi w_1(s) + \gamma_1(s) & -\xi(w_1^2(s)-1)^{\frac{1}{2}} \\ \xi(w_1^2(s)-1)^{\frac{1}{2}} & -\xi w_1(s) - \gamma_2(s) \end{bmatrix} \begin{bmatrix} k_1 & 0 \\ 0 & k_2 \end{bmatrix}}{\Delta(s)(\tau s + 1)}. \quad (43)$$

Knowing the transfer function, input and output, it can get the system functional representation (Figure 3).

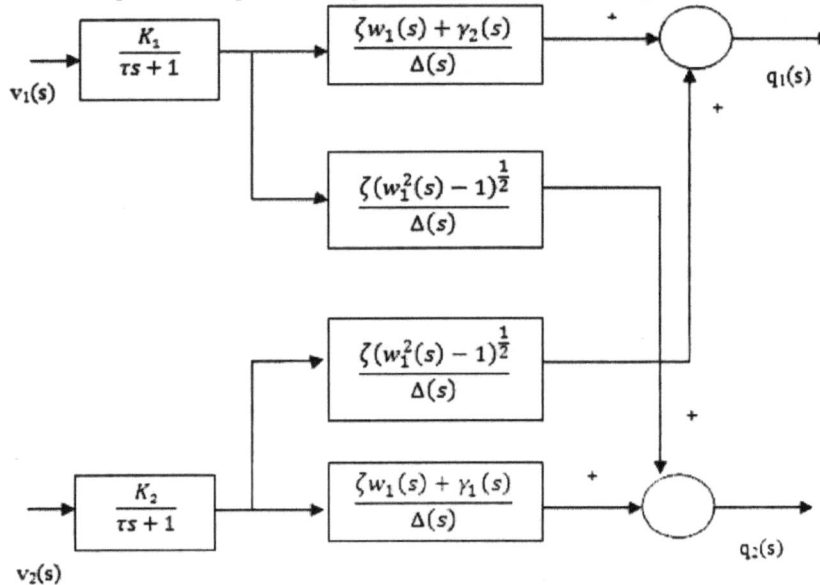

Figure 3 Functional representation system for automatic adjustment of the humidity and flow air, the MIMO type

The fan introducing the air in the greenhouse, and the extracted air from the greenhouse to provide a flow of air of about $K_1\%$, respectively $K_2\%$ these values being a function of the voltage applied to the fan. In addition, the air introduced into the greenhouse is conditioned so that the moisture thereof will be chosen by computer control, thus realizing to maintain a constant humidity within the greenhouse.

To achieve numerical simulation are chosen following values: $K_1 = 0.02124\ m^3V/s$; $K_2 = 0.01416\ m^3V/s$; $v_{1max} = 360\ V$; $v_{2max} = 360\ V$; $p_{1max} = 1.0342\ bar$, $p_{2max} = 1.0342\ bar$, $\xi = 0.2526\ m^{-2}$, $\Gamma(s) = s\sqrt{LC} = 0.0036\ s$, $l_1 = 30\ m$; $m_1 = 3.723$ kg/s; $m_2 = 4.407\ kg/s$; $f_1 = 1.281\ Ns/m^3$; $f_2 = 1.922\ Ns/m^3$; $a_2 = 0.371\ m^2$.

In this case, it will obtain the following transfer function:

$$H_1(s) = \frac{\xi w_1(s) + \gamma_2(s)}{\Delta(s)} = \frac{11.88s + 7.55}{125.84s^2 + 149,79s + 39.22}; \quad (44)$$

328

$$H_2(s) = H_2'(s) = \frac{\xi(w_1^2(s)-1)^{\frac{1}{2}}}{\Delta(s)} = \frac{2.3431}{125.84s^2 + 149.79s + 39.22} ; \quad (45)$$

$$H_3(s) = \frac{\xi w_1(s) + \gamma_1(s)}{\Delta(s)} = \frac{10.03s + 5.8}{125.84s^2 + 149.79s + 39.22} ; \quad (46)$$

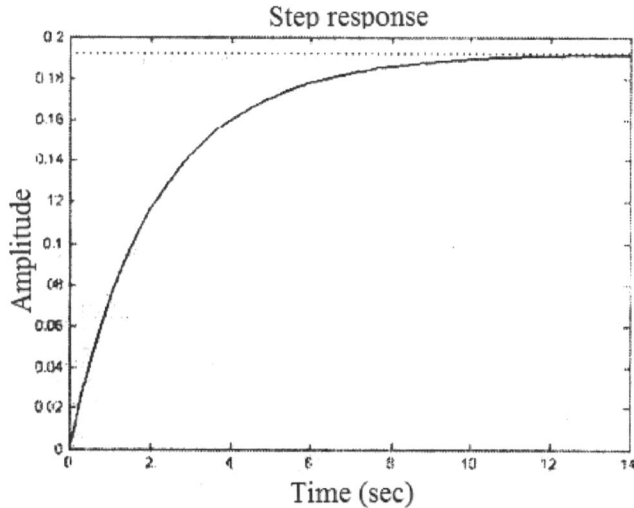

Figure 4 The system's response to an impulse type gear for the transfer function H_1

$$H_4(s) = \frac{K_1}{\tau s + 1} = \frac{0.02124}{10s + 1} ; \quad (47)$$

$$H_5(s) = \frac{K_2}{\tau s + 1} = \frac{0.01416}{10s + 1} . \quad (48)$$

With this transfer function can be built in Simulink control system model proposed but before that check the stability of each transfer functions in MATLAB and obtain:

For H_1 (s) we obtain:

$Z_1 = 0, -0.6355$; $p_1 = -0.8014, -0.3889$; $k_1 = 0.0944$,

where Z are zeroes system, p - poles of the system and k - the gain.

The system is completely observable and controllable as the transfer function poles and zeros are in the left half-plane. It is also clear that the system is stable; its step response is illustrated in Figure 4.

It is noted that the rise time is small, the output is about 5 times smaller than the input, the answer is weakened, leading to the grant of a controller to increase the output.

Analog for $H_2(s)$, $H_3(s)$, $H_4(s)$ şi $H_5(s)$ is obtained:

$Z_2 = 0$;	$p_2 = -0.8014, 0.3889$;	$k_2 = 0.0186$;
$Z_3 = 0; -0.5783$	$p_3 = -0.8014, 0.3889$;	$k_3 = 0.0797$;
$Z_4 = 0$;	$p_4 = -0.1000$;	$k_4 = 0.0021$;
$Z_5 = 0$;	$p_5 = -0.1000$;	$k_5 = 0.0014$;

2.2. Simulation automatically adjust the humidity and airflow

Transfer functions that describe the processes taking place are completely observable and controllable, so stable. However, it appears that the rise time is very high, so very hard to respond to a step input type and the output is vitiated by large errors.

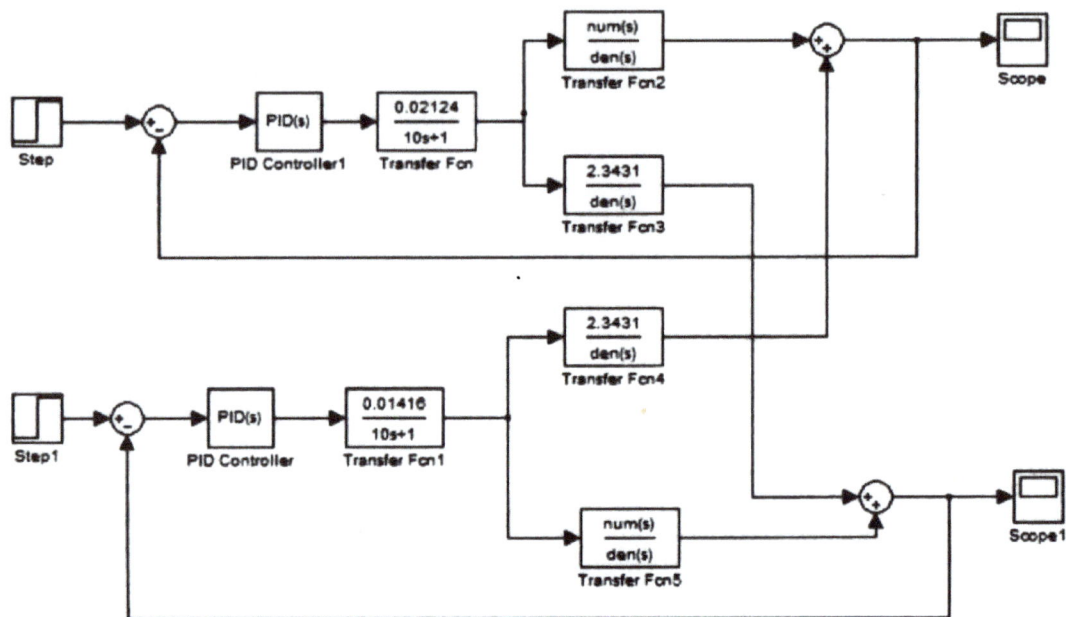

Figure 5 Scheme automatic adjustment MIMO system as a P-canonical structure

Further, the analysis is done M1MO system. It adopts a P-canonical structure, where H_1 (s) and H_3 (s) are the main transfer functions, and H_2(s) and H'2 (s) are transfer functions coupling

Figure 5 shows a schematic diagram of the system which is subject to automatic adjustment type input stage.

The calculation and optimization of Simulink controllers attached to this automatic adjustment have the following form:

$$H_{R1} = k_{p1} + k_{d1}s + k_{i1} / s ; \tag{49}$$

$$H_{R2} = k_{p2} + k_{d2}s + k_{i2} / s ; \tag{49}$$

where: k_{p1} = 433.15; k_{d1} = - 422.42; k_{i1} = 53.16; k_{p2}= 184.45; k_{d2} = 103.10; k_{i2} = 22.28;

With these values and transfer functions of regulators, this system exits the application forms for an entry level unit type are shown in Figures 6 and 7.

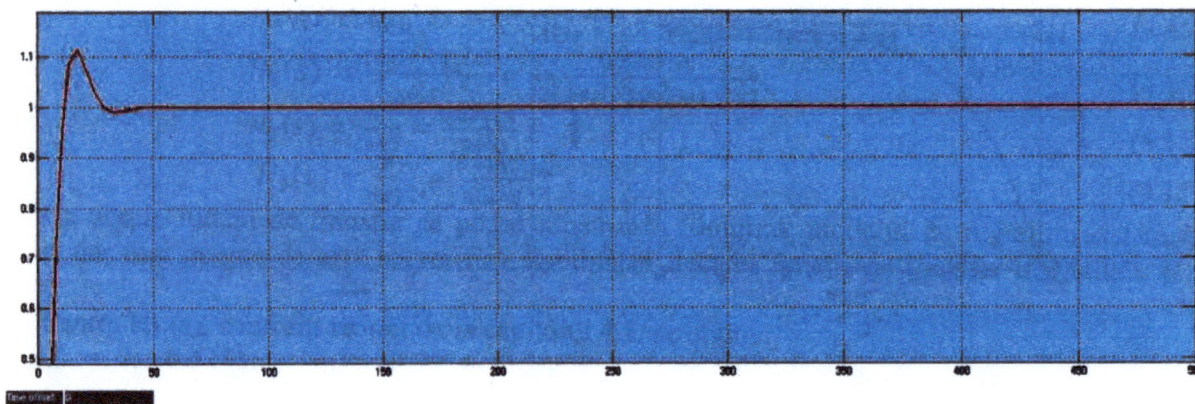

Figure 6 The response control system by granting regulators the transfer function HR_1

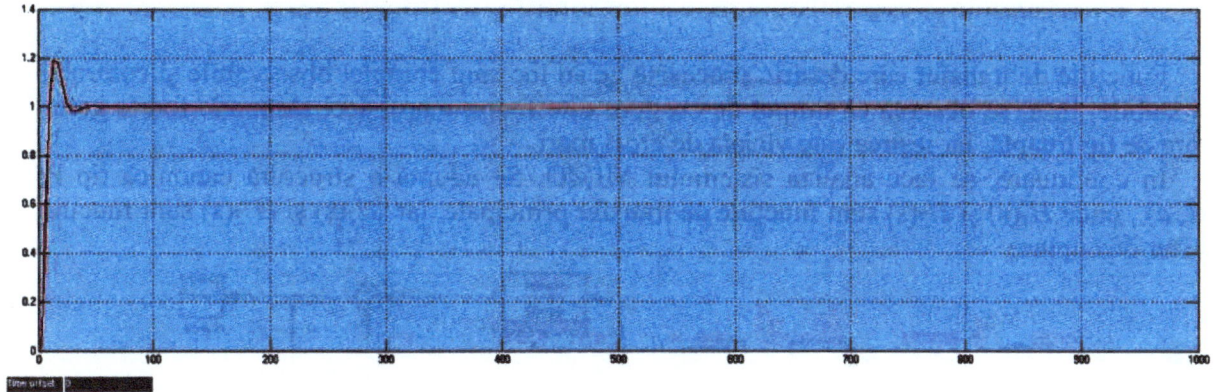

Figure 7 The response control system by granting regulators the transfer function HR_2

These two controllers attached automatic control system with two inputs and two outputs ensures stable behavior with a small over-adjustment with growth and transitional times relatively small, even if these shows high inertia.
These two controllers attached automatic control system with two inputs and two outputs ensures stable behavior with a small over-adjustment with growth and transitional times relatively small, even if these shows high inertia.

Controller parameters			Controller parameters	
	Tuned			Tuned
P	428.6914		P	184.4578
I	52.5069		I	22.2811
D	-430.2034		D	103.1047
N	0.17031		N	0.43692

Performance and robustness			Performance and robustness	
	Tuned			Tuned
Rise time (sec)	7.88		Rise time (sec)	7.68
Settling time (sec)	26.3		Settling time (sec)	28.8
Overshoot (%)	9.39		Overshoot (%)	9.15
Peak	1.09		Peak	1.09
Gain margin (db @ rad/sec)	Inf @ Inf		Gain margin (db @ rad/sec)	23.3 @ 0.869
Phase margin (deg @ rad/sec)	60 @ 0.17		Phase margin (deg @ rad/sec)	60 @ 0.161
Closed-loop stability	Stable		Closed-loop stability	Stable

Figure 8 The parameter values for the two regulators

Figure 8 presents the values of the controller parameters that control the performance and robustness, the described transfer function regulator HR_1 (s) and HR_2 (s).
From these two figures that rise time and during the transitional two regulators are about equal, the first regulator is higher rise time 7.88 seconds, compared to 7.68 seconds and 26.3 seconds during transient lower face 28.8 seconds, compared to the second controller. Also, the first over-control regulator is slightly higher, 9.39% to 9.15%, but both have given this peak as high performance over-adjustment is small and does not affect vegetables and flowers, these changes are within optimal conditions.

3. CONCLUSION

1. To transfer functions found functions which describe phenomena inside the greenhouse, actuators behavior, disturbances and constructive model proposed, were made automatic moisture control systems and ventilation closed loop that provides high performance adjusting both the oscillation values and as rise time and transient.
2. Satisfactory results can be obtained in the simulation automatically adjust the temperature and humidity in greenhouses for vegetables and flowers by using Fuzzy Logic controllers instead of PID. The advantage is clear from the lack of a precise mathematical model to describe physical phenomena [11], [12], [14].

REFERENCES

[1].Ashrae, P.: *Fundamental Handbook*, American Society of Heating, Refrigeration and Air Conditioning Engineers,2001.
[2].Augenbroe, G.: *Computer Models for the Building Industry in Europe*, Delft University of Technology Newsletter, 1994.
[3].Chua, K.J., Ho, J.C., Chou, S.K.:*A Comparative Study of Different Control Strategies for Indoor Air Humidity*, Energy and Buildings, 39, p. 537- 545, 2007.
[4].Haines, R.W., Hittle, D.C. : *Control System for Heating, Ventilation and Air Conditioning*, SpringerVerlag, Berlin, 2006.
[5].Jaymaha, L.: *Energy Efficient Building*, McGraw-Hill Publishing House,2006.
[6].Lovelay D.L., Virk, G.S. , Cheung, J.Y.M. : *Advanced Modeling for Control in Buildings*, In Proceedings of Clima 2000 Conference, London, 1993.
[7].Nichols, N.B. : *The Linear Properties of Pneumatic Transmission Lines*, Boulder Inst. Soc. American, Joint Auto Control, 1961.
[8].Palm, W. : *A Concise Introduction to Matlab*, McGraw-Hill Publishing House,2008.
[9].Păunescu, C.G., Brătucu, Gh.: *Research Regarding the Perfecting of Adjustment, Control and Monitoring Sstem for Humidity in Warehouses for Vegetable and Fruits*, In Journal of EcoAgriTurism, Vol.6 (2010), No. 3(20), p. 93-98, ISSN 1844-8577 (BIOATLAS 2010, International Conference: New Research in Food and Tourism, 28-30 May 2010, Braşov, Romania).
[10].Păunescu, C.G., Brătucu, Gh. : *Contribution to the Study of the Predictable Reliability for Humidity Automatic Adjustment Systems in Warehouses for Fruits and Vegetables*, In Journal of EcoAgriTurism, Vol.6 (2010), No. 4(21), p. 27-30, ISSN 1844-8577 (BIOATLAS 2010, International Conference: New Research in Food and Tourism, 28-30 May 2010, Braşov, Romania).
[11].Păunescu, C.G.: *Research Regarding the Temperature Automatic Control Simulation Using Fuzzy Logic*, In COMAT 2010 Bulletin Conference (International Conference Research and Innovation Engineering), 27-28 Oct. 2010. Vol. III, p. 192-196, Braşov, Romania, (under FISITA Aegis), ISSN 1844-9336.
[12].Păunescu, C.G.: *Research Regarding the Humidity Automatic Control Simulation Using Fuzzy Logic*, In The 4th International Conference Computational Mechanics and Virtual Engineering, COMEC 2011, 20-22 Oct. 2011. Vol. I, p. 244-248, Braşov, Romania.
[13].Schwartz, R., Friedland, B.: *Distributed Systems*, McGraw-Hill Publishing House, 1965.
[14].Sivanandam, S.N., Sumathi, S., Deepa, S.N.: *Introduction to Fuzzy Logic Using MATLAB*, Springer Verlag, Berlin - Heidelberg, 2007.
[15].Whalley, R., Ebrahhimi,M.: *Optimum Wind Tunnel Regulation*, In Proceedings Of the IMechE, Part G (4), p. 241-252, 2001.
[16].Zaheer-Udin, M.: *Temperature and Humidity Control in Indoor Environmental Spaces*, In Energy in Buildings, p. 275-284, 1993.

6th International Conference
"Computational Mechanics and Virtual Engineering "
COMEC 2015
15- 16 October 2015, Braşov, Romania

RESEARCH ON DESIGN FACTORS THAT INFLUENCE ENERGY CONSUMPTION FOR AIR-CONDITIONING IN GREENHOUSES FOR VEGETABLES AND FLOWERS

Bodolan Ciprian [1], Costiuc Liviu [2], Brătucu Gheorghe[3]
[1] Transilvania University of Braşov, Braşov, ROMANIA, bodolan.ciprian@gmail.com
[2] Transilvania University of Braşov, Braşov, ROMANIA, costiuc.liviu@gmail.com
[3]Transilvania University of Brasov, Braşov, ROMANIA, gh.bratucu@unitbv.ro

Abstract: *In greenhouses complex thermal phenomena take place for the exchange of hot or cold air, been signaled at the same time air circulation due to frequent venting. Air conditioning is one of the decisive moments in the scientific exploitation of greenhouses. By ventilation it avoids any accidents that may cause disturbance to the climate factors, especially in a hermetically closed greenhouse. Conditioning involves constant control and permanent supervising of heating, ventilation, moisture, shading and the concentration of carbon dioxide, using different methods, systems and devices. Automatic control ensure achievement of optimum parameters of the vegetation factors ensuring reliable obtaining productions. All the microclimate factors are influenced by the materials used in building of the greenhouse, the shape chosen for the design, the location and orientation of greenhouse, relief, exhibition ground, and soil composition.*
Keywords: *design factors, air-conditioning, energy consumption*

1. INTRODUCTION

In greenhouses complex thermal phenomena take place for the exchange of hot or cold air, been signaled at the same time air circulation due to frequent venting[1].
Solar energy that penetrates in greenhouses suffer significant changes, occurring largely with the passage of radiation through the material that covers the construction. Losses and accumulation of heat occurs, due to frequent air exchanges which strongly influences the greenhouse microclimate. All these factors are influenced by the materials used in building greenhouses, shape chosen for the design, location and orientation of greenhouses, relief, exhibition ground, soil composition. Air Conditioning seeks to ensure a status and composition of the air in greenhouse, in order to allow a normal growth and development of plants, simulating their fructifying ability. It has a crucial role in regulating inner temperature and water regime, to increase the exchange of air between the inner and the outside atmosphere, the reduction or increase of evapo-transpiration. By conditioning it seeks to maintain meteorological factors to the levels required by plants and foreseen in the greenhouse draft [2]. Conditioning involves constant control and permanent supervising of heating, ventilation, moisture, shading and the concentration of carbon dioxide, using different methods, systems and devices.

2. DESIGN FACTORS THAT INFLUENCE ENERGY CONSUMPTION FOR AIR-CONDITIONING

The elements which make part of skeleton structure of the greenhouse forms a surface where sunlight does not penetrate inside it. The shadows projected by them in different positions of the sun on the horizon are highly variable and in most cases above their real weight.

Representing graphically the influence of shading in greenhouses with structural elements, in Figure 1 highlight the effect of light loss interpreted in a cross-section of the building. Can be noticed that the most powerful shading is caused by the shaft of construction along its entire length, and it is all because of this phenomenon to two factors, namely:

➤ own shadow of the volume caused by non-transparent large pieces of structure generally crowded central longitudinal area;

➤ projected shadow bounded by lengthwise lines or the one that intersects surface separation, which moves in an average curve and are projected over the greenhouse crops.

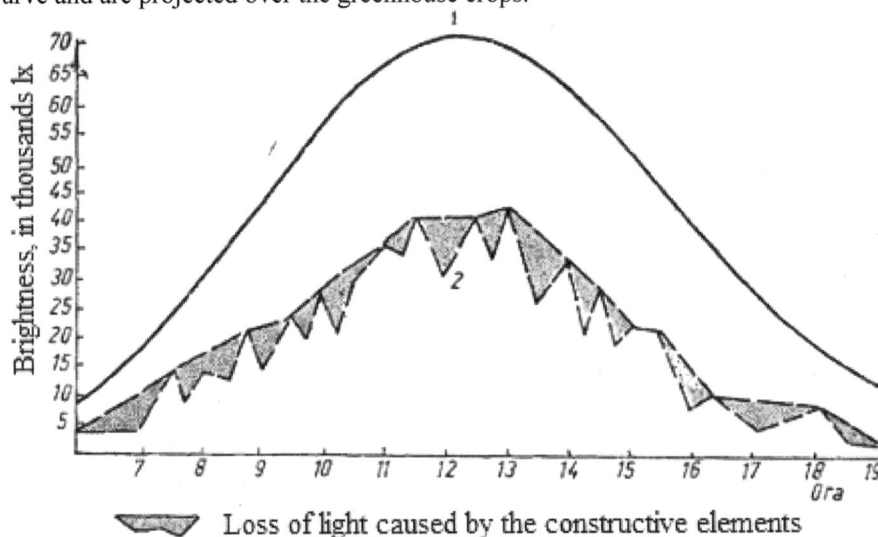

Figure 1. Influence of constructive elements of the intensity of greenhouses
1- outdoor light intensity curve; 2 - light intensity curve in greenhouse

In the cultivated area the shade is projected by greenhouse constructive elements such as gutter, mane and intermediate panels on the opening of over 3 meters (fig. 2). Greenhouses block water drainage trough between blocks raise significantly the percentage of shading. The length and width of the shadow of these design elements were designed following variations: as the sun descends on the horizon shadow grows in width and length, and the high position of the sun from the horizontal width and length of the shadow decreases [3].

Another series of pieces that shade the plants is formed by the side walls and the roof lath that limits the glass panels and supports them (Fig. 2). The lath is very common in the construction skeleton of greenhouses, and totalize the highest percentage of shadows on the crop. Thus, a greenhouse with many lath wood, with distances between them (120 x 52 cm) has a 40% up to 45% of shadows. A greenhouse with small wooden or metal lath, placed between them at a greater distance (220 x 120 cm) with wide openings between the pillars has a shading percentage of 25% -30%. From the point of view of using light, the better are the plastic greenhouses, having a shading percentage of 5-10%.

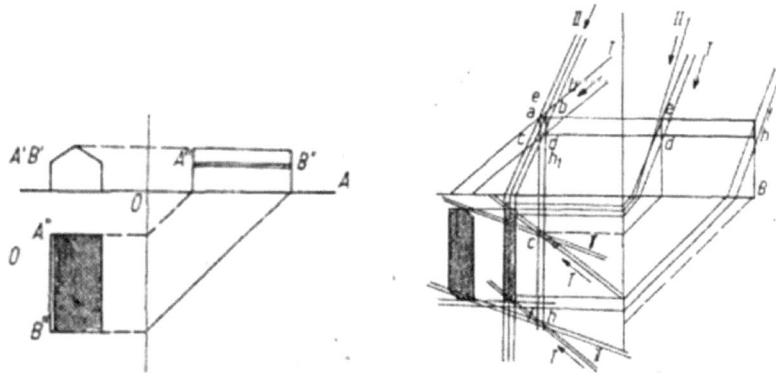

Figure 2. The shadow of the panels, ridges and mourning

Outside the main building elements indicated above in greenhouses with large openings there are helpful piece (transverse and longitudinal counterbraces) or wind braces girder which also contributes in reducing greenhouse light.

The level of shading in greenhouses apart from constructive elements are participating the ventilation and the heating elements that make up their entire technological function but by choosing operational solutions can occupy space outside of the interposer in the direction of penetration of sunlight.

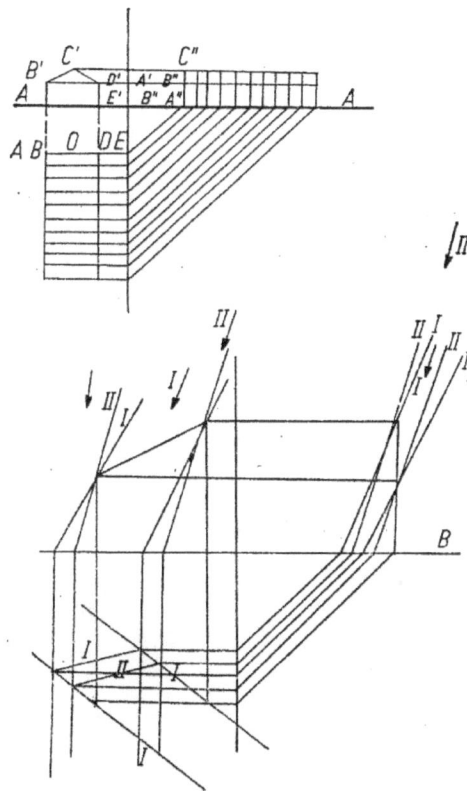

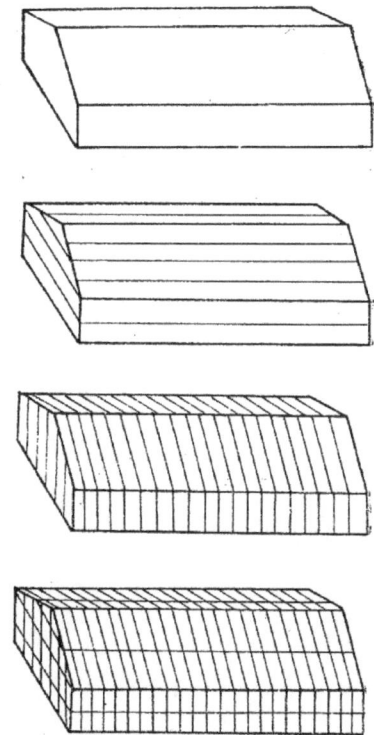

Figure 3. Lath shadow waged vertical and inclined:
A- position of parts in the construction of greenhouses in 3 projections (horizontal, front and side);
B- held various positions of lath shadow of the sun at different inclinations: I-first position of the sun II - second position of the sun; III- third position of the sun

Figure 4. Models greenhouses with different opaque and transparent areas (as Koza) at an angle of 32 ° and the thickness of 3 mm glass

Depending on the chosen constructive scheme there are known constructive designs based on the patterns showed in Figure 4 by Koza in 1972. Based on this constructive given schemes, in our country were made greenhouses, taking into account the construction materials used in model D. The C model is an improved construction and is made by take into account tha fact of increasing transparent areas and allowing better illumination. The B model excludes transversely lath involving expensive solutions without offering substantial extra light. The A model is the ideal solution achievable given that industry provides self-supporting transparent materials, such as polyesters.

Based on the above aspects can be deduced that light regimen in greenhouses depends largely in function of constructive particularities that are part of construction frame that covers the greenhouses. The percentage of shading has a variation limited by the size of structure elements according to their placement in the construction frame.

3. AIR CONDITIONING

Air conditioning is one of the decisive moments in the scientific exploitation of greenhouses. By ventilation it avoids any accidents that may cause disturbance to the climate factors, especially in a hermetically closed greenhouse.

Among the components of the environment, relative humidity and temperature are factors which are influenced by aeration and ventilation measures.

In the context of this paragraph two terms are used to achieve one or the same purpose, which refers to:

- aeration representing natural ventilation, spontaneous or infiltration of outside air (judging by heat loss) caused by the temperature gradient and wind speed amplified; This is the driving force of heat loss during operation of the heating system; it helps lower greenhouse temperature and CO_2 needed for photosynthesis supplement intake [4];
- ventilation or forced ventilation representing the mechanism that the air circulation inside the greenhouse is the result of some facilities and devices that require altering the composition of the environment in a short time.

The displacement air, where spontaneous ventilation through infrastructure leaks is given by the following equation (Okada and Takakura, 1972);

$$V = s \left(\frac{2g}{\gamma}\right)^{\frac{1}{n}} * \Delta P^{\frac{1}{n}} \qquad (1)$$

Where:

V – is the rate of the airflow in m^3 / h;
S – the effective area of orifice in m^2 (n = 2);
g – acceleration of gravity in m / sec^2;
γ – specific weight in kg/m^3;
n – experimental constant ranging between 1 and 2;
ΔP – the pressure difference, in kg/m^2.

In the greenhouse, as in any building through windows and other openings, there is a constant exchange of air when outside air penetrate into the greenhouse and indoor air comes out. Air circulation is caused by the action of wind and gravitational pressure. The last one occurs due to the difference between outside air temperature and inside air temperature. In greenhouses a low air circulation takes place especially along side of the roof. The air circulate upwardly towards vertical walls, along which it plays a leading role, so that air circulation occurs every frame. At some radiation an air exchange occurs between the frames [5].

From the data of the Agricultural Technical Institute Wageningen, Netherlands, it appears that from 15 cm above the ground, under sunny day with no wind, the air velocity increases with the increasing temperature difference between outside and inside. At 100 cm in height, along the south wall, growth is reduced due to the ascending air current braking (which arises with the increase of radiation) along the glass wall. It is formed due to the cold air moving gradually downwards.

At 1.9 m high, the influence of temperature difference between inside and outside is low, with the exception point situated at 100 cm high along the south wall, where is smaller as from 15 cm, but greater than 1.9 m.

The level of air velocity is low, between 5 cm / sec - 15 cm / sec. The speed of 20 cm / sec were recorded in exceptional situations. The taller greenhouses the measured air speeds are higher. In lower greenhouses horizontal component of the air velocity exceeds the vertical component of its speed.

In greenhouses with glass the airing involves certain rules that must be taken into account. Is indicated that for every section of the greenhouse that every slope to have its own system of opening and closing of windows or ventilation. For new greenhouses being built is recommended that the ventilation windows to be 20-25% of the total surface of the glass, depending on the respective climatic conditions, taking into account temperature and wind dominant. In spring and summer, can be used for air supply and exhaust both side windows and the windows of the ridge. Autumn and winter windows will be opened only from the ridge.

Ventilation in greenhouses covered with plastic requires special attention and care throughout the growing season. Block greenhouses requires a certain technique for ventilation. On sunny days is indicated to remove the side panels and part of the roof panels. If an increased frequency of wind the panels will not be removed from the side of dominant direction. If time is variable, with warm sunny days and cold cloudy days the ventilation should be made gradual by opening regular roof panels. On plastic greenhouses, ventilation take places by raising plastics at different heights of the side walls. In this case it should consider how to vent hot or cold days and on days with strong winds. Failure to follow these considerations may lead to plant damage due to excess heat greenhouses or building damage.

4. CONCLUSIONS

1. In greenhouses complex thermal phenomena take place for the exchange of hot or cold air, been signaled at the same time air circulation due to frequent venting.
2. Conditioning involves constant control and permanent supervising of heating, ventilation, moisture, shading and the concentration of carbon dioxide, using different methods, systems and devices.
3. The elements which make part of skeleton structure of the greenhouse form a surface where sunlight does not penetrate inside it. The shadows projected by them in different positions of the sun on the horizon are highly variable and in most cases above their real weight.
5. Air conditioning is one of the decisive moments in the scientific exploitation of greenhouses. By ventilation it avoids any accidents that may cause disturbance to the climate factors, especially in a hermetically closed greenhouse.

ACKNOWLEDGEMENTS

This work was partially supported by the strategic grant POSDRU/159/1.5/S/137070 (2014) of the Ministry of National Education, Romania, co-financed by the European Social Fund – Investing in People, within the Sectorial Operational Program Human Resources Development 2007-2013.

REFERENCES

[1] Ceauşescu, I., General and Special Horticulture, Didactic and Pedagogical Publishing, Bucharest, 1984;
[2] Ciofu, R. et al., Treated of Vegetable Growing, Ceres Publishing, Bucharest, 2004;
[3] Goian, M., et al., Horticulture, West Publishing, Timişoara, 2002;
[4] Oancea, I., Treated of Agricultural Technologies, Technical Publishing, Bucureşti, 1998;
[5] Stan, N., et al., Vegetable Farming, Ion Ionescu de la Brad Publishing, Iaşi, 2003

6th International Conference
" Computational Mechanics and Virtual Engineering"
COMEC 2015
15-16 October 2015, Braşov, Romania

EXPERIMENTAL RESEARCH ON THE MECHANICAL SOLICITATIONS OF THE GREENHOUSES OF VEGETABLES AND FLOWERS LOCATED ON ROOFTOPS

C.E. Badiu[1], M. Lateş[1], Gh. Brătucu[1], D.D Păunescu[1], D. Covaciu[1]

[1]Transilvania University of Braşov, Braşov, ROMANIA, eb@cebb.net

Abstract: This paper presents work algorithm and experimental research results related to the solicitations of the greenhouses for vegetables and flowers placed on rooftops in urban built environment. For research was designed and developed five models of possible forms of greenhouses, which were introduced in HM170 wind tunnel [8]. Air flow speeds to which they were exposed to the front and side surfaces, as well as the roofs of the layouts were 20, 25, 27.5, and 30 m/s. Were measured the pressures in 12 - 16 points on each of the surfaces specified, the data are necessary to design structures of greenhouses resistance of various shapes [2] .All the layouts had the same height and length, but were differentiated by the shapes of roofs and their inclination angles [1].

Keywords: Greenhouses for vegetables and flowers, mechanical solicitations, experimental researches.

1. INTRODUCTION

Resisting movement of an object relative to the air is proportional to the air density ρ, the front surface of the body S and the square of the velocity relative to the air v_a.
Aerodynamic drag force is defined by the formula:

$$F_a = \frac{1}{2} \cdot \rho \cdot c_x \cdot S \cdot v_a^2 \qquad (1)$$

where c_x is called the coefficient of aerodynamic resistance.
The coefficient of aerodynamic resistance c_x represents of the body form influence on the resistance to the force of the air and determined experimentally [3]. This factor is not a constant, but varies depending on the speed, air flow direction, the position and size of the object, density and viscosity of air. Speed, kinematics viscosity and a characteristic length scale of the object are incorporated into non dimensional coefficient called Reynolds number (Re). In compressible mediums is relevant and the speed of sound, and c_x depends on also on the Mach number (Ma). To some form of body drag coefficient of only depend on the number Re, Ma number and direction of the current. At low speeds is no longer dependent coefficient of Mach number (Ma) [4]. Also, the variation with Reynolds number is typically small for most areas of interest. For this reason the air current has the same direction relative to the examined body, it is considered to be constant coefficient c_x [5].
Aerodynamic drag coefficient determined experimentally for a spherical body is 0.47 [6]. Direction of the airflow in the case of a spherical body is not important, the body profile is the same in any direction. It should be noted that a building with the same floor area as a spherical body, but rectangular form, would have a drag coefficient of about 1, and the exposed surface would also significantly higher, so that the force of the wind presses on such construction would be higher (estimated 4 - 5 times higher).

2. MATERIALS AND METHOD

For experimental research on the influence of wind on roofs type greenhouse buildings from urban areas were built five models of greenhouses, which differ in shapes and angles of roofs (Figure 1) .In all frontal and lateral walls as well and the roof surfaces have been made holes with diameters of 3 mm, in which were introduced and were sealed heads gauge hoses connecting with tubes of multitube manometer by the equipment for measuring the pressure exerted at different wind speeds.

Figure 1. Greenhouse layouts built for experimental research

Wind tunnel HM170 used in experimental research (Figure 2) [7] is the type subsonic (air speed up to Mach 0.1), open circuit, outside air is taken in and expelled all outside, with increased speed.

Figure 2. HM 170 Educational Wind Tunnel manufactured by GUNT Gerätebau GmbH, Barsbütte, Germany.

www.gunt.de

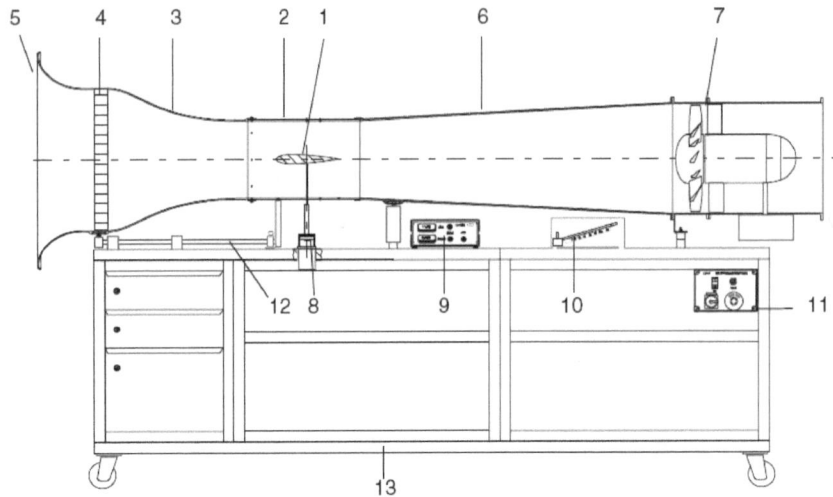

Figure 3. The components of wind tunnel HM170 [8]

The model 1, that is experimental investigated is fixed in the measurement section 2 (Figure3). Air is drawn into tunnel through the feeding hopper 5, and laminar flow is ensured by section 4 (possible cross components of the air flow is reduced to zero). The flowing of laminar air is accelerated by approximately 3.3 times in section 3 and in area 6 of the tunnel is made slower air speed that is pushed out through the fan 7.

Measurement of forces is achieved by means of the force transducer 8 which is integral with the experimental model 1 (Figure 4.a). This transducer inside the tunnel can perform measurements (after 2 directions - driving and port) on: strength, speed, pressure, aerodynamics coefficient to drive (drag) and lift (elevator). The measurements values of forces can be displayed on the screen the amplifier 9.

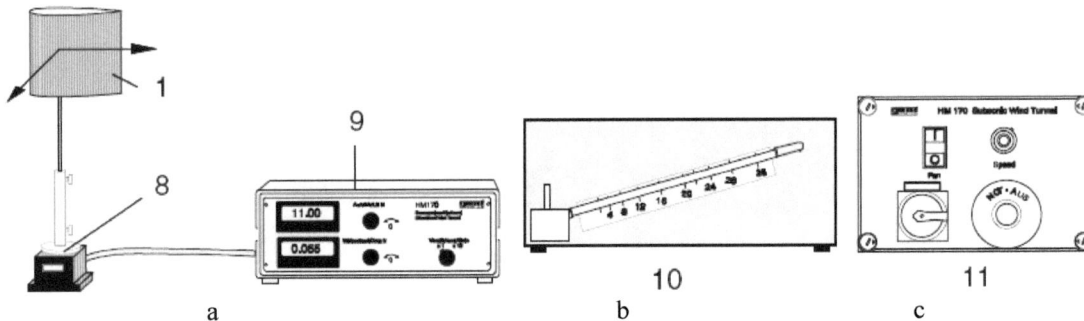

Figure 4. The measurement system (a), the manometer tube (b) and control panel (c) of the wind tunnel

Air speed in section for measurements 2 can be read on the manometer tube inclined 10 (Figure 4.b). The control panel 11 (Figure 4.c) contains a main switch ON / OFF power supply, emergency stop button, a button for adjusting the air speed (frequency converter) and an ON / OFF main switch for fan. Rail 12 allows the translation of wall to the measuring section and access inside section. System is placed within section 13 provided to the frame with rollers.

The air speed is measured with a thermal anemometer shown in Figure 5.

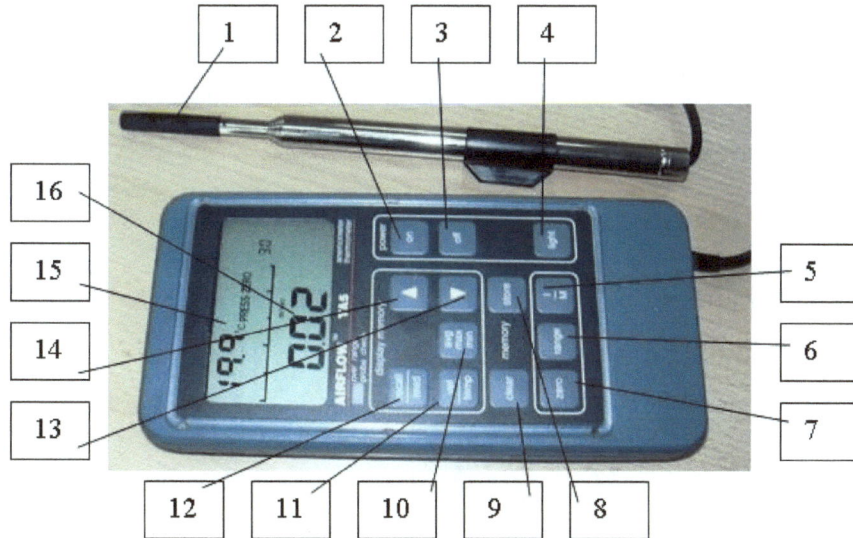

Figure 5. Thermal anemometer of the wind tunnel structure

Thermal anemometer (Figure 5) has the following features: 1 - type strain gauge sensor; 2 - button on; 3 - off button; 4 - light button of the screen; 5 - button for calculating the average measured; 6 - setting the unit of measurement; 7 - button calibration; 8 - Memory button; 9 - button to delete the stored value; 10 - button to display the minimum, maximum, average measured the activation button "on"; 11 - button measured temperature display; 12 - button display wind speed measured; 13 - down scroll button; 14 - scroll up button; 15 - display measured temperature value; 16 - display wind speed measured.

To experimental researches made in this paper was measured the pressure with which pushed the wind wich different speeds, using the multitube manometer presented in Figure 6.

Multitub manometer (Figure 6) [4] containing the 16 tubes of the type manometer with graduated scale 2, mounted on a hinged panel 1. Each tube is provided at the upper parts with connection nozzle 3. Water supply is achieved by means of tank 4, connected to link tube 5. The multitub manometer offers the possibility of measuring absolute or relative air pressure, static or dynamic pressure of the air in the flowing. The panel can tilt in 3 positions through lever 6, enabling the possibility to measurement of very low pressures. The angle can be read on the indicator panel 7: 1: 2 (63,4°), 1: 5 (78,7°) 1:10 (84,3°). Fixing the vertical direction of the panel is achieved by means of screws 8 taking account of the indicator 10. Panel fixing on stand 11 is achieved by tightening bolts 9.

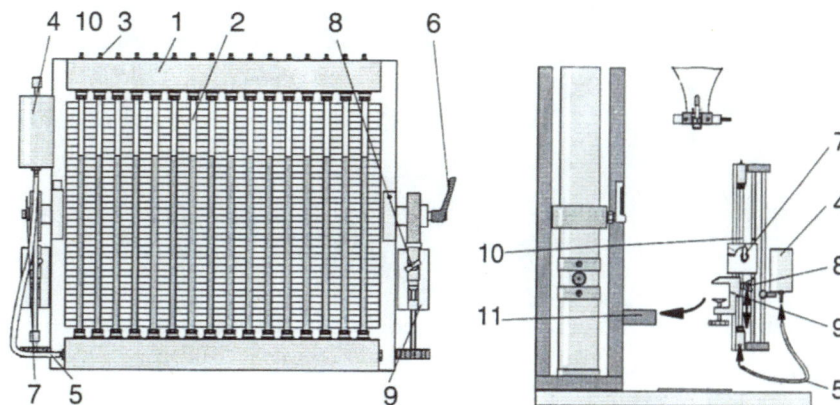

Figure 6. The mmultitub manometer of wind tunnel [4]

Multitub manometer (Figure 6) [4] containing the 16 tubes of the type manometer with graduated scale 2, mounted on a hinged panel 1. Each tube is provided at the upper parts with connection nozzle 3. Water supply is achieved by means of tank 4, connected to link tube 5. The multitub manometer offers the possibility of measuring absolute or relative air pressure, static or dynamic pressure of the air in the flowing. The panel can tilt in 3 positions through lever 6, enabling the possibility to measurement of very low pressures. The angle can be read on the indicator panel 7: 1: 2 (63,4°), 1: 5 (78,7°) 1:10 (84,3°). Fixing the vertical direction of the panel is achieved by means of screws 8 taking account of the indicator 10. Panel fixing on stand 11 is achieved by tightening bolts 9.

For water supply, the tank is fixed in the middle of manometers tubes and fed with water up to half the height of the tank (Figure 7 a).

At water supply, the nozzles upper of the manometers tubes are off, and on the principle of communicating vessels, the water level in the tank is the same in all tubes (Figure 7, b), taking into account atmospheric pressure.

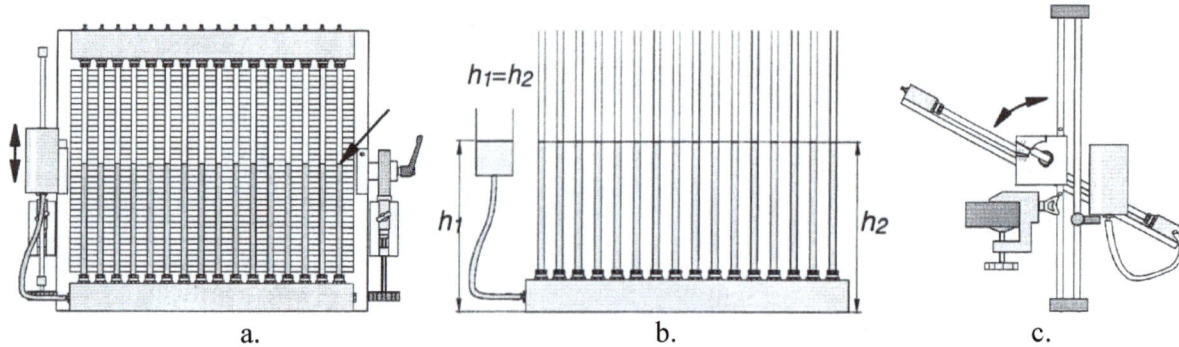

a.   b.   c.

Figure 7. Water supply (a), the water level (b) and adjusting the inclination (c) at the multitube manometer

For accurate measurements it is possible to adjust the inclination of the panel (Fig. 7 c) to 1: 2 (63,4°), 1: 5 (78,7°), 1:10 (84,3°) by operating the lever 6 (fig.17.10) and reading the indicator 7.

The models of greenhouses have been specially prepared to be subjected to experimental research in wind tunnel. In this respect they were performed on the frontal and lateral surfaces and on the roof the holes with diameters of 3 mm (Figure 8), which are inserted from the model inside the tubes connecting to the multitube manometer for measuring the pressure of the stream of air different speeds.

For entering the tubes was performed on the base plates of the models by a 30mm diameter hole. This hole served subsequently   and to layouts fixing on the bottom wind tunnel. The motherboards of layouts is detaches to model corps for introducing these tubes and fastening to the floor tunnel, after being mounted around the template, ready for experimental researches. The holes that have not participated in the current samples were coated with adhesive foil (Figure 9). For accurate recording of all results were numbered holes which participated in a particular sample.

Figure 8. Preparing the greenhouses models for experimental researches

Figure 9 The numbering to Holes on the layouts surfaces and cover those who do not participate in a particular sample

Figure 10. The wind tunnel with a greenhouse model located in a position of the airflow direction

Figure 10 is present the wind tunnel prepared for experimental research at a model of greenhouse, fixed by one of the airflow direction.

Figure 11 present the entire installation of greenhouse Prepared for an experimental research model and the a sequence of conducting experimental researches

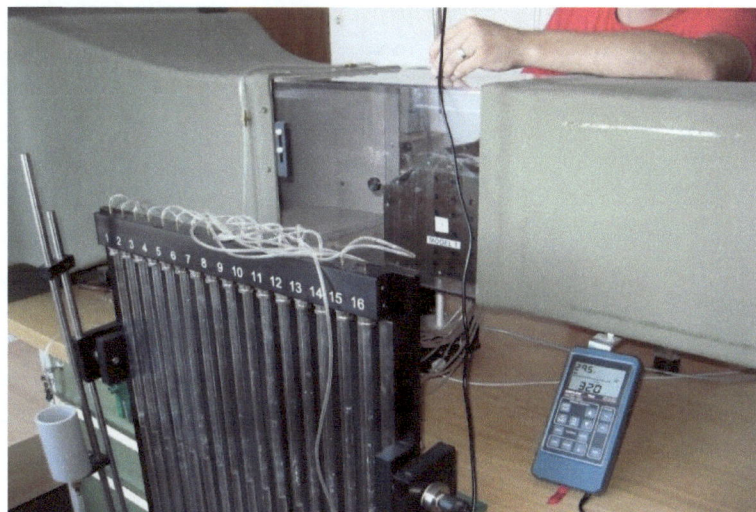

Figure 11 Moment of conducting to experimental researches

The wind speeds considered for experimental research had values of 20 m/s, 25m/s, 27.5/s and 30/s. For all the models were measured the airflow pressures on greenhouse surfaces where the frontal or lateral surfaces have been on the airflow direction.

3. RESULTS AND DISCUSSION

The force with the wind acting on greenhouse pending roofs of buildings in this research is assessed by pressures on their surfaces when blowing in different directions and at different speeds. Since lower wind speeds of 20 m/s is considered not dangerous to the stability of these greenhouses were investigated experimentally only the influences of winds with speeds of 20-30 m/s. Wind tunnel used in experimental research can not generate airflow speeds greater than 30m / s. For each greenhouse model were retained the pressures measured in the 12 to 16 holes on different surfaces when airflow direction has been frontal or lateral. The Gradation with the value 30 on the multitub manometer was considered the benchmark, so that higher values recorded for pressures means pushing forces and lower values are depressed forces.
In Tables 1, 2, 3 and 4 is illustrated the results of experimental research to model No 1 of greenhouse, where the roof slope was 30^0 from the horizontal

The pressure values on the front wall, the front direction of the air flow **Table 1**

MODEL: 1   FRONTAL ■   LATERAL: □   WALL: ■ROOF: □

	30 m/s	27,5 m/s	25 m/s	20 m/s
3	31,5	31,3	31	30,7
4	30,6	30,5	30,4	30,2
5	30,3	30,2	30,1	30
6	30	30	30	29,9
7	29,9	29,9	30	29,9
8	29,8	29,8	29,9	29,9
9	29,9	29,8	29,9	29,9
10	30,7	30,6	30,6	30,3
11	30,2	30	30	30,1
12	30	29,9	29,9	29,9
13	29,9	29,8	29,8	29,9
14	29,8	29,8	29,8	29,8
15	29,9	29,8	29,8	29,9

The pressure values on the lateral wall, the frontal direction of the air flow **Table 2**

MODEL: 1   FRONTAL ■   LATERAL: ■□   WALL□   ROOF: □

	30 m/s	27,5 m/s	25 m/s	20 m/s
3	-	-	-	-
4	-	-	-	-
5	30,3	30,4	30,4	30,2
6	30,1	30,2	30,1	30
7	30	30,1	30	30
8	29,9	30	30	30
9	30	30	30	30
10	30,9	30,8	30,8	30,5
11	30,5	30,7	30,5	30,3
12	33	33	32,5	31,5
13	30,4	30,4	30,4	30,2
14	30,7	30,6	30,6	30,4
15	-	-	-	-

The pressure values on the roof, the lateral direction of the air flow   **Table 3**

MODEL: 1   FRONTAL □   LATERAL: ■   WALL: □   ROOF: ■

	30 m/s	27,5 m/s	25 m/s	20 m/s
3	35,9	35	33,9	32,5
4	33,9	33,5	32,7	31,8
5	33,1	32,9	32,3	31,5
6	34,4	33,8	32,9	31,9
7	36,2	35,3	34,1	32,8
8	35,7	34,8	33,8	32,5
9	32,9	32,6	32	31,4
10	34,7	34,1	33,1	32,2
11	35,8	34,9	33,9	32,5
12	33,6	33,2	32,4	31,6
13	32,9	32,6	32	31,2
14	29	29,2	29,8	29,5
15	-	-	-	-

The pressure values on the lateral wall, the lateral direction of the air flow   **Table 4**

MODEL: 1   FRONTAL □   LATERAL: ■   WALL: ■   ROOF: □

	30 m/s	27,5 m/s	25 m/s	20 m/s
3	32,2	32	31,7	31
4	33	33	32,5	31,6
5	31	31	31	30,6
6	31	31	30,2	30,5
7	30,8	30,7	30,7	30,4
8	30,1	30,1	30,1	30
9	30	30	29,9	29,9
10	29,8	29,9	29,8	29,9
11	30,8	30,7	30,6	30,4
12	29,7	29,7	29,7	29,8
13	29,9	29,8	29,8	29,9
14	29,8	29,8	29,8	29,9
15	30,3	30,2	30,1	30

Similarly is presented the results of experimental research for the other four models of greenhouses. It highlights that at the same level of greenhouses the exposed surfaces to winds are even greater, as are smaller inclinations of roofs . It also the pressures on researched surfaces increased with the square of velocity of the air stream.

4. CONCLUSIONS

1. For the experimental research of the strength of greenhouses located on rooftops have produced five models, which have the same heights, but they differ in angles with that are inclined roofs. In this way, the exposed surfaces of pressing force on wind would be higher or lower depending on these angles.
2. The experimental investigations were conducted in an aerodynamic tunnel type HM170 manufactured by GUNT Educational Wind Tunnel Gerätebau GmbH, Barsbüttel, Germany. Air flow speeds to which they were exposed to the front and side surfaces, as well as the roofs of the models were 20, 25, 27.5, and 30 m / s
3 On the basis of experimental research may be completed formula (1), so as to establish the value of the pressing force of the air flow on an inclined surface with an angle to the direction of the current, respectively. Also these investigations are necessary to validate the simulation modeling software with finite element method for calculation of thrust and lift forces and overturning moment of the greenhouse by air currents.

REFERENCES

[1]. Badiu, C.E., Brătucu, Gh., 2014. The Effects of Wind on Roof Systems for Building. In Bulletin of the Transilvania University of Braşov, Vol. 7(56), No 1/2014, Series II, p 67-72, ISSN 2065-2135 (Print) ISSN 2065-2143 (CD-ROM).

[2]. Badiu, E.C., Brătucu, Gh., Păunescu, D.D., 2014. Types of Infrastructure Used for Crowing Plants in Greenhouses Located on the Roofs of Buildings. In The 5th International Conference COMAT2014, Braşov,Vol. 2, p.243-246, ISBN 978-606-19-0411-2.

[3]. Kijewski-Correa, T., A. Kareem, and M. Kochly. "Experimental Verification and Full-scale Deployment of Global Positioning Systems to Monitor the Dynamic Response of Tall Buildings. "Journal of Structural Engineering" 132.8 (2006): 1242-1253.

[4]. Lates, M.T., 2012. Wind Systems. Theory and Practice. Transilvania University of Brasov Publishing House, 2012. ISBN 978-606-19-0089-3. (in romanian). Lovse, J. W., et al. "Dynamic deformation monitoring of tall structure using GPS technology" Journal of Surveying Engineering" 121.1 (1995): 35-40.

[5] Ogaja, Clement, et al. "Toward the Implementation of On-line Structural Monitoring Using RTK-GPS and Analysis of Results Using the Wavelet Transform. "The 10th FIG International Symposium on Deformation Measurements". 2001.

[6]. Tamura, Yukio, et al. "Measurement of wind-induced response of buildings using RTK-GPS. "Journal of Wind Engineering and Industrial Aerodynamics" 90.12 (2002): 1783-1793.

[7]. * * *. www.gunt.de

[8] * * * Equipment for Engineering Education. Operating Instructions. HM170 Educational Wind Tunnel. G.U.N.T. Gerätebau GmBH. Barsbüttel, Germany.

The 6th International Conference on
"Computational mechanics and virtual engineering"
COMEC 2015
15-16 October 2015, Braşov, Romania

CURRENT LEGISLATIVE ISSUES REGARDING THE USE OF RENEWABLE ENERGY AND IMPLEMENTATION OF HOTEL UNITS ISOLATED IN THIS ISSUE

S. Blaj Brezeanu1, Gh. Brătucu1
[1] Transilvania University of Brasov, Brasov, ROMANIA, blaj.sebastian@yahoo.com

Abstract: The paper develops the main legislative restrictions of Romania and the European Union on the use of renewable energy as a substitute energy from non-renewable energy sources classics. It highlights the need for classification of global energy consumption in the phrase 20-20-20, by 2020 to reduce energy consumption compared to 2002 by 20% and 20% of the energy produced at that time come from sources renewable. It noted the expansion of wind turbines in Romania and the European Union introduction of green certificates, which take account of pollutants and the production of green energy.
Keywords: Renewable energy legislation, isolated hotel establishments.

1. INTRODUCTION

The explosive growth of the population of Earth and its need for energy, in the last century led to an uncontrolled development of pollution that power plants using fossil fuels coal, oil or natural gas have caused. Along with other pollutants it came to achieving a very large greenhouse, which in turn is considered the main factor in causing global warming we are witnessing at this time. Concern about the effect of global warming has made several civil organizations from different countries to draw alarm signals and then international bodies to take the initiative towards proposing preventive measures to be adopted voluntarily or later required by countries of the world . This point was achieved international conference in Kyoto that ended with a protocol which was subsequently joined several countries. EU countries have adopted this protocol, including proposals 20-20-20.

2. MATERIAL AND METHOD

Promoting Renewable Energy European Legal Framework
1997 - White Paper on the European Community Strategic Plan
Objective: By 2010, 12% of EC consumption will come from renewable energy
Features:
- Access to the electricity market discrimination;
- Development of new fiscal support measures;
- The use of biomass in transport and production of electrical and thermal energy;
Promoting RES in construction (buildings where rehabilitation and new construction)
Estimated results:
- 500 000 of roofs and front PV (ECU 1.5 billion)
- 10 000 MW of large wind power

- 10 000 MWh of biomass plants
- Integration of RES in 100 small communities, regions, islands [1].

2002 - The decision to adopt the Kyoto Protocol by the Member States (2002/358 / EC) [2]
Objective: In the period 2008-2012 the Member States will be reduced by at least 8% emissions of greenhouse gases compared to 1990 levels (Carbon dioxide, methane, nitrous oxide, fluorohydrocarbon, freon).

Notifications:
- Globally, the Protocol proposes a 5% reduction in emissions mentioned;
- May 31, 2002: The European Community ratified;
- Protocol entered into force on 16 February 2005.

Signed by:
- 188 countries worldwide have signed and ratified;
- The Protocol was signed by the US but has not been ratified by Congress;
- Do not express any position on the Protocol: China, Taiwan, Afghanistan, Somalia, Vatican, Palestinian Authority

2001 - The EC decision to use RES for producing electricity (2001/77 / EC)
2003 - Directive 2003/30 / EC of the European Parliament and of the Council promoting the use of biofuels and other renewable fuels for transport.
   Objective: biofuel use in transport in proportion of 10% the total consumption of petrol and diesel
2005 - Mandating new Member States to respect this decision
2006 - Expansion of the Directive by including Romania and Bulgaria
2008 - EC Communication on energy efficiency COM (2008) 772
   The objectives of EU development strategy:
- sustainable development
- security of supply of energy and raw materials
- economic competitiveness

Directive 2009/28 / EC of the European Parliament and of the Council on promoting the use of renewable energy:
Directive 20-20-20 [3]
   Purpose:
Respect for international commitments (Kyoto Protocol) by:
- Control of energy consumption in Europe
- The increased use of renewable energy with: energy savings and increased energy efficiency

Directive 20-20-20 Global Objectives, at EU level, By 2020 it will reach these goals simultaneously:
- 20% of global energy consumption will be provided from renewable sources;
- 10% of fuel used in transportation (gasoline and diesel) will be provided from renewable sources;
- Improving energy efficiency by 20%
- The 20% reduction of greenhouse gas emitted into the atmosphere effect.

   Application
Establish mandatory national targets in line with a 20% share of energy from renewable sources and a 10% share of renewable energy in transport in Community energy consumption by 2020.
The objective of 10% energy in transport SER is mandatory for each Member State.
The objective of 20% energy from RES will establish differentiated states.
Explaining the objective of 20% energy from RES, differentiated states take into account:
- The starting point (2005) including the existing level of energy from RES and energy mix SER potentials in the Member States;
- Gross domestic product (GDP) of each state;
- The gross final consumption of energy;
- Efforts to date of Member States in the use of energy from renewable sources.

Member States should adopt **National Action Plans** for renewable energy [2].
These plans should provide national targets set by Member States concerning the share of energy from renewable sources consumed in transport, electricity, heating and cooling in 2020.
Also, these plans must be harmonized with **National Energy Efficiency Plans**, so consider reducing energy consumption achieved from implementing energy efficiency measures.
Directive. 2010/31 / EU on the energy performance of buildings (**PEC**)
- It promotes the improvement of energy performance of buildings, taking into account outdoor climatic and local conditions, as well as indoor climate requirements and cost efficiency.

Directive 2012/27 / EU of the European Parliament and of the Council of 25 October 2012 on energy efficiency
        Objective: creating the legal framework for policy formulation and implementation National Energy Efficiency in order to achieve the national target of increasing energy efficiency. Measures of energy efficiency policy applies to the whole chain: primary resources production, distribution, supply, transportation and final consumption.
The most important renewable energy legislation is presented above [3].
Targets set at Member State level
Overall objective:% of total consumption of energy from RES

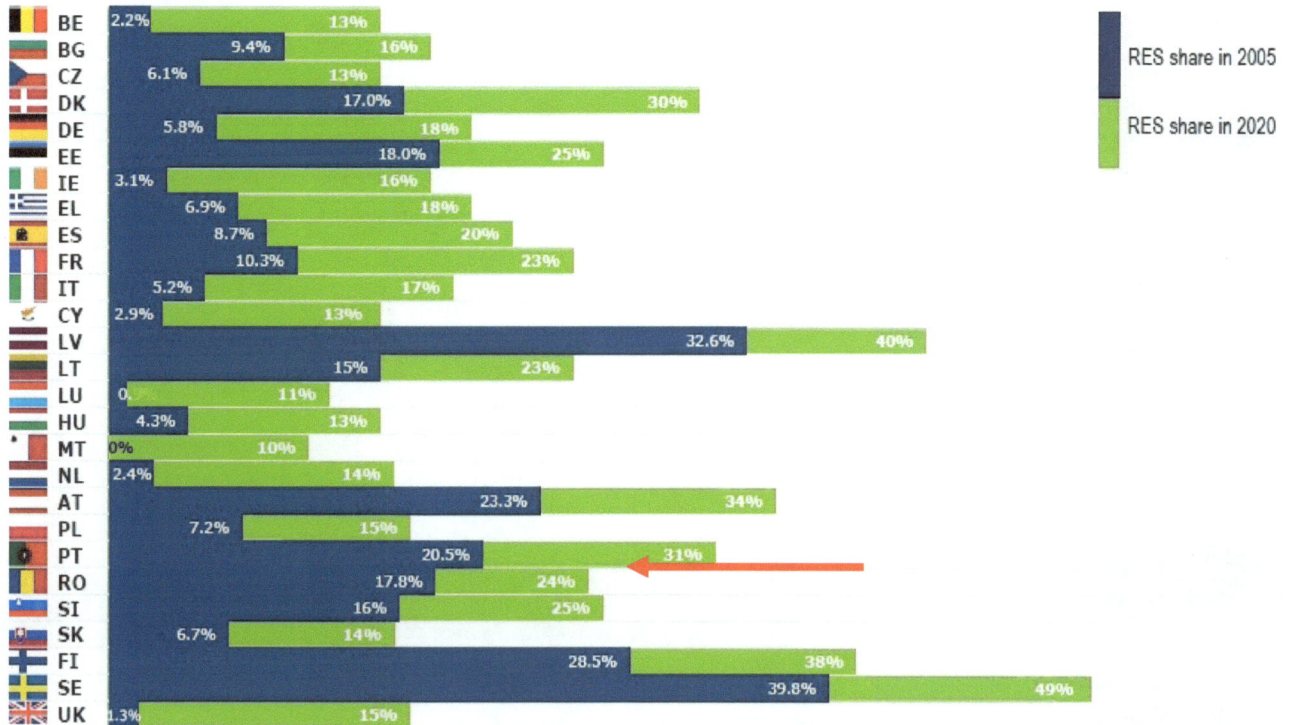

Based on 2005 starting point, recent progress and a balanced sharing of the effort, weighted by GDP/capita

Figure. 1 objective: 20% consumption SER

Country	According to Directive 1997, %	According to Directive 20-20-20, %	Relative growth, %	Notifications	Overall of total energy comes from
Hungary	4,3	13	202.3	She made national Plan	
Netherlands	2,4	14	**483.3**		
Austria	23,3	34	45.9		
Poland	7,2	15	108.3	She made national Plan	
Romania	**17,8**	**24**	**34.8**		
Slovenia	16	25	56.3		

[http://ec.europa.eu/energy/renewables/transparency_platform/action_plan_en.htm]

Table.1 Total consumption of energy from RES

349

Global sources of energy in 2006

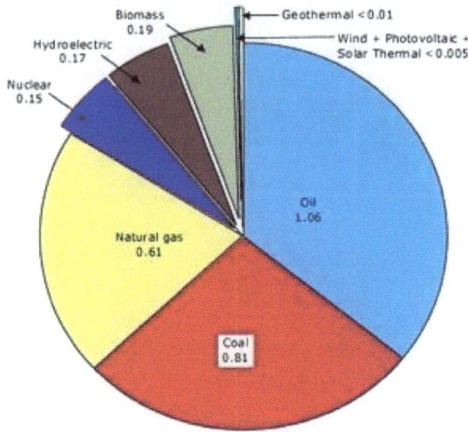

World Renewable Energy 2005

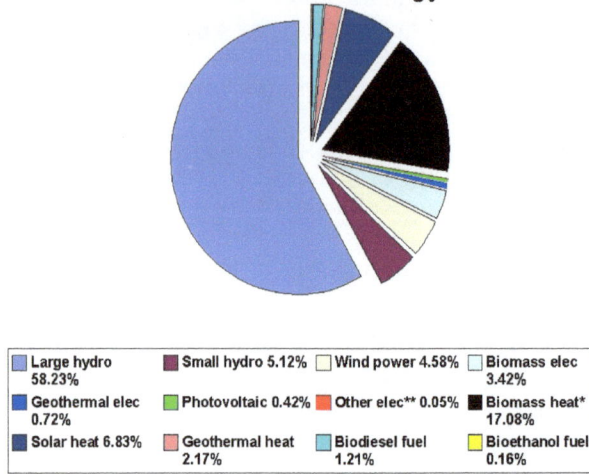

Figure 2 Participation in the world's
various energy sources
[www.energygroove.net]

Figure 3 Participation different sources
to renewable energy production
[www.energygroove.net]

World (2011)
15 tkWhrs/yr
Present Energy Distribution (Power)

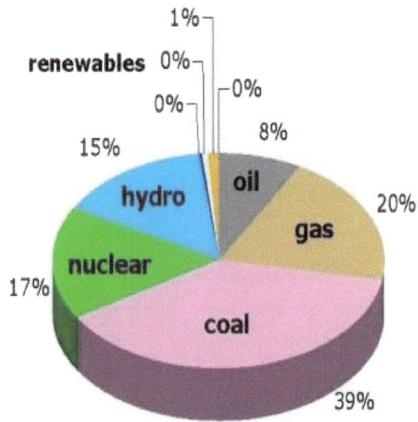

World (2040)
30 tkWhrs/yr
**A Target Sustainable Energy Distribution
by 2040 (Power)**

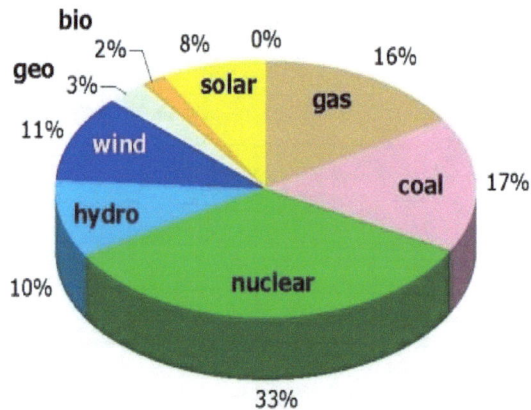

Figure 4 Global energy Demende by Region: 2011-2040
[http://www.forbes.com/sites/jamesconca/2012/05/13/what-is-our-energy-future/]

3 CONCLUSIONS

1. One of the main factors with negative influence on contemporary climate is considered system disorders energy production from conventional sources. This system annually emit significant quantities of CO_2 into the atmosphere, which by the greenhouse effect leading to dangerous climate change for the planet's future.
2. Creating energy independence isolated hotel units is part of a broader framework through which the weight increase energy derived from renewable energy sources in the national and global consumption of energy.

3. The main initiatives to reduce the negative impact of energy from conventional sources were Congress in Kyoto
4. At the moment Romania signed commitments, including the directive that the 20-20-20 by 2020 to reduce energy consumption compared to 2002, and 20% of energy obtained from renewable sources to energy

REFERENCES

[1] Brătucu Gh., Căpăţînă I., "Renewable energy and energy efficiency of rural tourism farms "in" Mechanization of agriculture ", nr. 9/2007, p.17-22, ISSN 1011-7296
[2] *** http://www.energie-solara.com.ro/politici_europene.html
[3] *** http://eur-lex.europa.eu/legal-content/RO
[4] *** http://ec.europa.eu/energy/renewables/transparency_platform/action_plan_en.htm
[5] *** www.energygroove.net
[6] *** http://www.forbes.com/sites/jamesconca/2012/05/13/what-is-our-energy-future/

6th International Conference
"Computational Mechanics and Virtual Engineering "
COMEC 2015
15-16 October 2015, Braşov, Romania

IMPACT RATE EVALUATION OF COMPOSITE SANDWICH PLATES USED IN SHIPBUILDING

Florentina Rotaru[1], Ionel Chirică[2], Elena F. Beznea[3]

[1]University "Dunarea de Jos", Galaţi, ROMANIA, Florentina.Rotaru@ugal.ro
[2]University "Dunarea de Jos", Galaţi, ROMANIA, Ionel.Chirica@ugal.ro
[3]University "Dunarea de Jos", Galaţi, ROMANIA, Elena.Beznea@ugal.ro

Abstract: *Low weight of ship hull structure is the target in the design process. Lower hull weight enables the possibility of low consumption and thus low emission of the ship. By using the last technologies in composite structures field, many structures have been built from light materials with high performances.*
Response of several composite sandwich plates to impact dynamic loads and to a quasi-static simulation of rigid spherical indenter has been evaluated and compared.
In this paper, the important aspect of dynamic loading of composite sandwich structures is presented. The dynamic response of composite sandwich plates is analyzed by impacting the plates at mid-skin surface with a steel sphere. The numerical analysis was performed by using AUTODYN 3D solver from ANSYS.
Keywords: *sandwich structures, numerical simulation, impact behavior, honeycomb core, foam core*

1. INTRODUCTION

In the last decades composite materials are being used extensively in the construction of ships and marine structures. Composites have a higher stiffness and strength by weight than other materials, such as steel or aluminum. Composite materials are used in various structures of the commercial or pleasure craft. So, the result is a lighter ship that can achieve a higher rate of low emission than the same type of ship built of aluminum or steel. Therefore, the lighter weight keeps fuel costs down, involving significant savings for a ship.

Problems of collision and crashing are very important for ship structures and sandwich structures that have shown good capabilities in absorbing energy. Therefore it is necessary to acquire more and more, better knowledge on the impact behavior of ship structures made out of composite sandwich. Core deformation and failure are important factors for the energy absorption capability of composite sandwich structures. After fracture of the skin, the impacting object may damage and penetrate into the core. In the case of honeycomb cores, damage consists of crushing or buckling of cell walls in the area surrounding the impact point. In foam cores, damage looks like a crack for low-energy impacts.

Impact behavior studies are performed to predict how composites respond to collisions with piers, loads from breaking waves, damage from running aground and debris from possible underwater explosions. The impact testing reveals important data, such as the ductile-to-brittle transition point and residual strength after contact with huge forces.

In [8] Abrate studies the needed speed of a projectile to penetrate the panels made out of layered composites and sandwich. Brenda L.Buitrago studies in [3] the impact behavior of the sandwich panels made out of carbon fibers AS4 and epoxy resin 8552 with core of aluminum 3003. A comparison between experimental and numerical tests produced a gap of 2%. Kilchert has studied impact with small and big speed on sandwich plates with various cores (ex.honeycomb, foldcore) with experimental and numerical methods with package software PAM Crush. The thesis investigates the numerical modeling of sandwich structures with aramid paper foldcore and fiber composite face sheets in quasi-static and impact load cases. For that purpose, existing approaches reproducing cellular sandwich structures on the basis of shell-based meso-models are adapted to aramid paper foldcores. The author focused on the strain rate effects in the material model in case of dynamic loading, on modeling and friction.

2. SANDWICH MATERIAL

The sandwich panels used in this study consist of three main parts:
- Two face sheets of composite glass fibres /Epoxy resin with the nominal face thickness of 1mm;
- A honeycomb polypropylene core and polystyrene core;

The sandwich panels have a square shape of 340mm x340mm. The total thickness of the panel is 22mm. The indenter is a steel sphere, with the diameter of 60mm.

3. RESULTS AND DISCUSSION

3.1 Results for composite sandwich plates with honeycomb core

In the tables 1 and 2 the types of specimens made out of composite sandwich with core of honeycomb used for collision with a steel indenter and impact conditions are presented so for low velocity and high velocity. The results such as impact energy, maximum displacement and total time are presented. As is it seen, the bigger velocity leads to bigger energy and bigger displacement for the same impact time.

Table 1: Low velocity impact for honeycomb polypropylene core

Specimen	Indenter type	Indenter mass [Kg]	Impact velocity [m/s]	Impact energy [J]	Displacement [mm]	Time [s]
1	Steel ball	1.5	3.65	10	8.73	0.0025
2	Steel ball	1.5	5.10	20	12.20	0.0025
3	Steel ball	1.5	6.32	30	15.07	0.0025
4	Steel ball	1.5	7.30	40	17.32	0.0025
5	Steel ball	1.5	8.16	50	19.29	0.0025
6	Steel ball	1.5	8.94	60	21.02	0.0025
7	Steel ball	1.5	9.66	70	22.61	0.0025

Table 2: High velocity impact for honeycomb polypropylene core

Specimen	Indenter type	Indenter mass [Kg]	Impact velocity [m/s]	Impact energy [J]	Displacement [mm]	Time [s]
1	Steel ball	1.5	30	675	70.91	0.0025
2	Steel ball	1.5	35	918	82.71	0.0025
3	Steel ball	1.5	40	1200	156.23/237.06	0.0025

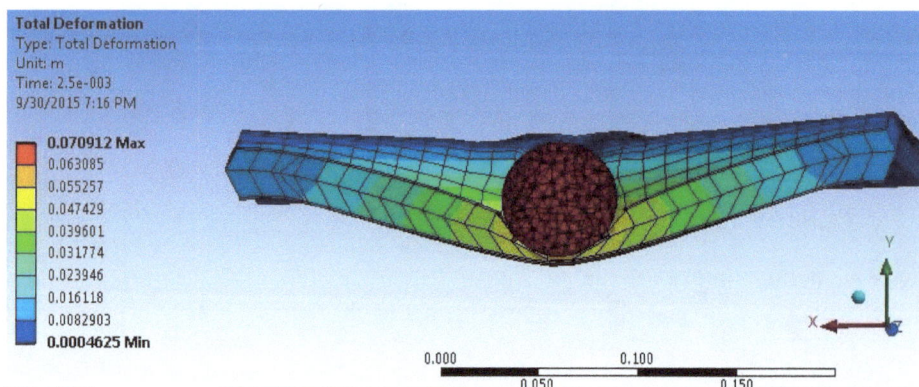

Figure 1: Plate deformation during collision. Velocity v= 30[m/s], honeycomb, E=675 [J]

In Figures 1 and 2 the deformation maps and deforming shapes of the composite plate with core made out of honeycomb, for speeds of 30[m/s] and 40[m/s] are illustrated.

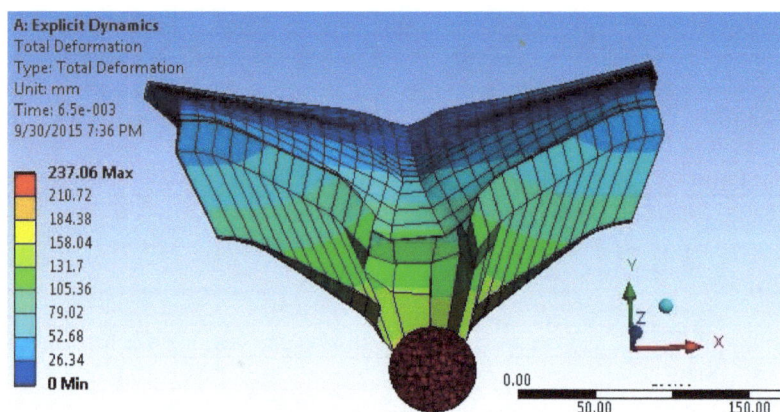

Figure 2: Plate deformation during collision. Velocity v= 40[m/s], honeycomb, E=1200 [J]

3.2 Results for composite sandwich plates with polystyrene core

In the tables 3 and 4 the types of specimens made out of composite sandwich with core of polystyrene used for collision with a steel indenter and impact conditions are presented so for low velocity and high velocity. The results such as impact energy, maximum displacement and total time are presented. As is it seen, the bigger velocity leads to bigger energy and bigger displacement for the same impact time.

Table 3: Low velocity impact for polystyrene core

Specimen	Indenter type	Indenter mass [Kg]	Impact velocity [m/s]	Impact energy [J]	Displacement [mm]	Time [s]
1	Steel ball	1.5	3.65	10	8.39	0.0025
2	Steel ball	1.5	5.10	20	11.46	0.0025
3	Steel ball	1.5	6.32	30	13.84	0.0025
4	Steel ball	1.5	7.30	40	15.65	0.0025
5	Steel ball	1.5	8.16	50	17.05	0.0025
6	Steel ball	1.5	8.94	60	18.40	0.0025
7	Steel ball	1.5	9.66	70	19.63	0.0025

Table 4: High velocity impact for polystyrene core

Specimen	Indenter type	Indenter mass [Kg]	Impact velocity [m/s]	Impact energy [J]	Displacement [mm]	Time [s]
1	Steel ball	1.5	60	2700	58.88	0.0025
2	Steel ball	1.5	65	3168	140.82	0.0025
3	Steel ball	1.5	70	3675	159.82	0.0025

In Figures 3 and 4 the deformation maps and deforming shapes of the composite plate with core made out of polystyrene, for speeds of 60[m/s] and 70[m/s] are illustrated.

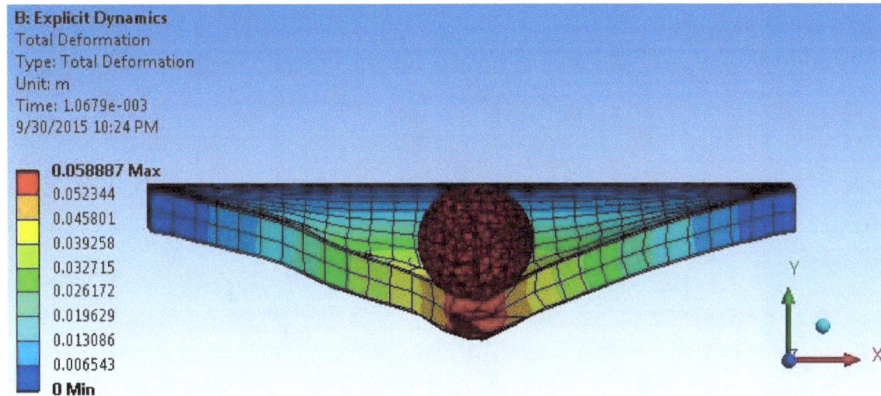

Figure 3: Plate deformation during collision. Velocity v= 60[m/s], core: polystyrene, E=2700 [J]

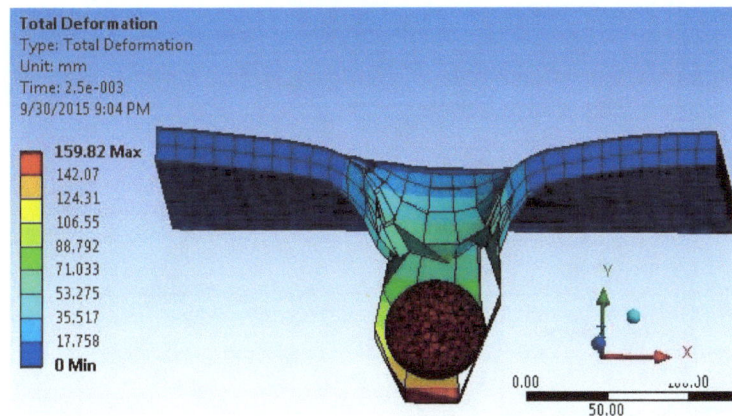

Figure 4: Plate deformation during collision, Velocity v= 70[m/s], core: polystyrene, E=3675 [J]

Variations of impact velocity versus maximum displacement of the sandwich panels for core made out of honeycomb and polystyrene are illustrated in Figure 5.
As it is seen, in the case of core made out of polystyrene the same maximum displacement is obtained for a bigger speed (almost double) than the speed for core made out of honeycomb. Also, the energy absorbed by the composite sandwich plate with core of polystyrene is bigger than the energy absorbed by the composite sandwich plate with core of honeycomb.

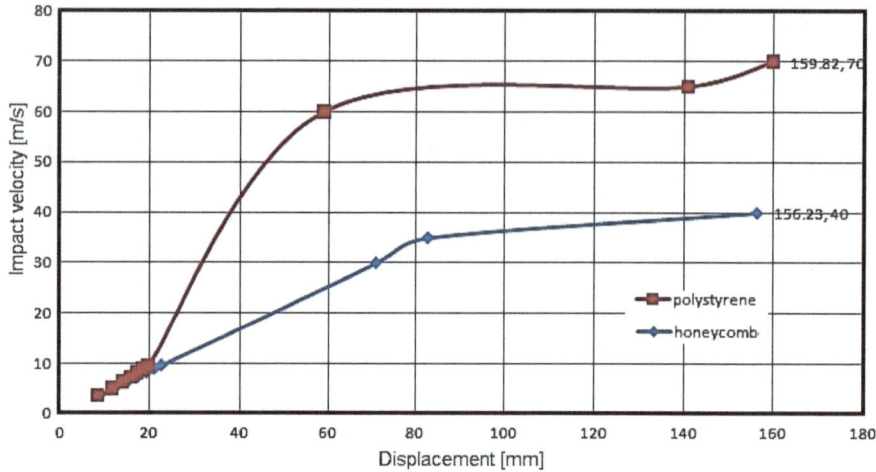

Figure 5: Variation of maximum displacement after collision versus indenter speed

3.3 Impact rate (Impact multiplier)

The deformation δ_d obtained in a plate loaded by impact (dynamic loading) is bigger than the deformation δ_s obtained in the same plate by loading with the same force but in static action.
The coefficient (ratio) ψ calculated as ratio between dynamic deformation δ_d and static deformation δ_s.
The ratio ψ shows also the values of the impact effects (stresses and strains).

$$\psi = \frac{\delta_d}{\delta_s} \tag{1}$$

In the case of the studied plates, the variations of the impact multiplier for the both types of cores versus impact velocity are illustrated in Figure 6.
As it is seen, the impact multiplier in the case of polystyrene is bigger than the impact multiplier in the case of honeycomb.

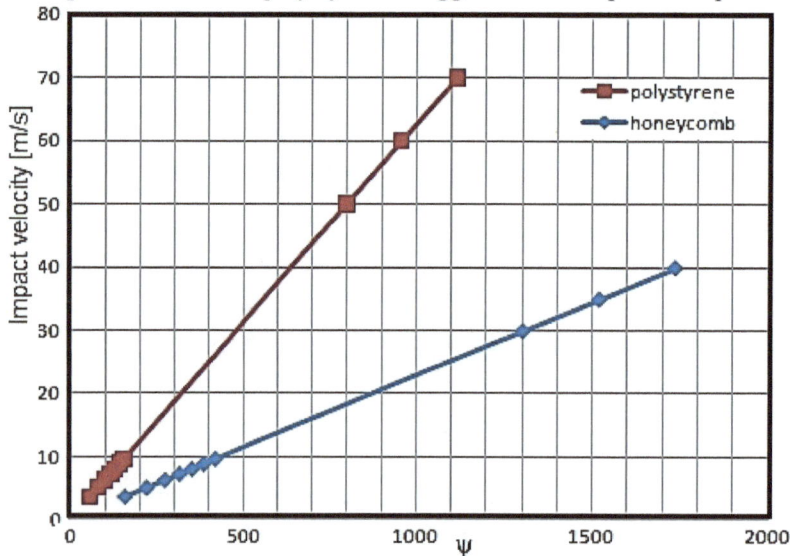

Figure 6: Variation of impact multiplier versus impact velocity for the both cores types

4. CONCLUSIONS

In the paper the response of two types of composite sandwich plates (with core made out of polystyrene and honeycomb) to dynamic loads and to a quasi-static simulation of rigid spherical indenter has been evaluated and compared.
The modality to crashing of the material and also the type of the damage are depending on the indenter mass, geometry and material structure. Important parameters for impact phenomena are the indenter speed and the indenter energy. The

two types of speeds have been used: low speeds (from 3.65m/s to 9.66m/s) and high speeds (greater than 30m/s). In static analysis the behavior of the panel with core made out of honeycomb was better than the behavior of the panel with core made out of polystyrene. That is for a force of 15N, the displacement for honeycomb was 0.054mm and for polystyrene was 0.401mm. In dynamic analysis the behavior of the both panel types was vice versa. The sandwich composite panel with core made out of polystyrene has a better dynamic behavior than the sandwich composite panel with core made out of honeycomb. In this case, the panel crashed for an indenter speed of 70m/s.

Also, the cut-out made by the indenter penetration is cleaner for the entrance face than the exit face. The exit cutout has in all cases an delamination. This delamination is extended on an area of couple cm from the main cutout, in fibers direction. Since the number of layers is increasing, the energy is increasing.

ACKNOWLEDGEMENTS

The work has been funded by the Sectoral Operational Programme Human Resources Development 2007-2013 of the Ministry of European Funds through the Financial Agreement POSDRU/159/1.5/S/132397

REFERENCES

[1] Callister W., Fundamentals of Materials Science and Engineering, John Wiley&Sons, New York, 2001.

[2] Nayeemuddin, Abdul Nazeer, To study the effect of low velocity impacts on composite material buy using Finite element analysis software LSDYNA, India, Volume 9, Issue 5 (Nov-Dec.3013), PP 01-09.

[3] Brenda L., Santiuste C., Sanchez-Saez S., Barbereo E., Navarro C., Modeling of composite sandwich structures with honeycomb core subjected to high-velocity impact, Madrid, Spain.

[4] Kilchert S.W., Nonlinear finite element modeling of degradation and failure in folded core composite sandwich structures, Doctoral thesis, University of Stuttgart 2013.

[5] Naresh,Ch., Numerical Investigation into Effect of Cell Shape on the Behavior of Honeycomb Sandwich Panel, International Journal of Innovative Research in Science, Engineering and Technology December 2013.

[6] Arabouli,J., Schmitt ,Y.,Pierrot,J.L., Royer F.X. Numerical Simulation and experimental bending behaviour of multi-layer sandwich structures, Laboratory de Physique des Milieux Denses, France 2014.

[7] Calciu Elena, Determinarea multiplicatorului de impact in cazul solicitarilor dinamice asupra unor calote subtiri realizate din material composite, revista Academiei Fortelor Terestre, 2000.

[8] Abrate,S., Ballistic Impacts on Composite and Sandwich Structures, Major Accomplishments in Composite Materials and Sandwich Structures, Vol.IV, pp. 465-501, 2010.

[9] Zienkiewicz, O.C. & Taylor, R.L., The Finite Element Method, Butterworth-Heinemann, Oxford, 2000.

[10] Tsai, S.W. & Wu, E.M., "A General Theory of Strength for Anisotropic Materials", Journal of Composite Materials, Vol. 5, pp. 58-80, 1971.

[11] Meyer-Piening, H.-R., "Sandwich Plates: Stresses, Deflection, Buckling and Wrinkling Loads. Proceedings of the 7th International Conference on Sandwich Structures, Aalborg, Denmark, 2005.

[12] Bathe, K.J., Finite Element Procedures in Engineering Analysis, Prentice-Hall Inc., Englewood Cliffs, 1982.

[13] Lok T.S. & Cheng Q.H., "Elastic Deflection of Thin-Walled Sandwich Panels", Journal of Sandwich Structures and Materials, Vol. 1, 279-298, 1999.

[14] Shenoi, R.A. & Wellicome, J.F., Composite Materials in Maritime Structures, Cambridge University Press, 1993.

6th International Conference
"Computational Mechanics and Virtual Engineering "
COMEC 2015
15-16 October 2015, Brașov, Romania

THE INFLUENCE OF THE KNOT TYPE IN THE HYBRID MODEL LINE-FISHING HOOK

Ștefan Dimitriu

Transilvania University Brasov, ROMANIA, e-mail sdimi@unitbv.ro

Abstract: Numerical simulation of the hybrid mechanical system line-fishing hook is based on some physical parameters, like the angle between the line and the shank of the hook. The main influence on this angle is done by the knot type, the hook shape and other factors. This article studies the influence of the knot type on the main angle line-fishing hook.

Keywords: fishing hook, knot, mechanical testing

1. INTRODUCTION

The fishing hook characteristics were discussed in a previous article [1] and some classifications were proposed to improve qualitative and quantitative analysis with finite element model. However, the geometry of this very interesting model line – fishing hook is influenced by many variables. Let's take a closer look on the main angle α, the angle between the line and the shank of the hook [1]. It seems to be simple to identify this angle, but in practice there are so many methods to tie the knot which introduces a lot of variables. It seems reasonable to fix some parameters, like the nature and the diameter of the line, and the type of the hook, in order to perform the study of the geometry.

A general situation is the case of a straight shank with plane end. This reduces considerable the study, as the braided knots used for the hook with eye end are many. Also, the hook with an eye at its end should be simulated as a spatial articulated body. If the hook end and is simple and plane, the line starts from the shank like from an articulation, but obviously there is not a perfect articulation. Here could be introduced another parameter, the nature of the hook. If the hook is made from wire, we should consider a circular shape of the cross section, but if the hook is forged, the cross section will be generally a rectangle. We did choose to start the study with the common hook made from wire.

2. ANALYTICAL MODEL

As it was presented in [1] and [2], assuming the load Q is applied in the worst case, exactly in the point of the hook, as in figure 1, the bending moment is a function of the parameter ε – the angle that indicates the current transverse section of the hook, starting from its point:

$$M(\varepsilon) = -Q \frac{d}{2} \left(\sin \varepsilon \sin \alpha + \cos \alpha - \cos \varepsilon \cos \alpha \right)$$

(1)

The maximum bending moment appears for the critical value of the ε angle

$$\varepsilon_{cr} = -arctg\left(tg\,\alpha\right) = -arctg\,\frac{d}{a}$$

(3)

For the second step, let's assume a more general value of the angle α as a complex function of material and knot type. The equations remain the same, but the value of the angle α will be obtained from experimental analysis and the angle α is no longer a function of the shape parameters d and a.

The variation of the function M(ε) for an unit force Q and unit distance d, assuming a value of ε angle of 1 decimal degree is presented in figure 2. The parameter α starts at 10 decimal degrees (0.15708 radians) and stops at 45 decimal degrees (0.70685 radians). This interval was observed in practice to cover, in general, the values of angle α.

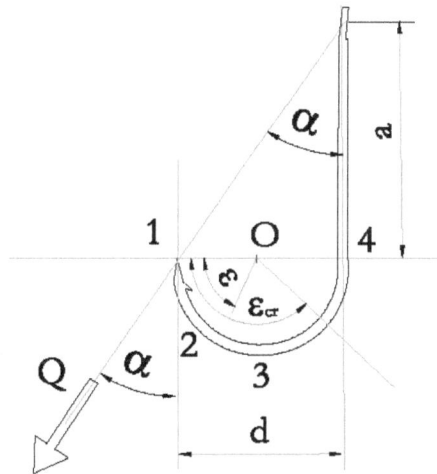

Figure 1: General geometry or the hook

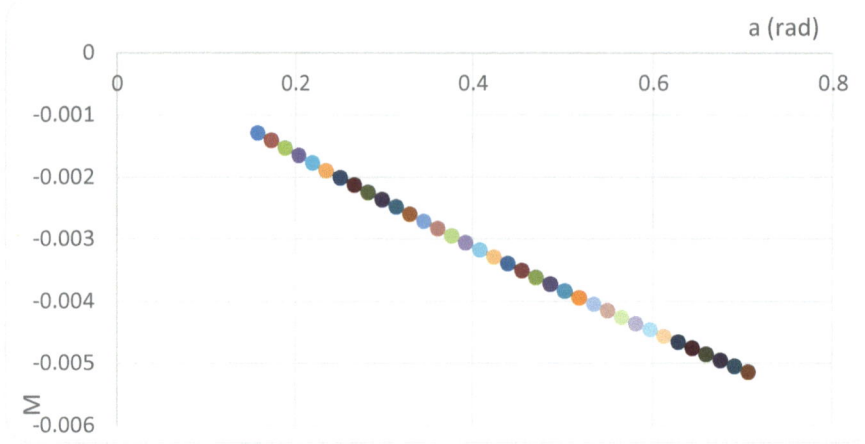

Figure 2: The variation of the bending moment M for $\alpha \in [10 \ldots 45]$ decimal degrees

In conclusion, the α parameter has an important influence on the bending moment. For this reason it is necessary to carefully observe its values in different situations.

Generally, knots are not as strong as the line. In fact, a knot weakens the line up to 50 %. The problem is to choose the right knot, but it seems there it should be compatibility between knot, line (function of nature of material and diameter) and the shape of the hook. In this study we tried to identify some mechanical parameters of this complex assembly line-knot-fishing hook.

3. EXPERIMENTAL SETUP

The method used to tie a simple knot for a plan end of a hook is presented in the figure 3 and 4. Note the position end the mode the line goes parallel to the shank and specially the mode it continues at the end of the knot.

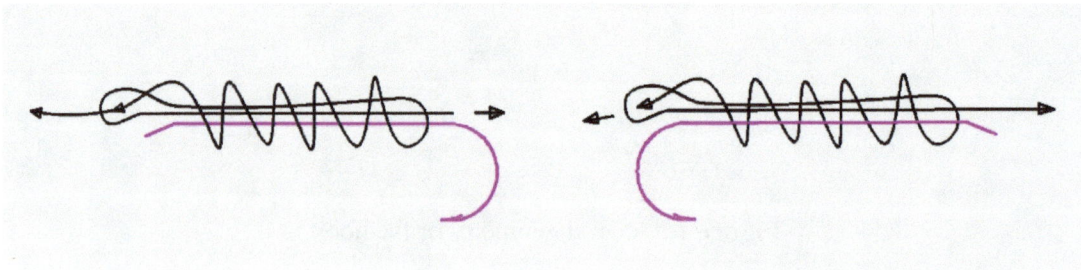

Figure 3: Knot K1   **Figure 4**: Knot K2

The knot named K1 present a curl at the end and the line pass trough, so finally the line will take a direction other than the shank (figure 5). The knot K2 allows the line to continue parallel to the shank. The difference is observed directly on the angle α values, as the line movement is restricted, depending on how the line starts out from the knot.

If we examine the knot, we can observe three relative positions of the line and the shank, as in figure 5.

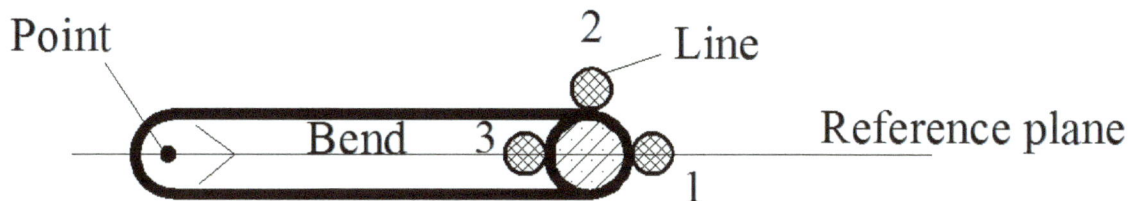

Figure 5: The relative position of the line and the shank of the hook

In order to measure the value of the angle α, the two knot K1 and K2 were tied, using three position of the line like in the figure 5. When the knot is tied, usually the angler puts the line in the position 1 or 3. The position 2 is also possible. It should be mentioned the knot is able to move around the shank, under great variable loading, especially when it is used a polymer line.

The line was a fluoropolymer, type monofilament, 0.30 mm, and the hook was a Mustad, 220 A-NI Hollow Point Crystal, size 6 [3].

The load was applied in two situations: firstly on the bend (considering this is a general case), and secondly, directly on the point of the hook (an extreme position, but also possible).

Some loading situation are presented in the figures 6-10.

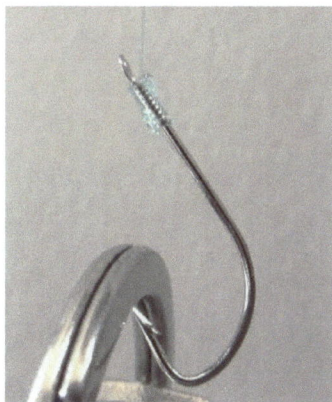

Figure 6: K1 position 1, point bend

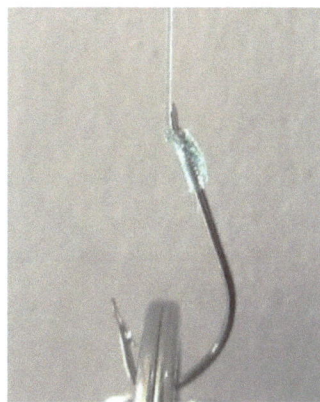

Figure 7: K1 position 3, bend

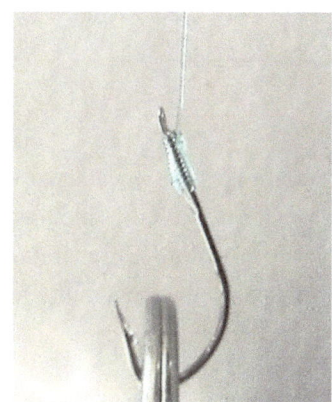

Figure 8: K2 position 1,

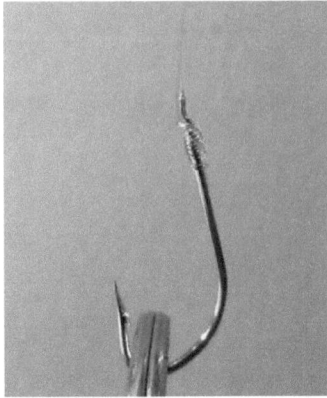

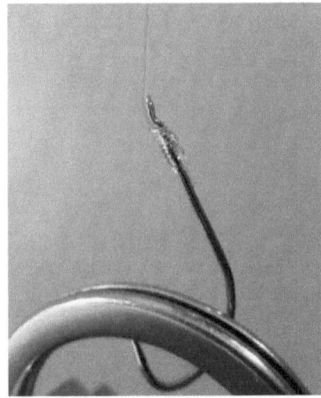

Figure 9: K2 position 3, bend   **Figure 10**: K2 position 3, point

The average α values measured are presented in the table 1.

Table 1: Measured values of , angle α

Node	Position	Applied load	Angle a decimal degree	Node	Position	Applied load	Angle a decimal degree
K1	1	on bend	22.70	K2	1	on bend	25.60
	2	on bend	19.40		2	on bend	19.30
	3	on bend	17.60		3	on bend	19.60
	1	on point	37.45		1	on point	36.30
	2	on point	27.75		2	on point	29.90
	3	on point	27.40		3	on point	27.70

The first position of the line allows the biggest value of the α. The second position leads to intermediary values. The third position force the line to start out over the plane end (or the eye, if exists) and this diminish the angle α. In general, the knot K2 generates greater values of α. Here was observed an exception, for loading applied on the point, position 1, when the angle is greater for the knot K1.

3. CONCLUSION

The main factor which influences the value of the bending moment for a fishing hook is the angle α .The values of the α parameter were experimentally observed for two types of simple knots. The angle α depends on many factors. In this study more parameters were fixed, in order to look carefully at the influence of the knot, the line positions and the point of application of the load. The next factor to study is the shape and the type of the hook and others.

In this experiment, , angle α varies from 17.60 degrees up to 37.45 degrees. This means a variation of the bending moment almost from simple to double (104 %). Note the analytical expression of the bending moment is deduced in pure bending hypothesis for a beam of constant transversal section. For the FEM simulations, a value of angle should be considered according to real use of the fishing devices. The higher is the value of the angle α, the higher is the value of the bending moment for the same hook.

REFERENCES

[1] Dimitriu St.,Hybrid mechanical structures with curved bars. Mechanical model of the fishing hook – line, 4th International Conference "Advanced Composite Materials Engineering "COMAT 2012, Brasov, Romania, vol.3 pp. 843-846

[2] Şerbu A., Dimitriu Şt., *Rezistenţa Materialelor*, Editura Universităţii Transilvania din Braşov, 2005, ISBN 973-635-430-X

[3] Catalog Mustad http://www.mustad.no/pdfcatalog/emea/064-077.pdf

TABLE OF CONTENTS

www.ingramcontent.com/pod-product-compliance
Lightning Source LLC
Chambersburg PA
CBHW050803220326
41598CB00006B/108

* 9 7 8 1 9 3 9 7 5 7 3 1 9 *